Lecture Notes in Computer Science 10135

Commenced Publication in 1973
Founding and Former Series Editors:
Gerhard Goos, Juris Hartmanis, and Jan van Leeuwen

More information about this series at http://www.springer.com/series/7409

Meikang Qiu (Ed.)

Smart Computing and Communication

First International Conference, SmartCom 2016
Shenzhen, China, December 17–19, 2016
Proceedings

 Springer

Editor
Meikang Qiu
Pace University
New York, NY
USA

ISSN 0302-9743   ISSN 1611-3349   (electronic)
Lecture Notes in Computer Science
ISBN 978-3-319-52014-8   ISBN 978-3-319-52015-5   (eBook)
DOI 10.1007/978-3-319-52015-5

Library of Congress Control Number: 2016962023

LNCS Sublibrary: SL3 – Information Systems and Applications, incl. Internet/Web, and HCI

Printed on acid-free paper

This Springer imprint is published by Springer Nature
The registered company is Springer International Publishing AG
The registered company address is: Gewerbestrasse 11, 6330 Cham, Switzerland

Preface

This volume contains the papers presented at SmartCom 2016: The International Conference on Smart Computing and Communication held during December 17–19, 2016, in Shenzhen City, Guangdong Province, China. SmartCom 2016 was organized and supported by Pace University and Shenzhen University. The conference received 210 submissions. Each submission was carefully reviewed by at least two, and mostly four, Program Committee members. The Program Committee decided to accept full 59 papers. The program also included four invited talks, which were given by Prof. Guoliang Chen (Chinese Academy of Sciences and Academy, China), Prof. Qing Yang (University of Rhode Island, USA), Dr. Shui Yu (Deakin University, Australia), and Prof. Meikang Qiu (Columbia University and Pace University, NY, USA).

We would like to thank all authors who submitted their papers to SmartCom 2016, and the conference attendees for their interest and support, which made the conference possible. We further thank the Organizing Committee for their time and efforts; their support allowed us to focus on the paper selection process. We thank the Program Committee members and the external reviewers for their hard work in reviewing the submissions; we thank Dr. Laizhong Cui for serving as the chair of the BigNetworking workshop; the conference would not have been possible without their professional reviews. We also thank the invited speakers for enriching the program with their presentations. We thank Prof. Guoliang Chen, honorary general chair of the SmartCom 2016, for his advice throughout the conference preparation process. Last but not least, we thank EasyChair for making the entire process of the conference manageable.

We hope you find these proceedings useful, educational, informative, and enjoyable!

December 2016

Meikang Qiu
Zhong Ming
Yang Xiang

Organization

Honorary General Chair

Guoliang Chen   Chinese Academy of Sciences and Academy, China

General Chairs

Zhong Ming   Shenzhen University, China
Yang Xiang   Deakin University, Australia
Meikang Qiu   Pace University, USA

Program Chairs

Tao Zhang   New York Institute of Technology, USA
Zongming Fei   University of Kentucky, USA
Haibo Zhang   University of Otago, New Zealand

Volume Editor

Meikang Qiu   Pace University, USA

Operation Chair/Web Chair

Keke Gai   Pace University, USA

Local Chairs

Rui Mao   Shenzhen University, China
Shuibin Cai   Shenzhen University, China

Publicity Chair

Hui Zhao   Henan University, China

Program Committee

Lixin Tao   Pace University, USA
Xiaofu He   Columbia University, USA
Wenjia Li   New York Institute of Technology, USA
Thomas Austin   San Jose State University, USA
Hui Zhao   Henan University, China

Wenyun Dai	Pace University, USA
Yu-Dong Yao	Stevens Institute of Technology, USA
Bo Luo	The University of Kansas, USA
Xin Li	Carnegie Mellon University, USA
Sang-Yoon Chang	Advanced Digital Sciences Center, Singapore
Jinjun Xiong	IBM Research, USA
Emmanuel Bernardez	IBM Research, USA
Yan Zhang	University of Oslo, Norway
Haibo Zhang	University of Otago, New Zealand
Suman Kumar	Troy University, USA
Yong Guan	Lowa State University, USA
Wei Yu	Towson University, USA
Zhiqiang Lin	University of Texas at Dallas, USA
Zhipeng Wang	China Electronics Standardization Institute, USA
Yong Zhang	The University of Hong Kong, SAR China
Fuji Ren	The University of Tokushima, Japan
Ukka Riekki	University of Oulu, Finland
Hao Hu	Nanjing University, China
Art Sedighi	Global Head of Cloud Architecture and Strategy, TD Bank, USA
Zhongming Fei	University of Kentucky, USA
Weigang Li	University of Brasilia, Brazil
Peng Sun	University of Huston, USA
Yujun Li	Xi'an University of Technology, China

Contents

Cost Reduction for Data Allocation in Heterogenous Cloud Computing Using Dynamic Programming

Hui Zhao[1], Meikang Qiu[2(✉)], Keke Gai[2], Jie Li[1], and Xin He[1]

[1] Software School, Henan University, Kaifeng 475000,
Henan, China
{zhh,jsjt9,hexin}@henu.edu.cn
[2] Department of Computer Science, Pace University,
New York City, NY 10038, USA
{mqiu,kg71231w}@pace.edu

Abstract. Heterogeneous clouds are helpful for improving performance when the data processing task becomes a challenge in big data within different operating environment. Non-distributive manner has some limitation, such as overload energy and low performance resource allocation mechanism. This paper address on this issue and propose an approach to find out the optimal data allocation plan for minimizing total costs of the distributed heterogeneous cloud memories in mobile cloud systems. In this paper, we propose a novel approach to find out the optimal data allocation plan to reduce data processing cost through heterogeneous cloud memories for efficient MaaS. The experimental results proved that our approach is an effective mechanism.

Keywords: Data allocation · Cost reduction · Heterogenous cloud computing · Dynamic programming

1 Introduction

Cloud computing is used broadly in recent years. Heterogeneous clouds are helpful for improving performance when the data processing task becomes a challenge in big data within different operating environment [1]. Combining heterogeneous embedded systems and could computing can receive lots of benefits within big data environments. Currently, cloud-based memories usually deploy non-distributive manner on cloud side. Non-distributive manner has some limitation, such as overload energy and low performance resource allocation mechanism [2]. These limitation restrict the implementation of cloud-based heterogeneous memories. This paper address on this issue and propose an approach

This work is supported by National Natural Science Foundation of China (No. U1304615) and the Science and Technology Research Key Project of Henan Province Science and Technology Department (No.162102210172).

M. Qiu (Ed.): SmartCom 2016, LNCS 10135, pp. 1–11, 2017.
DOI: 10.1007/978-3-319-52015-5_1

to find out the optimal data allocation plan for minimizing total costs of the distributed heterogeneous cloud memories in mobile cloud systems.

Processors and memories that provide processing services are hosted by individual cloud servers in current cloud infrastructure. The challenge of this type of deployment is that deploying distributed memories in clouds faces a few restrictions [3]. Allocating data to multiple cloud-based memories will meet obstacles because of the various impact factors and parameters [4]. The various configurations and capabilities can limit the performance of the whole systems because the naive task allocation is inefficient to operate the entire heterogeneous cloud memories. This means that optimizing the cost of using heterogeneous cloud memories is restricted by the multiple dimensional constraint conditions.

Addressing on this issue, we propose a novel approach that named *Cost-aware Multi Dimension Data Allocation Heterogeneous Cloud Memories* (CM2DAH). The goal of CM2DAH is to find out the optimal data allocation plan to reduce data processing cost through heterogeneous cloud memories for efficient *Memory-as-a-Service* (MaaS). The cost can be any resource that spend in the operation of MaaS, such as execution time, power consumption, etc. The basic idea of the proposed approach is using dynamic programming to minimize the processing costs via mapping data processing capabilities for different datasets inputs. For supporting the proposed approach, we propose an algorithm, which named *Cost-aware Multi Dimension Dynamic Programming* (CM2DP) Algorithm. This algorithm is designed to find out the optimal total costs by using dynamic programming.

The main contribution of this paper are twofold:

1. We propose an approach for solving the data allocation problem with multiple dimensional constraints for heterogeneous cloud memories. The proposed approach is an attempt in using heterogeneous cloud memories to minimize total cost by dynamically allocate data to various cloud resources.
2. We propose an algorithm to solve the problem of data allocations in heterogeneous cloud memories. The proposed algorithm can produce global optimal solutions that are executed by mapping local optimal solutions and using dynamic programming.

The rest of this paper are organized as follows. In Sect. 2, we reviewed the recent related works in data allocation in heterogeneous cloud. A motivation example is presented in Sect. 3. In additional, main algorithms are given in Sect. 4. Moreover, we display a number of experiment results in Sect. 5. Finally, conclusions are stated in Sect. 6.

2   Related Works

Some prior researches have addressed the improvement of the entire cloud systems' performances. The basic working manner of cloud computing is using distributed computing. Thus, exploring the optimization of cloud resources has been worked in a few different aspects.

First, interconnecting different cloud resources is a popular research topic in cloud computing. The relationships among cloud nodes are considered the crucial parts of the performance improvements in cloud systems. An example is *Internet-of-Things* (IoT) that requires intercommunications and interconnections among different infrastructure. For improving the system's performance, previous researches have attempted on a variety of dimensions. For instance, cost-aware cloud computing is a design target that saves energies by implementing scheduling algorithms on Virtual Machine (VM). The outcomes are highly associated with the cost requirements or other configured parameters.

Moreover, some other researches concentrated on integrating heterogeneous cloud computing with mobile wireless networks. For example, Huajian et al. [5] proposed an optimizing approach of the file management system that used mobile heterogeneous mobile cloud computing. This system applied transparent integrations to increase the efficiency of cloud computing. Next, Kostas Katsalis et al. [6] concentrated on wireless networking channels and proposed an approach using semantic Web for strengthening communications among multiple vendors.

Furthermore, consider the real-time services' requirements, Mahadev Satyanarayanan et al. [7] proposed a Cloudlet-based solution that was designed to reduce execution time by establishing a pool of VMs that are close to users. This design was proved that it was an efficient way to reduce the response time. Another research considered high-density computing resources and developed a cognitive management solution that was based on the mobile applications [8]. This solution can be used in the dynamic networking environment, which optimized the operations of mobile ends, Cloudlet, and computing resources in clouds. Next, geographical distributions can also impact on the performance of cloud systems. Hongbin Liang et al. [9] proposed an approach using Semi-Markov Decision Process (SMDP) to determine the service migrations. This approach could efficiently reduce the interrupts of services. Similarly, it has been assessed that a prediction-based approach could be helpful for planning the computing resource assignments based on the predictions [10]. However, most prior researches did not consider implementing heterogeneous cloud computing such that the potential optimizations were ignored.

3 Motivational Example

In this section, we give a motivational example to describe the operational processes of CM2DP in this scenario. Assume that there are four cloud providers offering MaaS with different performances, namely M1, M2, M3, and M4. Table 1 shows the costs for different cloud memory operations.

As shown in Table 1, we consider four main costs that include Read (R), Write (W), Communications (C), and Move (MV). R and W refer to the operation costs of reading and writing data. C refer to the costs happened to communication processes through the Internet. MV refer to the costs happened at the occasions when switching from one memory provider to another set.

Table 1. Cost for different cloud memories. Four types of memories are M1, M2, M3, and M4; main costs derive from Read, Write, Communications (Com.), and Move.

Operation		M1	M2	M3	M4
Read (R)		4	2	1	50
Write (W)		6	4	2	50
Comm. (C)		10	10	10	10
Move (MV)	M1	0	4	6	40
	M2	3	0	5	40
	M3	2	3	0	40
	M4	40	40	40	0

Table 2 shows the number of the memory accesses, including *Read* and *Write*. We assume there are seven input data that are A, B, C, D, E, F, and G. For example, data A require 7 reads and 6 writes according to the table.

Table 2. The number of the memory accesses.

Data	Read	Writes
A	7	6
B	6	3
C	8	6
D	1	1
E	3	1
F	6	6
G	2	3

Our example's initial status is to allocate $A \rightarrow M2, B \rightarrow M2, C \rightarrow M4, D \rightarrow M2, E \rightarrow M3, F \rightarrow M4$, and $G \rightarrow M3$. Moreover, The example configuration is that there are 2 M1, 2 M2, and 2 M3. We assume M4 is always available for data.

Based on the conditions given by Tables 1 and 2, we map the costs of data allocations to could memories in Table 3.

As shown in the Table 3, data A costs $7 * 4 + 6 * 6 + 10 + 4 = 77$ when it is allocated to M1, switched from M2. We call the Table 3 as a *B Table (BTab)*. Actually, we can use a 3-dimension array to refer to every row in Table 3. For instance, we use $\text{BTab}_a[x][y][z]$ to refer to the costs that data A allocated to M1, M2, M3, and M4. X, y, z shows the used number of cloud memories of M1, M2, and M3 respectively. In this case, $\text{BTab}_a[1][0][0] = 77$ indicate the costs of data A is allocated to M1, since we use 1 M1 here. $\text{BTab}_a[0][1][0] = 48$ indicate the costs of data A is allocated to M2. $\text{BTab}_a[0][0][1] = 34$ indicate the costs of

Table 3. Mapping the costs for cloud memories.

Data	M1	M2	M3	M4
A	77	48	34	700
B	55	34	27	500
C	118	90	70	710
D	23	16	18	150
E	30	23	15	250
F	110	86	68	610
G	38	29	18	300

data A is allocated to M3. $BTab_a[0][0][0] = 700$ indicate the costs of data A is allocated to M4 because either M1, M2, and M3 are not available here.

We are going to use Table 3 to calculate the total cost after the allocations and the optimal allocation plan. For obtaining the final optimal data allocation plan, we will produce a *D Table (DTab)* showing the optimal data allocation plan. The generation process is given as follows, which consists of a few steps.

First, we only consider one input data A. In this case, we just copy $BTab_a$ as $DTab_a$. Table 4 displays the costs of $DTab_a$.

Table 4. The costs of $DTab_a$.

DTab cell	Cost
$DTab_a[0][0][0]$	700
$DTab_a[0][0][1]$	34
$DTab_a[0][1][0]$	48
$DTab_a[1][0][0]$	77

As shown in Table 4, we mark the cloud memory to the corresponding cost. Then, we add data B into our sight and generate the $DTab_{ab}$. We calculate every cell of $DTab_{ab}$ to get the optimal costs and the optimal allocation plan of data A and data B, according to $BTab_b$ and $DTab_a$. The $DTab_{ab}$ is shown in Table 5. For example, $DTab_{ab}[0][1][1]=\min(BTab_b[0][0][0]+DTab_a[0][1][1], BTab_b[0][0][1])+DTab_a[0][1][0], BTab_b[0][1][0]) + DTab_a[0][0][1], BTab_b[1][0][0]) + DTab_a[-1][1][1])$. We just drop two elements, $BTab_b[0][0][0]) + DTab_a[0][1][1]$ and $BTab_b[1][0][0]) + DTab_a[-1][1][1]$, since $DTab_a[0][1][1]$ and $DTab_a[-1][1][1]$ are not valid. Thus $DTab_{ab}[0][1][1] = \min(BTab_b[0][0][1]) + DTab_a[0][1][0], BTab_b[0][1][0]) + DTab_a[0][0][1]) = \min(27 + 48, 34 + 34) = \min(75, 68) = 68$. That means the minimum cost is 68 when we have 1 M2 and 1 M3 available. The optimal plan is allocate data A to M2 and data B to M3 since 68 is generated by $BTab_b[0][1][0]) + DTab_a[0][0][1]$.

Table 5. The costs of DTab$_{ab}$.

DTab cell	Cost
DTab$_{ab}$[0][0][0]	1200
DTab$_{ab}$[0][0][1]	534
DTab$_{ab}$[0][0][2]	61
DTab$_{ab}$[0][1][0]	548
DTab$_{ab}$[0][1][1]	68
DTab$_{ab}$[0][2][0]	82
DTab$_{ab}$[1][0][0]	577
DTab$_{ab}$[1][0][1]	89
DTab$_{ab}$[1][1][0]	103
DTab$_{ab}$[2][0][0]	132

We can gain the data allocation plan for each data from the calculation method mentioned above. Finally, we get the last result: DTab$_{abcdefg}$[2][2][2] = 438 and the optimal allocation plan is A→M2, B→M2, C→M3, D→M4, E→M1, F→M3, G→M1.

4   Algorithms

In this section, we propose the algorithm of CM2DP, which is designed to find the optimal data allocation plan for N-dimension heterogeneous cloud memories by using N-dimension dynamic programming. Assume that there are n types of memories available. We define one type of the memory as one dimension. The definition of *N-Dimensional Heterogeneous Cloud Memories* is given by Definition 1.

Definition 1. N-Dimensional Heterogeneous Cloud Memories: ∃ n *types of the memory available for alternatives, we define "n" as the number of dimensions and the available memory set is called N-Dimensional heterogeneous Memories. Each memory type can have different number of memories.*

We define a few definitions used in our algorithm, including *B Table* defined by Definition 2, *D Table* defined by Definition 3 and *Plan* by Definition 4.

Definition 2. B Table: *We use a multiple-dimensional array to describe a table that stores the cost of each data at each memory. Inputs include each data's cost when it is assigned to a memory. The output will be a multiple-dimensional array. ∃ j types of memory, $\{M_j\}$, and each M_j has n_j memories available. The mathematical expression is $BTab[i]\langle(M_1,\ nb_1),\ (M_2,\ nb_2),\ \ldots,\ (M_j,\ nb_j)\rangle$, which represents the cost when data$_i$ use j types of memory and the number of each memory type.*

Definition 3. D Table: *We use a multiple-dimensional array to describe a table that shows the total cost of data allocations by storing all task assignment plans. Input is the B Table. The output will be a multiple-dimensional array storing all task assignments as well as their costs. $\exists$ j types of memory, $\{M_j\}$, and each M_j has n_j memories available. The mathematical expression is $DTab[d]\langle(M_1, nd_1), (M_2, nd_2), \ldots, (M_j, nd_j)\rangle$, which represents the cost of all d data when using j types of memory. The tuple shows the assignment plan.*

Definition 4. Plan: *We use a multiple-dimensional array to describe a table that shows the optimal data allocation plan. Input is the B Table. The output will be a multiple-dimensional array storing all task assignments as well as their costs. $\exists$ j types of memory, $\{M_j\}$, and each M_j has n_j memories available. The mathematical expression is $plan[d]\langle(M_1, nd_1), (M_2, nd_2), \ldots, (M_j, nd_j)\rangle$, which represents the allocation plan of all d data when using j types of memory.*

Algorithm 4.1. Cost-aware Multi Dimension Dynamic Programming (CM2DP) Algorithm

---

Require: The B Table $BTab$
Ensure: The optimal data allocation plan $PlanD$
1: input the B Table
2: initialize Plan
3: DTab[0] ← BTab[0]
4: FOR ∀ rest cell in DTab,
5:    /* DTab[i]$\langle(M_1, nd_1), (M_2, nd_2), \ldots, (M_j, nd_j)\rangle$ */
6:    minCost ← ∞
7:    minCostIndex ← *null*
8:    FOR ∀ cell in BTab[i],
9:       /* BTab[i]$\langle(M_1, nb_1), (M_2, nb_2), \ldots, (M_j, nb_j)\rangle$ */
10:       IF ∃ DTab[i-1]$\langle(M_1, nd_1 - nb_1), (M_2, nd_2 - nb_2), \ldots, (M_j, nd_j - nb_j)\rangle$
11:          sum ← BTab[i]$\langle(M_1, nb_1), (M_2, nb_2), \ldots, (M_j, nb_j)\rangle$ + DTab[i-1]$\langle(M_1, nd_1 - nb_1), (M_2, nd_2 - nb_2), \ldots, (M_j, nd_j - nb_j)\rangle$
12:          IF sum<minCost
13:             minCost ← sum
14:             minCostIndex ← $\langle(M_1, nb_1), (M_2, nb_2), \ldots, (M_j, nb_j)\rangle$
15:          ENDIF
16:       ENDIF
17:       IF minCostIndex != null
18:          plan[i]$\langle(M_1, nd_1), (M_2, nd_2), \ldots, (M_j, nd_j)\rangle$. add(Plan[i-1]$\langle(M_1, nd_1 - nb_1), (M_2, nd_2 - nb_2), \ldots, (M_j, nd_j - nb_j)\rangle$)
19:          plan[i]$\langle(M_1, nd_1), (M_2, nd_2), \ldots, (M_j, nd_j)\rangle$. add(allocate $data_i$ to $\langle(M_1, nb_1), (M_2, nb_2), \ldots, (M_j, nb_j)\rangle$)
20:    ENDFOR
21: ENDFOR
22: RETURN Plan.

The Algorithm 4.1 shows the proposed algorithm and the main phases of our algorithm include:

1. Input *B Table*.
2. Copy BTab[0] to DTab[0] to produce the first partial D Table, which represents add data[0] to D Table.
3. Find the minimal cost from the sums of BTab[1] cells and their supplemental cells in DTab[0] that generated by the last Step 2.
4. Assigning optimal data allocation plan for data 0 and data 1 according to the minimal costs.
5. Add all data by applying the same method used from Step 3 to Step 4 until the D Table is generated.
6. Find the optimal data allocation by searching the lowest cost in D Table according to the memory condition. Output the task assignment plan.

5   Experiment and the Results

In this section, we illustrated our experimental evaluations. Section 5.1 represented our experimental settings. Section 5.2 provided partial experimental results.

5.1   Experiments Settings

We use experimental evaluations to assess the performance of the proposed scheme. The evaluation is based on a simulation that compare CM2DP algorithm and FIFO algorithm, greedy algorithm, and MDPDA algorithm [11]. We implemented our experiments on a Host that ran Windows 8.1 64 bit OS. The main hardware configuration of the Host was: Intel Core i5-4210U CPU, 8.0 GB RAM. We have three main experimental settings that are designed to assess the proposed approaches performance. Three experiments settings are:

1. We configured 3 kinds of memories and 7 kinds of data, 3 available M1, 3 available M2, and 2 available M3.
2. We configured 4 kinds of memories and 8 kinds of data, 2 available M1, 3 available M2, and 2 available M3. M4 is always available for data.
3. We configured 4 kinds of memories and 10 kinds of data, 2 available M1, 3 available M2, and 3 available M3. M4 is always available.

5.2   Experiments Results

As shown in Fig. 1, our proposed approach has the less total cost than FIFO algorithm and greedy algorithm under setting 1. The FIFO algorithm has the biggest total cost because there is no technical choose to allocate the data to the memories that has less cost. The greedy algorithm usually has the larger total cost than CM2DP and MDPDA algorithm. But sometimes greedy algorithm can get the optimal solution, too. The MDPDA algorithm and our proposed algorithm have the same total cost result since they both can get optimal solution.

Figure 2 shows the comparison among FIFO algorithm, greedy algorithm, CM2DP algorithm, and MDPDA algorithm under setting 2. DM2DP algorithm

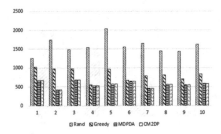

Fig. 1. Comparisons of costs under setting 1

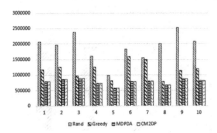

Fig. 2. Comparisons of costs under setting 2

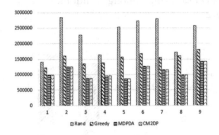

Fig. 3. Comparisons of costs under setting 3

and MDPDA algorithm always have better performance than FIFO and greedy algorithm under this setting.

Moreover, Fig. 3 shows showed another group of comparison results of FIFO algorithm, greedy algorithm, CM2DP algorithm, and MDPDA algorithm under setting 3. The figure represents that our proposed algorithm has the same optimal result with MDPDA algorithm, which is better than FIFO algorithm and greedy algorithm. As shown in the above three figures, we can find that usually the more data and memories we have, the more cost can be saved by using our proposed algorithm.

The Figs. 4, 5, and 6 show the comparison of execution time consumption of FIFO algorithm, greedy algorithm, CM2DP algorithm, and MDPDA algorithm under three settings respectively. From these figures we can find that our

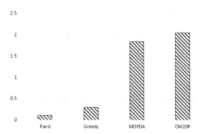

Fig. 4. Comparisons of execution time among Random, Greedy, CM2DP and MDPDA algorithms under setting 1

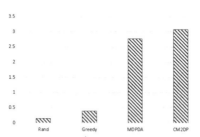

Fig. 5. Comparisons of execution time among Random, Greedy, CM2DP and MDPDA algorithms under setting 2

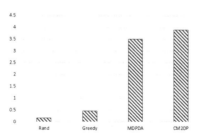

Fig. 6. Comparisons of execution time among Random, Greedy, CM2DP and MDPDA algorithms under setting 3

proposed algorithm has the less execution time than the MDPDA algorithm when they both can get the optimal solution under every setting.

In summary, our approach always can get the optimal data allocation plan, which is better than the results of FIFO algorithm and the greedy algorithm. Moreover, CM2DP has the better execution performance than MDPDA, another optimal algorithm.

6   Conclusions

In this chapter, we proposed a approach that named *Cost-aware Multi Dimension Data Allocation Heterogeneous Cloud Memories* (CM2DAH) to find out the optimal data allocation plan in heterogeneous cloud memories. To support the proposed approach, a N-dimension dynamic programming algorithm was designed. This algorithm is named *Cost-aware Multi Dimension Dynamic Programming* (CM2DP) Algorithm. The experimental results proved that our approach is an effective mechanism compare with FIFO algorithm, greedy algorithm, and MDPDA algorithm.

References

1. Hao, F., Min, G., Chen, J., Wang, F., Lin, M., Luo, C., Yang, L.: An optimized computational model for multi-community-cloud social collaboration. IEEE Trans. Serv. Comput. **7**(3), 346–358 (2014)
2. Basu, A., Gandhi, J., Chang, J., Hill, M., Swift, M.: Efficient virtual memory for big memory servers. In: ACM SIGARCH Computer Architecture News, vol. 41, pp. 237–248. ACM (2013)
3. Wei, Z., Pierre, G., Chi, C.: Cloudtps: Scalable transactions for web applications in the cloud. IEEE Trans. Serv. Comput. **5**(4), 525–539 (2012)
4. Qiu, M., Chen, L., Zhu, Y., Hu, J., Qin, X.: Online data allocation for hybrid memories on embedded tele-health systems. In: 2014 IEEE International Conference on High Performance Computing and Communications, 2014 IEEE 6th International Symposium on Cyberspace Safety and Security, 2014 IEEE 11th International Conference on Embedded Software and System, pp. 574–579. IEEE (2014)
5. Mao, H., Xiao, N., Shi, W., Lu, Y.: Wukong: a cloud-oriented file service for mobile internet devices. J. Parallel Distrib. Comput. **72**(2), 171–184 (2012)
6. Katsalis, K., Sourlas, V., Korakis, T., Tassiulas, L.: A cloud-based content replication framework over multi-domain environments. In: 2014 IEEE International Conference on Communications (ICC), pp. 2926–2931. IEEE (2014)
7. Satyanarayanan, M., Bahl, P., Caceres, R., Davies, N.: The case for vm-based cloudlets in mobile computing. IEEE Pervasive Comput. **8**(4), 14–23 (2009)
8. Chen, M., Zhang, Y., Li, Y., Mao, S., Leung, V.: EMC: emotion-aware mobile cloud computing in 5G. IEEE Network **29**(2), 32–38 (2015)
9. Liang, H., Cai, L., Huang, D., Shen, X., Peng, D.: An smdp-based service model for interdomain resource allocation in mobile cloud networks. IEEE Trans. Veh. Technol. **61**(5), 2222–2232 (2012)
10. Sood, S., Sandhu, R.: Matrix based proactive resource provisioning in mobile cloud environment. Simul. Model. Pract. Theory **50**, 83–95 (2015)
11. Qiu, M., Chen, Z., Ming, Z., Qin, X., Niu, J.: Energy-aware data allocation with hybrid memory for mobile cloud systems. IEEE Syst. J. **99**, 1–10 (2014)

Minimizing Bank Conflict Delay for Real-Time Embedded Multicore Systems via Bank Mapping

Zhihua Gan[1,2]([✉]), Mingquan Zhang[1], Zhimin Gu[1], Jizan Zhang[1], and Hai Tan[1]

[1] School of Computer Science and Technology, Beijing Institute of Technology,
Beijing, China
800521@bit.edu.cn
[2] School of Software, Henan University, Kaifeng, China

Abstract. Multi-core architectures may meet the increasing performance requirement of real-time systems. However, it is harder to compute the WCET estimation in multi-core platforms due to inter-task interference that tasks suffer when accessing shared hardware resources. In this paper, we propose a finer grained approach to analyze the inter-task interference for multi-core platforms with the TDMA policy and bank-column cache partitioning, and our approach can reasonably estimate inter-task interference delays. Moreover, we make bank-to-core mapping to optimize the interference delays, and develop an algorithm for finding the best bank-to-core mapping. The experimental results show that our interference analysis approach can improve the tightness of interference delays by 14.68% on average compared to Upper Bound Delay (UBD) approach, and the optimized bank-to-core mapping can achieve the WCET improvement by 9.27% on average.

Keywords: Bank conflict · Multicore system · Worst case execution time

1 Introduction

The increasing demand for new functionality in real-time embedded system is driving an increment in the performance requirement of embedded processors. Multi-core processors are believed to be one of major solutions for real-time embedded systems [1–3], since they can provide high performance with low cost and relatively simple design [17]. However, applying multi-cores for real-time system is difficult due to inter-task interference accessing hardware shared resources [4]. These interferences complicate the behavior of a system, resulting in difficulties for performing a tight worst case execution time(WCET) analysis for a given task.

Previous solutions [5–9] have been focusing on the effect of inter-task interference caused by on-chip shared resources. For example Chattopadhyay et al. [5] employed the maximum bus access delay to estimate WCET according to execution contexts instead of aligning each loop head execution to the first TDMA

© Springer International Publishing AG 2017
M. Qiu (Ed.): SmartCom 2016, LNCS 10135, pp. 12–21, 2017.
DOI: 10.1007/978-3-319-52015-5_2

slot. In [8], the authors improved the treatment of loop structures. Kelter et al. [9] improved analysis efficiency via bounding the upper bound of TDMA offsets. However, these researches mainly concentrated on the bus access interference and cache storage interference, and ignored the effect of bank conflict on the WCET estimation. In multi-core architecture, the shared cache usually consists of multiple banks that can be accessed in parallel, i.e., different cache requests can access different banks simultaneously. A bank can only handle one cache request at a time, when two or more cache requests try to access the same bank at the same time, the bank conflict takes place. The bank conflict brings extra execution time for the task, and its influence on WCET estimation has to be taken into account for ensuring safety of WCET. To the best of our knowledge, only Paolieri et al's work [12] and Yoon et al's work [13] considered bank conflict for WCET estimation. But they all employed the Upper Bound Delay(UBD) method to estimate the interference delay, in which a potential maximum delay (i.e., upper bound delay) that each cache request suffers is bounded, then, this delay is added to each request during WCET analysis. However, not all requests can suffer from bank conflict, even though bank conflicts occur among a group of requests, the delay of bank conflict suffered by each request is different. Therefore, this method causes pessimistic WCET.

The goal of this paper is to minimize the bank conflict delay and obtain the tighter WCET estimation for embedded multicore systems with TDMA policy. We make the following major contributions. (1) We propose a finer-grained approach to analyze bank conflict and bus access interference based on request timing, which can improve the tightness of interference delays. (2) We make bank mapping to optimize conflict delays, and develop an algorithm for finding the best bank mapping according to the queue of cores, such that the effect of inter-core bank conflict on the WCET estimation is minimized.

The rest of the paper is organized as follows. Section 2 describes the system model. The bank conflict analysis is described in Sect. 3. Section 4 presents the optimization algorithm of bank mapping, and experimental results are provided in Sect. 5. The conclusion is presented in Sect. 6.

2   System Model

2.1   Embedded Multi-core Architecture

We consider an embedded multi-core architecture consisting of N_{core} homogeneous core, $\mathbb{C} = \{C_1, C_2, \cdots, C_{(N_{core})}\}$. Each core has its own private L1 instruction cache and L1 data cache. All the cores share an L2 combined cache $\mathbb{B}$ which is partitioned into N_{bank} banks $\{B_1, B_2, \cdots, B_{(N_{bank})}\}$, and each bank is subdivided into N_{column} columns. Bank access latency is L_M cycles (same for read/write operations for all banks). The real-time shared bus connecting cores and the shared L2 cache adopts TDMA policy and full-duplex as assumed by [12]. Every bus round has L_{round} equal time slots, $\mathbb{R} = \{S_1, S_2, \cdots, S_{(L_{round})}\}$.

The length of one slot is L_B cycles, which is equal to the time of the bus completing one request. The core-to-slot mapping is one-to-one mapping, i.e., the $C_i(\in \mathbb{C})$ is mapped to $S_i(\in \mathbb{R})$. In addition, the penalty of L2 cache miss is $L2_{penalty}$ cycles.

2.2 Task Model

A hard real-time set comprises multiple independent hard real-time tasks(HRTs). All HRTs are partitioned to N_{core} cores in advance. The tasks allocated to the same core are executed sequentially and task migration is not allowed. In multicore systems, multiple tasks can be executed simultaneously in different cores. Let Γ_{sim} be the set of HRTs mapped to core $C_i(\in \mathbb{C})$, $HT_i = \{HRT_1, HRT_2, \cdots, HRT_{n_i}\}$, n_i be the number of the HRTs in HT_i, and the demanded L2 cache of $HRT_i(\in HT_i)$ be $size_{HRT_j}$ columns. Thus, the demanded L2 cache size of C_i is $size_{C_i} = max(size_{HRT_j}|1 \leq j \leq n_j)$ columns.

3 Bank Conflict Delay Analysis

In this section, we analyze the bank conflict delay suffered by a HRT. Assuming that $m(\leq N_{core})$ cores $\{C_1, C_2, \cdots, C_m\}$ sharing one bank try to access the bank in the lth bus round, and they do not miss their own bus slots. The value of $C_i(1 \leq i \leq m)$ is the index of the corresponding bus slot. Let $bcd_{ij}, i \neq 1$, be the bank conflict delay suffered by HRT running on C_i in the lth bus round shown in Fig. 1, which can be expressed by formulation (2).

$$bcd_{ij} = Max\{bcd_{(i-1)j} + L_M - (C_i - C_{i-1}) \cdot L_B, 0\} \tag{1}$$

In formulation (1), the $bcd_{(i-1)j}$ denotes the bank conflict delay suffered by the HRT in lth -1 bus round. If $i = 1$ and $l = 1$, $bcd_{11} = 0$, otherwise, the bcd_{1l} can be computed by formulation (2). In formulation (2), C_{pre} denotes the predecessor core of C_1 and bcd_{pre_k} be the bank conflict delay suffered by the HRT running on C_{pre} in the kth bus round, which are shown in Fig. 2

$$bcd_{ij} = Max\{bcd_{pre\_k} + (j-k-1) \cdot N_{core} \cdot L_B - (N_{core} - (N_{core} - C_1) \cdot L_B, 0\} \tag{2}$$

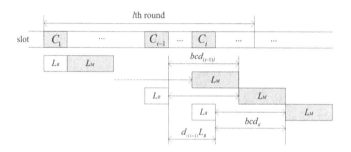

Fig. 1. The bank conflict delay suffered by HRT running on C_i in the lth bus round

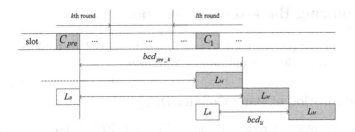

Fig. 2. The bank conflict delay suffered by C_1 in the lth bus round

The formulation (2) can be simplified as formulation (3):

$$bcd_{ij} = Max\{bcd_{pre\_k} + L_M + (j - k) \cdot N_{core} \cdot L_B - (C_{pre} - C_1) \cdot L_B, 0\} \quad (3)$$

In formulation (3), the bank conflict delay suffered by C_1 is transferred from the precious bus round, which is the initial bank conflict delay of the shared bank in the current bus round. There are two factors to affect the initial bank conflict delay, i.e., the number of requests to access bank B_i in one bus round and the distribution of the corresponding slots. A bus round can contain at most $\left\lfloor \frac{L_{round} \cdot L_B}{L_M} \right\rfloor$ requests without any bank access conflict, for one L2 cache access needs be at least L_M cycles. Let $N_{c\_b}^i$ be the number of cores sharing bank B_i. In the worst case, the total requests in one bus round to access one bank is $N_{c\_b}^i$. If $N_{c\_b}^i > \left\lfloor \frac{L_{round} \cdot L_B}{L_M} \right\rfloor$ and $N_{c\_b}^i \cdot L_M > L_{round} \cdot L_B$, the total time of access L2 cache is greater than the length of a bus round, and the initial bank conflict delay in the following bus round is $N_{c\_b}^i \cdot L_M - L_{round} \cdot L_B$ cycles.

In addition, the distribution of the corresponding slots affect the initial bank conflict delay in the following bus round even. The initial bank conflict delay cannot be propagated across bus rounds if $N_{c\_b}^i \leq \left\lfloor \frac{L_{round} \cdot L_B}{L_M} \right\rfloor$. Otherwise, the initial bank conflict delay cannot be transmitted in bus rounds if more than $\left\lfloor \frac{L_{round} \cdot L_B}{L_M} \right\rfloor$ cores have requests to access L2 cache in several sequent bus rounds. In this paper, we define transferred bank conflict delay of one bank in one bus round as the initial bank conflict delay of the bank in the bus round if the initial bank conflict delay can be propagated across bus rounds.

Transferred bank conflict delay brings more negative impact to the predictability of the hard real-time multi-core system for it can be propagated across bus rounds. In order to minimize the bank conflict delay on one bank, the number of cores sharing one bank need be less than or equal to $\left\lfloor \frac{L_{round} \cdot L_B}{L_M} \right\rfloor$ to guarantee no transferred bank conflict delay. If not, the number of the cores shared one bank is minimum to decrease transferred bank conflict delay.

4   Optimizing Bank-to-core Mapping

In this section, we present bank-to-core mapping to reduce bank conflict, and design an algorithm for this optimization problem.

4.1   Optimization of Bank Conflict Delay

Let N_c^i be the total bus rounds demanded by the HRT running on C_i, and D_c^i be the total bank conflict delay suffered by the HRT, which can be expressed as $D_c^i = \sum_{l=1}^{N_c^i} bcd_{il}$. Let D_{jl} be the total bank conflict delay on bank B_j in the lth bus round, which can expressed by $D_{jl} = \sum_{i=1}^{K_{jl}} bcd_{il}$, where K_{jl} is the total count of the requests to access bank B_j in the lth bus round. Let C_{bj} be the core set mapped to bank B_j, N_b^j be the maximum number of bus rounds of the HRTs in the core set C_{bj}, and D_b^j be the total bank conflict delay on B_j. The D_b^j can be expressed as follows:

$$D_b^j = \sum_{l=1}^{N_b^j} D_{jl} = \sum_{l=1}^{N_b^j} \sum_{i=1}^{K_{jl}} bcd_{il} = \sum_{i=1}^{N_{c\_b}^i} D_c^i = \sum_{i=1}^{N_{c\_b}^i} \sum_{l=1}^{N_c^i} bcd_{il} \tag{4}$$

As bank conflict delay suffered by one HRT is nonnegative, minimizing the total bank conflict delay for each HRT is equivalent to minimizing the total bank conflict delay on each bank, and that can be expressed by $min(D_c^i | \forall C_i \in \mathbb{C}) \Leftrightarrow min(D_b^j | \forall B_j \in \mathbb{B})$. The total bank conflict delay on one bank also is nonnegative for the bank conflict delay suffered by one HRT is nonnegative, therefore, we can derive the following formulation $min(\sum_{j=1}^{N_{bank}} D_b^j) \Leftrightarrow min(\sum_{j=1}^{N_{bank}} \sum_{l=1}^{N_b^j} \sum_{i=1}^{K_{jl}} bcd_{il})$. Let x_{ij} be the mapping from $C_i(\in \mathbb{C})$ to $B_j(\in \mathbb{B})$, and n_{ij} be the column number of C_i mapped to B_j. If $C_i \in \mathbb{C}$, x_{ij} is equal to 1, otherwise, x_{ij} is 0. As the demanded cache of HRTs are exclusively mapped to columns, n_{ij} is integer and $0 \leq n_{ij} \leq min\{N_{column}, Size_{c_i}\}$. If $x_{ij} = 1$, then $n_{ij} \geq 0$. Otherwise, $n_{ij} = 0$. x_{ij} and n_{ij} are the decision variables. The optimization model to minimize the bank conflict delay suffered by each HRT can be described as follows:
   Objective function:

$$min(\sum_{j=1}^{N_{bank}} \sum_{l=1}^{N_b^j} \sum_{i=1}^{K_{jl}} bcd_{il}) \tag{5}$$

Constraints:

$$N_{bank} \cdot N_{column} \geq \sum_{i=1}^{N_{core}} size_{C_i} \tag{6}$$

$$\forall C_i \in \mathbb{C} \quad size_{C_i} = \sum_{j=1}^{N_{bank}} n_{ij} \cdot x_{ij} \tag{7}$$

$$\forall B_i \in \mathbb{B} \quad N_{column} \geq \sum_{i=1}^{N_{core}} n_{ij} \cdot x_{ij} \tag{8}$$

In this optimization problem, the objective function describes that the overall bank conflict delay is minimum. In bank-column cache partitioning, a HRT exclusively accesses its allocated columns, therefore, the size of shared L2 cache need meet the total demand of N_{core} cores, which is expressed as constraint (6). Constraint (7) describes that the model need meet the demanded caching of each core. Constraint (8) is the size constraint of one bank.

4.2 Design of the Proposed Optimization Algorithm

Based on above conflict analysis, the optimization problem is transformed to find the optimal bank mapping, and the bank conflict delay suffered by the cores sharing the bank is minimum. In this subsection, we design algorithm for this optimization problem without any specific initial bank-to-core mapping. Algorithm 1 shows our algorithm for bank mapping optimization. It iteratively calls itself, until the column requirement of all tasks is satisfied. Algorithm 1 takes the number of cores and the column requirement of each tasks as an input. Line 1 initializes the initial minimal total conflict delays to infinity, and initializes decision variables used[] to *false* for performing recursion call. A recursive function $FindOptimalMapping()$ is defined to search the optimal bank mapping in the solution space in lines 2∼31. Lines 5∼15 generate the mapping of bank to core $BtoCmapping[][]$ based on the core queue $c\_seq[]$. Then, based on bank mapping $BtoCmapping[][]$, we calculate the conflict delays of each HRT in line 16. In line 17, we compute the conflict delays suffered by all HRTs. In lines 18∼22, the best bank mapping is saved. The recursion of the algorithm is practiced in lines 23∼30. In line 27, a core queue is generated in the recursive walk, and the generated core queue is stored in the array $c\_seq[]$.

5   Evaluation

In this section, we implement above approaches, and set up experiments to demonstrate the effectiveness of our optimization.

5.1   Experimental Setup

In our experiment, the multi-core architecture consists of 6 cores, each of which implements an in-order 5-stages pipeline without branch prediction. The instruction fetch queue size is 4, fetch width is 2 and the instruction window size is 8. Each core has 64B private instruction and data L1 cache (1-bank, 2-way, 8-byte per line, 1 cycle access and LRU replacement policy). The L2 cache is shared among all cores, the total size is 4KB, 4 banks (each of which is 1KB), 4-way, 32-byte per line, 4 cycles access (i.e., cycles), and LRU replacement policy. Each bank is partitioned into 8 columns and the size of each one is 128B. Bus access

---

Algorithm 1. Optimizing bank-to-core mapping

Require: N_{core}, $size_{C_i}$ $(C_i \in \mathbb{C})$
Ensure: the total conflict delays($MinDelay$),the bank mapping ($OptimalMap[][]$)
1: $MinDelay$ =Infinity, $used[j] = false(1 \leq j \leq N_{core})$;
2: **function** $FindOptimalMapping(N)$
3: **if** $N > N_{core}$ **then**
4:    $ncol = N_{column}$; $nbank = 1$;
5:    **for** each core C_i in $c\_seq[]$ **do**
6:       **if** $size_{C_i} \geq ncol$ **then**
7:          **while** $size_{C_i} \geq ncol$ **do**
8:             $BtoCmapping[i][nbank] = ncol$;
9:             $size_{C_i} = size_{C_i} - ncol$; $nbank + +$; $ncol = N_{column}$
10:            **end while**
11:            $BtoCmapping[i][nbank] = size_{C_i}$; $ncol = ncol - size_{C_i}$;
12:         **else**
13:            $BoCmapping[i][nbank] = size_{C_i}$; $ncol = ncol - size_{C_i}$;
14:         **end if**
15:      **end for**
16:      Computing $conflict\_delay[]$ suffered by each HRT_i based on formulation (3) ;
17:      $Total\_delay = \sum_{\forall C_i \in \mathbb{C}} Interference\_Delay[i]$;
18:      **if** $MinDelay > Total_{delay}$ **then**
19:         $MinEnergy = Total_{delay}$;
20:         $OptimalMap[][] = BtoCmapping[][]$;
21:      **end if**
22: **end if**
23: **for** $j = 1$; $j \leq N_{core}$; $j + +$ **do**
24:    **if** $!used[j]$ **then**
25:       $c\_seq[N] = C_j$;
26:       $used[j] = true$;
27:       $FindMinMapping(N + 1)$;
28:       $used[j] = false$;
29:    **end if**
30: **end for**
31: **end function**

---

latency is 2 cycles (i.e., $L_M = 4$ cycles). All benchmarks used in this section are part of Mälardalen wcet benchmarks [14] as shown in Table 1 including the byte size Bytes and the lines of code LOC.

In order to get the L2 cache size of all benchmarks demanded, we use Chronos [15] to measure their WCET that the L2 cache size is 128 B, 256 B, 512 B, 1 KB, 2 KB and 4 KB respectively. The configurations of measurement are: the instruction and data L1 cache are 64 B (1-bank, 2-way, 8-byte per line, LRU replacement policy), respectively. L2 cache is 4-way, 32-byte per line and LRU replacement policy. According to these results, we adopt the L2 cache size provided in Table 1 (column 3). In addition, columns 6–9 in Table 1 present the initial bank-to-core mapping.

5.2 Experimental Results

5.2.1 Our Conflict Analysis Approach Versus UBD Approach

In this experiment, our focus is the effectiveness of conflict analysis approach. Therefore, we assume that the bank mapping is used for HRTs in Table 1 (i.e., the order of core in queue $c\_seq[]$ is $\{C_1, C_2, C_3, C_4, C_5, C_6\}$). We compare our approach with UBD approach where the upper bound delay can be expressed as $UBD = (N_{core} - 1) \cdot Max(L_B, L_M)$. Figure 3 illustrates this comparison

Table 1. The benchmarks

Benchmark	CodeSize (bytes)	Columns	LOC	Core	B_1	B_2	B_3	B_4
prime	797	1	47	C_1	1	0	0	0
bsort100	2779	16	128	C_2	7	8	0	0
cnt	2880	8	267	C_3	0	0	8	0
fibcall	3499	1	72	C_4	0	0	0	1
insertsort	3892	4	92	C_5	0	0	0	4
expint	4288	2	157	C_6	0	0	0	2

in bank conflict delays for all benchmarks in Table 1. The bank conflict delay values are normalized to the delays obtained by UBD approach. We can see that the bank conflict delays obtained by our approach is less than the delays obtained by UBD approach for HRTs, which is because our approach takes request timing into consideration on runtime inter-task conflict on shared cache, and can reasonably estimate the bank conflict delay. The proposed approach can improve the tightness of bank conflict delays by 14.68% on average compared to the UBD approach.

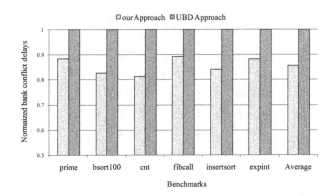

Fig. 3. Comparison of bank conflict delays of different HRT for two approaches

5.2.2   Impact of Bank-to-core Mapping on WCET

The sum of bank conflict delays suffered by all tasks in task set under different bank-to-core mapping are shown in Fig. 4. We can observe that the solution space of bank mapping is 720, and the total bank conflict delays suffered by all tasks have significant difference under different bank mapping. The bank conflict delays are 29836 cycles under the worst mapping. In contrast, the bank conflict delays suffered by all tasks are only 20242 cycles under the best mapping. In order to compare the effect of bank mapping on WCET. We use the mapping in Table 1 as the non-optimized mapping. One of the mappings with the minimum conflict delays is shown in Table 2. In this bank-to-core mapping, the WCET for all tasks under the optimized bank-to-core mapping improved by 9.27% on average.

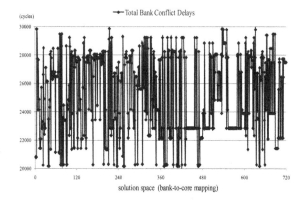

Fig. 4. The total bank conflict delays suffered by all HRTs

Table 2. The optimal bank-to-core mapping

Benchmark	Core	B_1	B_2	B_3	B_4
prime	C_1	0	1	0	0
bsort100	C_2	0	7	8	1
cnt	C_3	4	0	0	4
fibcall	C_4	0	1	0	0
insertsort	C_5	4	0	0	0
expint	C_6	0	0	0	2

6   Conclusion

In this paper, we presented a finer-grained approach to analyze the bank conflict. This approach can reasonably estimate conflict delays. Furthermore, we optimize the conflict delays through bank-to-core mapping and design the optimizing algorithms to find the optimal mapping. Using our analysis approach, the bank conflict delays can be improved by 14.68% on average compared to existing method. The experimental results show that bank mapping has great impact on WCET estimation, which can achieve a 9.27% improvement in WCET on average.

Acknowledgements. This work is supported by the National Natural Science Foundation of China(No.61370062,61462004). We thank the anonymous reviewers for their feedback.

References

1. ARC Advisory Group, Process Safety System Worldwide Outlook, Market Analysis and Focrcast through (2015)
2. Qiu, M., Zhong, M., Li, J., Gai, K., Zong, Z.: Phase-change memory optimization for green cloud with genetic algorithm. IEEE Trans. Comput. **64**(12), 3528–3540 (2015)
3. Gai, K., Qiu, M., Zhao, H., Tao, L., Zong, Z.: Dynamic energy-aware cloudlet-based mobile cloud computing model for green computing. J. Netw. Comput. Appl. **59**, 46–54 (2016)
4. Wilhelm, R., Engblom, J., Ermedahl, A., et al.: The worst-case execution-time problem - overview of methods and survey of tools. ACM TECS **7**(3), 1–53 (2008)
5. Chattopadhyay, S., Chong, L.K., Roychoudhury, A., Kelter, T., Marwedel, P., Falk, H.: A unified WCET analysis framework for multi-core platforms. ACM Trans. Embed. Comput. Syst. **13**(4), 1–29 (2014)
6. Yan, J., Zhang, W.: WCET analysis for multi-core processors with shared L2 instruction caches. In: Proceedings of the 14th IEEE Real-Time and Embedded Technology and Applications Symposium, pp. 80–89 (2008)
7. Zhang, W., Yan, J.: Accurately estimating worst-case execution time for multi-core processors with shared direct-mapped instruction caches. In: Proceedings of the 15th IEEE International Conference on Embedded and Real-Time Computing Systems and Applications, pp. 455–463 (2009)
8. RadojkoviC, P., Girbal, S., Grasset, A., Quinones, E., Yehia, S., Cazorla, F.J.: On the evaluation of the impact of shared resources in multithreaded COTS processors in time-critical environments. ACM Trans. Arch. Code Optim. **8**(4), 1–25 (2012)
9. Kelter, T., Falk, H., Marwedel, P., Chattopadhyay, S., Roychoudhury, A.: Bus-aware multicore WCET analysis through TDMA offset bounds. In: Proceedings of the 2011 Euromicro Conference on Real-Time Systems, pp. 3–12 (2011)
10. Gai, K., Qiu, M., Tao, L., Zhu, Y.: Intrusion detection techniques for mobile cloud computing in heterogeneous 5G. Secur. Commun. Netw. **9**, 1–10 (2015)
11. Qiu, M., Gai, K., Thuraisingham, B., Tao, L., Zhao, H.: Proactive user-centric secure data scheme using attribute-based semantic access controls for mobile clouds in financial industry. Future Gener. Comput. Syst. (2016, in press)
12. Paolieri, M., Quiñones, E., Cazorla, F.J., Bernat, G., Valero, M.: Hardware support for WCET analysis of hard real-time multicore systems. In: Proceedings of the 36th IEEE/ACM International Symposium on Computer Architecture, pp. 57–68 (2009)
13. Yoon, M.K., Kim, J.E., Sha, L.: Optimizing tunable WCET with shared resource allocation and arbitration in hard real-time multicore systems. In: Proceedings of the 32th IEEE Real-Time Systems Symposium, pp. 227–238 (2011)
14. Gustafsson, J., et al.: The Malardalen WCET benchmarks-past, present and future. In: WCET workshop
15. Li, X., Liang, Y., Mitra, T., Roychoudhury, A.: Chronos: a timing analyzer for embedded software. Sci. Comput. Program. **69**(1), 56–67 (2007)
16. RapiTime: Worst-case execution time analysis. User Guide. Rapita Systems (2014)
17. Gai, K., Li, S.: Towards cloud computing: a literature review on cloud computing and its development trends. In: The 4th International Conference on Multimedia Information Networking and Security, pp. 142–146, Nanjing, China (2012)

A Hybrid Algorithm Based on Particle Swarm Optimization and Ant Colony Optimization Algorithm

Junliang Lu[1,2], Wei Hu[1,2(✉)], Yonghao Wang[3], Lin Li[1,2], Peng Ke[1,2], and Kai Zhang[1,2]

[1] College of Computer Science and Technology,
Wuhan University of Science and Technology, Wuhan, China
`ljllujunliang@foxmail.com`,
`{huwei,lilin,ke_peng,zhangkai}@wust.edu.cn`
[2] Hubei Province Key Laboratory of Intelligent Information Processing
and Real-Time Industrial System, Wuhan, China
[3] The DMT Lab, Birmingham City University, Birmingham, UK
`yonghao.wang@bcu.ac.uk`

Abstract. Particle swarm optimization (PSO) and Ant Colony Optimization (ACO) are two important methods of stochastic global optimization. PSO has fast global search capability with fast initial speed. But when it is close to the optimal solution, its convergence speed is slow and easy to fall into the local optimal solution. ACO can converge to the optimal path through the accumulation and update of the information with the distributed parallel global search ability. But it has slow solving speed for the lack of initial pheromone at the beginning. In this paper, the hybrid algorithm is proposed in order to use the advantages of both of the two algorithm. PSO is first used to search the global solution. When it maybe fall in local one, ACO is used to complete the search for the optimal solution according to the specific conditions. The experimental results show that the hybrid algorithm has achieved the design target with fast and accurate search.

Keywords: Particle swarm optimization · Ant colony optimization · Optimal solution

1 Introduction

Many engineering problems are proven to be NP complete or NP hard. In recent years, the intelligent heuristic optimization algorithms become more and more attractive for they can be used to search for the optimal solutions to solve NP problem partially [1]. However, due to the particularity and complexity of the various problems, each algorithm shows its advantages and disadvantages. The key challenge is the tradeoff between the time performance and optimization effect. If the algorithm is designed to find the optimal solution in limited time, it has to compromise in the optimization effectiveness and vice versa. One of the promising approach is to adopt the ideas of different algorithms and find a mixed way to be a better solution.

Particle swarm optimization (PSO) and ant colony optimization (ACO) are two important algorithms to find optimal solutions for the NP problems. PSO was proposed

© Springer International Publishing AG 2017
M. Qiu (Ed.): SmartCom 2016, LNCS 10135, pp. 22–31, 2017.
DOI: 10.1007/978-3-319-52015-5_3

as an evolutionary calculation method based swarm intelligence [2]. It is an optimization method based on iteration, which is similar to genetic algorithm. PSO algorithm has adopted the concepts of group and evolution, which is derived from the research on the behaviors of the birds' feed [3]. It operates mainly based on individual adaptive values. This algorithm has simple design concept and it is easy to implement. It has strong global search ability with less experienced parameters. However, it also has obvious disadvantage that this algorithm is easy to fall into local optimal solution though it has fast global search capability with fast initial speed.

ACO is a different type of intelligent optimization algorithm. This algorithm was derived from the study of path finding behaviors of ants' activity of looking for food [4]. It also adopts the concepts of group and evolution and is based on iterative optimization. It uses a positive feedback mechanism. This algorithm could converge to the optimal solution through the pheromone that updating continuously. However, it has a slow convergence speed due to the lack of pheromone at the beginning [5].

PSO and ACO are popular algorithms. They are used to solve many problems. These two algorithms can be used individually or jointly [6–10]. Many researches focused on the performance and convergence of the two algorithms [11–14] that showing both of PSO and ACO have their advantages and disadvantages. In this paper, our design is to adopt their concepts to obtain the optimal solutions. PSO has a better performance to search the optimal solutions. This algorithm is used as the initial processing. When premature convergence is emerging, ACO is used to complete the rest of the optimization process. The key is how to identify the premature convergence. Our design focuses on the aggregation of the particles. It helps our approach to switch PSO to ACO.

This paper is organized as the follows. Section 2 provides the PSO based optimization. Section 3 describes the ACO based optimization. Section 4 depicts the hybrid algorithm. The experiments and result analysis are discussed in Sect. 5. And at last, we give the conclusions in Sect. 6.

2 Basic Principle of PSO and Its Optimization

PSO simulates birds' feeding behaviors. Imagine that a flock of birds search for food in the area randomly, in which there is only a piece of food. All the birds do not know the location of the food, but they know the distances between their own current positions to the food. The simplest and most effective method is to search the area, in which a bird is the nearest to the food. PSO algorithm was inspired from this kind of thought. It is used to solve the optimization of the problem. Each bird in this search space may be the solution for the problem. The bird can be considered as an idealized particle without quality and volume in this space. Each particle has an adaptive value that is determined by the optimization function, and there is a velocity that determines the direction and distance of the flight. Located in a S dimension of the target search space, there are m particles to form a group, where the current position of the particle i is represented as $X_i(x_{i1}, x_{i2},..., x_{iS})$, current flight velocity $V_i(v_{i1}, v_{i2},..., v_{iS})$ and the position of $P_i(p_{i1}, p_{i2}, ..., p_{iS})$ in which the best position P_{pbest}(that is, with the best fitness value of the position, which is an individual extreme value). The current space of all particles have experienced the best position P_{gbest}(global extreme value). In each iteration, the particle updates itself by tracking the two extreme values. Particle i in S dimensional space update its velocity and position according to the following computation:

$$v_{is}(t+1) = w * v_{is}(t) + c_1 r_1 \left(P_{pbest}(t) - x_{is}(t)\right) + c_2 r_2 \left(P_{gbest}(t) - x_{is}(t)\right) \tag{1}$$

$$x_{is}(t+1) = x_{is}(t) + v_{is}(t+1) \tag{2}$$

Among them, $i \in [1, m]$ and $s \in [1, S]$; inertia weight w is non-negative number to control the influence from the previous velocity on the current velocity, which has very big effect on balancing the global search ability and local search ability of the algorithm. When w is small, the previous velocity has little effect on the local search ability of PSO algorithm. When the w is large, the previous velocity has great influence on the global search ability of PSO algorithm. c_1 and c_2 are learning factors, which are non-negative values. r_1 and r_2 are independent pseudo-random numbers, which obey the uniform distribution on [0, 1]. $v_{is} \in [-v_{max}, v_{max}]$, and vmax is constant. In the process of updating, the maximum velocity in each dimension of a particle is restrained as vmax, and the coordinates in each dimension of a particle is also restrained in the permitted range. At the same time, P_{pbest} and P_{gbest} are constantly updated in the iterative process. The final output is P_{gbest}, which is the optimal solution output by the algorithm.

Standard particle swarm algorithm completes the search for the optimal solution through the individual extremum and global extremum. The operations are simple with fast convergence. However, when the number of iterations increases, the particles become similar with the population convergence. It may result into local optimal solution. Such approach to track the particles' positions is replaced by other methods. The optimal solution is searched through the crossover of the individual extremum and the global extremum, or the mutation of the particles.

3   Principle of ACO Algorithm and Its Model

3.1   Optimization Principle of ACO

Ants have the ability to find the shortest path from their nest to the food without any cues. They can avoid the obstacles appropriately according to the terrain and be adapt to search a new path as the different choice. The nature of such phenomenon is that the ants will release a special kind of secretion called pheromone. Pheromone will disappear gradually over time. The remains can represent the distance of the path. And then, the ants can change their paths according the concentration of the rest pheromone. If the probability of choosing the path also high, more ants will choose this path. They will release more pheromone. When a path is chosen by more ants, it will keep more pheromone. This forms a positive feedback mechanism. Through such positive feedback mechanism, the ants can find its nest to food source of the shortest path finally. In particular, when there is an obstacle between the ants and food source, the ants can not only pass around the obstacles, and they can find the shortest path after a period of time of the positive feedback through the changes of the pheromone in different paths.

3.2 Elitist Ant System (EAS)

EAS is a relatively well global optimization of ACO algorithm [15]. Assumes that a path $e(i, j)$ has the $\tau_{ij}(t)$ as the concentration of the pheromone track at time t. At the initial moment, the different paths have the same pheromone. Ant k $(k = 1, 2, 3, ..., m)$ determines its direction during the movement according to the concentration of the pheromone in each path. $p_{ij}^k(t)$ represents the probability that the ant k shifts from position i to position j at t time. Then the probability can be obtained as follows:

$$p_{ij}^k = \begin{cases} \dfrac{\tau_{ij}^\alpha(t) * \eta_{ij}^\beta(t)}{\sum_{s \in allowed_k} \tau_{is}^\alpha(t) * \eta_{is}^\beta(t)}, & j \in allowed_k \\ 0, & others \end{cases} \tag{3}$$

In (3), $allowed_k = \{0, 1, ..., n - 1\}$. In addition, $tabu_k$ is defined as a taboo list and indicates the positions that ant k can choose in the next step. $tabu_k(k = 1,2,3,...,m)$ is used to record the current position of ant k. η_{ij} represents the visibility of a path $e(i, j)$, which is general set as $\eta_{ij} = \dfrac{1}{d_{ij}}$. d_{ij} represents the distance between position i and position j. α represents the relative importance of the tracks; β indicates the relative importance of visibility; ρ represents the persistent of the track; $1-\rho$ is the disappear degree of the information. After the ants complete a cycle, the amount of the pheromone in each path are adjusted according to the follows:

$$\tau_{ij}(t + n) = \rho \tau_{ij}(t) + \sum_{k=1}^{m} \Delta \tau_{ij}^k(t) \tag{4}$$

3.3 Max-Min Ant System

MMAS (Max-Min Ant System) was proposed as a general-purpose ant colony algorithm [16]. It improved ant colony algorithm in the following areas. Firstly, only the ants in the shortest path could update the pheromone after an iteration. The update methods was same to EAS. Secondly, the concentration of the pheromone was set in the range $[\tau_{min}, \tau_{max}]$ to improve the efficiency. Any values beyond this range would be forced to set in this range as τmin or τmax. Thirdly, the initial values of the pheromone in each path were set as the τmax for better search. Furthermore, a smaller evaporation coefficient was set in order to find more search paths.

4   HOA: Hybrid Optimization Algorithm

4.1   HOA Architecture

The basic idea of HOA is derived from the advantages of PSO and ACO without their defects. The process of HOA has two different stages. The former is to use PSO for the efficiency, the global search ability and the fast convergence. When the premature is emerging, ACO is used to replace PSO. PSO will output the basic information of the paths. They are the initial parameters for ACO, which means ACO can obtain a better initialization compared with the original one. In the process of the algorithm, ant colony algorithm is used for the positive feedback. Its overall framework is shown in Fig. 1.

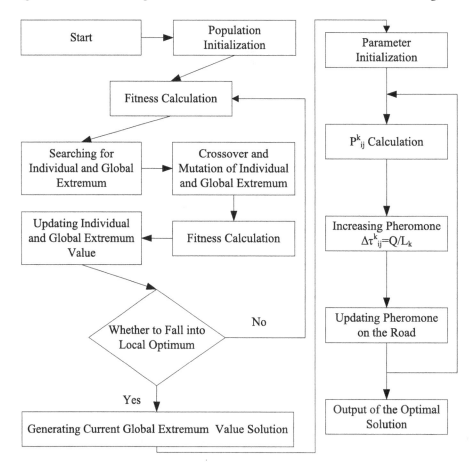

Fig. 1. HOA framework

4.2 Premature Phenomena and Early Maturity of PSO

PSO is similar to many heuristic optimization algorithms just like the other heuristic algorithms. This algorithm has no operations such as selection, crossover and mutant. It means this algorithm is relatively simple with faster convergence. However, if a particle finds a current optimal node, the other particles will quickly move closer to this particle during the processing. If this node is the local optimal node, the particles will fall into the local optimal solution without escape. And the search cannot be restart. The output is a local optimal solution, which is called premature convergence phenomenon.

The existing works showed that particles in PSO will aggregate whether or not the algorithm is premature convergence or global convergence. The particles will aggregate at a special position or several special positions, which is determined by the problem itself and the selection of the fitness function. In our design, the heuristic starts from the normal distribution and fitness variance is adopted as the judgment condition to the premature convergence.

If particle swarm particle number is n, F_i represents the fitness of the i-th particle, F_{avg} represents the average fitness of the PSO, fitness variance σ^2 can be obtained as the follows:

$$\sigma^2 = \sum_{i=1}^{n} \left(\frac{F_i - F_{avg}}{F} \right)^2 \tag{5}$$

In (5), F is the normalized calibration factor and it is used as the limit to the size of the σ^2 variance. F can be obtained as the follows:

$$F = \begin{cases} \max\left|F_i - F_{avg}\right|, \max\left|F_i - F_{avg}\right| > 1, \ among \ i \in [1, n] \\ 1, \qquad\qquad\qquad\qquad\qquad\qquad\quad others \end{cases} \tag{6}$$

The variance gives the degree of aggregation of the particles in the particle swarm. The smaller the σ^2 variance is, the aggregation degree of particle swarm is big. If the algorithm does not meet the end condition and the aggregation is too large, the particle swarm algorithm falls into the so-called premature phenomenon. So when $\sigma^2 \leq h$ (h for a given constant), this algorithm will process using ACO.

5 Experiments and Results Analysis

5.1 Implementation

In order to overcome the problem that PSO is easy to fall into local optimal solution and ACO is lack of initial pheromone, HOA is designed to take use of the advantages of both PSO and ACO. The steps of HOA are as follows. First, PSO is chosen as the initial algorithm. Its parameters are set as the follows. The total number of particles is 100. The parameter h, which is used to judge whether PSO falls into a local optimal solution, is set to 15 as a constant. And then when the PSO falls into a local optimal solution, the output of PSO algorithm will be recorded as bestPath. And then ACO is initialized with the 5 times of bestPath as the initial pheromone concentration. And then ACO starts with 4 as the number of ants. At last ACO gives the output.

5.2 Experimental Results and Analysis

TSP (Travelling Salesman Problem) is an important combinatorial optimization problem, which has been proved to be NP complete. In order to verify HOA, TSP is adopted as the target problem.

Taking the urban plan as an example, the theoretical optimal value is 423.7406. The parameters of PSO algorithm are set as the follows: the total number of particles is 100; the number of cities is 30; The parameter h, which is used to judge whether PSO falls into a local optimal solution, is set to 15 as a constant. The parameters of ACO is set as the follows. α and β are set random values in [1, 2] (here $\alpha = \beta = 2$ as usual); $\rho = 0.9$ (usually from [0, 1]); the number of ants is 4.

The experimental results are shown in Table 1. PSO still contributes most iterations in HOA. PSO will compact the search space as an intermediate one and provide enough initial information for PSO. It also means that when PSO may fall into the local optimal solution during the process. At that moment, the premature is detected and PSO is switched to ACO. ACO only provides relatively less iterations. ACO can complete the whole algorithm in limited iterations. When h is suitable, HOA can complete the process quickly.

Table 1. Experimental results of HOA

α	β	h	HOA (PSO + ACO)	
			Shortest path length	Iterative times
2	2	8	431.9335	165 + 2
2	2	5	425.2667	312 + 0
2	2	10	425.9887	149 + 38
2	2	4	427.8107	167 + 16
2	2	4	425.5095	598 + 3
2	2	12	430.8101	167 + 0
2	2	7	423.7406	267 + 11
2	2	7	429.3803	153 + 5
1	2	8	427.1752	302 + 205
1	2	8	425.6807	164 + 0
1	2	7	448.0349	162 + 44
1	2	7	423.7406	285 + 6
1.5	2	7	423.7406	269 + 4

Table 2 shows the experimental results of the HOA for the length of the shortest paths and the iterations. HOA has less iterations. Furthermore, HOA can iterate repeatedly. And it can also avoid the local optimal solution trap with higher accuracy. The experimental results also show that this algorithm can complete the processing in 30 s, while the average execution time of the hybrid particle swarm algorithm exceeds 100 s. In a word, HOA has better optimization performance on the accuracy and efficiency.

Table 2. Experimental results of PSO

α	β	PSO	
		The shortest path length	Iterative times
2	2	434.8224	598
2	2	429.3803	588
2	2	424.6918	965
2	2	430.1988	867
2	2	446.2989	610
2	2	439.3706	107
2	2	445.9569	588
2	2	445.1756	783
1	2	448.6918	608
1	2	443.7406	253
1.5	2	453.1411	998
1.5	2	434.4903	615
2	2	434.8224	598

Figure 2 shows the average value comparison of HOA and the hybrid particle swarm algorithm after running 20 times. When PSO is running, HOA and hybrid PSO almost have the same curve. It means they almost have the same convergence speed. However, when the convergence is going to the end, HOA is faster. ACO provides more accurate solution.

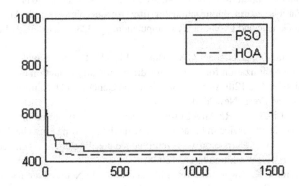

Fig. 2. Comparison of the iterative process of PSO and HOA

6   Conclusions

This paper presents the HOA algorithm based on the particle swarm algorithm and ant colony algorithm. This algorithm is designed to take use of the advantages of both PSO and ACO whereas to avoid their disadvantages. PSO improves the convergence speed and provides the input to ACO for further processing. ACO is used to avoid the premature convergence. The switch time is determined by the fitness variance. ACO helps HOA to provide the accuracy of processing. Our algorithm is verified through the experiments. TSP is used as the target problem. The experimental results shows that our algorithm can provide faster convergence speed and higher accuracy.

References

1. Amudhavel, J., Kumar, K.P., Monica, A., Bhuvaneshwari, B., Jaiganesh, S., Kumar, S.S.: A hybrid ACO-PSO based clustering protocol in VANET. In: The 2015 International Conference on Advanced Research in Computer Science Engineering & Technology. ACM Press, New York (2015). Articles 25
2. Lam, H.T., Nicolaevna, P.N., Quan, N.T.M.: A heuristic particle swarm optimization. In: The 9th Annual Conference on Genetic and Evolutionary Computation, p. 174. ACM Press, New York (2007)
3. Snyman, J.A., Kok, S.: A strongly interacting dynamic particle swarm optimizational method. In: The 9th Annual Conference on Genetic and Evolutionary Computation, p. 183. ACM Press, New York (2007)
4. Wu, C., Zhang, C., Wang, C.: Topology optimization of structures using ant colony optimization. In: The First ACM/SIGEVO Summit on Genetic and Evolutionary Computation, pp. 601–608. ACM Press, New York (2009)
5. Chen, Y., Wong, M.L.: Optimizing stacking ensemble by an ant colony optimization approach. In: The 13th Annual Conference Companion on Genetic and Evolutionary Computation, pp. 7–8. ACM Press, New York (2011)
6. Al-Rifaie, M.M., Bishop, M.J., Blackwell, T.: An investigation into the merger of stochastic diffusion search and particle swarm optimization. In: The 13th Annual Conference on Genetic and Evolutionary Computation, pp. 37–44. ACM Press, New York (2011)
7. Khosla, A.: Particle swarm optimization for fuzzy models. In: The 9th Annual Conference Companion on Genetic and Evolutionary Computation, pp. 3283–3296. ACM Press, New York (2007)
8. Sinnott-Armstrong, N.A., Greene, C.S., Moore, J.H.: Fast genome-wide epistasis analysis using ant colony optimization for multifactor dimensionality reduction analysis on graphics processing units. In: The 12th Annual Conference on Genetic and Evolutionary Computation, pp. 215–216. ACM Press, New York (2010)
9. Cao, S., Qin, Y., Liu, J., Lu, R.: An ACO-Based user community preference clustering system for customized content service in broadband new media platforms. In: The 2008 IEEE/WIC/ACM International Conference on Web Intelligence and Intelligent Agent Technology, pp. 591–595. IEEE Press, Washington, DC (2008)
10. Rajini, A., David, V.K.: Swarm optimization and Flexible Neural Tree for microarray data classification. In: The Second International Conference on Computational Science, Engineering and Information Technology, pp. 261–268. ACM Press, New York (2012)
11. Chen, S., Montgomery, J.: A simple strategy to maintain diversity and reduce crowding in particle swarm optimization. In: The 13th Annual Conference Companion on Genetic and Evolutionary Computation, pp. 811–812. ACM Press, New York (2011)
12. Ugolotti, R., Cagnoni, S.: Automatic tuning of standard PSO versions. In: The Companion Publication of the 2015 Annual Conference on Genetic and Evolutionary Computation, pp. 1501–1502. ACM Press, New York (2015)

13. Abdelbar, A.M.: Is there a computational advantage to representing evaporation rate in ant colony optimization as a gaussian random variable? In: The 14th Annual Conference on Genetic and Evolutionary Computation, pp. 1–8. ACM Press, New York (2012)
14. Chira, C., Pintea,C.M., Crisan, G.C., Dumitrescu, D.: Solving the linear ordering problem using ant models. In: The 11th Annual Conference on Genetic and Evolutionary Computation, pp. 1803–1804. ACM Press, New York (2009)
15. Hemmatiana, H., Fereidoona, A., Sadollahb, A., Bahreininejad, A.: Optimization of laminate stacking sequence for minimizing weight and cost using elitist ant system optimization. Adv. Eng. Softw. **57**, 8–18 (2013)
16. Wang, G., Gong, W., Kastner, R.: Instruction scheduling using MAX-MIN ant system optimization. In: The 15th ACM Great Lakes symposium on VLSI, pp. 44–49. ACM Press, New York (2005)

Understanding Networking Capacity Management in Cloud Computing

Haihui Zhao[1]([⊠]), Yaoguang Qi[1], Hongwei Du[1], Ningning Wang[1],
Guofu Zhang[1], Hailong Lu[2], and Wenbao Liu[2]

[1] College of Mechanical and Electronic Engineering,
China University of Petroleum (East China), .
Qingdao 266555, Shandong, China
zhaohaihui@upc.edu.cn
[2] Energy Equipment and Research Institute
of Lanshi Group Ltd., Lanzhou, China

Abstract. The suction and operation parameters remarkably influence the performance and efficiency of the pumping unit system, the theoretical and experiential model selection methods being verified to be not effective on parameters adoption. The FKM cluster algorithm model selection is put forward to deal with huge data processing in model selection, the efficiency curves comparison indicates: the pumping unit model selection on a oilfield block keep more stable and higher efficiency. The efficiency membership to operation parameters combination is revealed to be the approach to best performance and high efficiency. The reasonable match of equipment and process saves power rather than pure equipment.

Keywords: Pumping unit · Model selection · FKM algorithm · Cloud computing · Cluster

1 Introduction

The conventional pumping unit model selection depends on the personal experience and the general model selection diagram, which are both inaccurate while compared with the practice. Especially for the permeable oil fields, the rule of the producing fluid change being different from the general oilfield, selecting pumping equipment with the general methods will lead to the low utilization rate of load and electrical power, and the low system efficiency. Conventional selection method being on the basis of the model selection diagram [1,2], the

This work was supported by grants from National Major Special Project of Oil and Gas "Study and Promotion of the Self-Adaptive Control Technology of Drainage Based on Shaft Flow Field" (2016ZX05042003-001); National Major Special Project of Oil and Gas "Key Equipment Development of Integrated Development of Three Kind of Unconventional gas in One Well" (2016ZX05066004-002); "Fundamental Research Funds for the Central Universities" (16CX02004A); NSFC (51174224).

© Springer International Publishing AG 2017
M. Qiu (Ed.): SmartCom 2016, LNCS 10135, pp. 32–41, 2017.
DOI: 10.1007/978-3-319-52015-5_4

diagram is suitable for the general oilfield block, but not for the unconventional reservoir. The unconventional reservoir differs from the conventional reservoir on the aspects of permeability, pressure and other undetermined factors.

Moreover, the factors possess some fuzzy characteristics [3], which are not involved in the diagram method and experience method [4]. The test data consists of the basis on which the characteristics can be embodied and the rule of the load change can be induced while the reasonable algorithm is used, The model selection based on production data processing can provide the approximation selection. The cloud computing has been verified to be able to process the huge data processing and engineering project well in recent years. This research attempts to find a method about selecting model of pumping units on the production and reservoir parameters [1,5].

The main contributions of this work are threefold:

1. A new thinking of the data classification is provided, and the FKM algorithm is induced into the application of pumping unit model selection. The fuzzy membership is solved to embody the efficiency sensitivity to the load utilization rate, the torque utilization rate and other suction parameters etc.
2. Based on the general computing platform analysis, the cloud computing platform special for model selection and efficiency evaluation is set up, and it can be used to compute and analyze the test data in distributed computing mode.
3. We put forward the innovative use of the cloud computing method by means of the FKM algorithm on the basis of the huge test data.

2   Related Work

2.1   Pumping Unit Model Selection Principle

The model selection was usually implemented in this technical route: determining the pump depth and the sucker rod combination; analyzing the balance situation and the load utilization rate, torque utilization rate and system efficiency; contrasting and analyzing different parameters combination and determine the mode selection.

2.1.1   Pump Depth

The reasonable flow pressure is determined by Eq. (1).

$$p_{fp} = \frac{1}{1-n}(\sqrt{n^2 p_s^2 + n(1-n)p_s p_{ef}} - n p_s) \tag{1}$$

Among them, $n = \frac{0.1033\alpha z t(1-f)}{293.15B}$. p_{fp} is the minimum allowable flowing pressure; p_s is the saturation pressure; p_{ef} is the effective formation pressure; α is the crude oil solubility coefficient; f is the water cut in oil well, B is the crude oil volume factor, dimensionless; T is the temperature of oil layer. The pump depth is:

$$L_p = H - \frac{p_{pf} - \rho g H_s}{\rho g} \tag{2}$$

2.1.2 Reducer Torque

According to the determined sucker rod combination, the operating parameter and experimental formula, the reducer torque computation of several typical pumping units is provided as the followings:

Conventional beam-pumping unit:

$$M_N = \overline{TF}[P - B - W_b\frac{c}{a}(1 - \frac{ca_A}{ag})]\eta_b^{kl} - M_c sin(\theta - \tau) \tag{3}$$

Dual horsehead beam-pumping unit:

$$M_N = \overline{TF}(P - B)\eta_b^{kl} - M_0 sin\theta \tag{4}$$

Derrick framed pumping unit:
Up stroke:

$$M_N = (P - Q)R + J\frac{d\omega}{dt} \tag{5}$$

Down stroke:

$$M_N = (Q - P)R + J\frac{d\omega}{dt} \tag{6}$$

Depth is a main factor determining the polished rod load, which is the variable considered firstly while selecting the pumping unit model, the polished rod load thereby being the basic parameter of the reducer torque. Both of them are the precondition of model selection in diagram method. In the condition of general reservoir suction parameters, the diagram method is extensively used. But in the unconventional reservoir, the formation parameters vary remarkably from them of the conventional reservoir. Simultaneously, this method is out of the advantage of the test data's accuracy in prediction of the load data and model selection. Test data's processing and analyzing require the advanced computing technology [6,7].

2.2 Cloud Computing in Data Processing

Huge data is generated during the test data acquisition, and the data processing in model selection requires high computing and analyzing ability. The cloud computing has the advantage for big data processing, having been verified in many fields. To process the big data, the cloud computing task is distributed into a large number of computer resource pool, enabling all kinds of application system obtain the computing power, storage space [8–10] and all kinds of software service [11–13]. The large amount of data cannot be computed before because of the relative slow processing speed of the computer and the server. Now, the cloud computing and the concurrent computing technology are mature and introduced in the huge data, the data computation will be distributed to enough servers [14,15]. As shown in Fig. 1, this platform is composed of 3 layers:

1. Physical layer includes hardware resources, such as computers, servers, hard disks and software system to control these resources. Resource pool is the virtual resources virtualized from the physical resource [16], in order to facilitate the unified management [17,18].
2. Data processing layer is responsible for task scheduling [19], resource management [20], user management and other administrative tasks [21].
3. The application layer encapsulates cloud services into standard service to the costumers and provides computation and analysis [22,23].

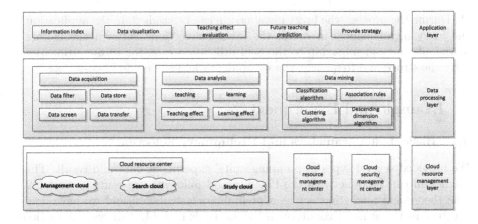

Fig. 1. Three layers of the platform.

3 FKM Clusters on Suction Parameters

To process the huge data of the test well, the FKM cluster analysis method is introduced according to the efficiency membership to the performance parameters combination. The basic principle of FKM method is to cluster the positions data into several classes, and the data (1<k<n) is divided into k classes. The u_{ij} represents the x_j membership coefficient to the c_i class. The fuzzy matrix $u = \{u_{ij}\}_{n\mathbb{x}k}$ represents the sample $\{x_j\}_n$ membership to every fuzzy cluster $c = \{c_j\}_k$, so every fuzzy cluster center is solved, shown as $m = \{m_i\}_k$.

Equation (7) is the objective function. to make the objective function reach the min value. Equations (8) and (9) should be fitted.

$$u_{ij} = \begin{cases} 1 & \|x_i - m_j\| = 0(i - j) \\ 0 & \|x_i - m_j\| \neq 0(i \neq j) \\ (\sum_{r-1}^{k} \frac{\|x_i-m_j\|^{2/(q-1)}}{\|x_i-m_r\|^{2/(q-1)}})^{-1} & \|x_i - m_j\| \neq 0 \end{cases} \tag{7}$$

$$J_w(x,c) = \sum_{j=1}^{k} J_j = \sum_{j=1}^{k}\sum_{i=1}^{n} u_{ij}^q \|x_i - m_j\|^2 \tag{8}$$

$$m_j(s+1) = \frac{\sum\limits_{i=1}^{n} [u_{ij}(s)]^q x_j}{\sum\limits_{i=1}^{n} [u_{ij}(s)]^q} \quad \text{where } i \in [1,k] \tag{9}$$

$$\sum_{i=1}^{n} u_{ij} = 1 \text{ where } \forall i, i \in [1,n], j \in [1,k]$$

$$0 < \sum_{i=1}^{n} u_{ij} < n \text{ where } j, i \in [1,n], j \in [1,k] \tag{10}$$

During the process, the fuzzy membership can be studied. In the application, the fuzzy membership can reveal the inner connections between the factors, or the efficiency membership to the load utilization rate, the torque utilization rate, the stroke and frequency, the bump depth and the crude viscosity etc.

4 Example and Experiment

4.1 Basic Test Parameters and Preliminary Selection

We provided several types of pumping units' experiment data and result. This prediction can be done with the Eqs. 1–6 and following the route in Sect. 2.2. The theoretical prediction act as a?preliminary?calculation and we have done the test in the general condition listed in Table 1. Table 2 shows the suction parameters and the combination parameters that we solved according to the theoretical model (the two tables of data are about the CYJ10-4-73HY pumping unit). To indicate the effect on model selection on the basis of FKM clusters result, the three types of classical pumping units were selected and every type of pumping units were respectively chosen of 10 sets, being tested on the aspects of load utilization, power utilization, performance and efficiency etc. The results are shown in Table 3, Figs. 2 and 3.

4.2 Experiment Result of the FKM Data Processing

The system efficiency and the pump efficiency, both related with the pumping unit stroke and frequency, Table 4 shows the basic suction parameters and operating parameters in cloud computing, and Table 5 shows the three classical beam-pumping units model selection on FKM data processing in cloud computing.

Table 1. The basic data of 19# block.

Daily production (t)	Oil layer depth (m)	Saturation pressure (MPa)	Formation pressure (MPa)	Crude oil viscosity (mPa.s)	Crude oil density (t/m3)	Moisture content (%)	Following pressing (MPa)	Submerge depth (m)
2.2	2000	3.19	19.50	14.5	0.8396	91	2	297

Table 2. The model parameters selection

Motor related power (t)	Stroke (m)	Frequency (min⁻¹)	Pump diameter (mm)	Maximum load (kN)	Minimum (kN)	Max torque (kN.m)	Sucker combination (mm)
73	3.2	4	38	77	55	19.1	$\varphi19 \times \varphi22$

Table 3. The three classical pumping units model selection result. (DPC: Day Power Consumption)

Model	Polished rod load (kN)		Load (%)	Utilization	Reducer torque (kN.m)		Torque rate (%)	Utilization	DPC (kw.h)
	Average	Max	Average	Max	Average	Max	Average	Max	
CYJ10-4-73YH	21.7	26.2	50.9	67.5	62.3	116.5	45.3	91.5	62.8
WCYJJ10-6-12Z	25.5	39.7	51.1	71.1	76.3	129.7	41.6	82.4	79.7
WCYJD12-06-40Z	36.3	43.1	46.2	72.3	85.1	169.1	32.7	65.5	91.2

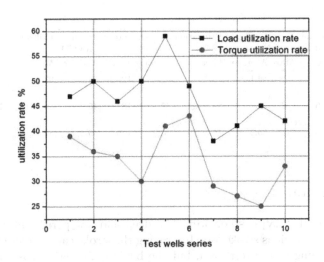

Fig. 2. Utilization curves of theoretical selection l model.

Table 4. Basic suction parameters and operating parameters in cloud computing

Motor related power (t)	Stroke (m)	Frequency (min⁻¹)	Pump diameter (mm)	Maximum load (kN)	Minimum (kN)	Max torque (kN.m)	Sucker combination (mm)
70	5.1	2.5	38	90	62	25.1	$\varphi19 \times \varphi22$

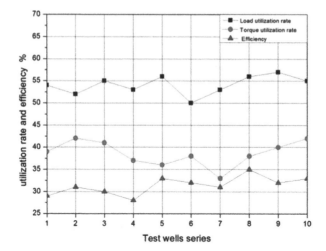

Fig. 3. Efficiency and utilization curves of optimized selection model.

Table 5. The three classical pumping units model selection result. (DPC: Day Power Consumption)

Model	Polished rod load (kN)		Load (%)	Utilization	Reducer torque (kN.m)		Torque rate (%)	Utilization	DPC (kw.h)
	Average	Max	Average	Max	Average	Max	Average	Max	
CYJ10-4-73YH	24.5	28.5	56.1	76.7	65.5	125.1	51.4	97.6	55.1
WCYJJ10-6-12Z	29.6	45.5	57.7	80.5	82.7	137.6	45.9	86.3	73.6
WCYJD12-06-40Z	39.6	502	55.5	76.3	90.9	175.7	36.5	82.1	80.8

5 Discussion

1. **Pumping unit type selection:** The deep and low permeability reservoir can result in obvious stroke loss. Improving the stroke can decrease the stroke loss, according to the test data, but the beam pumping unit is not suitable for the reservoir because it cannot afford the large polished rod load and long stroke. But it can be known that the long stroke chain pumping unit efficiency is 6–10% higher than that of the beam-pumping unit, which can be known by comparing the curves in Figs. 2 and 3.

2. **Motor type selection:** As for the pumping units with new types of motors, although their system efficiency is higher (for example some pumping units equipped with permanent magnet motor), the motor is not reliable, and the cost is much more than the three-phase asynchronous motor. The tree-phase asynchronous motor has the performance stability and reliability. So, in this paper, the tree-phase asynchronous motor is proposed to be selected prior in beam pumping unit and chain pumping unit.

3. **Efficiency:** The study shows that the utilization rate of the load and torque are both lower than expected, and the utilization rate of load and torque has the direct relation with the system efficiency. Quite some wells' efficiency being low, they are lower than the theoretical efficiency by 5–11%. As the study result, the Fig. 1 shows the system efficiency of the pumping units whose model selection is based on the preliminary selection of formula (1–6 etc.), and the Fig. 2 shows the system efficiency of the model selection on FKM algorithm in cloud computing. It can be seen that the efficiency is not stable and vary from different suction parameters in Fig. 1, but the efficiency curves in Fig. 2 is stable and higher than that in Fig. 1. The average increase is as following: beam pumping unit of 7.7–9.1; motor commutation pumping unit of 3.5–6.2; chain pumping unit of 3.3–6.1. The efficiency membership to the operating parameters combination is revealed in the clusters result. It indicates: the high efficient wells operating parameters act as the cluster centers in the cloud computing data processing, based on the formula 7–10, the high efficient suction parameters combination can be deduced. In theory, the FKM cluster centers can be regarded as the maximum value point in well system efficiency.

6 Conclusion

The pumping unit theoretical model selection is not accurate, whose efficiency cannot reach the high value, and the pumping units' efficiency is not stable, the theoretical pumping unit model selection method has limit usage. Moreover, the FKM cluster algorithm leads to the more accurate result, and the pumping units' efficiency is stable and higher than that of the formula model selection. The high efficiency well's suction parameters acting as the original clusters, the efficiency membership to operating parameters combination is solved. It provides a new idea in pursuing high efficiency and reasonable equipment-process match. The equipment cannot save power, and the key of saving power point is the reasonable match of the equipment and the process. Finally, the chain pumping unit and motor commutation pumping unit are more reasonable with long stroke and low frequency than beam-pumping unit, having 6–10% higher efficiency than that of the beam pumping unit.

References

1. Zhao, H., Qi, Y., Du, H., Wang, N., Zhang, G., Liu, W., Lu, H.: Cloud computation processing for oilfield block data and chain drive pumping unit polished rod motion model. J. Signal Process. Syst. **85**, 1–10 (2016)
2. Yin, H., Gai, K.: An empirical study on preprocessing high-dimensional class-imbalanced data for classification. In: The IEEE International Symposium on Big Data Security on Cloud, pp. 1314–1319, New York, USA (2015)
3. Yin, H., Gai, K., Wang, Z.: A classification algorithm based on ensemble feature selections for imbalanced-class dataset. In: The 2nd IEEE International Conference on High Performance and Smart Computing, pp. 245–249, New York, USA (2016)

4. Gai, K., Qiu, M., Zhao, H., Dai, W.: Anti-counterfeit schema using monte carlo simulation for e-commerce in cloud systems. In: The 2nd IEEE International Conference on Cyber Security and Cloud Computing, pp. 74–79. IEEE, New York, USA (2015)
5. Zhao, H., Qi, Y., Du, H., Wang, N., Zhang, G., Liu, W., Lu, H.: Running state of the high energy consuming equipment and energy saving countermeasure for Chinese petroleum industry in cloud computing. Concurr. Comput. Pract. Exp. **28**, 1 (2016)
6. Zhu, J.: Pumping unit selection and energy saving optimization design method research. Oil Prod. Eng. **3**(2), 42–45 (2013)
7. Zheng, B.: Duplex permanent magnet motor pumping unit. Oil-Gas Field Surf. Eng. **29**(8), 109–110 (2010)
8. Li, Y., Gai, K., Qiu, L., Qiu, M., Zhao, H.: Intelligent cryptography approach for secure distributed big data storage in cloud computing. Inf. Sci. **PP**(99), 1 (2016)
9. Gai, K., Qiu, M., Zhao, H.: Security-aware efficient mass distributed storage approach for cloud systems in big data. In: The 2nd IEEE International Conference on Big Data Security on Cloud, pp. 140–145, New York, USA (2016)
10. Gai, K., Qiu, M., Chen, L., Liu, M.: Electronic health record error prevention approach using ontology in big data. In: 17th IEEE International Conference on High Performance Computing and Communications, pp. 752–757, New York, USA (2015)
11. Gai, K., Li, S.: Towards cloud computing: a literature review on cloud computing and its development trends. In: 2012 Fourth International Conference on Multimedia Information Networking and Security, pp. 142–146. Nanjing, China (2012)
12. Gai, K., Qiu, M., Thuraisingham, B., Tao, L.: Proactive attribute-based secure data schema for mobile cloud in financial industry. In: The IEEE International Symposium on Big Data Security on Cloud; 17th IEEE International Conference on High Performance Computing and Communications, pp. 1332–1337, New York, USA (2015)
13. Qiu, M., Gai, K., Thuraisingham, B., Tao, L., Zhao, H.: Proactive user-centric secure data scheme using attribute-based semantic access controls for mobile clouds in financial industry. Future Gener. Comput. Syst. **PP**, 1 (2016)
14. Gai, K., Qiu, M., Tao, L., Zhu, Y.: Intrusion detection techniques for mobile cloud computing in heterogeneous 5G. Secur. Commun. Netw. **9**, 1–10 (2015)
15. Gai, K., Steenkamp, A.: A feasibility study of platform-as-a-service using cloud computing for a global service organization. J. Inf. Syst. Appl. Res. **7**, 28–42 (2014)
16. Gai, K., Qiu, M., Zhao, H.: Cost-aware multimedia data allocation for heterogeneous memory using genetic algorithm in cloud computing. IEEE Trans. Cloud Comput. **PP**(99), 1–11 (2016)
17. Zhang, S., Zhang, S., Duan, H.: FKM-based cluster analysis method for intelligent form selection of high-rise structures. In: The 17th National Conference on Computer Application of Engineering Construction, p. 1 (2014)
18. Gai, K., Qiu, M., Zhao, H., Xiong, J.: Privacy-aware adaptive data encryption strategy of big data in cloud computing. In: 2016 IEEE 3rd International Conference on Cyber Security and Cloud Computing (CSCloud), The 2nd IEEE International Conference of Scalable and Smart Cloud (SSC 2016), pp. 273–278. IEEE, Beijing, China (2016)
19. Gai, K., Du, Z., Qiu, M., Zhao, H.: Efficiency-aware workload optimizations of heterogenous cloud computing for capacity planning in financial industry. In: The 2nd IEEE International Conference on Cyber Security and Cloud Computing, pp. 1–6. IEEE, New York, USA (2015)

20. Gai, K., Qiu, M., Zhao, H., Liu, M.: Energy-aware optimal task assignment for mobile heterogeneous embedded systems in cloud computing. In: 2016 IEEE 3rd International Conference on Cyber Security and Cloud Computing (CSCloud), pp. 198–203. IEEE, Beijing, China (2016)
21. Gai, K., Qiu, M., Zhao, H., Tao, L., Zong, Z.: Dynamic energy-aware cloudlet-based mobile cloud computing model for green computing. J. Netw. Comput. Appl. **59**, 46–54 (2015)
22. Gai, K., Qiu, M., Chen, M., Zhao, H.: SA-EAST: security-aware efficient data transmission for ITS in mobile heterogeneous cloud computing. ACM Trans. Embed. Comput. Syst. **PP**, 1 (2016)
23. Li, Y., Gai, K., Ming, Z., Zhao, H., Qiu, M.: Intercrossed access control for secure financial services on multimedia big data in cloud systems. ACM Trans. Multimed. Comput. Commun. Appl. **12**(4s), 67 (2016)

Cloud Learning Community of Engineering Drawing

Fenna Zhang$^{(\boxtimes)}$, Yaoguang Qi, Long Pan, Yong Yang, Hao Zhang,
and Yao Yao

College of Mechanical and Electronic Engineering,
China University of Petroleum (East China),
Qingdao 266555, Shandong, China
zhangfn@upc.edu.cn

Abstract. General online teaching system cannot afford the huge information data exchange, held back in the modern teaching method and technology support. This paper aims at establishing the structure of the cloud learning community on big data. The Oracle server on cloud computing is selected to provide the data processing support. Based on the investigation on the students' browse online and the homework completion situation, the existing teaching resource is integrated, and the frame work of the cloud learning community on big data is established to improve the communication and integration. Cloud platform layers and the key data processing technology are analyzed. The cloud learning community can match the data processing technology and expose the students in the advanced cloud teaching stimulate the study enthusiasm.

Keywords: Cloud learning community · Engineering drawing · Big data · Resource integration · Online teaching

1   Introduction

"Engineering Drawing" aims at training the students on drawing and analyzing, especially emphasizing the space imagination ability. It was usually found that some students studied other course well but could not follow up in this course. This phenomenon results from the absence of the space imagination ability, thereby the efficient teaching and learning mode is required. The multi-media teaching and learning community are being used widely, having been mature in technology [1–3]. But now, in the big data and cloud era, the information and material transferred being much larger, the original online teaching cannot fit

This work was supported by grants from National Major Special Project of Oil and Gas "Study and Promotion of the Self-Adaptive Control Technology of Drainage Based on Shaft Flow Field" (2016ZX05042003-001); National Major Special Project of Oil and Gas "Key Equipment Development of Integrated Development of Three Kind of Unconventional gas in One Well" (2016ZX05066004-002); "Fundamental Research Funds for the Central Universities" (16CX02004A); NSFC (51174224).

M. Qiu (Ed.): SmartCom 2016, LNCS 10135, pp. 42–50, 2017.
DOI: 10.1007/978-3-319-52015-5_5

the requirement of big data [4, 5]. Learning community online requiring vast data processing, the recent development of new technologies and the implementations have enabled a variety of innovative network-based approaches, such as cloud computing [6–10] and big data. Many improvements have been made by previous researches for the purpose of the high performance achievements [11, 12], and it can benefit the huge amount data exchange in the learning community. This paper addresses the issue of cloud learning community on big data solution.

2 Related Work

2.1 Technology Supports

In distributed cloud storage system, the management, access control [13], data retrieval, and the huge data processing require high-end server to ensure the efficiency, high availability and high robustness [14, 15]. Oracle is by far the most popular in the large-scale relational database management system. As an open distributed database, it can work properly under all kinds of heterogeneous systems and has the high transaction processing speed. The advantages make it be the optimized server in the cloud teaching of this research.

Moreover, database management system includes resource library, database and library management system. The resource library is the integration of the model files, video files and graphics files; database stores the directory of all files in the resource library; the library management system is responsible for managing the resource library and the database, its main function being to add, delete, update and invoke database [16, 17].

Next, multi-media teaching material and models online can be browsed in students' online learning, and the exercises, tests can be done online [18, 19]. The teaching video is abundant after accumulation and the model online provide the operations such as rotating, moving and cutting etc. Updating and recombination is the approach to the cloud teaching under cloud teaching conditions [20].

2.2 Investigation on Browse Online

The teacher is traditionally the inculcator but the organizer in online teaching. The constructivism deemed: the students' experience should be stimulated to be the new growth points of knowledge. Online learning encourages students' enthusiasms and has been investigated on the browse purpose, as shown in Fig. 1. (Note: the y value represents the ratio between the student amount on an aspect browse and the total sample student amount.) The chart indicates: the teaching video is the most popular; the guide, analysis and group discussion are often viewed; the independent learning is the mode students prefer.

2.3 Investigation on Homework Completion

An investigation had been taken during 5 months, in order to testify the effect of the independent learning. To show the pertinence, the sample students were

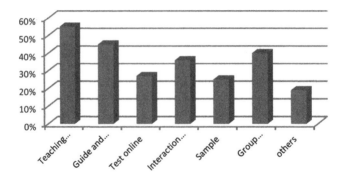

Fig. 1. Students attention and online purpose distribution.

divided into 3 levels: level 1, students with 80–100 score; level 2, students with 60–80 score; level 3, students with score below 60. The chart shown in Fig. 2 indicates: the cost time on homework will descend by about 2 hours for level 2 and level 3. For level1, the time increase a little because of the further thinking independently. From the chart it can be known that the independent learning effect is obvious.

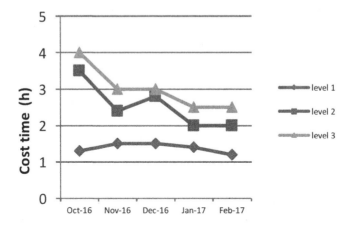

Fig. 2. The home work cost time of 3 levels students.

3   Content of Learning Community for Mechanical Drawing

Learning community is a virtual teaching and learning structure, being for independent learning, emphasizing the independence and the interaction. According to the subject construction in the UPC teaching renovation, the structure model is given as Figs. 3 and 4. The learning community mainly includes the teachers and students, but to evaluate the effect of the teaching and learning, the educational evaluation committee is involved in. Figure 3 shows the members construct relationship. The community structure is shown in Fig. 4. As the preliminary work, the multi-media online teaching needs to be led further and supplemented:

Firstly, the learning community is the update for the online teaching. The teaching video contains huge data to ensure the resolution ratio, but the original background servers were usually Access servers, not suitable for huge data transfer and exchange. The Oracle background server should be taken for larger data base, which is mentioned in Sect. 2. Simultaneously, our national quality curriculum (Engineering Drawing) needs to be turned perfected and the test bank online should be supplemented too. Secondly, the modeling room online will be in use, which is in VR mode. In this VR environment, the roam mode for viewing is emphasized, and the students can study the model in any angle as they wish. The individualized learning and collaborative learning can be realized. Thirdly, the discussion group online can run on interaction. It can hold meeting for discussion online and exchange the information freely. The students who do not want to ask teacher questions in reality can freely find the answer. The learning community is a salon online, being not limited of time.

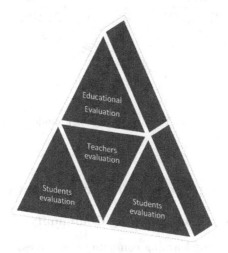

Fig. 3. The learning community academic evaluation.

Community is characterized with the interaction to some extent, the interaction being the basis. The teacher, the student, the teaching resource and the organizing strategy compose the main body. The interaction model is shown in Fig. 5. In this model, the teaching management platform is the support for the interaction. The interaction sent by teacher contains: teacher's guide to student study; announcing course message; assigning homework; organizing discussion

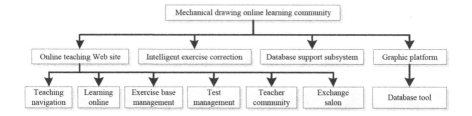

Fig. 4. The learning community structure.

and answering questions. The interaction from student includes: asking questions; self testing; exercise online and submission for correction; discussing and cooperating [21,22].

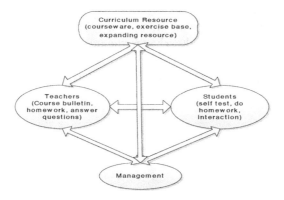

Fig. 5. Teaching interaction mode.

4   Cloud Learning Community in Big Data

4.1   Significance of Cloud Learning Community

In the era of big data, the learning community on Access server is not suitable of high data processing, thereby cloud computing and big data technology is adopted with the advantage of big data storage, computing, big data analysis and data mining to support the learning community [23]. Cloud learning community uses the cloud storage and compute technology to provide a more diversified and individual character of teaching and learning activities. The teachers and students can not only understand the graphics teaching status quo according to the data, but also can predict the future in teaching, so that they can make the right forecast and decisions according to the present situation. All above talked put the teachers and students in an active state.

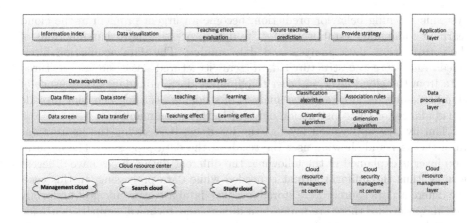

Fig. 6. Cloud computing platform construction.

4.2 Big Data Origin of Cloud Learning Community

Cloud data source of learning community mainly includes the following aspects: (1) Sakai database, mainly storing the basic information of teachers and students; (2) the mongo database recording time duration about the students learning; (3) the record about the number of teachers and students' post/replies; (4) the information records of the user's role, function modules.

4.3 Cloud Learning Community Layer Construction

The basis of the cloud learning community on big data is shown in Fig. 6. Known from this figure, the learning community is composed of 3 layers: the cloud resource management layer, the data processing layer and the application layer [22].

Firstly, the cloud resource management layer contains cloud resource center, cloud resource management center and cloud security management center. The cloud resource center is to preserve and update the resource of management cloud, search could and learning cloud. The cloud resource management center is to manage the resources of monitoring, distribution, recycling, updating and maintenance etc. The existing data resources include the cloud classroom, the number of teachers and students in the classroom, the total number of courses, student's learning behavior records and other data sources.

Secondly, with the increase of the content about the cloud learning community, it will produce a lot of data. But the current teaching support system is short of the data analysis, and there is no system analysis to feedback the data on teaching work. Thus the data cannot play the role it should be. So, providing the method of using the existing data to forecast and analyze it is necessary. And then, how to feedback the data and promote teachers' teaching quality, how to integrate the existing teaching resources, and how to analyze the systemic teaching behavior prediction, become an urgent problem in the cloud learning community establishment. The data processing layer has functions of data acquisition, data analysis and data mining. The data acquisition filter, screen and store the structured data, semi-structured data and unstructured data from the cloud layer, in order to eliminate data redundancy. The usage of cloud teaching resource and the analysis of students' learning behavior provide the guidance for the construction of cloud learning community.

Thirdly, the application layer is the interface, it receiving the instructions from the user and showing user the result of the data processed. The application layer is explained in the following: the different data sources have different format, thus the uniform format is needed while data processing. In this case, the different data sources were stored in Sakai database, convenient of uniformly invoking. Because there are many data dimensions, technical support is needed to analyze and find out the data dimensions for implementing further data mining.

4.4   Technical Supports

It is difficult to use cloud computing technology to analyze the connection between the various data, bring obstacle in maximize the application effect of different education resources. While constructing the platform, the technology is selected on purpose to weaken the data difference effect. In the teaching resources sharing platform, the data processing layer is the most important, and the parallel loading of the data storage module becomes the core of the whole platform. The Hadoop distributed technology provides the platform, the model, the method of data storage and the data processing, thereby the technology of the platform based on the big data has the following core:

1. Using the Hadoop distributed file system to store huge amounts of teaching source data;
2. Dealing with the huge amounts of data with the Map Reduce distributed computing model;
3. Using Sakai distributed database to store the processed huge amounts of data, in order to realize the mass of teaching data storage management.

The learning community is integrated into the application layer, and this system can integrate the students and the teachers in interaction with huge data exchange which can transfer freely on cloud computing.

5   Conclusions

Integration is the most relied function in the learning community, the exchanged huge data processing requiring the Oracle server on big data and cloud computing to upgrade the original teaching resource. The frame of the learning community is put forward, and the existing resource should be updated to fit the huge data exchanging requirement, the key technology to integrate the modules is the distributed cloud computing technology. The cloud computing platform construction is provided, treating the existing database as the part of the application layer. It provided the connection between the existing module and the basic cloud layer, being the approach to cloud learning community realization.

References

1. Gai, K., Qiu, M., Zhao, H., Tao, L., Zong, Z.: Dynamic energy-aware cloudlet-based mobile cloud computing model for green computing. J. Netw. Comput. Appl. **59**, 46–54 (2015)
2. Gai, K., Qiu, M., Zhao, H.: Cost-aware multimedia data allocation for heterogeneous memory using genetic algorithm in cloud computing. IEEE Trans. Cloud Comput. **99**, 1–11 (2016)
3. Gai, K., Qiu, M., Chen, M., Zhao, H.: SA-EAST: security-aware efficient data transmission for ITS in mobile heterogeneous cloud computing. ACM Trans. Embed. Comput. Syst. **PP**, 1 (2016)
4. Li, Y., Gai, K., Ming, Z., Zhao, H., Qiu, M.: Intercrossed access control for secure financial services on multimedia big data in cloud systems. ACM Trans. Multimed. Comput. Commun. Appl. **12**(4s), 67 (2016)
5. Li, Y., Gai, K., Qiu, L., Qiu, M., Zhao, H.: Intelligent cryptography approach for secure distributed big data storage in cloud computing. Inf. Sci. **PP**(99), 1 (2016)
6. Gai, K., Qiu, M., Zhao, H.: Security-aware efficient mass distributed storage approach for cloud systems in big data. In: The 2nd IEEE International Conference on Big Data Security on Cloud, pp. 140–145, New York, USA (2016)
7. Gai, K., Li, S.: Towards cloud computing: a literature review on cloud computing and its development trends. In: 2012 Fourth International Conference on Multimedia Information Networking and Security, pp. 142–146, Nanjing, China (2012)
8. Gai, K., Qiu, M., Thuraisingham, B., Tao, L.: Proactive attribute-based secure data schema for mobile cloud in financial industry. In: The IEEE International Symposium on Big Data Security on Cloud; 17th IEEE International Conference on High Performance Computing and Communications, pp. 1332–1337, New York, USA (2015)
9. Gai, K., Du, Z., Qiu, M., Zhao, H.: Efficiency-aware workload optimizations of heterogenous cloud computing for capacity planning in financial industry. In: The 2nd IEEE International Conference on Cyber Security and Cloud Computing, pp. 1–6. IEEE, New York, USA (2015)
10. Qiu, M., Zhong, M., Li, J., Gai, K., Zong, Z.: Phase-change memory optimization for green cloud with genetic algorithm. IEEE Trans. Comput. **64**(12), 3528–3540 (2015)
11. Yin, H., Gai, K.: An empirical study on preprocessing high-dimensional class-imbalanced data for classification. In: The IEEE International Symposium on Big Data Security on Cloud, pp. 1314–1319, New York, USA (2015)
12. Gai, K., Qiu, M., Chen, L., Liu, M.: Electronic health record error prevention approach using ontology in big data. In: 17th IEEE International Conference on High Performance Computing and Communications, pp. 752–757, New York, USA (2015)
13. Ma, L., Tao, L., Zhong, Y., Gai, K., RuleSN: research and application of social network access control model. In: IEEE International Conference on Intelligent Data and Security, pp. 418–423, New York, USA (2016)

14. Gai, K., Qiu, M., Zhao, H., Xiong, J.: Privacy-aware adaptive data encryption strategy of big data in cloud computing. In: 2016 IEEE 3rd International Conference on Cyber Security and Cloud Computing (CSCloud), The 2nd IEEE International Conference of Scalable and Smart Cloud (SSC 2016), pp. 273–278. IEEE, Beijing, China (2016)
15. Gai, K., Qiu, M., Zhao, H., Liu, M.: Energy-aware optimal task assignment for mobile heterogeneous embedded systems in cloud computing. In: 2016 IEEE 3rd International Conference on Cyber Security and Cloud Computing (CSCloud), pp. 198–203. IEEE, Beijing, China, (2016)
16. Gai, K., Steenkamp, A.: A feasibility study of Platform-as-a-Service using cloud computing for a global service organization. J. Inf. Syst. Appl. Res. **7**, 28–42 (2014)
17. Gai, K., Steenkamp, A.: Feasibility of a Platform-as-a-Service implementation using cloud computing for a global service organization. In: Proceedings of the Conference for Information Systems Applied Research ISSN, vol. 2167, p. 1508 (2013)
18. Gai, K., Qiu, M., Zhao, H., Dai, W.: Anti-counterfeit schema using monte carlo simulation for e-commerce in cloud systems. In: The 2nd IEEE International Conference on Cyber Security and Cloud Computing, pp. 74–79. IEEE, New York, USA (2015)
19. Hoi, S., Wang, J., Zhao, P.: Libol: a library for online learning algorithms. J. Mach. Learn. Res. **15**(1), 495–499 (2014)
20. Steenkamp, A., Alawdah, A., Almasri, O., Gai, K., Khattab, N., Swaby, C., Abaas, R.: Teaching case enterprise architecture specification case study. J. Inf. Syst. Educ. **24**(2), 105 (2013)
21. Agrawal, D., Das, S., El Abbadi, A.: Big data, cloud computing: current state and future opportunities. In: Proceedings of the 14th International Conference on Extending Database Technology, pp. 530–533. ACM (2011)
22. Demirkan, H., Delen, D.: Leveraging the capabilities of service-oriented decision support systems: putting analytics and big data in cloud. Decis. Support Syst. **55**(1), 412–421 (2013)
23. Yin, H., Gai, K., Wang, Z.: A classification algorithm based on ensemble feature selections for imbalanced-class dataset. In: The 2nd IEEE International Conference on High Performance and Smart Computing, pp. 245–249, New York, USA (2016)

Energy Saving Method for On-Chip Data Bus Based on Bit Switching Activity Perception with Multi-encoding

Mingquan Zhang[1,2(✉)], Zhihua Gan[1], Zhimin Gu[1], and Jizan Zhang[1]

[1] School of Computer Science and Technology, Beijing Institute of Technology,
Beijing, China
ncepu_zmq@163.com
[2] Department of Computer, North China Electric Power University,
Baoding, China

Abstract. The dynamic energy consumption of the multi-core on-chip data bus is more and more large in the whole system energy consumption. With the reduction of the technology size the bus dynamic energy consumption, which is brought by the coupling switching activity (SA) is increasing, and the effect of one single bus encoding on the bus energy saving is not significant. To settle the problems, we propose a new method for energy saving of on-chip data bus, which is based on bit SA perception, and four bus encoding schemes are introduced. By means of hardware structures and algorithms, the number of bit SA in each encoding scheme is perceived, and the encoding scheme with the minimum SA number is automatically selected to encode the value to be transferred. The simulation results show that the method can effectively optimize dynamic energy saving of on-chip data bus.

Keywords: Bus energy saving · Switching activity · Coupling capacitance · Invert encoding

1 Introduction

With the development of the program complexity and the gradually increased integration of the chip, the energy consumption of per unit area and per unit time is significantly increased, which brings many problems. Firstly, the rapid increase of energy consumption is bound to hinder the performance further improved, studies have shown that now in multi-core especially in on-chip bus design, the urgent problem to be solve is dynamic energy consumption control [1,2]. Secondly, the high on-chip bus energy consumption will increase the cost of the chip design, such as additional cooling devices and heating protection circuits are need to bring in, making the package costs of the chip increased significantly [3]. Finally, the increase of the bus energy consumption makes the temperature of the chip higher, which leads to a series of problems, such as the reduction of system reliability, the shortening of the battery power supply time and the increase

© Springer International Publishing AG 2017
M. Qiu (Ed.): SmartCom 2016, LNCS 10135, pp. 51–61, 2017.
DOI: 10.1007/978-3-319-52015-5_6

of the execution cost [4]. So optimizing the data bus dynamic energy saving becomes the research object for software and hardware designers.

Switching activity (SA) on data transmission gives the number of times the bus lines are discharged and recharged between 0 and 1, and is directly responsible for the dynamic energy consumption on the data bus. The low power encoding techniques [5–7] are widely used to reduce SA for dynamic energy consumption and the effects of crosstalk (signal noise, delay) during data transmission on buses. They aim to transform the data being transmitted on buses in such a manner so that the self-SA and coupling SA on buses are reduced. The bus encoding schemes can be classified three types [8]: Algebraic encoding, which refers to the transformation of the original code into other forms of code, such as bus invert encoding; Permutation encoding, which is the permutation of the original code; Probability encoding, which is using statistical analysis of the program, getting the probability of continuous data or instructions, then encoding the data or instruction with high probability to make the number of SA as small as possible.

Bus invert encoding [9] proposed by M. Stan et al. is the most well-known to perform very well in the arena of energy reduction over the data bus. In bus invert encoding, a value is sent as it is, or in a bit-inverted form, depending on the state of the bus during the previous transaction. The main idea is to reduce the dynamic energy consumption by reducing the SA of the bus. Based on this, the researchers presented the odd/even invert encodings, the partial invert encoding and other extended forms [10–12]. However, an encoding scheme is used alone will cause the inactive or unrelated bits unnecessary switching, which reduces the effects of data bus energy saving. Due to the data characteristic is different in different stages of applications and changing with different applications, the single encoding scheme is insufficient to meet the goal of optimizing the data bus energy saving. These encoding schemes have some flaws as follows: (a) They aren't considered the influence of coupling SA, which accounts for a large proportion of total SA and leads to large number of bus energy consumption in deep sub-micron (DSM) technology. (b) The ratio of energy incomings to the introducing cost is not high with one encoding scheme alone. (c) The scope of the encodings is limited.

To solve these problems, in this paper we design a method named SAPME (Switching Activity Perception Multi-Encoding), which can perceive the SA number of the bus. The SAPME method introduces four encoding schemes: swap encoding, invert encoding, odd invert encoding and even invert encoding, which has some advantages compared with other bus energy saving encodings as follows: (a) Its structure is simple and the energy consumption of itself is low; (b) The coupling SA and energy consumption introduced by it are considered; (c) It can automatically select the encoding scheme according to the SA number of different applications and different stages of the same application.

The remainder of the paper is organized as follows. Section 2 describes the platform and energy model. The SAPME encoding scheme is described in Sect. 3. Section 4 presents the experiments and evaluation. We conclude this work in Sect. 5.

2   The Used Platform and Bus Energy Model

2.1   Multi-core Platform with SAPME

Figure 1 shows our on-chip multi-core platform with SAPME connected by bus. The details of the structure as follows: four cores, each core has a private 4-way set associative first level data (DL1) cache and first level instruction (IL1) cache with a size of 32 KB, respectively, 16-way second level (L2) shared cache with a size of 1 MB, the cache line size of both cache levels is 64 B, five SAPME modules - one at each of the cores and one at the L2 end as shown in Fig. 1. The data bus is alternately used by different cores, so as to achieve the purpose of access the shared L2 cache. When different cores access the L2 cache at the same time, a conflict is produced, then an appropriate arbitration mechanism is needed to select the access order to ensure the consistency of the data.

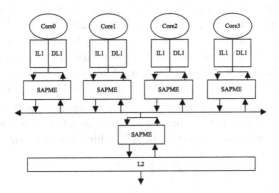

Fig. 1. On-chip multi-core platform with SAPME

2.2   Dynamic Energy Model

The DSM bus capacitance model is shown in Fig. 2, where C_L is the self-capacitance between a bit line and ground, and C_I is the coupling capacitance between adjacent bit lines. The capacitance factor λ is defined as the ratio of the coupling capacitance to the self-capacitance; that is, $\lambda = C_I/C_L$, it highly depends on the manufacturing and layout details. The value of λ becomes high when the technology shrinks. As the technology continuing scaled-down, the value of λ will become larger in the future [13]. In this paper, we investigate the bus with 70 nm technology and the λ is about 5.

As mentioned earlier SA makes the bus lines discharge and recharge between 0 and 1, and is responsible for the bus dynamic energy consumption. In this paper we mainly use the value optimization of data transmission to optimize the

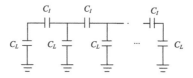

Fig. 2. Capacitance model of a DSM bus

on-chip bus data dynamic energy consumption. The SA, which is the source of dynamic energy consumption, mainly consists of two parts: the self-SA (caused by self-capacitance) and the coupling SA (caused by coupling capacitance). The on-chip data bus dynamic energy consumption E can be calculated by (1). Equations (2) and (3) describe the energy consumption due to self-SA (E_S) and coupling SA (E_I), respectively. In (2), X (named vertical distance) is the corresponding total number of self-SA, which can be calculated by summing up $X_{i,j}$. Thus, X is given by: $X = \sum_{i=1}^{n} \sum_{j=1}^{t-1} X_{i,j}$, where $X_{i,j}$ represents the SA for bit line i from cycle j to cycle $j+1$. It is equal to one if there is a 0 to 1 switching, otherwise, $X_{i,j}$ is equal to zero; t is the total number of clock cycles needed for the data transmission, n is the bit-width of the bus and V_{DD} is the supply voltage. In (3), Y (named horizontal distance) is the total number of coupling SA, which can be calculated by summing up $Y_{(i,i+1),j}$. Thus, Y is given by: $Y = \sum_{i=1}^{n-1} \sum_{j=1}^{t-1} Y_{(i,i+1),j}$, where $Y_{(i,i+1),j}$ represents the coupling SA from cycle j to cycle $j+1$ between bit line i and its adjacent bit line $i+1$. We use the similar coupling bus model employed in [14], which is summarized in Table 1.

$$E = E_S + E_I \tag{1}$$

$$E_S = C_L \cdot V_{DD}^2 \cdot X \tag{2}$$

$$E_I = C_I \cdot V_{DD}^2 \cdot Y \tag{3}$$

Put (2), (3) and $\lambda = C_I/C_L$ into (1), we can get

$$E = (X + \lambda \cdot Y) \cdot C_L \cdot V_{DD}^2 \tag{4}$$

We denote D_{enc} is the total SA distance and D_{enc} is given by

$$D_{enc} = X + \lambda \cdot Y \tag{5}$$

From (4) and (5), to obtain a minimum value of E, we just make the value of D_{enc} minimum.

3   SAPME Encoding Scheme

3.1   Design of the Proposed Encoding Scheme

Our encoding method is added to multi-core system as shown in Fig. 1. There are five SAPME modules - one at each of the cores and one at the L2 end. According

Table 1. The value of $Y_{(i,i+1)}$

$Y_{(i,i+1)}$	Bit line i at time $j \rightarrow j+1$				
Bit line $i+1$ at time $j \rightarrow j+1$	SA	$0 \rightarrow 0$	$0 \rightarrow 1$	$1 \rightarrow 0$	$1 \rightarrow 1$
	$0 \rightarrow 0$	0	1	0	0
	$0 \rightarrow 1$	1	0	2	0
	$1 \rightarrow 0$	0	2	0	1
	$1 \rightarrow 1$	0	0	1	0

to their functions, SAPME module can be divided into two parts: encoding and decoding (Figs. 3 and 4 show their structures, respectively). SAPME method, using algebraic and permutation encoding schemes, introduces four encoding schemes: Swap encoding, Invert encoding, Odd invert encoding and Even invert encoding, which can effectively reduce the data bus SA total distance of a data transmission, and so the bus dynamic energy consumption is effectively reduced. For the value entering the SAPME encoder, the encoder calculates the total SA distance according to (5) and automatically selects the encoding scheme, which can obtain the minimum SA distance, and generates the encoded value and coding number to be sent. In order to ensure the receiver can correctly decode the transmitted data, our method needs two extra control indication lines, which indicate the scheme to be used.

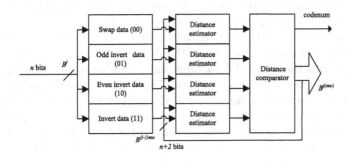

Fig. 3. SAPME encoder schematic diagram

3.2 SAPME Encoding Structure and Algorithm

We denote B^j is the value to be sent on the n-bit width data bus at cycle j, and it can be expressed as $B^j = (b_n^j, b_{n-1}^j, b_{n-2}^j, \cdots, b_1^j)$. $B^{(j-1)enc}$ represents the encoded value at cycle $j - 1$. Four encoding schemes of SAPME are expressed as: Swap $B^{(swap)}$, Invert $B^{(inv)}$, Odd Invert $B^{(odd)}$, Even Invert $B^{(even)}$ with the coding number 00, 11, 01, 10, respectively. Swap the adjacent bits of B^j and append its coding number, we get the encoded value $B^{j(swap)}$. The encoded value that all the bits of B^j are inverted then appended its coding number is

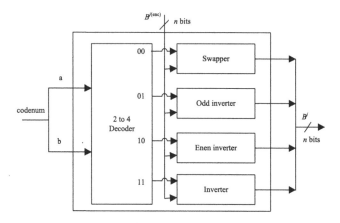

Fig. 4. SAPME decoder schematic diagram

defined as $B^{j(inv)}$. The encoded value that the odd bits of B^j are inverted then appended its coding number is defined as $B^{j(odd)}$. The encoded value that the even bits of B^j are inverted then appended its coding number is defined as $B^{j(even)}$. The coding number indicates which encoding scheme is used to encode the sending value. The function $ST_{\_n}(d1, d2)$ is used to calculate the vertical distance X between the two n-bit width values, the function $CT_{\_n}(data)$ is used to calculate the horizontal distance Y of n-bit width value, and D_{total} represents the minimum total SA distance of the encoded value.

Algorithm 1 is SAPME encoding algorithm, which demonstrates how the encoder selects the encoding scheme with minimum total SA distance according to (5) and returns the encoded data value and its coding number. Figure 3 shows the SAPME encoder schematic diagram. The input value with n-bit width is entered the four encoding components at the same time and they produce four types $(n + 2)$-bit encoded values. And subsequently, the encoded values are sent to distance estimator, in which obtained their total SA distance, respectively. In order to reduce the delay, each encoding process and distance evaluation are treated with at the same time. Finally, the distance comparator is responsible for selecting the encoded value $B^{j(enc)}$ which has the minimum SA distance D_{total} and producing coding number *codenum*.

3.3 SAPME Decoding Structure and Algorithm

Algorithm 2 is SAPME decoding algorithm, which demonstrates how the decoder returns the original value using coding number. Figure 4 shows the SAPME decoder schematic diagram. The SAPME decoder is mainly consisted of one 2–4 decoder and four decoding circuits. The input of the 2–4 decoder is the control signal coding numbers a and b. The output of the 2–4 decoder produces one of the four decoding signals controlling one following decoding component to be valid. The valid component decodes to obtain the n-bit original value B^j.

Algorithm 1. SAPME Encoding Algorithm

Require: Input $value$, n;
Ensure: Output $encoded\_value$, $codenum$.
1: $B^{(swap)} \leftarrow swap(value)$;
2: $B^{(inv)} \leftarrow invert(value)$;
3: $B^{(odd)} \leftarrow odd\ invert(value)$;
4: $B^{(even)} \leftarrow even\ invert(value)$;
5: $D_{swap} \leftarrow ST\_n(B^{(swap)}, B^{(j-1)enc}) + \lambda \cdot CT\_n(B^{(swap)})$;
6: $D_{inv} \leftarrow ST\_n(B^{(inv)}, B^{(j-1)enc}) + \lambda \cdot CT\_n(B^{(inv)})$;
7: $D_{odd} \leftarrow ST\_n(B^{(odd)}, B^{(j-1)enc}) + \lambda \cdot CT\_n(B^{(odd)})$;
8: $D_{even} \leftarrow ST\_n(B^{(even)}, B^{(j-1)enc}) + \lambda \cdot CT\_n(B^{(even)})$;
9: $D_{total} \leftarrow \min(D_{swap}, D_{inv}, D_{odd}, D_{even})$;
10: **if** $D_{total} == D_{swap}$ **then**
11:     $encoded\_value \leftarrow B^{(swap)}$;
12:     $codenum \leftarrow 00$;
13: **end if**
14: **if** $D_{total} == D_{odd}$ **then**
15:     $encoded\_value \leftarrow B^{(odd)}$;
16:     $codenum \leftarrow 01$;
17: **end if**
18: **if** $D_{total} == D_{even}$ **then**
19:     $encoded\_value \leftarrow B^{(even)}$;
20:     $codenum \leftarrow 10$;
21: **end if**
22: **if** $D_{total} == D_{inv}$ **then**
23:     $encoded\_value \leftarrow B^{(inv)}$;
24:     $codenum \leftarrow 11$;
25: **end if**
26: **return** $encoded\_value$, $codenum$;

Algorithm 2. SAPME Decoding Algorithm

Require: Input $encoded\_value$, $codenum$;
Ensure: Output $original\_code$.
1: **if** $codenum == 00$ **then**
2:     $original\_code \leftarrow swap(encoded\_value)$;
3: **end if**
4: **if** $codenum == 01$ **then**
5:     $original\_code \leftarrow odd\ invert(encoded\_value)$;
6: **end if**
7: **if** $codenum == 10$ **then**
8:     $original\_code \leftarrow even\ invert(encoded\_value)$;
9: **end if**
10: **if** $codenum == 11$ **then**
11:     $original\_code \leftarrow invert(encoded\_value)$;
12: **end if**
13: **return** $original\_code$.

The four decoding components are: Swapper for swapping adjacent bits of the encoded value $B^{j(swap)}$, Inverter for inverting all the bits of the encoded value $B^{j(inv)}$, Odd inverter for inverting all the odd bits of the encoded value $B^{j(odd)}$ and Even inverter for inverting all the even bits of the encoded value $B^{j(even)}$.

4   Evaluation

4.1   Simulation Environment and Benchmarks

To quantify the effect of dynamic bus energy saving with our method, we have carried out simulation experiments under the structure of shown in Fig. 1 on the Archimulator [15], which is a multi-core architectural simulator of our research group. The default simulation parameters that have been used in the simulation experiments are given in Table 2. We have used three memory-intensive benchmarks with helper threads (ht) from the Olden [16] and CPU2006 [17] suites: mst_ht, em3d_ht, and 429.mcf_ht, representing typical tasks that might be present in a multi-core system. All applications were executed with the default input sets provided with the benchmarks suites. All three benchmarks are cross-compiled using GCC at O3 optimization level.

Table 2. Default simulation parameters

Parameter	Value
Number of cores	4
IL1, DL1 size	32 KB
L1 associativity	4-way
Bus width	$32 + 2$
Bus energy/line	11.6 pJ
Coupling factor λ	5

4.2   Analysis of Dynamic Energy Saving Effect

Through the simulation experiments, we compare the effect of reducing bus energy consumption under different measures. The measures we have conducted

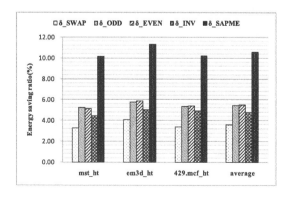

Fig. 5. Dynamic energy saving ratio comparison chart under different measures ($\lambda = 5$)

Table 3. The ratio of dynamic energy saving with different measures (%)

	δ_SWAP	δ_ODD	δ_EVEN	δ_INV	δ_SAPME
mst_ht	3.24	5.21	5.11	4.40	10.12
em3d_ht	4.02	5.72	5.86	4.96	11.30
429.mcf_ht	3.35	5.31	5.34	4.87	10.18
Average	3.54	5.41	5.44	4.74	10.53

are: exclusively using swap encoding (SWAP); exclusively using invert encoding (INV); exclusively using odd invert encoding (ODD); exclusively using even invert encoding (EVEN) and our method (SAPME). We use (6) to measure the energy saving effect of each measure. With 70 nm technology, we compare the effect of the above various measures on the dynamic energy saving when coupling factor λ is equal to 5.

$$\delta = \left(1 - \frac{E_{enc}}{E_{org}}\right) \times 100\% \tag{6}$$

The results are shown in Fig. 5. From Fig. 5 we can see that the effect of energy saving is not obvious when each of other measures is used, whose maximum ratio of energy saving is 5.44%, this is because the incomings from using a single measure offsets its own energy consumption. However, when the SAPME method is used, we can get the maximum energy saving ratio is 11.3% (em3d_ht), and the average energy saving ratio can be reached 10.53%. This is because the SAPME encoder can perceive the SA number and automatically select the encoding scheme for minimizing the total SA distance, which makes the SA be reduced more greatly. According to (4) and (5), the percentage of the reduced SA distance directly determines the dynamic energy saving effect of the data bus. And the SAPME method, which has obtained the maximum reduction of bus total SA distance, makes the energy saving effect the best. In this paper we also carry out some other experiments, which are introduced more encoding schemes than SAPME, the results show that their effect is not significantly improved than SAPME on bus dynamic energy saving. However, they need to add more hardware units and control lines than SAPME which will increase the on-chip area cost and their own energy consumption. So the SAPME composed of four encoding schemes is a better option for on-chip data bus energy saving. The energy saving detailed results are shown in Table 3.

5   Conclusion

In this paper we investigate the dynamic energy saving of on-chip multi-core data bus with the influence of coupling capacitance in DSM technology. We design an on-chip multi-core structure with bus energy saving modules, and present a new method called SAPME to optimize the dynamic energy saving of data bus. The SAPME method can perceive the SA number and automatically select the encoding scheme for minimizing the total SA distance. The results of simulation experiments show that our method can significantly reduce the total SA distance and effectively improve the bus dynamic energy saving ratio by about 10.53% in 70 nm CMOS technology. Compared with other methods, the SAPME can be always obtained a better effect on data bus dynamic energy saving.

Acknowledgements. We thank the anonymous reviewers for their valuable feedback. This work has been supported by the National Science Foundation of China (No. 61370062).

References

1. Jafarzadeh, N., Palesi, M., Khademzadeh, A., Afzali-Kusha, A.: Data encoding techniques for reducing energy consumption in network-on-chip. IEEE Trans. Very Large Scale Integr. (VLSI) Syst. **22**(3), 675–685 (2014)
2. Wang, S.N., Luo, B., Shi, W.S., Tiwari, D.: Application configuration selection for energy-efficient execution on multicore systems. J. Parallel Distrib. Comput. **87**(1), 43–54 (2016)
3. Tang, J., Thanarungroj, P., Liu, C., et al.: Pinned OS/services: a case study of XML parsing on intel SCC. J. Comput. Sci. Technol. **28**(1), 3–13 (2013)
4. Niu, L.W.: Energy efficient scheduling for real-time embedded systems with QoS guarantee. Real-Time Syst. **47**(2), 75–108 (2011)
5. Chang, K.C.: Reliable network-on-chip design for multi-core system-on-chip. J. Supercomputing **55**(1), 86–102 (2011)
6. Sankaran, H., Katkoori, S.: Simultaneous scheduling, allocation, binding, re-ordering, and encoding for crosstalk pattern minimization during high-level synthesis. IEEE Trans. Very Large Scale Integr. Syst. **19**(2), 217–226 (2011)
7. Kaushik, B.K., Agarwal, D., Babu, N.G.: Bus encoder design for reduced crosstalk, power and area in coupled VLSI interconnects. Microelectron. J. **44**(9), 827–833 (2013)
8. Verma, S.K., Kaushik, B.K.: Novel bus encoding scheme for RC coupled VLSI interconnects. In: Wyld, D.C., Wozniak, M., Chaki, N., Meghanathan, N., Nagamalai, D. (eds.) NeCoM/WeST/WiMoN -2011. CCIS, vol. 197, pp. 435–444. Springer, Heidelberg (2011). doi:10.1007/978-3-642-22543-7_44
9. Stan, M.R., Burleson, W.P.: Bus-invert coding for low-power I/O. IEEE Trans. Very Large Scale Integr. (VLSI) Syst. **3**(1), 49–58 (1995)
10. Fang, C.-H., Fan, C.-P.: Novel low-power bus invert coding methods with crosstalk detector. J. Chin. Inst. Eng. **34**(1), 123–139 (2011)
11. Yoon, M.: Achieving maximum performance for bus-invert coding with time-splitting transmitter circuit. IEICE Trans. Fundam. Electron. Commun. Comput. Sci. **E95A**(12), 2357–2363 (2012)
12. Chiu, C.-T., Huang, W.-C., Lin, C.-H., et al.: Embedded transition inversion coding with low switching activity for serial links. IEEE Trans. Very Large Scale Integr. (VlSI) Syst. **21**(10), 1797–1810 (2013)
13. International Technology Roadmap for Semiconductors. http://www.itrs.net
14. Wong, S.-K., Tsui, C.-Y.: Dynamic reconfigurable bus encoding scheme for reducing the energy consumption of deep sub-micron instruction bus. In: 2004 IEEE International Symposium on Circuits and Systems - Proceedings, pp. II321–II324. Institute of Electrical and Electronics Engineers Inc. (2004)

15. http://github.com/mcai/Archimulator/
16. Rogers, A., Carlisle, M.C., Reppy, J.H., Hendren, L.J.: Supporting dynamic data-structures on distributed-memory machines. ACM Trans. Program. Lang. Syst. **17**(2), 233–263 (1995)
17. Henning, J.L.: SPEC CPU2006 benchmark descriptions. SIGARCH Comput. Archit. News **34**(4), 1–17 (2006)

A Novel PSO Based Task Scheduling Algorithm for Multi-core Systems

Jia Tian[1,2], Wei Hu[1,2(✉)], Yonghao Wang[3], Lin Li[1,2], Peng Ke[1,2], and Kai Zhang[1,2]

[1] College of Computer Science and Technology,
Wuhan University of Science and Technology, Wuhan, China
springreef@foxmail.com,
{huwei,lilin,ke_peng,zhangkai}@wust.edu.cn
[2] Hubei Province Key Laboratory of Intelligent Information Processing
and Real-Time Industrial System, Wuhan, China
[3] The DMT Lab, Birmingham City University, Birmingham, UK
yonghao.wang@bcu.ac.uk

Abstract. Multi-core processors have been the mainstream in computer architecture. It also provides the enhancement of the parallelism degree of multiple tasks. An emerged challenge is how to schedule the multiple tasks to the cores for high efficiency. In this paper, a novel task scheduling algorithm is proposed for multi-core systems. This algorithm is based on optimized particle swarm algorithm, which is used to find the optimal solution for the task scheduling. The experimental results have showed that the proposed algorithm can improve the efficiency of task scheduling for multi-core systems.

Keywords: Optimized particle swarm algorithm · Multi-core systems · Task scheduling introduction

1 Introduction

More and more transistors are integrated onto the single chip, which provides huge potential to improve the performance of the processors [1]. When CPU with a single core confronts with the great challenge in system performance and power-consuming, multi-core processors have been taken as a promising diagram [2–4]. More cores on a single chip can improve the performance and cut down the power consumption by reducing the frequency of each core. Multiple tasks can execute in parallel on multiple cores. The high parallelism of the tasks can enhance the performance of the systems. However, a new challenge is emerging for multi-core processors [5]. When more tasks are running in parallel, they may communicate with some other cores. It plays an important role in the system performance on how to schedule these tasks to the different cores. The scheduler should consider the performance of the multiple cores while making the scheduling decision. When the tasks are scheduled to different cores, the communication relationship will also be different. The key is to reduce the communication penalty, scheduling time and improve the performance of the multi-core processors [6].

© Springer International Publishing AG 2017
M. Qiu (Ed.): SmartCom 2016, LNCS 10135, pp. 62–71, 2017.
DOI: 10.1007/978-3-319-52015-5_7

Recent works have proposed to address this challenge of the scheduling schemes for multi-core processors. Partitioned hierarchical real-time scheduling was proposed for multi-core processors [7]. In this design, the applications are considered as well-defined components in a hierarchical manner. The scheduling could be operated in the different hierarchies. Preemptibility-aware scheduling (PAS) was proposed as a responsive scheduling algorithm to reduce the scheduling latency in multi-core systems [8]. One core prepared for the urgent interrupt by both of interrupt-enabled and being in preemptible sections. And other approaches were also proposed to schedule the tasks for high performance. In this paper, a PSO (Particle Swarm Optimization) based task scheduling algorithm is proposed for multi-core processors. The optimized particle swarm algorithm is used to find the optimal solution for the task scheduling. It provides a novel approach to improve the performance of multi-core processors.

This paper is organized as the follows. Section 2 provides the related works. Section 3 describes the system model and the PSO based task scheduling algorithm. The experiments and result analysis are discussed in Sect. 4. And at last, we give the conclusions in Sect. 5.

2 Related Work

When there is only a single core on chip, the scheduling algorithm aims to manage this core with high efficiency. The system cannot have the parallelism in thread level. No threads can share this core at the same time. However, the tasks can run in parallel on multiple cores on the same chip. The communication among the tasks will consume more time. How to map the tasks to the cores is one of the key paths to solve the above problem. There are existing works on this problem to provide better solutions [9–12].

There are different types of scheduling algorithms. When multiple tasks are mapped to the multiple cores, the efficiency is the main target. Static scheduling algorithms were proposed to complete the mapping between the tasks and the cores. Heuristic mapping algorithm [13], genetic mapping algorithm [14], QoS guaranteed mapping algorithm [15] and multi-target mapping method for mesh network [16] were typical such algorithms. The scheduling was determined before the execution of the tasks. It means that such algorithms could obtain a well optimization of the scheduling. However, the scheduling could not be adjusted at run-time. When the situations were changed, the algorithms had to re-calculate the scheduling approach. Dynamic scheduling algorithms could manage the scheduling of tasks at run-time according to the instant proofing [17–20]. Such algorithms could reduce the traffic on-chip. Some specific hardware or software units could be added for the efficient management. When the traffic was heavy, these units were the bottleneck. A different algorithm was proposed in [21], which was a scenario based mapping method. The scenarios were set as the states in a state machine. When the states changed, the scheduling would be triggered.

When more cores are available, many algorithms focuses on both of the scheduling and the power consumption. DVS/DVFS (Dynamic Voltage Scaling/Dynamic Voltage Frequency Scaling) are traditionally used on single core chip. Now they are also used in multi-core processors for both efficiency and power consumption [22–24]. [25]

focused on how to detect the idle time of the cores to put the shared memory into sleep to save energy. [26] proposed an approach for task scheduling based on the run-time characteristics of individual tasks, which were considered as the critical factor in making decisions for the scheduling. [27] presented the concept that the tasks with the same run-time features could be dispatched a clustered region of the cores. Such algorithms could provide new designs to take into account the efficiency and energy consumption. However, it also means the compromise of the efficiency.

The task scheduling itself should be improved to find a fast approach to complete the task dispatching. The potential optimal scheduling sequence should be provided to the system for the improvement of the performance. In this paper, PSO is used for this target.

3 PSO Based Task Scheduling Algorithm

3.1 System Model

The technical advantages of multi-core processor is to support the multiple tasks running in parallel on multiple cores. It is parallelism in thread level. The cores have relatively simple structure. The tasks will communication with each other through the bus, wires between cores or shared memory. A system model is proposed to abstract and simplify the multi-core systems.

For a multiprocessor system, assuming that there are n identical processors. The number of tasks to be scheduled is m. The directed acyclic graph (DAG) is used to represent the tasks in the system. DAG can be represented by quintuple $G = (V, E, R, C, W)$.

$V = \{V_i\}$ is the set of vertices, which is the collection of the tasks. Vertex V_i is used to represent a task. (V_i, V_j) is used to represent the edge between the two vertices V_i and V_j.

$E = \{e\}$ is the set of the edges. $e(i, j)$ is the edge between the two vertices V_i and V_j, i.e. (V_i, V_j). All edges in E are directed ones.

$R = \{R_i\}$ is the set of execution time of the tasks. R_i is used to represent the execution time of a task V_i.

$C = \{c\}$ is the set of the weight of the edges. The weight is a compound factor, which is the coupling degree of the traffic and control between the tasks.

$W = \{w(V_i, V_j)\}$ is the set of communication overhead between the tasks V_i and V_j. W is an additional information of the edges. When the two tasks are running on the same core, their communication overhead is at the minimum and the corresponding w is also at the minimum.

Figure 1 shows the DAG with ten tasks. The values of the edges are the overhead of communication between tasks. They are just the weights of the edges.

3.2 Algorithm Design

PSO algorithm is proposed for the simulation of a simple social model. In the PSO algorithm, each optimization problem of potential solutions is search in the space of a bird, known as "particles". All particles have an adaptive value (Value Fitness) determined by the optimized function, and each particle has a velocity that determines the direction of their flight and the displacement of each step. Then the particles follow the

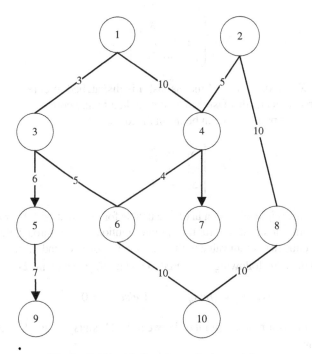

Fig. 1. DAG with direction and edge weights

current optimal particle in the solution space. PSO algorithm needs to initialize a group of random particles (random solution), and then to find the optimal solution through the iteration. In each iteration, the particle tracks two "extreme" to update them. The first one is the optimal solution of the particle itself, which is called the individual extremum. The other is the optimal solution for the whole population, which is called the global extremum. The two basic formulas of particle swarm optimization algorithm are as follows.

$$v_{ij}^{k+1} = \text{w} * v_{ij}^{k} + c_1 * r_1 * \left(pBest_{ij}^{k} - x_{ij}^{k} \right) + c_2 * r_2 * (gBest_{ij}^{k} - x_{ij}^{k}) \tag{1}$$

$$x_{ij}^{k+1} = x_{ij}^{k} + v_{ij}^{k+1} \tag{2}$$

In (1) and (2), the particle velocity is v_{ij}^{k+1}; w is the inertia weight, and x_{ij}^{k} is the position of the particle. $pBest_{ij}^{k}$ and $gBest_{ij}^{k}$ are defined as optimal solutions of the particle itself and the whole group respectively. r_1 and r_2 is random values (0 or 1). c_1 and c_2 is the learning factors. k is the number of iterations; i is the number of particles; and j is the dimension of the target space.

This algorithm is designed for multi-core processors with multiple tasks. In this optimization, the positions of the particles are represented by X which is a binary value:

$$X = \begin{bmatrix} X_{11} & \cdots & X_{1n} \\ \vdots & \ddots & \vdots \\ X_{m1} & \cdots & X_{mn} \end{bmatrix} \qquad (3)$$

Each X_{ij} in X is 0 or 1. $X_{ij} = 1$ means task i is dispatched to core j. If a row in the matrix X is zero, it means this task can be dispatched to any core.

The speed of the particles S can be normalized as:

$$S = \begin{bmatrix} S_{11} & \cdots & S_{1n} \\ \vdots & \ddots & \vdots \\ S_{m1} & \cdots & S_{mn} \end{bmatrix} \qquad (4)$$

At this time, the original location in (2) are unable for the situations. After the initialization, the algorithm starts to search for the optimal solution. It goes through the V matrix of each line, to find the maximum value of the location of each row, and the location of the x_{ij} is set to 1. And then, the following (5) is used to replace the position in (2):

$$\text{if}(\text{rand}() < sig(s_{ij}))\, x_{ij} = 1 \; else \; x_{ij} = 0 \qquad (5)$$

In (5), *rand ()* is a random number between [0,1]; $sig(s_{ij})$ is the Sigmoid function shown as follows:

$$sig(s_{ij}) = 1/(1 + e^{-s_{ij}}) \qquad (6)$$

In order to ensure that the group can move evenly, the speed range of particles is set as [−4,4]; and the $s(v_{ij})$'s range is closer to the middle value, rather than near from 0 to 1.

The weight w in (1) has a great influence on the global search ability and local search ability of the algorithm. When w is very small, the PSO algorithm is easy to fall into local extreme value. When the w is large, the convergence of PSO algorithm is low. So the weight w range is set in [0.6, 1.0]. In this range, the balance of the algorithm is suitable, and it is easy to find the global optima through the iterations.

The weight w is got by using the linear decreasing:

$$w = w_{max} - \frac{w_{max} - w_{min}}{k_{max}} * k \qquad (7)$$

In (7), w_{max} is the maximum weight; w_{min} is the minimum weight; k_{max} is the number of iterations of the algorithm; k is the number of iterations of the algorithm. Such approach can improve the convergence rate of the algorithm, and has a significant good effect to find the optimal solution.

After each update of the position, the fitness values will be recomputed. The edge sets are traversed. If two tasks are scheduled to the same core, the adaptive value is set to 1, which is the minimum value. The communication overhead is represented by $F(i)$, which the communication time of each pair of tasks. The number of edges at system initialization is represented by num. Then F(i) can be calculated as follows:

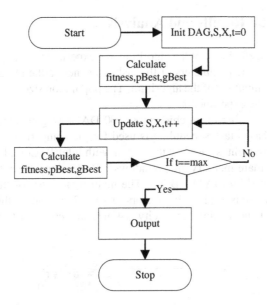

Fig. 2. Algorithm flow diagram

$$F(i) = \begin{cases} b * w(i) * c(i), & \text{different cores.} \\ b * c(i) * 1, & \text{on the same core.} \end{cases} \tag{8}$$

In (8), b is the coefficient of the scheduling time; $w(i)$ is the weight on the edge of the DAG; $c(i)$ is the data control correlation between tasks. The fitness function of the system is:

$$f(s) = \sum_{i=1}^{num} F(i) \tag{9}$$

The flow of discrete particle swarm algorithm is as follows:

Step 1. In this step, the following works will be completed including the initialization of the original DAG, the initialization of the particle position and velocity, and setting the number of cycles $t = 0$.
Step 2. In this step, the particles' fitness values are calculated, and the *pBest* and *gBest* are set up.
Step 3. In this step, the particles' velocity V and position X are updated.
Step 4. In this step, the particles' fitness values are calculated, the *pBest* and *gBest* are setup, and $t = t + 1$.
Step 5. If the maximum cycle times are completed, the optimal solution is output; otherwise the algorithm goes to step 3.

The diagram of this algorithm is shown in Fig. 2.

4   Experimental Results and Analysis

In order to obtain the efficiency of the algorithm, the experiments are set up with a series of random characteristic parameters of the tasks to generate the DAG. The generated DAG is used as the input set of the algorithm. The population size and iteration parameters are set up for the experiments.

For each given number of on-chip cores, 100 DAG are generated randomly. The genetic algorithm based task scheduling is used for the comparison. The setup of the experimental environment is based on the system with Intel(R) Core(TM) i3-3240 CPU (3.40 GHz), 8 GB main memory and Windows 7 Professional operating system. The tests are implemented by MATLAB 7.0. The main parameters of the algorithm are: particle population size is 30; learning factors c1 and c2 are both 2; the maximum iteration number is 100; the maximum velocity is 4; $w(i)$ and $c(i)$ are 1 to 10; b is set as the natural coefficient e.

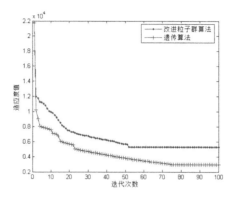

Fig. 3. Particle swarm optimization algorithm curve

Table 1. Comparison of simulation results

Number of processor cores	Number of threads	The shortest time to obtain the global optimal solution (ms)	
		DPSO	GA
4	30	39	48
	60	71	81
	80	66	97
6	30	32	40
	60	58	73
	80	61	89
8	30	26	31
	60	52	61
	80	59	82

As Table 1 shows, the efficiency of the algorithm will be improved when the number of cores increases. But the amplitude of the improvement is reduced until the algorithm is stable. when the number of tasks increases, the efficiency of the algorithm is gradually improved. Figure 3 shows another improvement of this algorithm. This algorithm is stable in the iteration number of 27 times, and this PSO based algorithm can find the global optimal solution quickly (Fig. 4).

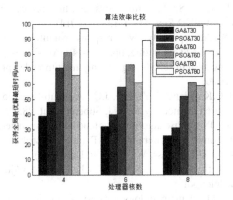

Fig. 4. Performance comparison of GA algorithm and improved particle swarm optimization algorithm under different processing cores and number of threads

5 Conclusions

When more cores are integrated onto a single chip, the tasks can run in parallel to achieve better performance and cut down the power consumption. However, how to improve the scheduling efficiency is emerging as a new challenge. In this paper, a novel PSO based task scheduling algorithm is proposed for this problem. Optimized particle swarm algorithm is used to find the optimal scheduling solution based the system model. The experimental results show that this algorithm can improve the efficiency of the task scheduling for multi-core processors.

References

1. Khaira, M.S.: Micro-2010: lead performance microprocessor of the year 2010-myth or reality. In: Twelfth International Conference on VLSI Design, pp. 157–163. IEEE Press, Washington, DC (1999)
2. Olukotun, K., Nayfeh, B.A., Hammond, L., Wilson, K., Chung, K.: The case for a single-chip multiprocessor. ACM SIGOPS Operating Syst. Rev. **30**(5), 2–11 (1996)
3. Nayfeh, B.A., Olukotun, K.: A single-chip multiprocessor. IEEE Comput. **30**(9), 79–85 (1997)
4. Cesario, W., Baghdadi, A., Gauthier L., Lyonnard, D., Nicolescu, G., Paviot, Y., Yoo, S., Jerraya, A.A., Diaz-Nava, M.: Component-based design approach for multicore SoCs. In: Design Automation Conference, pp. 789–794. ACM Press, New York (2002)

5. Cong, J., Yuan, B.: Energy-efficient scheduling on heterogeneous multi-core architectures. In: 2012 ACM/IEEE International Symposium on Low Power Electronics and Design, pp. 345–350. ACM Press, New York (2012)
6. Das, R., Ausavarungnirun, R., Mutlu, O., Kumar, A., Azimi, M.: Application-to-core mapping policies to reduce memory interference in multi-core systems. In: The 21st International Conference on Parallel Architectures and Compilation Techniques, pp. 455–456. ACM Press, New York (2012)
7. Åsberg, M., Nolte, T., Kato, S.: Towards partitioned hierarchical real-time scheduling on multi-core processors. SIGBED Rev. **11**(2), 13–18 (2014)
8. Lee, J., Lim, G., Suh, S.: Preemptibility-aware responsive multi-core scheduling. In: The 2011 ACM Symposium on Applied Computing, pp. 748–749. ACM Press, New York (2011)
9. Fedorova, A.: Operating system scheduling for chip multithreaded processors. Ph.D. thesis, Harvard University (2006)
10. Zhuang, Y.C., Shieh, C.K., Liang, T.Y., Lee, J.Q.: A group-based load balance scheme for software distributed shared memory systems. J. Supercomput. **28**, 295–309 (2004)
11. Attiya, G., Hamam, Y.: Two phase algorithm for load balancing in heterogeneous distributed systems. In: 12th Euromicro Conference on Parallel, Distributed and Network-Based Processing, pp. 434–439. IEEE Press, Washington, DC (2004)
12. Aas, J.: Understanding the Linux 2.6.8.1 CPU scheduler. http://josh.trancesoftware.com/linux/linux_cpu_scheduler.pdf
13. Ho, W.H., Pinkston, T.M.: A methodology for designing efficient on-chip interconnects on well-behaved communication patterns. In: The 9th International Symposium on High-Performance Computer Architecture, pp. 377–388. IEEE Press, Washington, DC (2003)
14. Srinivasan, K., Chatha, K.S.: ISIS: a genetic algorithm based technique for custom on-chip interconnection network synthesis. In: 18th International Conference on VLSI Design, pp. 623–628. IEEE Press, Washington, DC (2005)
15. Murali, S., Benini, L., Micheli, G.: Mapping and physical planning of networks-on-chip architectures with quality-of-service guarantees. In: Conference on 2005 Asia and South Pacific Design Automation Conference, pp. 27–32. IEEE Press, Washington, DC (2005)
16. Ascia, G., Catania, V., Palesi, M.: Multi-objective mapping for mesh-based NoC architectures. In: 2004 International Conference on Hardware/Software Codesign and System Synthesis, pp. 182–187. ACM Press, New York (2004)
17. Barcelos, D., Brião, E.W., Wagner, F.R.: A hybrid memory organization to enhance task migration and dynamic task allocation in NoC-based MPSoCs. In: The 20th Annual Conference on Integrated Circuits and Systems Design, pp. 282–287. ACM Press, New York (2007)
18. Chou, C., Marculescu, R.: User-aware dynamic task allocation in networks-on-chip. In: The Conference on Design, Automation and Test in Europe, pp. 1232–1237. IEEE Press, Washington, DC (2008)
19. Hu, J., Marculescu, R.: Energy-Aware communication and task scheduling for network-on-chip architectures under real-time constraints. In: The Conference on Design, Automation and Test in Europe, vol. 1, pp. 234–239. IEEE Press, Washington, DC (2004)
20. Briao, E.W., Barcelos, D., Wronski, F., Wagner, F.R.: Impact of task migration in NoC-based MPSoCs for soft real-time applications. In: IFIP International Conference on Very Large Scale Integration, pp. 296–299. IEEE Press, Washington, DC (2007)
21. Schor, L., Bacivarov, I., Rai, D., Yang, H., Kang, S.H., Thiele, L.: Scenario-based design flow for mapping streaming applications onto on-chip many-core systems. In: 2012 International Conference on Compilers, Architectures and Synthesis for Embedded Systems, pp. 71–80. ACM, New York (2012)

22. Wang, Y., Liu, H., Liu, D., Qin, Z.W., Shao, Z.L., Sha, E.H.M.: Overhead-aware energy optimization for real-time streaming applications on multiprocessor System-on-Chip. ACM Trans. Des. Autom. Electron. Syst. 16(2) (2011). Article 14

23. Zhang, D.S., Guo D.K., Chen, F.Y., Wu, F., Cao, T., Jin, S.Y.: TL-plane-based multi-core energy-efficient real-time scheduling algorithm for sporadic tasks. ACM Trans. Archit. Code Optim. 8(4) (2012). Article 47

24. Liu, C., Li, J., Rubio, J., Speight, E., Lin, X.Z.: Power-efficient time-sensitive mapping in heterogeneous systems. In: The 21st International Conference on Parallel Architectures and Compilation Techniques, pp. 23–32. ACM Press, New York (2012)

25. Fu, C., Zhao, Y., Li, M., Xue, C.J.: Maximizing common idle time on multi-core processors with shared memory. In: The 2015 Design, Automation & Test in Europe Conference & Exhibition, pp. 900–903. IEEE Press, Washington, DC (2015)

26. Dhiman, G., Kontorinis, V., Tullsen, D., Rosing, T., Saxe, E., Chew, J.: Dynamic workload characterization for power efficient scheduling on CMP systems. In: The 16th ACM/IEEE International Symposium on Low Power Electronics and Design, pp. 437–442. ACM Press, New York (2010)

27. Ma, K., Li, X., Chen, M., Wang, X.R.: Scalable power control for many-core architectures running multi-threaded applications. SIGARCH Comput. Archit. News 39(3), 449–460 (2011)

Artificial Bee Colony Algorithm with Hierarchical Groups for Global Numerical Optimization

Laizhong Cui, Yanli Luo, Genghui Li$^{(\boxtimes)}$, and Nan Lu

College of Computer Science and Software Engineering,
Shenzhen University, Shenzhen, People's Republic of China
{cuilz,lunan}@szu.edu.cn, meidaile210@163.com,
li_genghui@126.com

Abstract. Artificial Bee Colony (ABC) algorithm is a relatively new swarm-based optimization algorithm, which has been shown to be better than or at least competitive to other evolutionary algorithms (EAs). Since ABC generally performs well in exploration but poorly in exploitation, ABC often shows a slow convergence. In order to address this issue and improve its performance, in this paper, we present a novel artificial bee colony algorithm with hierarchical groups, named HGABC. In employed bee phase of HGABC, the population is divided into three groups based on the fitness values of the food source positions, and three solution search strategies with different characteristics are correspondingly employed by different groups. Moreover, in onlooker bee phase, onlooker bees conduct exploitation in the most promising area of search space, instead of around some good solutions. In order to demonstrate the performance of HGABC, we compare HGABC with four other state-of-the-art ABC variants on 22 benchmark functions with 30D. The experimental results show that HGABC is better than other competitors in terms of solution accuracy and convergence rate.

Keywords: Artificial bee colony algorithm · Hierarchical group · Exploitation in the most promising area · Global numerical optimization

1 Introduction

Global optimization problems (GOPs) always arise in almost all of science research and engineering fields. population-based random optimization algorithms, such as genetic algorithm (GA) [1, 2], ant colony optimization (ACO) [3], particle swarm optimization (PSO) [4] and artificial bee colony algorithm (ABC), have been becoming a popular and promising way to handle these GOPs. ABC was developed by Karaboga [5] firstly, inspired by the collective foraging behavior of honey bee colony. The performance of ABC was demonstrated by comparing ABC with other evolutionary algorithms (EAs). Due to its simple structure, easy implementation and good performances, ABC has successfully attracted numerous researcher's attention and been applied to solve many practical engineering optimization problems [6–9].

© Springer International Publishing AG 2017
M. Qiu (Ed.): SmartCom 2016, LNCS 10135, pp. 72–85, 2017.
DOI: 10.1007/978-3-319-52015-5_8

However, like other EAs, ABC often shows a slow convergence speed [10] since its solution search equation does well in exploration but poorly in exploitation. The search equation is the core operator of ABC, which significantly affects the performance of ABC. Therefore, in order to keep a better balance between exploration and exploitation, many new search equations were proposed. Inspired by PSO, Zhu and Kwong [11] introduced the information of the global best solution into the solution search equation to improve the exploitation ability of ABC (GABC). The experimental results showed that GABC is better than ABC on most benchmark functions. Karaboga and Akay [13] introduced two new parameters i.e., modification rate (MR) and scaling factor (SF), into the solution equation to control frequency and magnitude of perturbation, respectively. In order to combine the advantage of different solution search equations, Kiran et al. [14] proposed a new method, which integrates five search equations to generate candidate solutions by the way of cooperation and competition. Moreover, Wang et al. [12] proposed the MEABC algorithm to improve the local and global search capability of the ABC, in which a pool of three distinct solution search strategies coexists throughout the search process and produces new solutions competitively. Recently, Karabaga et al. [15] proposed a new search equation for onlooker bees (qABC), which uses the valuable information of the best solution among the neighbors to improve the search efficiency of ABC. At the same time there are some improvements that blend with other operations [16, 17], and so on.

According to above considerations, the performance of ABC mainly depends on its solution search equation. Therefore, it is a promising way to improve the performance of ABC by introducing new search equation or integrating multiple search equations. In this paper, we follow this basic idea and propose an improved ABC algorithm, named HGABC. In employed bee phase of HGABC, all employed bees are divided into three groups according to the quality of their food source positions (fitness values), and different groups use different solution search equations. Moreover, to enhance the local exploitation ability in a promising area, in onlooker bee phase of HGABC, the most promising area is firstly recognized based on the quality of all food source positions, and onlooker bees conduct exploitation only around the positions located in the most promising area. The experimental results on 22 benchmark functions show that HGABC performs more competitively and effectively when it is compared with the other ABC variants.

The rest of this paper is organized as follows. Section 2 introduces ABC algorithm briefly. The proposed algorithm is presented detailedly in Sect. 3. Section 4 discusses and analyzes the experimental results. Finally, Sect. 5 concludes this paper.

2 Artificial Bee Colony Algorithm

Inspired by the waggle dance and foraging behaviors of honey bee colony, ABC algorithm has been developed. In ABC algorithm, the position of a food source represents a possible solution to the optimization problem, and the nectar amount of a food source position corresponds to the quality (fitness value) of the associated solution. The number of the employed bees or the onlooker bees is equal to the number of food sources. The basic ABC algorithm consists of four basic phases, namely initialization phase, employed bee phase, onlooker bee phase and scout bee phase.

2.1 Initialization Phase

In the initialization phase, the necessary parameters, *i.e.*, the number of food source position *SN*, the termination condition and the parameter *limit*, should be initialized firstly. Then, the initial food source positions are randomly produced in the whole search space by Eq. (1) as follows,

$$x_{i,j} = x_{\min,j} + rand(0,1)(x_{\max,j} - x_{\min,j}) \tag{1}$$

where $i = 1, 2, \cdots, SN$, $j = 1, 2, \cdots, D$, *SN* is the population size, and $x_{i,j}$ is the *j*th dimension of the *i*th solution. $x_{\min,j}$ and $x_{\max,j}$ are the lower and upper bounds of the *j*th dimension of the problem, respectively. $rand(0,1)$ is a random number in the range of [0,1]. The fitness value of the food source positions are calculated as follows,

$$fit(x_i) = \begin{cases} 1/(1+f(x_i)) & \text{if} (f(x_i) \geq 0) \\ 1 + abs(f(x_i)) & \text{else} \end{cases} \tag{2}$$

where $f(x_i)$ is the objective function value of the *i*th food source position, and $fit(x_i)$ is the fitness value of the *i*th food source position.

2.2 Employed Bee Phase

In this phase, each employed bee flies to a distinct food source position to search for better food source position, and the candidate food source position is generated as follows,

$$v_{i,j} = x_{i,j} + \phi_{i,j}(x_{i,j} - x_{k,j}) \tag{3}$$

where $i = 1, 2, \cdots, SN$ and $j = 1, 2, \cdots, D$; $k \in \{1, 2, \ldots, SN\}$ and it is different from i; *D* is the dimension of the problem; $\phi_{i,j}$ is a random number in the range of $[-1,1]$. After the generation of the candidate solution v_i, if the candidate solution is better than the old one, the old solution will be replaced by the candidate solution. Otherwise, the old solution will be kept.

2.3 Onlooker Bee Phase

After all employed bees complete their search process, they will share the information (quality and position of food source) of their food source position to onlooker bees by assigning each food source position a selection probability, which is calculated as follows,

$$p_i = fit(x_i) \Big/ \sum_{i=1}^{SN} fit(x_i) \tag{4}$$

where p_i is the selection probability of the ith food source position, and each onlooker bee selects a food source position to perform search according to the selection probability of each food source position. The same search strategy and greedy selection method are employed by onlooker bees to perform further exploitation.

2.4 Scout Bees Phase

In the scout bee phase, if a certain food source position (solution) fails to be updated during a predetermined cycle (defined as "*limit*"), the corresponding employed bee becomes a scout bee and the food source position should be replaced by a new one, which is generated randomly according to Eq. (1).

After the initialization, ABC enters a loop of employed bee phase, onlooker phase and scout bee phase until the terminal condition is satisfied.

3 Artificial Bee Colony Algorithm with Hierarchical Groups (HGABC)

In the original ABC or other ABC variants [11], only one search strategy is employed by employed bee and onlooker bee, which may result in that the search ability of these methods are limited. Inspired by the observation in the team work of human being, since each member in the team has different characteristics, such as knowledge, attitude and skill, the whole team usually is divided into multiple groups according to their abilities, and each group takes different responsibilities or tasks. By this way, the work efficiency can be significantly improved. In original ABC, although the colony contains three types of bees, *i.e.*, employed bee, onlooker bee and scout bee, different types of bees are responsible for different search abilities. However, ABC treats all employed bees (or onlooker bee) equally because all employed bees (or onlooker bees) employ the same search strategy. While in real bee colony, each employed bee (or onlooker bee) is a unique individual, and the search ability of them may be different from each other. Therefore, different employed bees (and onlooker bees) may adopt different search strategies in fact.

According to above consideration, in this paper, we propose a novel artificial bee colony algorithm with hierarchical groups, named HGABC. To be specific, in HGABC, the employed bees are divided into three groups based on the quality of their food source positions, and different groups employ different search strategies so as to be responsible for different search abilities. Moreover, in order to pay more attention to the exploitation in the most promising area, all onlooker bees only search around the food source positions which locate in the most promising area. Similarly, three search strategies could be used by onlooker bees in a random manner. The proposed strategies are described in detail as follows.

3.1   Division of Employed Bees and Search Strategies

In ABC, each employed bee occupies a food source position. Since each employed bee has distinct ability and should adopt different search strategy, in order to differentiate employed bees, we firstly divide the employed bees into three group based on the quality of their own food source positions. To be specific, the employed bees firstly sort from best to worst based on the quality of their food source positions. The first $a \cdot SN$ employed bees, the medium $b \cdot SN$ employed bees and the last $c \cdot SN$ employed bees, respectively constitute the high group, medium group and low group, where $a, b, c \in [0, 1]$ and $a + b + c = 1$.

The high group includes some current good solutions, which may be located in the local optimal areas or the global optimal area. Therefore, its employed bees should learn the beneficial information from the current best solution and conduct exploitation toward the current best solution. The employed bees belonging to the high group adopt the search strategy as follows,

$$v_{i,j} = x_{k,j} + \varphi_{i,j}(xbest_j - x_{k,j}) \tag{5}$$

where $xbest$ is the current best solution; x_k is randomly selected from the population, which is different from x_i and $xbest$; $\varphi_{i,j}$ is a random number in the range of $[0,1]$; j is a randomly selected dimension.

With respect to the medium group, it consists of some neither better nor worse solutions that are not far from or close to the global optimal area. Their employed bees should take the responsibility of obtaining balance between the exploitation and exploration. Therefore, the employed bees in the medium group use the search strategy as follows,

$$v_{i,j} = x_{k,j} + \phi_{i,j}(x_{k,j} - x_{q,j}) + \varphi_{i,j}(xbest_j - x_{k,j}) \tag{6}$$

where $xbest$ is the current best solution, and x_k and x_q are randomly selected from the population, which are distinct from each other and different from x_i and $xbest$. $\varphi_{i,j}$ is a random number in the range of $[0,1]$, and $\phi_{i,j}$ is a random number in the range of $[-1, 1]$. j is a randomly selected dimension.

Regarding to the low group, it contains the current bad solutions that may be far from the local optimal areas or the global optimal area with a high probability, and its employed bees should be responsible for exploration by exploiting new areas randomly. Therefore, the third kind of employed bees employ the search strategy as follows,

$$v_{i,j} = x_{k,j} + \phi_{i,j}(x_{k,j} - x_{q,j}) \tag{7}$$

where x_k and x_q are randomly selected from the population, which are distinct from each other and different from x_i. $\phi_{i,j}$ is a random number in the range of $[-1, 1]$. j is a randomly selected dimension.

Overall, in our proposed algorithm, all employed bees are divided into three groups, namely the high group, the medium group and the low group. The employed bees in different groups adopt different search strategies and undertake different search tasks. More specifically, the high group's employed bees pay more attention to exploitation, the low group's employed bees focus on exploration, and the medium group's employed bees are responsible to balance between exploration and exploitation.

3.2 Search Strategy of Onlooker Bee

In original ABC, after all employed bees complete their search tasks, the onlooker bees start to work depending on the information provided by the employed bees. To be specific, each onlooker bee will select a food source position to conduct exploitation by the roulette wheel method, which is a time-consuming procedure. Moreover, the better the quality of the food source position is, the bigger the selection probability is. In order to pay more attention to the promising area and accelerate the convergence, in this paper, we present a most promising area search strategy for onlooker bee. The details are described as follows.

Algorithm 1. Procedure of onlooker bee phase	
01:	Locate in the most promising area
02:	**for** i=1:SN
03:	Randomly select a position x_s in the most promising area
04:	u=rand(0,1);
05:	**if** $u \leq s_1$
06:	Generate the candidate solution v_s use Eq. (5)
07:	**else if** $u \leq s_2$
08:	Generate the candidate solution v_s use Eq. (6)
09:	**else**
10:	Generate the candidate solution v_s use Eq. (7)
11:	**end if**
12:	**if** $f(v_s) \leq f(x_s)$
13:	Replace x_s by v_s, *counter(s)*=0
14:	**else**
15:	*counter(s)*=*counter(s)*+1;
16:	**end if**
17 :	**end for**

Fig. 1. The pseudo-code of onlooker bee phase

In order to recognize the most promising area, each food source position denotes an area. To be specific, for the ith food source position, if the Euclidean distance between food source position x_i and x_j ($j = 1, 2, \cdots, SN$ and $j \neq i$) is less than the radius r, the position x_j belongs to the area located by the position x_i. Moreover, the radius r is calculated as follows,

$$r = \frac{\sum_{i=1}^{SN-1} \sum_{j=i+1}^{SN} d(x_i, x_j)}{SN(SN-1)/2} \tag{8}$$

where $d(x_i, x_j)$ is the Euclidean distance between x_i and x_j, and SN is the number of the food source positions.

Obviously, there are SN areas in the search space and the best quality area based on the average fitness value of its members is treated as the most promising area. After the most promising area is identified, the onlooker bees only fly to a randomly selected food source position located in the most promising area to search.

	Algorithm 2.The procedure of HGABC
01:	**Initialization**: Generate SN solutions according to Eq. (1)
02:	**while** $FES < maxFES$
03:	Sort the population and divide the population into three groups
04:	**for** i =1 to SN // **employed bee phase**
05:	**if** $i \leq a \cdot SN$ // **the high group**
06:	Generate a new solution v_i^G use Eq. (5).
07:	**else if** $i \leq (a+b) \cdot SN$ // **the medium group**
08:	Generate a new solution v_i^G use Eq. (6).
09:	**else** // **the low group**
10:	Generate a new solution v_i^G use Eq. (7).
11:	**end if**
12:	Evaluate the new solution v_i^G
13:	**if** $f\left(v_i^G\right) \leq f\left(x_i^G\right)$
14:	Replace x_i^G by v_i^G , $counter(i)$=0
15:	**else**
16:	$counter(i)$= $counter(i)$+1
17:	**end if**
18:	**end for**
19:	**Algorithm 1** // **onlooker bee phase**
20:	FES=FES+2SN
21:	Select X_{max}^G with max $counter$ value // **scout bee phase**
22:	**if** $counter$(max) > $limit$
23:	Replace X_{max}^G by a new solution generated according to Eq.(1)
24:	FES=FES+1, $counter$(max)=0
25:	**end if**
26:	**end while**
	Output: The food source (solution) with the smallest objective value

Fig. 2. The pseudo-code of HGABC

Moreover, to make the onlooker bees show different search abilities and keep a better balance between exploration and exploitation, the above three search equations (Eqs. (5), (6) and (7)) are employed by onlooker bees in a random manner based on two control parameters s_1 and s_2.

Table 1. Benchmark functions in experiments

Name	Function	Range	Min	Accept
Sphere	$f_1(x)=\sum_{i=1}^D x_i^2$	$[-100,100]^D$	0	1×10^{-8}
Elliptic	$f_2(x)=\sum_{i=1}^D (10^6)^{\frac{i-1}{D-1}} x_i^2$	$[-100,100]^D$	0	1×10^{-8}
SumSquare	$f_3(x)=\sum_{i=1}^D i x_i^2$	$[-10,10]^D$	0	1×10^{-8}
sumPower	$f_4(x)=\sum_{i=1}^D |x_i|^{(i+1)}$	$[-1,1]^D$	0	1×10^{-8}
Schwefel 2.22	$f_5(x)=\sum_{i=1}^D |x_i| + \prod_{i=1}^D |x_i|$	$[-10,10]^D$	0	1×10^{-8}
Schwefel 2.21	$f_6(x)=max\{|x_i|,\ 1\le i\le n\}$	$[-100,100]^D$	0	1×10^0
Step	$f_7(x)=\sum_{i=1}^D (\lfloor x_i+0.5\rfloor)^2$	$[-100,100]^D$	0	1×10^{-8}
Exponential	$f_8(x)=\exp\left(0.5*\sum_{i=1}^D x_i\right)$	$[-10,10]^D$	0	1×10^{-8}
Quartic	$f_9(x)=\sum_{i=1}^D i x_i^4 + random[0,1]$	$[-1.28,1.28]^D$	0	1×10^{-1}
Rosenbrock	$f_{10}(x)=\sum_{i=1}^{D-1}\left[100(x_{i+1}-x_i^2)^2+(x_i-1)^2\right]$	$[-5,10]^D$	0	1×10^{-1}
Rastrigin	$f_{11}(x)=\sum_{i=1}^D \left[x_i^2-10\cos(2\pi x_i)+10\right]$	$[-5.12,5.12]^D$	0	1×10^{-8}
NCRastrigin	$f_{12}(x)=\sum_{i=1}^D \left[y_i^2-10\cos(2\pi y_i)+10\right]$ $y_i=\begin{cases} x_i & |x_i|<\frac{1}{2} \\ \frac{round(2x_i)}{2} & |x_i|\ge\frac{1}{2} \end{cases}$	$[-5.12,5.12]^D$	0	1×10^{-8}
Griewank	$f_{13}(x)=1/4000\sum_{i=1}^D x_i^2 - \prod_{i=1}^D \cos\left(\frac{x_i}{\sqrt{i}}\right)+1$	$[-600,600]^D$	0	1×10^{-8}
Schwefel2.26	$f_{14}(x)=418.98288727243380*D - \sum_{i=1}^i x_i\sin\left(\sqrt{|x_i|}\right)$	$[-500,500]^D$	0	1×10^{-8}
Ackley	$f_{15}(x)=20+e-20\exp\left(-0.2\sqrt{\frac{1}{D}\sum_{i=1}^D x_i^2}\right)$ $-\exp\left(\frac{1}{D}\sum_{i=1}^D \cos(2\pi x_i)\right)$	$[-50,50]^D$	0	1×10^{-8}

(continued)

Table 1. (*continued*)

Name	Function	Range	Min	Accept
Penalized1	$f_{16}(x) = \dfrac{\pi}{D}\left\{10\sin^2(\pi y_1) + \sum_{i=1}^{D-1}(y_i-1)^2\left[1+10\sin^2(\pi y_{i+1})\right] + (y_D-1)^2\right\} + \sum_{i=1}^{D}u(x_i,10,100,4)$ $y_i = 1+1/4(x_i+1), \quad u_{x_i,a,k,m} = \begin{cases} k(x_i-a)^m & x_i > a \\ 0 & -a \le x_i \le a \\ k(-x_i-a)^m & x_i < a \end{cases}$	$[-100,100]^D$	0	1×10^{-8}
Penalized2	$f_{17}(x) = \dfrac{1}{10}\left\{\sin^2(\pi x_1) + \sum_{i=1}^{D-1}(x_i-1)^2\left[1+\sin^2(3\pi x_{i+1})\right] + (x_D-1)^2\left[1+\sin^2(2\pi x_{i+1})\right]\right\} + \sum_{i=1}^{D}u(x_i,5,100,4)$	$[-100,100]^D$	0	1×10^{-8}
Alpine	$f_{18}(x) = \sum_{i=1}^{D-1}|x_i\cdot\sin(x_i)+0.1\cdot x_i|$	$[-10,10]^D$	0	1×10^{-8}
Levy	$f_{19}(x) = \sum_{i=1}^{D-1}(x_i-1)^2\left[1+\sin^2(3\pi x_{i+1})\right]+\sin^2(3\pi_1) + |x_D-1|\left[1+\sin^2(3\pi x_D)\right]$	$[-10,10]^D$	0	1×10^{-8}
Weierstrass	$f_{20}(x) = \sum_{i=1}^{D}\left(\sum_{k=0}^{k_{max}}\left[a^k\cos(2\pi b^k(x_i+0.5))\right]\right) - D\sum_{k=0}^{k_{max}}\left[a^k\cos(2\pi b^k 0.5)\right], a=0.5, b=3, k_{max}=20$	$[-1,1]^D$	0	1×10^{-8}
Himmelblau	$f_{21}(x) = 1/D\sum_{i=1}^{D}\left(x_i^4-16x_i^2+5x_i\right)$	$[-5,5]^D$	-78.3	-78
Michalewicz	$f_{22}(x) = -\sum_{i=1}^{n}\sin(x_i)\sin^{20}\left(\dfrac{i\times x_i^2}{\pi}\right)$	$[0,\pi]^D$	$-D$	$-D+1$

According to the proposed modifications, the pseudo-code of onlooker bee phase is shown in Fig. 1 and the completed pseudo-code of the proposed algorithm HGABC is shown in Fig. 2.

4 Experiments

In order to demonstrate the performance of our proposed algorithm HGABC, we compare HGABC with four ABC methods, *i.e.*, the basic ABC [5], GABC [11], qABC [15] and MEABC [12] on 22 benchmark functions with 30D, which are listed in Table 1. To make a fair comparison, for all compared algorithms, SN and *limit* are set to 50 and $SN \cdot D$, respectively. Other parameters are set the same as the original papers. For HGABC, a, b and c are respectively set to 0.2, 0.3 and 0.5; s_1 and s_2 are set to 0.25 and 0.75, respectively. The maximal number of function evaluation (*maxFES*) is used as the termination condition, which is set to 5000 $\cdot$ D. All algorithms conduct 25 times independent runs on each function. The experimental results are given in Table 2. For the sake of clarity, the best results are marked in **boldface**. Moreover, the Wilcoxon's rank sum test at 5% significance level on results gained by two competing algorithms is also conducted to show the significant differences between HGABC and other ABC methods. The results of the test are represented as "+", "−", "=", which mean that the compared algorithm is significantly better than, worse than, equal to HGABC, respectively.

As shown in Table 2, the metric of *mean* and *std* respectively denote the average value and standard deviation of the best objective function value of 25 independent runs. According to these metrics, HGABC successfully gets the best results on all functions except that f_4, f_{10} and f_{14}. To be specific, HGABC is better than ABC, GABC, qABC and MEABC on 18, 12, 18 and 9 functions, respectively. On the contrary, HGABC is only beaten by GABC and MEABC on 1 and 1 function, respectively. Moreover, ABC and qABC is unable to perform better than HGABC on any cases.

In addition, in order to clearly show the convergence speed and robustness of different algorithms, more experimental results about the average FES (AVEN) and success rate (SR) are also given in Table 1. AVEN represents the average FES needed to reach the threshold defined in Table 1. In Table 1, "NAN" denotes that the algorithm cannot get any solutions, whose objective function is smaller than the acceptable value in 25 independent runs. SR represents the ratio of the number of success runs in the 25 independent runs. The success run means that algorithm can find the solution, whose objective function value is less than the acceptable value. Obviously, the search accuracy of HGABC is better than or equal to other algorithms on all functions, excluding f_4, f_{10} and f_{14}. Similarly, the SR of HGABC is 100% on all functions except f_{10}, on which all algorithms are unable to get a 100% success rate. Overall, HGABC is better than the competitors in terms of solution accuracy, convergence speed and robustness.

Table 2. Comparison results on 22 test functions with 30D

Alg	ABC mean(std) AVEN(SR)	GABC mean(std) AVEN(SR)	qABC mean(std) AVEN(SR)	MEABC mean(std) AVEN(SR)	HGABC mean(std) AVEN(SR)
f_1	8.05e−18 (6.08e−18) − 100/82934	6.97e−33 (4.93e−33) − 100/50130	1.60e−15 (1.32e−15) − 100/72618	3.45e−40 (4.66e−40) − 100/44910	**2.18e−57** **(2.33e−57)** 100/32314
f_2	4.77e−10 (3.76e−10) − 100/135230	1.92e−26 (2.12e−26) − 100/76150	1.53e−10 (3.87e−10) − 100/123870	6.17e−37 (6.37e−37) − 100/55908	**7.59e−55** **(1.12e−54)** 100/40498
f_3	1.55e−19 (1.31e−19) − 100/75366	2.98e−34 (2.38e−34) − 100/45478	3.14e−16 (2.92e−16) − 100/63946	2.74e−41 (2.06e−41) − 100/41710	**2.91e−58** **(2.97e−58)** 100/30186
f_4	2.41e−31 (9.09e−31) − 100/23266	1.83e−52 (6.33e−52) − 100/14106	3.01e−21 (1.31e−20) − 100/13342	**4.93e−86** **(1.20e−85)** + 100/12014	5.32e−76 (1.81e−75) − 100/10022
f_5	6.55e−11 (2.12e−11) − 100/125030	5.95e−18 (1.76e−18) − 100/77478	1.09e−08 (3.89e−09) − 48/148340	1.47e−21 (6.87e−22) − 100/66784	**5.85e−30** **(3.45e−30)** 100/48882
f_6	4.35e+00 (8.60e−01) − 0/NAN	2.55e−01 (1.30e−01) − 100/109060	9.36e−02 (1.79e−02) − 100/35898	3.00e+00 (1.37e+00) − 4/131500	**4.57−03** **(3.39−03)** 100/58034
f_7	**0.00e+00** **(0.00e+00) =** 100/11314	**0.00e+00** **(0.00e+00) =** 100/10314	**0.00e+00** **(0.00e+00) =** 100/6482	**0.00e+00** **(0.00e+00) =** 100/18974	**0.00e+00** **(0.00e+00)** 100/11962
f_8	**7.18e−66** (4.37e−73) = 100/150	**7.18e−66** (9.22e−77) = 100/150	**7.18e−66** (2.98e−72) = 100/150	**7.18e−66** (3.63e−79) = 100/100	**7.18e−66** **(7.66e−80)** 100/150
f_9	6.42e−02 (1.37e−02) − 100/93186	2.80e−02 (6.51e−03) − 100/41966	2.78e−02 (8.01e−03) − 100/11018	2.98e−02 (8.07e−03) − 100/45748	**1.26e−02** **(2.78e−03)** 100/20674
f_{10}	**6.79e−02** **(5.93e−02) =** 72/120030	8.21e−01 (3.73e+00) = 68/77515	5.56e−01 (6.12e−01) − 36/**75828**	9.34e−02 (1.17e−01) = 80/115880	1.99e−01 (2.77e−01) 56/98729
f_{11}	2.68e−14 (1.03e−13) − 100/99214	**0.00e+00** **(0.00e+00) =** 100/68134	1.23e−10 (1.68e−10) − 100/112510	**0.00e+00** **(0.00e+00) =** 100/51876	**0.00e+00** **(0.00e+00)** 100/36926
f_{12}	4.25e−13 (1.57e−12) − 100/110050	**0.00e+00** **(0.00e+00) =** 100/76642	4.95e−10 (5.78e−10) − 100/119730	**0.00e+00** **(0.00e+00) =** 100/55650	**0.00e+00** **(0.00e+00)** 100/39794
f_{13}	3.08e−04 (1.54e−03) − 96/96783	4.51e−08 (2.25e−07) = 96/61688	2.48e−12 (6.35e−12) − 100/95790	**0.00e+00** **(0.00e+00) =** 100/53762	**0.00e+00** **(0.00e+00)** 100/38790
f_{14}	4.51e−12 (1.59e−12) − 100/84338	**2.18e−13** (6.03e−13) + 100/65670	3.88e−10 (1.46e−09) − 100/112170	2.76e−12 (1.50e−12) − 100/53292	3.64e−12 (0.00e+00) 100/40506

(*continued*)

Table 2. (*continued*)

Alg	ABC mean(std) AVEN(SR)	GABC mean(std) AVEN(SR)	qABC mean(std) AVEN(SR)	MEABC mean(std) AVEN(SR)	HGABC mean(std) AVEN(SR)
f_{15}	3.83e−09 (2.27e−09) − 96/144000	1.49e−14 (2.92e−15) − 100/89178	1.61e−06 (8.36e−07) − 0/NAN	6.79e−15 (1.97e−15) − 100/76954	**3.84e−15** **(9.84e−16)** **100/55678**
f_{16}	1.29e−18 (1.76e−18) − 100/79398	**1.57e−32** **(5.59e−48) =** 100/45786	4.12e−15 (7.77e−15) − 100/63282	**1.57e−32** **(5.59e−48) =** 100/40080	**1.57e−32** **(5.59e−48)** 100/28266
f_{17}	8.19e−18 (1.71e−17) − 100/84730	4.06e−33 (2.30e−33) − 100/49750	1.83e−15 (1.51e−15) − 100/75322	**1.50e−33** **(0.00e+00) =** 100/44876	**1.50e−33** **(0.00e+00)** 100/30854
f_{18}	3.15e−06 (1.85e−06) − 0/NAN	3.88e−07 (6.54e−07) − 16/129980	1.43e−05 (3.92e−05) − 0/NAN	1.78e−17 (6.15e−17) − 100/68064	**7.80e−31** **(1.20e−30)** **100/49198**
f_{19}	8.23e−14 (1.25e−13) − 100/91734	1.39e−31 (1.41e−32) = 100/50934	9.29e−10 (9.59e−10) − 100/123530	**1.35e−31** **(2.23e−47) =** 100/42450	**1.35e−31** **(2.23e−47)** 100/32006
f_{20}	3.06e−02 (3.75e−02) − 0/NAN	3.60e−02 (4.19e−02) − 0/NAN	8.71e−03 (8.44e−03) − 0/NAN	**0.00e+00** **(0.00e+00) =** 100/89724	**0.00e+00** **(0.00e+00)** 100/68534
f_{21}	−7.83e+01 (4.10e−15) = 100/26934	−7.83e+01 (5.02e−15) = 100/15986	−7.83e+01 (7.11e−15) = 100/6838	−7.83e+01 (5.80e−15) = 100/12760	−7.83e+01 (2.90e−15) 100/7666
f_{22}	−2.999e+01 (8.26e−04) − 100/25362	−2.999e+01 (1.01e−03) − 100/21778	**−3.00e+01** **(1.12e−05) =** 100/2310	**−3.00e+01** **(2.19e−07) =** 100/17126	−3.00e+01 (3.29e−06) 100/9530
+/ =/−	0/4/18	1/9/12	0/4/18	1/12/9	

To clearly show the advantages of HGABC, the convergence curves of the *mean* on some representative functions are plotted in Fig. 3. It can be seen from Fig. 3 that HGABC converges faster than ABC, GABC, qABC and MEABC on both unimodal functions and multimodal functions. In conclusion, the experimental results demonstrate that our modifications of employed bee phase and onlooker bee phase can ontain a better balance between exploration and exploitation, and effectively improve the performance of ABC.

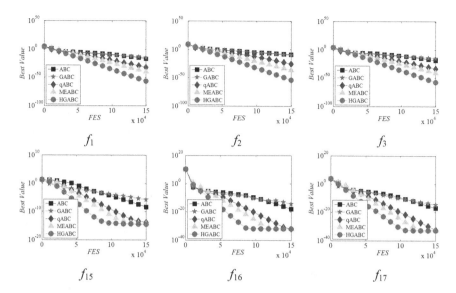

Fig. 3. Convergence curve of all ABCs on some representative functions

5   Conclusion

This paper presents a new ABC algorithm, called HGABC. In HGABC, in order to differentiate the employed bees, the employed bees are divided into three groups according to the quality of their food source positions. The employed bees belonging to different groups employ different search strategies and are responsible for different search abilities. Moreover, to speed up convergence and pay more attention to the most promising area, the onlooker bees using three search strategies in a random manner only exploit in the most promising area. The comparison results on 22 benchmark functions show that HGABC can significantly improve the performance of ABC and outperform other ABC methods in terms of solution accuracy, convergence speed and robustness. In future, we can apply HGABC to handle real world engineering problems.

Acknowledgments. This work is supported by the National Natural Science Foundation of China under Grant 61402294, Major Fundamental Research Project in the Science and Technology Plan of Shenzhen under Grants JCYJ20140509172609162, JCYJ20-140828163633977, JCYJ20140418181958501, and JCYJ20160310095523765.

References

1. Holland, J.H.: Adaptation in natural and artificial systems: an introductory analysis with applications to biology, control, and artificial intelligence. U Michigan Press (1975)
2. Yang, C., Gui, W., Kong, L., et al.: A genetic algorithm based optimal scheduling system for full-filled tanks in the processing of starting materials for alumina production. Can. J. Chem. Eng. **86**(4), 804–812 (2008)
3. Dorigo, M., Birattari, M., Stutzle, T.: Ant colony optimization. IEEE Comput. Intell. Mag. **1**(4), 28–39 (2006)
4. Kennedy, J.: Particle swarm optimization. In: Encyclopedia of Machine Learning, pp. 760–766. Springer US, Heidelberg (2011)
5. Karaboga, D.: An idea based on honey bee swarm for numerical optimization. Technical report-tr06, Erciyes university, engineering faculty, computer engineering department (2005)
6. Karaboga, D., Basturk, B.: A powerful and efficient algorithm for numerical function optimization: artificial bee colony (ABC) algorithm. J. Global Optim. **39**(3), 459–471 (2007)
7. Karaboga, D., Akay, B.: A modified artificial bee colony (ABC) algorithm for constrained optimization problems. Appl. Soft Comput. **11**(3), 3021–3031 (2011)
8. Singh, A.: An artificial bee colony algorithm for the leaf-constrained minimum spanning tree problem. Appl. Soft Comput. **9**(2), 625–631 (2009)
9. Li, G., Niu, P., Xiao, X.: Development and investigation of efficient artificial bee colony algorithm for numerical function optimization. Appl. Soft Comput. **12**(1), 320–332 (2012)
10. Karaboga, D., Akay, B.: A comparative study of artificial bee colony algorithm. Appl. Math. Comput. **214**(1), 108–132 (2009)
11. Zhu, G., Kwong, S.: Gbest-guided artificial bee colony algorithm for numerical function optimization. Appl. Math. Comput. **217**(7), 3166–3173 (2010)
12. Wang, H., Wu, Z., Rahnamayan, S., et al.: Multi-strategy ensemble artificial bee colony algorithm. Inf. Sci. **279**, 587–603 (2014)
13. Akay, B., Karaboga, D.: A modified artificial bee colony algorithm for real-parameter optimization. Inf. Sci. **192**, 120–142 (2012)
14. Kiran, M.S., Hakli, H., Gunduz, M., et al.: Artificial bee colony algorithm with variable search strategy for continuous optimization. Inf. Sci. **300**, 140–157 (2015)
15. Karaboga, D., Gorkemli, B.: A quick artificial bee colony (qABC) algorithm and its performance on optimization problems. Appl. Soft Comput. **23**, 227–238 (2014)
16. Qiu, M., Ming, Z., Li, J., et al.: Phase-change memory optimization for green cloud with genetic algorithm. IEEE Trans. Comput. **64**(12), 3528–3540 (2015)
17. Gai, K., Qiu, M., Zhao, H.: Cost-aware multimedia data allocation for heterogeneous memory using genetic algorithm in cloud computing. IEEE Trans. Comput. (2016) doi:10.1109/TCC.2016.2594172

An Buffering Optimization Algorithm for Cooperative Mobile Service

Lei Hu[1,2], Huan Shen[1,2(✉)], Qingsong Shi[3], Jiajia Xu[1,2], Wei Hu[1,2], and Peng Ke[1,2]

[1] College of Computer Science and Technology,
Wuhan University of Science and Technology, Wuhan, China
{huwei, ke_peng}@wust.edu.cn,
ShenHuan201620@outlook.com,
{lylehu, double2hao}@foxmail.com
[2] Hubei Province Key Laboratory of Intelligent Information Processing and
Real-Time Industrial System, Wuhan, China
[3] College of Computer Science, Zhejiang University, Hangzhou, China
zjsqs@zju.edu.cn

Abstract. With the development of wireless network technology and embedded technology, mobile devices have more powerful hardware and are used wildly than before. However, mobile services have different features compared with the traditional desktop services. Different types of wireless networks can provide different network widths for the mobile services. When images are transferred to mobile devices. The performance will be affected by network circumstances. This makes it necessary to adjust the buffering strategy for local data to improve the user experiences. In this paper, a novel buffering optimization algorithm is proposed for cooperative mobile service. The buffering optimization model is constructed as basis and the algorithm can adjust the buffering strategy according to the real-time network width. The traffic will be reduced to achieve better response to users and save the limited network width. The experimental results show that this algorithm can improve the performance without loss of the user experiences.

Keywords: Mobile devices · Buffering optimization algorithm · Network width model

1 Introduction

With the rapid development of wireless network technology and embedded technology, mobile devices have been widely used than ever [1]. Mobile service is flourishing in different areas for its portability and flexibility. Multimedia is one of the most important data form in more and more mobile services [2, 3]. Images are almost the basic data form in mobile applications. The users of mobile services need fast response for their requests. The time-lasting responses will deplete the users' patience. However, there are also some limitations when images are used under different wireless network circumstances. Firstly, the basic networks are determined before the mobile services. There are different types of the wireless networks [4]. When the mobile devices are

© Springer International Publishing AG 2017
M. Qiu (Ed.): SmartCom 2016, LNCS 10135, pp. 86–94, 2017.
DOI: 10.1007/978-3-319-52015-5_9

moving, they will change among these networks. Different networks will provide different network widths to these devices. Secondly, wireless network can not provide the same stable network width as the traditional network with lines. The mobile devices may have to tolerate the changing network width. The above limitations make it necessary to adjust the buffering strategy for local data to improve the user experiences.

The existing researches focused on different impact factors to improve the performance [5–9]. As a typical optimization design, on-chip memory was used to provide the fast access to memory to obtain fast responses [10]. And some other optimization approaches were proposed to improve the performance of the wireless network [11, 12]. However, what the wireless networks provide to the related devices can not guarantee the stable widths under different circumstances. In this paper, a novel buffering optimization algorithm is proposed to improve the performance of the image buffering on mobile devices. The buffering model is constructed as the basis. The traffic is adjusted by compressing the images according the network widths.

This paper is organized as the follows. Section 2 depicts the background for the optimization. Section 3 describes this algorithm. Section 4 provides experimental results and analysis. And at last, we give the conclusions in Sect. 5.

2 Background

2.1 Basic Mobile Data Access Structure

When wireless networks become the mainstream, B/S (Browser and Server) structure is also used as the main one to construct the service model for the mobile services. The typical structure is showed in Fig. 1. The mobile devices send their http requests to the web server including a data block with the request in information. The request has three parts including the request approach with URI protocol and its version, the request header and the request body. The web server will send the data block back to the mobile devices as the response if the network is available. The data in the response will be analyzed and showed to the users. Mobile devices will also send data to the servers for long-term storage.

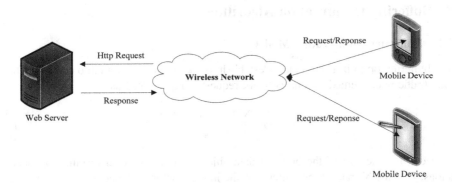

Fig. 1. Mobile data access structure

When mobile devices fetch data from the servers frequently, such processes will consume much time. The data being fetched will be buffered in the local memory for the upcoming utilization. The network width plays an important role in the response speed. Wireless networks provide shared widths among many mobile devices. A single mobile device may obtain enough width for the transmission or it can only obtain limited width according to the total number of mobile devices shared the same wireless network. The images with bigger size will be affected by the network width.

2.2   Basic Buffering Mechanism Analysis

Mobile devices fetch data from web server. However, fetching data through network will consume much time especially the frequent accessing and the limited network width. All the fetched data should be buffered for the future utilization. Buffering mechanism is used to cache the fetched data and reduce the network operations to improve the performance and cut down the power consumption.

There are two typical buffering strategies. They are memory buffering and file buffering respectively. Memory buffering uses the main memory as the buffer. The system will assign some special areas to the applications for the data buffering. However, mobile devices have limited main memory compared with the desktop computers. The buffers always have not enough space. Furthermore, memory buffering may result in the memory leakages. File buffering uses the spare hard disk space as the buffer. When the applications need the data, they can fetch them from the buffer without network accesses. However, hard disk is very slow compared with the main memory. When more data are placed on such space, it's slow to fetch them. The compromise is to combine the memory buffering and file buffering, which is as Fig. 2 shows.

Data buffering mechanism for mobile devices can improve the performance. However, such mechanism is used to increase the local efficiency. The network traffic is not reduced and it has close relationship with the network and the size of the buffered images. If the images can be pre-processed on web server, the size of the images can be adjusted for the transmission. The combination can have better performance through local buffering on mobile devices and dynamic adjusting on web server.

3   Buffering Optimization Algorithm

3.1   Buffering Optimization Model

The buffering optimization model is establish to take various factors into account. The total traffic w consumed by one network request is represented as:

$$w = \alpha * \frac{\sum_1^n f_n * (1 - P_t)}{N} \tag{1}$$

Here, f is the size of the buffered data object; N is the number of buffering operations; P_t is the hit rate of the buffer; α is the impact factor of the network state. If w is the target for the reduction, three factors should be considered according to (1).

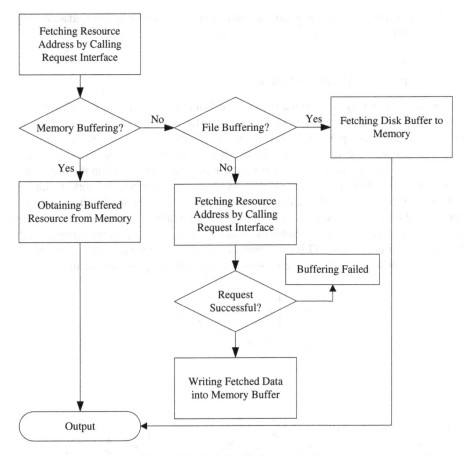

Fig. 2. Combined buffering mechanism

The replacement algorithm will affect the hit rate. There are four popular algorithms including soft reference/weak reference algorithm, first in first out algorithm, least recent used algorithm and least frequent used algorithm. Soft reference/weak reference is used as a lazy replacement approach. It is applicable to the following circumstance. When some data objects having long lifecycle have occupied large memory space, this algorithm is adopted to avoid the memory leakage. But it will consume more traffic. First in first out algorithm is simplest one. It uses one queue to track all the data objects with high efficiency. But the data objects don't have the same features. Least recent used algorithm is the best one. But it is almost impossible to obtain the related information before the operations. This algorithm is difficult to implement in real systems. In this paper, least frequent used algorithm is adopted as the basic replacement algorithm.

P_t and α have close relationship. When the size of the buffered data object is large enough and the network cannot provide available width, the response of the data fetching will be slow. The network cannot be adjusted by the mobile devices. The network state can be obtained at run-time. The size of the buffered data object can be adjusted according to the network state to achieve better performance.

3.2 Optimization Algorithm Design

The network state may be changed under the different circumstances. But it cannot be adjusted by web server or the mobile devices. The optimization algorithm aims to change the image compression algorithms according to the network state. The general wireless networks are divided into three types including robust network, common network and weak network. Robust network can provide broad width to the users with almost no delay. Wifi based wireless network is the typical robust network. Common network can provide enough width for texts and general media data. But it can support images and other multimedia data well for the shared width. 3G and 4G based wireless network is the typical common network. Weak network cannot provide enough width for most multimedia data. 2G based wireless network is the typical weak network.

The compressed quality of the images is represented as β as showed in (2).

$$\beta = \left\{ 0.9 - \frac{\frac{1}{\sqrt{p_1^2 * n_1 + p_2^2 * n_2}}}{|0.5 - p_3|} \right. \tag{2}$$

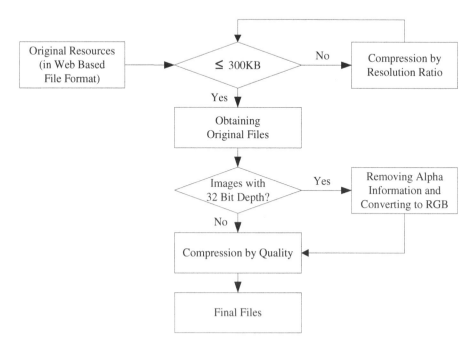

Fig. 3. Processing flow of the images

Here, p_1, p_2 and p_3 is the ratios of request times of data access to the total request times under robust network, common network and weak network respectively. n_1 and n_2 are the total requests of the former two network types respectively. The request times are counted through the network interfaces statistically. If it's robust network, $\alpha = 0$ and the images will not be compressed. If it's common network, α and β are linear correlation. Namely the smaller compression is, the lower traffic loss is. The value of β is adjusted dynamically. And the advantage of such operations is that the compression of the images can reduce the network traffic when the users send more requests to the web server. When β is higher, it means the images can have broader width and they can avoid the compression. The process is showed in Fig. 3. The compression will lose some information from PNG file with 32 bit depth for Alpha channel to JPEG. The Alpha information should be removed from the original images. At the same time, if the size of the images is less than 20 KB, these images will not be compressed to avoid the following situation: the size of the images may increase by the forced compression and the network traffic also increases.

4 Experimental Results and Analysis

The compression quality parameter of the image resources is ranged from 0 to 1. It represents the compression quality of the related images. 100% means the best compression effect. The lower the compression quality is, the smaller the file size is. At the same time, the display effect of the images on mobile devices changes smoothly with the reduction of the file size. When the algorithm is tested, an image with 212 KB size is selected as the test object. The effects of the compression are showed in Fig. 4. The new file sizes are 212 KB, 40 KB and 10 KB respectively. Though the file size is reduced, the display effect of the images has no obvious decline in quality.

The buffering optimization algorithm is tested under different network states. The compression quality parameters are obtained through providing the parameters to the network interface. The web server will calculate the related parameters and return them back. Each request will fetch 20 images from the web server. The results will be

a) $\beta=1$ b) $\beta=0.8$ c)$\beta=0.1$

Fig. 4. Effect after compression

analyzed. 15 images are uploaded images to the web server and the total size of the images is 2.05 MB. The values of β are 1, 0.9 and 0.31 for three different types of the network. The comparison is based on the four larger images in the 15 ones. The size of the compressed images is compared with the original size as showed in Fig. 5 (RNET means the robust network; CNET means the common network and WNET means the weak network).

As Fig. 5 shows, when there is enough width provided by the network for the transmission of the images with different sizes, the images will not be compressed. When the width is not enough, the images will be compressed and then transferred through the network. The compressed images are small enough to be transferred under the corresponding network states. And the compressed images will be buffered quickly by local applications in mobile devices.

Figure 6 Shows the traffic comparison of three types of network. The images aren't compressed under RNET for the width enough. Though the images may have large size, RNET can support the fast transmission and the local applications can process the images. The images are compressed under CENT and WNET. Though the images have been compressed for the fast transmission, the image quality will not affect the display effect according to the quality effect experiments. The compression improves the transmission speed and local buffering.

Figure 7 shows the compression ratios of the different network states. The compression ratio is determined by the value of β. When the value of β is small, it means the network cannot afford enough width for recent requests. And the images will be compressed for the responses. However, the images cannot be compressed without limitation. So when the value of β is very small, the images are compressed almost under a fixed compression ratios.

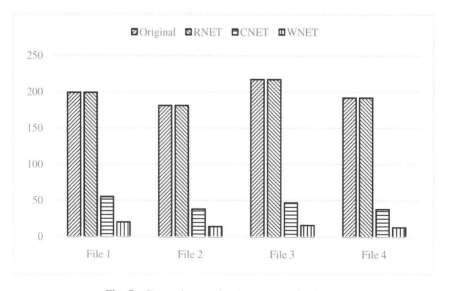

Fig. 5. Comparison under three types of network

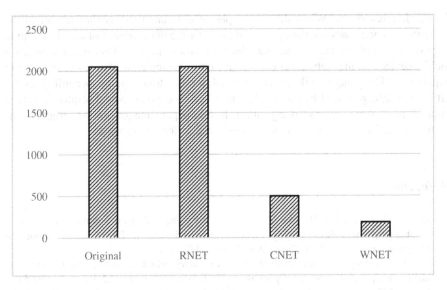

Fig. 6. Traffic comparison of three types of network

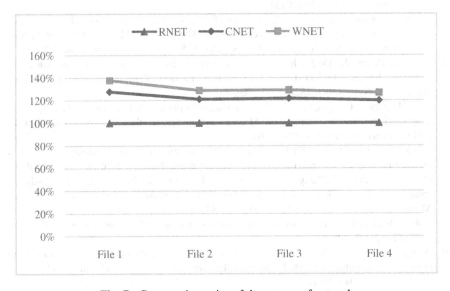

Fig. 7. Compression ratios of three types of network

5 Conclusions

With the development of wireless technology and embedded technology, mobile devices are more powerful than before. However, their resources are still limited compared with the traditional desktop computers. Such devices are connected to the servers through wireless network. The users of mobile devices have more requirements for the fast responses. When different types of data are used as the support to the contents, it is necessary for the systems to consider the optimization of the algorithm to improve the performance of the data fetching and buffering. This paper presents a mobile device and the web server cooperative optimization algorithm to ensure the user experience. The images will be compressed for the transmission according to the different widths provided by the wireless network. The experimental results shows the improvement of the optimized algorithm. In this paper, images are the target for the optimization. More data types can be considered in the future work.

References

1. Jia, R., Hu, W., Li, Z.H., Ma, Z.W., Wang, J.P.: Rapid display optimization based on on-chip programmable memory for mobile intelligent devices. In: 2015 Chinese Automation Congress, pp. 13–17. IEEE Press, Washington, D.C. (2015)
2. van Der Schaar, M., Sai Shankar, N.: Cross-layer wireless multimedia transmission: challenges, principles, and new paradigms. IEEE Wirel. Commun. 12(4), 50–58 (2005)
3. Sheng, Z.G., Fan, J., Liu, C.H., Leung, V.C.M., Liu, X., Leung, K.K.: Energy-efficient relay selection for cooperative relaying in wireless multimedia networks. IEEE Trans. Veh. Technol. 64(3), 1156–1170 (2015)
4. Biswas, S., Bicket, R., Wong, E., Musaloiu-E, R., Bhartia, A., Aguayo, D.: Large-scale measurements of wireless network behavior. In: The 2015 ACM Conference on Special Interest Group on Data Communication, pp. 153–165. ACM Press, New York (2015)
5. Kang, K., Benini, L., De Micheli, G.: Cost-effective design of mesh-of-tree interconnect for multicore clusters with 3-D stacked L2 scratchpad memory. IEEE Trans. Very Large Scale Integr. VLSI Syst. 23(9), 1828–1841 (2015)
6. Mao, H., Zhang, C., Sun, G., Shu, J.: Exploring data placement in racetrack memory based scratchpad memory. In: Non-volatile Memory System and Applications Symposium, pp. 1–5. IEEE Press, Washington, D.C. (2015)
7. Khalili, S., Simeone, O.: Inter-layer per-mobile optimization of cloud mobile computing: a message-passing approach. Trans. Emerg. Telecommun. Technol. 27(6), 814–827 (2016)
8. Zheng, K., Yang, Z., Zhang, K., Chatzimisios, P., Yang, K., Xiang, W.: Big data-driven optimization for mobile networks toward 5G. IEEE Netw. 30(1), 44–51 (2016)
9. Ahmed, E., Gani, A., Sookhak, M., Hamid, S.H.A., Xia, F.: Application optimization in mobile cloud computing: motivation, taxonomies, and open challenges. J. Netw. Comput. Appl. 52, 52–68 (2015)
10. Wang, Z., Gu, Z., Shao, Z.: WCET-aware energy-efficient data allocation on scratchpad memory for real-time embedded systems. IEEE Trans. Very Large Scale Integr. VLSI Syst. 23(11), 2700–2704 (2015)
11. Liang, C., Yu, F.R., Zhang, X.: Information-centric network function virtualization over 5G mobile wireless networks. IEEE Netw. 29(3), 68–74 (2015)
12. Meng, T., Wu, F., Chen, G.: Code-based neighbor discovery protocols in mobile wireless networks. IEEE/ACM Trans. Netw. 24(2), 806–819 (2016)

SDN Protocol Analysis with Process Algebra Method

Chen Fu$^{(\boxtimes)}$ and Fu Bing

Department of Computer Science, Beijing Foreign Studies University,
Beijing, Haidian 100089, China
chenfu@servst.com

Abstract. With so many users and applications handled by the Internet infrastructure, the Internet has to evolve continuously to meet the requirement. Separation of control plane and management plane has been studies as a new architecture model in recent years by Internet community. Software Defined Network (SDN is just one of the most influential models in this field. But there lack the rigor formal analysis of the protocol behavior with SDN. In this paper, a formalization method with algebra to analysis SDN protocol is proposed. The method is self-contained and universal to any protocol analysis. With the method we can formally analysis the detail of SDN protocol to find the uncertain and shortcoming aspects.

Keywords: Software Defined Network · Protocol analysis · Process algebra · Formalization method

1 Introduction

With the overwhelming success in mobile app and Internet service, the Internet has become a huge cyberspace. Mass network services have a deep influence on our daily life. But the popularity of Internet breeds huge challenge to the transmission capability, network security, and network management. With billions of transactions streaming on Internet, the Internet has to deal all of these efficiently. A trustworthy network and services infrastructure are extremely important. Many future network or next generation network projects has been going on in the past ten years. For example, the 7th Framework Programme for research and technological development in EU. OpenFlow-based Software Defined Network comes into being under this background. Just as the other network protocol, there lack the rigor formal analysis of the protocol behavior with SDN. Formalization method with process algebra to analysis SDN protocol is necessary under this situation. Using process algebra as formalization is different from other means because of its mathematical rigor. So, a formalization method with process algebra to analysis SDN protocol is proposed in this paper. The method is self-contained and universal to any protocol analysis. With the method we can formally analysis the detail of SDN protocol to find the uncertain and shortcoming aspects.

The rest of the paper is organized around our main topic, which include the following: Sect. 2 reviews related works and discusses some background materials about

© Springer International Publishing AG 2017
M. Qiu (Ed.): SmartCom 2016, LNCS 10135, pp. 95–101, 2017.
DOI: 10.1007/978-3-319-52015-5_10

the process algebra protocols analysis; Sect. 3 presents operational semantics of the method with process algebra; Sect. 4 gives conclusions and future work.

2   Protocol Analysis with Process Algebra

Process Algebra is an algebraic approach that is used for formulating problems [1]. Hoare's Communicating Sequential Processes [3], Bergstra & Klop's Algebra of Communicating Processes [4]. Labeled transition systems, behavioral equivalences, operational semantics, congruence, and bisimulation equivalence are the most important concepts in the Process Algebra. The important one in these concepts is behavioral equivalences. In this paper we focus on the specification and verification of SDN protocols.

Protocol validation and analysis with process algebra is studied for a long time [5–9]. Process algebra can be used to analyze and evaluate protocols or support protocol design from the beginning [10]. Analysis with process algebra is unambiguous, which can dramatically avoid the misunderstandings, clarify the protocol details. In recently years the process algebra is mostly used to analysis of Mobile Ad Hoc Wireless Networks (MANETs) and their protocols [11]. For example, W-calculus is presented for formally modeling and reasoning about MANETs [12]. Generally speaking, in recently year Process algebra is just used as a tool for supporting protocol design, verification and analysis.

3   SDN Protocol Analysis Based on Process Algebra

3.1   SDN Architecture

In this section we will discuss the problem of the SDN protocol analysis based on process algebra. SDN is an approach to computer networking that allows network administrators to manage network services through abstraction of lower-level functionality [13–17]. SDN architectures decouple network control and forwarding functions, enabling network control to become directly programmable and the underlying infrastructure to be abstracted from applications and network services [18]. This means that network programming languages is the hot topic to support the SDN. Now, in network community, the emerging SDN specific programming languages include Frenetic [19], NetCore [20], Pyretic [21], and NetKat [22]. The common SDN protocols include OpenFlow, Open Network Environment, and Network virtualization platform et al. And the OpenFlow is used by many platforms. With these program languages sand protocols, we can define the behavior of the network infrastructure with abstract model, just like advanced programming languages to the computer architecture. So the rigorous semantic description and analysis of the protocol are dramatically important to the network security and network management ability of these program languages.

3.2   SDN Protocol Analysis

Why should we analyze the SDN Protocol? Formal verification of the SDN with rigorous mathematics logic can examine the requirements and properties system should meet. Especially for SDN, as programmable network architecture, we should design the SDN protocol with detail enough. Misunderstanding or malicious using of SDN based platform can break-down the underlying networks infrastructure. For example, Frenetic-OCaml and Pyretic are designed with a plenty of the formal verification [23–33].

3.3   Removing the Flow in SDN Flow-Table

According OpenFlow protocol, there are three methods to delete entries stream:

- The controller issued a delete request flow table entries;
- OpenFlow switch entries expire mechanism;
- Flow entry remaining lifetime.

Among them, the first two methods to remove the switch in terms of a certain passivity. The following description of the flow table entries deletion process is given.

Set 1:Flow_S={f$_i$| f$_i$ is flow table entry} ;

Set 2:Switch_S={Si|Si is Switch status} ;

Set 3:Flow_Removed={ expiry, delete or eviction } ;

⊔⊔⊔⊔ : Switch flow table entry tag ;

hard_timeout: Expiration of flow table entry tag ;

idle_timeout: Flow table entry idle time ;

$OF\_I \in Switch\_S$, $hard\_timeout \in Switch\_S$,

$idle\_timeout \in Switch\_S$;

$\alpha.Control\_Re$: controller process;

$\alpha.Switch$: Switch process;

Flow table deletion relation $FR = \{\alpha.Control\_Re, \alpha.Switch\} \times Flow\_S$; Flow   table deletion behavior Flow_Remove1 :

$(\overline{\alpha}.Control\_Re \mid \alpha.Switch) \vee s_i.Switch$;

$s_i.Switch \equiv (OF\_I).Switch \vee$

(hard_timeout+idle_timeout).Switch;

Flow_Remove1a: Delete all match table entries ;

Flow_Remove1b: Delete all match table entries ;

Flow table deletion behavior, Flow_Remove2 :

$\alpha.Switch \mid s_i.Switch;$

$$\equiv \frac{s_i.Switch \xrightarrow{\text{hard\_timeout}} \alpha.Switch}{\alpha.Switch \mid s_i.Switch \to \alpha.Switch}$$

$$\equiv \frac{s_i.Switch \xrightarrow{\text{idle\_timeout}} \alpha.Switch}{\alpha.Switch \mid s_i.Switch \to \alpha.Switch}$$

Flow table deletion behavior, Flow_Remove3 :

$\alpha.Switch \mid s_i.Switch;$

$$\equiv \frac{s_i.Switch \xrightarrow{OF\_1} \alpha.Switch}{\alpha.Switch \mid s_i.Switch \to \alpha.Switch}$$

Flow_Remove3a: Delete table entry according to the weight ;

Flow_Remove3b: Delete table entry by duration ;

Flow_Remove3c: Delete table entry by ther conditions ;

Flow_Remove3d: Delete table entry by Synthetic condition ;

Flow table deletion process description :

(1Ttable deletion request process Controller :

$\overline{(\text{Modify-State}.Control\_Re|\text{Modify-State.Switch})}$
$\wedge (\text{OFPFC\_DELETE} \leftrightarrow 1)$
$\Rightarrow \text{Flow\_Remove1a}$

$\overline{(\text{Modify-State}.Control\_Re|\text{Modify-State.Switch})}$
$\wedge (\text{OFPFC\_DELETE\_STRICT} \leftrightarrow 1)$
$\Rightarrow \text{Flow\_Remove1b}$

$((\text{Flow\_Remove1a} \vee \text{Flow\_Remove1a}) \leftrightarrow 1)$
$\wedge (\text{OFPFF\_SEND\_FLOW\_REM} \leftrightarrow 1)$
$\Rightarrow \overline{\text{Flow\_Removed.Switch}}|\text{Flow\_Removed}.Control\_Re;$

(2) Time constraint threshold Flow table deletion :

(Hard_timeout !=0 $\wedge$ Time_Threshold $\leftrightarrow$ 1) $\vee$

(Idle_timeout !=0 $\wedge$ Packet_Num_Threshold $\leftrightarrow$ 1)

$\wedge$(OFPFC_DELETE $\leftrightarrow$ 1 $\vee$ OFPFC_DELETE_STRICT $\leftrightarrow$ 1)   (3) Switch defined

$\Rightarrow$ Flow_Remove2;

(Flow_Remove2 $\leftrightarrow$ 1) $\wedge$ (OFPFF_SEND_FLOW_REM $\leftrightarrow$ 1)

$\Rightarrow$ $\overline{\text{Flow\_Removed.Switch}}$|Flow_Removed.*Control_Re*;

sources recovery

{ (*OFPTMPEF _ IMPORTANCE* $\leftrightarrow$ 1) $\wedge$

(*OFPTMPEF _ LIFETIME* $\leftrightarrow$ 0) $\wedge$

(*OFPTMPEF _ OTHER* $\leftrightarrow$ 0)}

$\Rightarrow$ Flow_Remove3a;

{ (*OFPTMPEF _ IMPORTANCE* $\leftrightarrow$ 0) $\wedge$

(*OFPTMPEF _ LIFETIME* $\leftrightarrow$ 1) $\wedge$

(*OFPTMPEF _ OTHER* $\leftrightarrow$ 0)}

$\Rightarrow$ Flow_Remove3b;

{ (*OFPTMPEF _ IMPORTANCE* $\leftrightarrow$ 0) $\wedge$

(*OFPTMPEF _ LIFETIME* $\leftrightarrow$ 0) $\wedge$

(*OFPTMPEF _ OTHER* $\leftrightarrow$ 1)}

$\Rightarrow$ Flow_Remove3c;

{(*OFPTMPEF _ IMPORTANCE* $\vee$ *OFPTMPEF _ LIFETIME*

$\vee$ *OFPTMPEF _ OTHER*) $\leftrightarrow$ 1}

$\Rightarrow$ Flow_Remove3d;

((Flow_Remove3a $\vee$ Flow_Remove3b $\vee$ Flow_Remove3b

$\vee$ Flow_Remove3d) $\leftrightarrow$ 1) $\wedge$

(OFPFF_SEND_FLOW_REM $\leftrightarrow$ 1)

$\Rightarrow$ $\overline{\text{Flow\_Removed.Switch}}$|Flow_Removed.*Control_Re*;

4 Conclusion and Future Work

We give a formal operational semantics of SDN protocol in detail with labeled transition systems. After a succinct description of the SDN protocol, we prove the correctness and analyses the complexity of SDN protocol with process algebra. The area of formal methods to SDN protocol is very difficult because of no mature tools and techniques available. How to assure the correct behavior of SDN network protocol can be summarized them below:

- Using the universal mathematical calculus symbol to all aspect of SDN protocol is very important.
- Large scale network data streams should be verified by formal operational semantics.
- A variety of network scenarios should be modeled to validate.

All these directions should be pursued in next work.

Acknowledgements. This work is supported in part by the National Science Foundation of China under No. 61672104, 61170209, 61502038, U1509214; Program for New Century Excellent Talents in University No. NCET-13-0676. Key Program of BFSU 2011 Collaborative Innovation Center No. BFSU2011-ZD04.

References

1. http://theory.stanford.edu/ ~ rvg/process.html
2. Milner, R.: Communication and Concurrency. Prentice Hall, Englewood Cliffs (1989)
3. Hoare, C.A.R.: Communicating Sequential Processes. Prentice Hall, Englewood Cliffs (1985)
4. Baeten, J.C.M., Weijland, W.P.: Process Algebra. Cambridge University Press, Cambridge (1990)
5. http://theory.stanford.edu/ ~ rvg/
6. De Nicola, R.: Behavioral Equivalences. In: Padua, D. (ed.) Encyclopedia of Parallel Computing, pp. 120–127. Springer, New York (2011)
7. Bourke, T., Van Glabbeek, R.J., Höfner, P.: Mechanizing process algebra for network protocols. J. Autom. Reason. **56**, 309–341 (2016). Springer Verlag
8. Simonak, S., Hudak, Š., Korecko, Š.: Protocol specification and verification using process algebra and petri nets. In: 2009 International Conference on Computational Intelligence. Modelling and Simulation, Brno, pp. 110–114 (2009)
9. Robert, J.: Colvin: modelling and analysing neural networks using a hybrid process algebra. Theor. Comput. Sci. **623**, 15–64 (2016)
10. Höfner, P.: Using process algebra to design better protocols. Forum"Math-for-Industry" (2015)
11. Singh, A., Ramakrishnan, C.R., Smolka, S.A.: A process calculus for mobile Ad Hoc networks. In: Coordination, pp. 296–314 (2008)
12. Singh, A., Ramakrishnan, C.R., Smolka, S.A.: A process calculus for Mobile Ad Hoc Networks. Sci. Comput. Program. **75**(6), 440–469 (2010)
13. Software-defined networking (SDN) definition. Opennetworking.org. Accessed 26 Oct 2014
14. Gai,K., Li, S.: Towards cloud computing: a literature review on cloud computing and its development trends. In: 2012 Fourth International Conference on Multimedia Information Networking and Security, Nanjing, China, pp. 142–146 (2012)
15. Gai, K., Qiu, L., Zhao, H., Qiu, M.: Cost-aware multimedia data allocation for heterogeneous memory using genetic algorithm in cloud computing. IEEE Trans. Cloud Comput. **PP**(99), 1 (2016)
16. Gai, K., Qiu, M., Zhao, H., Tao, L., Zong, Z.: Dynamic energy-aware cloudlet-based mobile cloud computing model for green computing. J. Netw. Comput. Appl. **59**, 46–54 (2016)
17. Qiu, M., Zhong, M., Li, J., Gai, K., Zong, Z.: Phase-change memory optimization for green cloud with genetic algorithm. IEEE Trans. Comput. **64**(12), 3528–3540 (2015)
18. http://www.frenetic-lang.org. Accessed 12 Sept (2013)
19. Monsanto, C., Foster, N., Harrison, R., Walker, D.: A compiler and run-time system for network programming languages. In: ACM SIGPLAN Notices, vol. 47, pp. 217–230. ACM (2012)
20. Monsanto, C., Reich, J., Foster, N., Rexford, J., Walker, D.: Composing software defined networks. In: NSDI, April 2013
21. Anderson, C.J., Foster, N., Guha, A., Jeannin, J.-B., Kozen, D., Schlesinger, C., Walker, D.: NetKAT: semantic foundations for networks (2013)

22. Bjorner, N., Foster, N., Brighten Godfrey, P., Zave, P.: Formal foundations for networking (Dagstuhl Seminar 15071). Dagstuhl reports **5**(2), 44–63 (2015)
23. Casado, M., Foster, N., Guha, A.: Abstractions for Software-Defined Networks. CACM **57** (10), 86–95 (2014)
24. McClurg, J., Hojjat, H., Foster, N., Cerny, P.: Event-driven network programming. In: ACM SIGPLAN Conference on Programming Language Design and Implementation (PLDI), Santa Barbara, CA, June 2016
25. Reitblatt, M., Canini, M., Guha, A., Foster, N.: FatTire: declarative fault tolerance for software-defined networks. In: ACM SIGCOMM Workshop on Hot Topics in Software Defined Networking (HotSDN), Hong Kong, China, August 2013
26. Gai, K., Qiu, M., Zhao, H., Liu, M.: Energy-aware optimal task assignment for mobile heterogeneous embedded systems in cloud computing. In: 2016 IEEE 3rd International Conference on Cyber Security and Cloud Computing (CSCloud), Beijing, China, pp. 198–203. IEEE (2016)
27. Gai, K., Du, Z., Qiu, M., Zhao, H.: Efficiency-aware workload optimizations of heterogenous cloud computing for capacity planning in financial industry. In: The 2nd IEEE International Conference on Cyber Security and Cloud Computing, New York, USA, pp. 1–6. IEEE (2015)
28. Gai, K., Qiu, M., Tao, L., Zhu, Y.: Intrusion detection techniques for mobile cloud computing in heterogeneous 5G. Secur. Commun. Netw. 1–10 (2015)
29. Li, Y., Gai, K., Qiu, L., Qiu, M., Zhao, H.: Intelligent cryptography approach for secure distributed big data storage in cloud computing. Inf. Sci. **PP**(99), 1 (2016)
30. Qiu, M., Gai, K., Thuraisingham, B., Tao, L., Zhao, H.: Proactive user-centric secure data scheme using attribute-based semantic access controls for mobile clouds in financial industry. Fut. Gener. Comput. Syst. **PP**, 1 (2016)
31. Li, Y., Dai, W., Ming, Z., Qiu, M.: Privacy protection for preventing data over-collection in smart city. IEEE Trans. Comput. **65**, 1339–1350 (2015)
32. Gai, K., Qiu, M., Zhao, H., Xiong, J.: Privacy-aware adaptive data encryption strategy of big data in cloud computing. In: The 2nd IEEE International Conference of Scalable and Smart Cloud (SSC 2016), Beijing, China, pp. 273–278. IEEE (2016)
33. Gai, K., Qiu, M., Zhao, H., Dai, W.: Anti-counterfeit schema using monte carlo simulation for ecommerce in cloud systems. In: The 2nd IEEE International Conference on Cyber Security and Cloud Computing, New York, USA, pp. 74–79. IEEE (2015)

Unsupervised Pre-training Classifier Based on Restricted Boltzmann Machine with Imbalanced Data

Xiaoyang Fu[✉]

Department of Computer Science and Technology,
Zhuhai Key Laboratory of Symbolic Computation and Knowledge Engineering
of Ministry of Education, Zhuhai College of Jilin University,
Zhuhai 519041, China
dvndavidfu@vip.163.com

Abstract. Many learning algorithms can suffer from a performance bias for classification with imbalanced data. This paper proposes the pre-training the deep structure neural network by restricted Boltzmann machine (RBM) learning algorithm, which is pre-sampled with standard SMOTE methods for imbalanced data classification. Firstly, a new training data set can be generated by a pre-sampling method from original examples; secondly the deep neural network structure is trained on the sampled data and all unlabelled data sets by RBM greedy algorithm, which is called "coarse tuning". Then the neural networks are fined tuned by BP algorithm. The effectiveness of the RBM pre-training neural network (RBMPT) classifier is demonstrated on a number of benchmark data sets. Compared with only BP classifier, pre-sampling BP classifier and RBMPT classifier, it has shown that pre-training procedure can learn more representations of data better with unlabelled data and has better classification performance for classification with imbalanced data sets.

Keywords: Semi-supervised learning · Classification · Deep learning · Restricted boltzmann machine · Deep neural network

1 Introduction

Deep learning is a new field of Machine learning technology and powers many aspects of modern society. Deep learning methods allow multiple processing layers neural networks structure to learn intricate representations with levels of abstraction, which have dramatically improved the state-of-the-art in speech recognition [2], visual object recognition [1], nature language processing [18] and many other domains [8, 10, 17].

Although the most common form of machine learning is supervised learning, the semi-supervised or unsupervised learning had a catalytic effect in reviving interest in deep learning. In the late 1990s, neural nets and backpropagation being largely forsaken is mainly due to the commonly thought that simple gradient descent would get trapped in poor local minima and hard being trained in the deep neural networks. However, the expression Deep Learning was actually breakthrough around 2006 [6],

© Springer International Publishing AG 2017
M. Qiu (Ed.): SmartCom 2016, LNCS 10135, pp. 102–110, 2017.
DOI: 10.1007/978-3-319-52015-5_11

when unsupervised pre-training of deep FNNs helped to accelerate subsequent supervised learning through BP algorithm [4], which introduced more efficient learning methods RBM that allow it to pre-train deep nets by layer-by-layer way.

Another problem suffering researchers more years is that data sets are imbalanced when at least one class is represented only a small number of training examples (called the minority class) while the other classes made up the majority. Imbalanced data learning is of great important and challenge in many real application, such as image recognition [10] and oil spills detection [11], etc.

In general, imbalanced data learning involves two main aspects [19]. The first uses various sampling techniques to create an artificially balanced distribution of class examples for training. Among these techniques, random over-sampling [12] and random under-sampling [20] are the simplest ones to be applied by duplicating or eliminating instances randomly. SMOTE is proposed by Chawla [13], which creates synthetic instances by interpolating between similar known examples. It is a better one in pre-sampling approaches not losing some important information in the original data sets.

Negative Correlation Learning (NCL) has been introduced by Liu and Yao [15] with the aim of negatively correlate the error of each network within the ensemble. In this method, instead of training each network separately, a penalty term is introduced to minimize the correlation between the error of network and the error of the rest of the ensemble [14].

The second aspect uses cost adjustment(penalty term) within the learning algorithms so as to tend minority training, such as instead of the overall classification accuracy, using the weighted average accuracy of the minority and majority classes for training target function, or adding a negative correlation penalty term (NCL) in the cost function, etc.

Furthermore, for smaller data sets, unsupervised pre-training procedure helps to prevent over-fitting [7], leading to significantly better generalization when the number of labelled data is small.

In this paper, we study the pre-training the deep structure neural network by restricted Boltzmann machine (RBM) learning algorithm, which is pre-sampled with standard SMOTE methods for imbalanced data classification. First, a new training data set can be generated by a pre-sampling method from original examples; secondly the deep neural network structure is trained on the sampled data and all unlabeled data sets by RBMs, which is called "coarse tuning". Then the neural networks is fine-tuned by BP from top three layers while the regularized negative correlation penalty terms were participated into the cost function of neural network in order to overcome over-fitting and promote generalization ability.

The rest of this paper is organized as follows. Section 2 introduces the pre-sampling method, BP-MLP classifier, neural nets with RNCL algorithm and RBM learning process. Section 3 provides the results of the RBMPT classifier on a benchmark data sets and compared with no-sampling and standard SMOTE-sampling BP-MLP classifier. Finally, Sect. 4 concludes the paper.

2 The Model

2.1 Pre-sampling Algorithm

Let the training data set,

$$X = \{x_i, t_i\}_{i=1}^{n}, n = n_p + n_N$$

Where n is the overall number of examples, n_p and n_N is the number of minority class and majority class respectively, t_i is the target value corresponding x_i. After pre-sampling, generating a set of new examples $X' = \{x_i', t_i'\}_{i=1}^{n}, n' = n_p' + n_N'$,

Where n' is the number of pre-sampling training data set, n_p' and n_N' is the number of minority and majority class respectively, t_i' is target value corresponding x_i'.

The pre-sampling algorithm can be included as following steps:

(a) Extract samples randomly from majority class sets according to the rate of a, make $n_N' = a \cdot n_N$;
(b) Extract a sample x_i randomly from minority class n_p;
(c) Find out the k number of samples nearest the x_i and randomly extract a sample x_n from the k samples and generate a random number $vran \in [0, 1]$;
(d) Generate required synthetic instances x_s by interpolating between the seeds and minority nearest neighbor examples $x_s = x_i + vran \cdot (x_n - x_i)$;
(e) do repeat from (b) to (d) until generating $n_s = b \cdot n_p$; b is the rate of minority data, n_s is number of man-made data sets.
(f) Randomly mix up the instances of to n_N', n_p, n_s, generate the new training data sets as following:$x' = \{x_i', t\}_{i=1}^{n'}, n' = n_N' + n_p + n_s$.

2.2 BP-MLP Classifier

Three-layers neural network is used for based classifier, it is the most widely used neural network model, in which the connection weights training is normally completed by back-propagation (BP) learning algorithm [3]. In BP algorithm, the error is minimized when the network outputs match the desired outputs. The mean square error (MSE) for the neural network is defined as:

$$MSE = \sum_{i=1}^{n} O(x_i) - T_i \qquad (1)$$

Where $O(x_i)$ is the network output for input vector x_i, while x_i as its target value.

The cost function (*MSE*) is the function of the connection weights and is minimized when the network outputs match the desired target values.

2.3   Semi-supervised Pre-training Procedure

The Restricted Boltzmann Machine (RBM) is a network of symmetrically coupled stochastic binary units [9]. The undirected graphical model of an RBM is illustrated in Fig. 1, showing that the h_i are independent of each other when conditioning on v and the v_j are independent of each other when conditioning on h. RBM contains a set of visible units $v \in \{0,1\}^D$, and a set of hidden units $h \in \{0,1\}^P$, the energy of the state $\{v,h\}$ is defined as:

$$E(v, h) = -b'v - c'h - v^T W h \tag{2}$$

The probability that the model assigns to a visible vector v is:

$$p(v; \theta) = \frac{1}{Z(\theta)} \sum_h \exp(-E(v, h; \theta)) \tag{3}$$

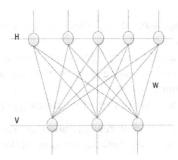

Fig. 1. A restricted Boltzmann machine with no hidden-to-hidden and no visible-to-visible connections

$$Z(\theta) = \sum_v \sum_h \exp(-E(v, h; \theta)), \tag{4}$$

$Z(\theta)$ is the partition function. $\theta = \{W, b, c\}$ are the model parameters. The conditional distributions over the hidden and visible units are given by:

$$p(h_j = 1|v, h_{-j} = \sigma(\sum_{i=1}^{D} W_{ij} v_i + c_j), \tag{5}$$

$$p(v_i = 1|h, v_{-i}) = \sigma(\sum_{j=1}^{P} W_{ij} h_j + b_i), \tag{6}$$

where $\sigma = 1/(1 + \exp(-x))$ is the sigmoid logistic function. The parameters updates, which are needed to perform gradient ascent in the log-likelihood, can be obtained from function (3).

$$\Delta \mathrm{W} = \alpha(E_{data}\left[vh^T\right] - E_{model}\left[vh^T\right], \tag{7}$$

$$\Delta b_i = \alpha(E_{data}\left[vv^T\right] - E_{model}\left[vv^T\right], \tag{8}$$

$$\Delta b_j = \alpha(E_{data}\left[hh^T\right] - E_{model}\left[hh^T\right], \tag{9}$$

Where α is a learning rate, E_{data} denotes an expectation with respect to the completed data distribution, and E_{model} is an expectation with respect to the distribution defined by the model(Eq. 3). We will sometimes refer to E_{data} as the data-dependent expectation and E_{model} as the model's expectation.

Although exact maximum likelihood learning in RBM's is still intractable, learning can be carried out efficiently using Contrastive Divergence. We use Gibbs sampling to approximate both expectations.

Now, the RBMPT classifier can be described as following:

(a) Generate a set of new data sets using an improved SMOTE over-sampling method;
(b) Initialize each individual network weights, negative correlation term λ and regularization parameter α_m;
(c) Use an RBM unsupervised learning algorithm to learn every level initial weights.
(d) For each network and corresponding training example set, adjust the weight W_m and regularization parameter α_m.
(e) Repeat from (c) for a desired number of iterations (epochs).

In above algorithm, weights and regularization of each network are evolved simultaneously and converged to the optimal solution while the error function of the networks is minimized.

The neural network we use is trained by the greedy layer-wise procedure in which each added layer is trained as RBM. The algorithm is showed as follows:

Algorithm:

```
TrainDNN(v_data, α, layers, W, b, c, isMeanField)
  for k=1 to layers do
    normalize(v_data)
    initializeW^k=0, b^k=0, c^k=0
      while not stopping criterion do
        sampleh⁰=v_data from datasets
        fori=1 to k-1 do
          ifisMeanField then
            Q(h_j^i = 1|h^{i-1}) = h_j^i, jall feature elements of h^i
          Else
            Sample h_j^i from Q(h_j^i h^{i-1})
          End if
        End for
        UpdateRBM(h^{k-1}, α, W^k, b^k, c^k)
      End while
  End for
```

3 Experimental Results

3.1 Datasets Selection

In this section, we will study some experimental results of RBMPT classifier and compared with no-sampling and standard SMOTE sampling BP-MLP classifier on a number of benchmark data sets. Six imbalanced data sets come from the UCI machine learning repository [16]. In these data sets, the maximum imbalanced ratio ranges from 9:1 to 19:1, the feature dimensions from 4 to 10, involving many applications. Their main characteristics are shown as Table 1.

The main implementation parameters of the based classifier are presented in Table 2. The neural networks have two hidden layers, each layer with 20 nodes. The BP training epochs are 1000 while the pre-training epochs with RBM are from 1000 to 1500 for different class samples.

Table 1. Main characters of samples

Data sets	Examples	Features	Imb. ratio	Application
Abalone_7	4175	8	9.7:1	Life
Balance	625	4	12:1	Social
Yeast_ME3	1480	8	9.1:1	Life
Flare	1065	10	19:1	Physics

Table 2. Main parameters of the based classifier

Data sets	Learning rate	Hidden Nodes	BP_epochs	RBM_epochs
Abalone_7	0.001	20	1000	1500
Balance	0.01	20	1000	1300
Yeast_ME3	0.001	20	1000	1300
Flare	0.001	20	1000	1000

3.2 Performance Comparison of Classifiers

Compared RBMPT classifier with no sampling and standard SMOTE-sampling BP-MLP classifier, the average values of G-mean and F-measure are shown in Table 3.

Table 3. Main parameters of the based classifier

Data sets	RBMPT		SMOTE-sample		MLP	
	G	F_{min}	G	F_{min}	G	F_{min}
Abalone_7	0.483	0.334	0.483	0.359	0.218	0.011
Balance	0.926	0.762	0.789	0.444	0.703	0.356
Yeast_ME3	0.874	0.788	0.838	0.784	0.846	0.739
Flare	0.693	0.286	0.506	0.200	0.196	0.067

Table 4. Balance dataset test analysis

Test item	Different classification tests				
	T1	T2	T3	T4	T5
Allscore	96.00	95.2	92.00	92.00	91.20
$F_{MAXCLASS}$	0.9782	0.9737	0.9558	0.9569	0.9500
$F_{MINCLASS}$	0.7619	0.7273	0.5833	0.4444	0.3556
G-mean	0.9264	0.8787	0.7462	0.7886	0.7028

Table 4 Notes:

T1:   RBMPT use pre-sampling and pre-training with all data;

T2:   RBMPT without pre-sampling; T3: RBMPT pre-training without unlabelled data;

T4:   Only pre-sampling test with BP-MLP; T5: BP-MLP;

Seen from Table 3, $F_{MINCLASS}$ is F-measure of small class datasets. According to G-mean value, the performance of pre-sampling neural network is better than BP-MLP neural network. Specially, the BP-MLP is hard to classify the small class datasets of Abalone_7 and Flare. Except the equal value between RBMPT and SMOTE-sampling classifiers on Abalone_7, the performance of RBMPT is the best of all. Being compared with according to F-mean can get the same results except Abalone_7 data set.

The experimental results show that pre-sampling method is beneficial to the imbalance datasets classification. By 'pre-training' several layers of progressively more complex feature detectors using learning every layer by RBM greedy learning method, the weights of the network could be initialized to sensible value. Seen from Table 4, the Balance dataset was classified with five situations by three classifiers. Being compared with test results, we can easily find the only BP-MLP classifier is the worst of all test targets, the RBMPT classifier test results without using unlabeled data because pre-training are unsurprisingly similar with that of pre-sampling BP-MLP classifier for semi-supervised learning classifier should be fed in enough data to learn the representation of network structure. It is very interesting findings that the RBMPT classifier can still get enough good results through only pre-training procedure without pre-sampling procedure. Using pre-sampling data and unlabeled data together to pre-train the networks and find the sensible globe initial weights shows the best test results of all. However, we also find sometimes the RBMPT classifier without pre-sampling does not classify on some datasets when the ratio of class samples is considerable large and pre-training data are not enough.

Another finding in the test results shows that the classification performance depends on the proportion of majority class datasets and man-made datasets to the extents, which majority dataset parameter (a) is set from 0.5 to 1.0 while man-made parameter (b) is set from 0.1 to 0.5. This is caused by the reasons that turning the parameter (a) smaller will lost the classification information of the original datasets while turning the parameter (b) larger will touch on the performance majority class with excessively unnecessary man-made data.

4 Conclusion

In this paper, we have study an RBMPT classifier with pre-sampling and its application for imbalanced data classification. Firstly a new training data set can be generated by pre-sampling from original examples and it improves the imbalanced ratio in training examples, the RBMPT is pre-trained on the both new training examples and unlabeled data to get sensible initial weights of nets. The unsupervised pre-training procedure here can learn the representation of intricate network structure and prevent from over-fitting, leading to significantly better generalization for the imbalance datasets. We propose a hybrid pre-sampling and pre-training classifier for imbalanced data classification. The effectiveness of RBMPT classifier is demonstrated on a number of benchmark data sets. Compared with no-sampling MLP and standard SMOTE-sampling BP-MLP classifier, it has shown that the RBMPT classifier has better classification performance for classification with imbalanced data.

References

1. Krizhevsky, A., Sutskever, I., Hinton, G.: ImageNet classification with deep convolutional neural networks. In: Proceedings of the Advances in Neural Information Processing Systems, vol. 25, pp. 1090–1098 (2012)
2. Hinton, G., et al.: Deep neural networks for acoustic modeling in speech recognition. IEEE Signal Process. Mag. **29**, 82–97 (2012)
3. Rumelhart, D.E., Hinton, G.E., Williams, R.J.: Learning representations by back-propagating errors. Nature **323**, 533–536 (1986)
4. Hinton, G.E., Osindero, S., Teh, Y.-W.: A fast learning algorithm for deep belief nets. Neural Comput. **18**, 1527–1554 (2006)
5. Bengio, Y., Lamblin, P., Popovici, D., Larochelle, H.: Greedy layer-wise training of deep networks. In: Proceedings of the Advances in Neural Information Processing Systems, vol. 19, pp. 153–160 (2006)
6. Hinton, G.E., Salakhutdinov, R.: Reducing the dimensionality of data with neural networks. Science **313**, 504–507 (2006)
7. Bengio, Y., Courville, A., Vincent, P.: Representation learning: a review and new perspectives. IEEE Trans. Pattern Anal. Mach. Intell. **35**, 1798–1828 (2013)
8. Mnih, V., et al.: Human-level control through deep reinforcement learning. Nature **518**, 529–533 (2015)
9. Salakhutdinov, R., Hinton, G.: Deep Boltzmann machines. In: Proceedings of the International Conference on Artificial Intelligence and Statistics, pp. 448–455 (2009)
10. Munder, S., Gavrila, D.: An experimental study on pedestrian classification. IEEE Trans. Pattern Anal. Mach. Intell. **28**(11), 1863–1868 (2006)
11. Japkowicz, N., Stephen, S.: The class imbalance problem: A systematic study. Intell. Data Anal. **6**(5), 429–449 (2002)
12. Barandela, R., Sanchez, J.S., Garcia, V., Rangel, E.: Strategies for learning in class imbalance problems. Pattern Recogn. **36**(3), 849–851 (2003)
13. Chawla, N.V., Bowyer, K.W., Hall, L.O., Keglmeyer, W.P.: SMOTE: Synthetic minority over-sampling technique. J. Artif. Intell. Res. **16**(1), 321–357 (2002)

14. Fu, X., Zhang, S.: Evolving neural network ensembles using variable string genetic algorithm for pattern classification. In: Proceedings of the 6th International Conference on Advanced Computational Intelligence, pp. 81–85 (2013)

15. Chen, H., Yao, X.: Regularized negative correlation learning for neural network ensembles. IEEE Trans. Neural Netw. **20**(12), 1962–1979 (2009)

16. Asuncion, A., Newman, J.: UCI Machine learning Repository (2007). http://www.ics.uci.edu/ ~ learn/MLRespository.html

17. Leung, M.K., Xiong, H.Y., Lee, L.J., Frey, B.J.: Deep learning of the tissue-regulated splicing code. Bioinformatics **30**, i121–i129 (2014)

18. Collobert, R., et al.: Natural language processing (almost) from scratch. J. Mach. Learn. Res. **12**, 2493–2537 (2011)

19. Bhowan, U., Johnston, M., Zhang, M., Yao, X.: Evolving diverse ensembles using genetic programming for classification with unbalanced data. IEEE Trans. Evol. Comput. **17**(3), 368–386 (2013)

20. Weiss, G.M., Provost, F.: Learning when training data are costly: The effect of class distribution on tree induction. J. Artif. Intell. Res. **19**, 315–354 (2003)

A Secure Homomorphic Encryption Algorithm over Integers for Data Privacy Protection in Clouds

Jyh-Haw Yeh[✉]

Department of Computer Science, Boise State University, 1910 University Drive,
Boise, Idaho 83725, USA
jhyeh@boisestate.edu

Abstract. If a secure and efficient fully homomorphic encryption algorithm exists, it should be the ultimate solution for securing data privacy in clouds, where cloud servers can apply any operation directly over the homomorphically encrypted ciphertexts without having to decrypt them. With such encryption algorithms, clients' data privacy can be preserved since cloud service providers can operate on these encrypted data without knowing the content of these data. Currently only one fully homomorphic encryption algorithm proposed by Gentry in 2009 and some of its variants are available in literature. However, because of the prohibitively expensive computing cost, these Gentry-like algorithms are not practical to be used to securing data in clouds. Due to the difficulty in developing practical fully homomorphic algorithms, partially homomorphic algorithms have also been studied in literature, especially for those algorithms homomorphic on arithmetic operations over integers. This paper presents a secure variant algorithm to an existing homomorphic algorithm over integers. The original algorithm allows unlimited number of arithmetic additions and multiplications but suffers on a security weakness. The variant algorithm patches the weakness by adding a random padding before encryption. This paper first describes the original algorithm briefly and then points out it's security problem before we present the variant algorithm. An efficiency analysis for both the original and the variant algorithms will be presented at the end of the paper.

Keywords: Homomorphic encryption · Non-deterministic encryption · Cipher equality test · Big data privacy · Data privacy in clouds

1 Introduction

The amount of information that companies and governments are collecting is growing exponentially: 90% of all the world's digital data were collected in just the last two years [1]. The term "big data" refers to a new generation of massive datasets, which contain information that can be searched for commercial purposes or stockpiled for subsequent mining. Storing these datasets in clouds allows users to exploit the data without having to maintain their own large systems.

© Springer International Publishing AG 2017
M. Qiu (Ed.): SmartCom 2016, LNCS 10135, pp. 111–121, 2017.
DOI: 10.1007/978-3-319-52015-5_12

However, storing valuable data on another entity's equipment exposes it to theft and unauthorized use. This threat can slow down companies from realizing the benefits of cost-efficient cloud computing and big data.

Most cyber-security efforts aim to prevent security breaches by outsiders; indeed, most cloud service providers do have greater resources to protect against outsider attacks than do their clients. Ironically, authorized system administrators inside the cloud pose a more significant threat. Because of curiosity or financial motivation, insiders who actually manage the data could become malicious and steal sensitive information. Recent surveys and research suggest that insider attacks are a widespread threat to cloud computing [2–4].

To protect data privacy against insider attacks, the ideal solution will be to have a Fully Homomorphic Encryption (FHE) algorithm that allows cloud servers (insiders) operate on the encrypted data without having to decrypt the data. However, for more than 30 years, cryptographers have tried to find FHE algorithms, in which any operation can be directly applied to ciphertexts without messing up the results. Mathematically, an FHE algorithm can be described as $f(m_1, m_2, \ldots, m_n) = D(f(E(m_1), E(m_2), \ldots, E(m_n)))$, where each m_i is a plaintext $\forall 1 \leq i \leq n$, f is any operation with n inputs, and E and D are the FHE encryption and decryption functions, respectively. Not until recently, Gentry was the first person to propose a workable FHE algorithm in 2009 [5]. Following Gentry's paper, several other variant algorithms [6–9] were also proposed in literature. However, none of these Gentry-like algorithms is practical for real world applications because of their expensive computational cost.

One can imagine how difficult it is to design an efficient FHE algorithm that is homomorphic on any operation. Rather than developing a practical FHE algorithm, researchers have tried to develop partially homomorphic encryption algorithms that target at specific data types with homomorphic property on specific operations. The author has proposed an efficient probabilistic homomorphic encryption (PHE) algorithm over integers in [10]. Comparing to Gentry's algorithm, the PHE algorithm is simple and very efficient. The trade-off is that the PHE algorithm is not homomorphic to all operations but only on arithmetic additions and multiplications. Though the PHE algorithm has only homomorphic property on arithmetic operations, it is still very useful for protecting numeric data in clouds. Unfortunately the PHE algorithm has a security weakness. A variant PHE algorithm is proposed in this paper to patch the security problem by applying random padding to the plaintext before encryption. Let's name this variant algorithm PPHE (Padded PHE). Comparing to the PHE algorithm, the PPHE algorithm is more secure but losing the capability of testing the equality of two ciphers. Two ciphers are said "equal" if they are ciphers of the same plaintext. Without the security weakness suffered by the PHE, the proposed PPHE algorithm is highly useful for real world applications in cloud computing.

Both the PHE and PPHE algorithms use a random number as an input to the encryption process. Thus, these two algorithms are non-deterministic, which is a property that can hide the equality relationship among ciphertexts and thus enhance the data privacy protection against adversaries who are able to eavesdrop encrypted data. The difference between PHE and PPHE is that PHE ciphers can be tested whether they are "equal" if someone knows the equality test key, whereas the PPHE ciphers cannot be tested for equality.

The paper is organized as follows. Section 2 describes related work in literature - partially homomorphic encryption algorithms on arithmetic additions and multiplications, followed by the Gentry-like algorithms and their computational cost. Section 3 describes the original PHE algorithm and its security weakness. Section 4 presents the PPHE algorithm and argues that why it is free from the security weakness suffered by the PHE algorithm. Section 5 analyzes the efficiency of both algorithms. Finally the paper is concluded in Sect. 6.

2   Related Work

Unlike the PHE and the proposed PPHE algorithms, some traditional encryption algorithms are partially homomorphic on either additions or multiplications but not both. For example, the unpadded RSA [11] and ElGamal [12] algorithms are homomorphic on multiplications, whereas the encryption algorithms of Paillier [13], and Okamoto and Uchiyama [14] are homomorphic on additions.

In 2009, Gentry proposed the first FHE algorithm using lattice-based cryptography [5]. Following his paper, many different implementations or optimizations to Gentry's algorithm have been proposed [6–9]. However, the reported results are far away from being practical. In their reports, each bit of plaintext requires ciphertext with thousands or even millions of bits [6–8,15]. An implementation of Gentry's algorithm that makes the system actually working needs a public key with 2.3 GB, and taking two hours to generate it. In addition, the re-encryptions took 30 min each [8,15]. Re-encryption (called recrypt) is an operation in the Gentry-like algorithms to reduce noises on ciphertexts after homomorphic operations. Currently the most efficient implementation of a Gentry-like algorithm has been done recently by Lauter, Naehrig and Vaikuntanathan [9] who managed to reduce the key sizes to roughly a megabyte!

3   PHE Algorithm

This section describes the PHE algorithm proposed in [10]. We extract some sections from [10] and present them in this paper concisely. The security of the PHE algorithm relies on a hard integer factorization problem.

Integer Factorization Problem: *Given a product N of two large primes p and q, the problem is to find the primes p and q from N.*

No efficient algorithm is available in literature that is able to solve the problem in a reasonable amount of time. The security of the famous RSA algorithm [11] is also based on this hard problem. The security model for both the PHE and the PPHE algorithms are the same and described below.

Security Model: Two parties are involved in the PHE or PPHE algorithm:

1. A data owner D_o owns the data. To protect data privacy, only the data owner can perform data encryption and decryption using an encryption key only known to D_o.
2. An authorized agent D_a can perform homomorphic additions, multiplications and/or equality tests on ciphertexts. D_o issues an operational key to D_a. With the operational key, D_a can only perform the homomorphic operations over ciphertexts, but cannot use it to decrypt data. From data owner's perspective, the agent D_a is actually an adversary and the homomorphic encryption algorithms used should preserve data privacy from the agent. Thus, even if the operational key is revealed to public, the data privacy is not compromised.

3.1   PHE Parameter Setup

If a data owner D_o wants to use the PHE algorithm to encrypt data, D_o needs to set up several PHE parameters and keys.

1. D_o randomly picks two large primes p_1 and p_2 repeatedly until $q = 2p_1 + 1$ is also a prime. D_o needs to choose p_1 large enough so that all his application data $m < p_1$. Otherwise the cipher of m cannot be decrypted back to m.
2. D_o computes $N = p_1 \times p_2$. The encryption key of the algorithm is the pair (p_1, N), which will be used by D_o to perform data encryption/decryption.
3. D_o randomly picks another large prime p_3 and computes $T = q \times p_3$. D_o also randomly picks a set of integers $h_1, h_2, \ldots, h_k \in Z_T^*$, and computes

$$g_i = h_i^{2(p_3 - 1)} \bmod T, \forall i = 1, 2, \ldots, k \qquad (1)$$

The equality test key is the set $(g_1, g_2, \ldots, g_k, T)$. The provided equality test algorithm in Sect. 3.4 later is a probabilistic algorithm. Thus, if more g_i are provided by D_o, the accuracy of the equality test is higher.
4. The homomorphic operational key of the algorithm is $(N, g_1, g_2, \ldots, g_k, T)$, where authorized agents can use N to perform homomorphic add/multiply operations and use $(g_1, g_2, \ldots, g_k, T)$ for equality testing.

3.2   PHE Encryption and Decryption

Only the data owner D_o knowing the encryption key (p_1, N) can encrypt and decrypt data. The PHE encryption only consists of one modular operation.

$$E_{PHE}(m, p_1, N) = rp_1 + m \bmod N \qquad (2)$$

where $m < p_1$ is the plaintext and r is a random positive integer.

To decrypt a ciphertext c, again only one modular operation is required. Equation (3) below described the decryption.

$$D_{PHE}(c, p_1) = c \bmod p_1 \tag{3}$$

Since p_1 is a factor of N, based on the Chinese Reminder Theorem, Eq. (3) can recover the plaintext $m < p_1$ from the ciphertext c, since

$$D_{PHE}(c, p_1) = c \bmod p_1 = (rp_1 + m \bmod N) \bmod p_1 = (rp_1 + m) \bmod p_1 = m$$

3.3 PHE Homomorphic Additions and Multiplications

The authorized agent D_a knowing the homomorphic operational key can perform the PHE addition, which is just a modular addition as described in Eq. (4) below.

$$H_+(c_1, c_2, N) = (c_1 + c_2) \bmod N \tag{4}$$

where $H_+(c_1, c_2, N)$ denotes the homomorphic sum of two ciphers c_1 and c_2, computed using an operational key N. We can verify the correctness of the homomorphic addition in Eq. (4) if the decryption of the sum $H_+(c_1, c_2, N)$ is $m_1 + m_2$, given that $m_1 + m_2 < p_1$. By Eq. (3), we have

$$D_{PHE}(H_+(c_1, c_2, N), p_1) = H_+(c_1, c_2, N) \bmod p_1 = (c_1 + c_2 \bmod N) \bmod p_1$$
$$= (c_1 + c_2) \bmod p_1 = (r_1 p_1 + m_1 + r_2 p_1 + m_2) \bmod p = m_1 + m_2$$

The authorized agent D_a knowing the operational key N can perform the homomorphic multiplication, which is just a modular multiplication of two ciphers c_1 and c_2 as described in Eq. (5) below.

$$H_\times(c_1, c_2, N) = (c_1 \times c_2) \bmod N \tag{5}$$

We can show the correctness of the homomorphic multiplication in Eq. (5) if $H_\times(c_1, c_2, N)$ can be decrypted to $m_1 \times m_2$, given that $m_1 \times m_2 < p_1$.

$$D_{PHE}(H_\times(c_1, c_2, N), p_1) = H_\times(c_1, c_2, N) \bmod p_1 = (c_1 \times c_2 \bmod N) \bmod p_1$$
$$= (c_1 \times c_2) \bmod p_1 = ((r_1 p_1 + m_1) \times (r_2 p_1 + m_2)) \bmod p_1$$
$$= (r_1 r_2 p_1^2 + r_1 m_2 p_1 + r_2 m_1 p_1 + m_1 m_2) \bmod p_1 = m_1 \times m_2$$

3.4 PHE Homomorphic Equality Test

Since PHE uses a random integer in each encryption, the algorithm is non-deterministic, where a same plaintext can be encrypted to different cipher each time. This is a feature that the equality relationship among ciphers can be hidden if only ciphers are presented. However, the agent D_a, on behalf of the data owner D_o, might need to compare the equality of two ciphers in some applications. The PHE provides a homomorphic equality test algorithm for D_a (if knowing the equality test key) to test whether two ciphers are actually "equal".

The PHE's equality test is an algorithm based on Euler's theorem. It is very similar to the famous primality test algorithm which is based on Fermat's Theorem. We describe both the Euler's and Fermat's theorems below.

Euler's Theorem: $\forall g \in Z_N^*$, $g^{\phi(N)} = 1 \mod N$, *where $\phi(N)$ is a totient of N.*

Fermat's Theorem: *If p is prime and for all $1 < g < p$, $g^{p-1} = 1 \mod p$.*

By the Fermat's theorem, p is not prime if $g^{p-1} \neq 1 \mod p$. However, p is not necessary but most likely a prime if $g^{p-1} = 1 \mod p$. By [16], for a hundred digits number p, the probability that p is not prime when $g^{p-1} = 1 \mod p$ is about $\frac{1}{10^{13}}$. If this false positive risk is not acceptable, the primality test of p can be more accurate by testing the Fermat's formula with multiple g's. For example, if we use g_1 and g_2 to test the primality of p twice by Fermat's formula and both tests return 1, the chance of p not being prime will be reduced to $\frac{1}{10^{26}}$.

Similarly, the accuracy of the PHE equality test below can be improved by a few tests. Based on the PHE parameter setup in Sect. 3.1, the equality test key $T = q \times p_3 = (2p_1 + 1) \times p_3$. Thus, the totient of T is $\phi(T) = 2p_1(p_3 - 1)$.

PHE Equality Test Theorem: Given PHE ciphers c_1 and c_2, and an equality test key (g, T) defined in Sect. 3.1, if c_1 and c_2 are homomorphically equal (i.e., the corresponding plaintexts $m_1 = m_2$), the following test returns true.

$$g^{|c_1 - c_2|} \stackrel{?}{=} 1 \mod T \qquad (6)$$

Proof: Based on Eq. (2), let $c_1 = (r_1 p_1 + m_1) \mod N$ and $c_2 = (r_2 p_1 + m_2) \mod N$. Since

$$|c_1 - c_2| \mod p_1 = |r_1 p_1 + m_1 - r_2 p_1 - m_2| \mod p_1 = |m_1 - m_2|$$

Thus, $|c_1 - c_2| = r p_1 + |m_1 - m_2|$ with a random integer r. If c_1 and c_2 are homomorphically equal (i.e., $m_1 = m_2$), by the Euler's theorem, we have

$$g^{|c_1-c_2|} = g^{r p_1 + |m_1 - m_2|} \mod T = g^{r p_1} \mod T = (h^{2(p_3-1)})^{r p_1} \mod T$$
$$= (h^r)^{2 p_1 (p_3 - 1)} \mod T = (h^r)^{\phi(T)} \mod T = 1$$

From the above theorem, we can conclude that two ciphers $c_1 \neq c_2$ if the test in Eq. (6) returns false. However if the test returns true, c_1 and c_2 are not necessary homomorphically equal but most likely they are! Similar to the primality test, the chance of falsely reporting equality of two ciphers is very small, probably comparable to the false positive chance in a primality test (i.e., 1 in 10^{13}). If the data owner cannot accept this false positive chance, k such tests can reduce the risk to 1 in 10^{13k}. In order for the agent D_a to perform k equality tests, the data owner needs to provide k equality test keys $g_1, g_2, \ldots, g_k$. The following procedure describes the equality test algorithm with multiple tests.

PHE-Equality-Test $(c_1, c_2, g_1, g_2, \ldots, g_k, T)$
{   for $i \leftarrow 1$ to k
        do if $g_i^{|c_1 - c_2|} \neq 1 \mod T$
                then   return false;
    return true; }

3.5   Security Weakness of PHE

In this section, we describe the security weakness of the PHE algorithm.

PHE Security Weakness: If the agent D_a collects two PHE homomorphically equal ciphers c_1 and c_2, D_a can derive the encryption key p_1 by

1. D_a computes $|c_1 - c_2|$, which is actually a PHE ciphertext of 0 and thus $|c_1 - c_2| = 0 \bmod p_1$. In other words, $|c_1 - c_2| = rp_1$ for some integer r.
2. D_a computes $\gcd(|c_1 - c_2|, N)$, where the gcd result is actually the encryption key p_1 since both $|c_1 - c_2| = rp_1$ and $N = p_1 p_2$ have a prime factor p_1 unless r is multiple to the large prime p_2.

---

4   Padded PHE (PPHE)

In response to the security weakness of the PHE algorithm described in the previous section, a padded PHE or PPHE algorithm is proposed as a variant algorithm. PPHE is more secure, but the equality test algorithm will be no longer working. It allows an unlimited number of arithmetic operations until the sum/product $m \not< 2^w$ or $m' \not< p_1$, where m and m' are the original plaintext and the padded plaintext respectively, and w is a predefined maximal bit-size of allowed data (represented as integers) in the application.

4.1   PPHE Parameter Setup

The PPHE has the same parameter setup as the PHE except two additional parameters w and z. The selection of w and z has impact to the selection of p_1.

1. w should be selected based on the application requirement. If the maximum data (integer) in an application is I_{max}, then w must be big enough so that $I_{max} < 2^w$. That is, w is the maximum bit size of all data in an application.
2. z is the number of random bits to be left padded to the data so that two encryptions of the same data m will be encrypting two different padded data. Typically, $z = 64$ would be big enough for real world applications.
3. The padded data m' is the size of $w + z$ bits. Recall that $m < p_1$ in PHE encryption. Similarly, in PPHE encryption, $m' < p_1$. This implies that if the data owner would like to have k consecutive homomorphic multiplications on ciphertexts, then choosing p_1 with a size at least $(k + 1)(w + z)$ bits.

4.2   PPHE Encryption and Decryption

The PPHE algorithm encrypts the padded m' rather than m. We use an example to demonstrate the padding procedure. Given a data $m = 13$, the padded m' is $m' = r_{z-1} \ldots r_1 r_0 | 0 \ldots 00 | 1101$, where $m = 13 = 1101_2$ with leading 0's to the w-th bit, followed by z padded random bits. Mathematically,

$$m' = R \times 2^w + m \tag{7}$$

where R is the value of the z padding bits. The equation below shows how to PPHE encrypt a data m.

$$E_{PPHE}(m, p_1, N) = rp_1 + m' \bmod N \tag{8}$$

where (p_1, N) is the encryption key and r is a random number.

To decrypt a PPHE cipher c, the decryption process in Eq. (3), i.e., $c \bmod p_1$, would recover the padded data m'. The original data m can then be recovered by $m = m' \bmod 2^w$. From the above two steps of PPHE decryption, in order to recover the original data m, the following two conditions must hold: (1) The padded data $m' < p_1$; and (2) The original data $m < 2^w$.

4.3 PPHE Homomorphic Additions and Multiplications

The homomorphic additions/multiplications in the PPHE are the same as those in the PHE algorithm as described in Eqs. (4) and (5) respectively.

4.4 The Equality Test Does Not Work in the PPHE

The PHE equality test algorithm described in the previous section does not work for the PPHE. Consider two PPHE ciphers c_1 and c_2 of plaintexts m_1 and m_2. If $m_1 = m_2$, then c_1 and c_2 are homomorphically equal. However, the test in Eq. (6), $g_i^{|c_1 - c_2|} \bmod T$, most likely will not return 1 since $|c_1 - c_2| \bmod p_1$ is now equal to $|m'_1 - m'_2|$ rather than $|m_1 - m_2|$. This implies $|c_1 - c_2| = rp_1 + |m'_1 - m'_2|$ for some random number r. Thus,

$$
\begin{aligned}
g_i^{|c_1 - c_2|} &= (g_i^{rp_1 + |m'_1 - m'_2|}) \bmod T = ((h_i^{2(p_3 - 1)})^{rp_1 + |m'_1 - m'_2|}) \bmod T \\
&= (h_i^{r\phi(T) + 2(p_3 - 1)|m'_1 - m'_2| \bmod \phi(T)}) \bmod T = (h_i^{2(p_3 - 1)|m'_1 - m'_2| \bmod \phi(T)}) \bmod T \\
&= 1 \quad \text{if } m'_1 = m'_2
\end{aligned}
$$

With z random padding bits, if $m_1 = m_2$, there is only 1 in 2^z chance that $m'_1 = m'_2$ and thus the above test most likely will not return 1.

Even without the equality test procedure available in the PPHE, the algorithm is still a valuable homomorphic encryption algorithm since it is secure and can be used for applications without the need for equality matching.

4.5 A Walk Through Example for the PPHE

1. Choose $p_1 = 32451533$, $q = 2p_1 + 1 = 64903067$, $p_2 = 103$ and $p_3 = 179$, where p_1 is $2 \times (w + z) = 24$-bit prime and it is large enough to support $w = 8$-bit data, $z = 4$-bit random padding and one homomorphic multiplication. For demonstration purpose, we use small primes p_2 and p_3.
2. Compute $N = p_1 \times p_2 = 3342507899$ and $T = q \times p_3 = 11617648993$.
3. The encryption key is $(p_1, N) = (32451533, 3342507899)$.

4. Let $h = 6$ be a random number selected. The equality test key is (g, T), where $g = h^{2(p_3-1)} \bmod T = 6^{356} \bmod T = 2628786245$. We give the equality test key used in PHE here to show that it is no longer working in PPHE.

Assume the PPHE algorithm encrypts two data $m_1 = 13$ and $m_2 = 8$ to two ciphers c_1 and c_2. First, the algorithm constructs two padded data $m_1' = 1010\ 00001101 = 2573$ and $m_2' = 0010\ 00001000 = 520$, where the four rightmost bits are random padding. Secondly, the algorithm encrypts m_1 and m_2 by Eq. (8) as follows: $c_1 = 12571 \times p_1 + 2573 \bmod N = 162260238$ and $c_2 = 8431 \times p_1 + 520 \bmod N = 2855735424$, where 12571 and 8431 are random numbers.

Now let's verify the PPHE decryption on c_1. The decryption has two mod operations, i.e., $(c_1 \bmod p_1) \bmod 2^w$. In this example, $(162260238 \bmod 32451533) \bmod 2^8 = 13$, which recovers m_1. Thus, the decryption works correctly.

For homomorphic addition $(c_1 + c_2) \bmod N = 3017995662$ and multiplication $(c_1 \times c_2) \bmod N = 1656366143$, PPHE works correctly since the decryptions of the sum and product, $(3017995662 \bmod p_1) \bmod 2^w = 21$ and $(1656366143 \bmod p_1) \bmod 2^w = 104$, equal to the sum and product of $m_1 = 8$ and $m_2 = 13$.

Finally, let's show that the equality test does not work in PPHE. Let $m_1'' = 0110\ 00001101 = 1549$ be another padded data of m_1. Assume the encryption of m_1'' is $c_1' = 2881 \times p_1 + 1549 \bmod N = 3245154849$. The equality test $g^{|c_1 - c_1'|} = g^{3082894611} \bmod T = 1041103470 \neq 1$, though both c_1 and c_1' are ciphers of m_1.

4.6   Security of the PPHE

In this section, we argue that the PPHE will not have the same security weakness suffered by the PHE algorithm. Assume the agent D_a or an adversary has two PPHE homomorphically equal ciphers c_1 and c_2, in which $m_1 = m_2$. Using the same attack described in the previous section, the adversary performs

$$|c_1 - c_2| = rp_1 + |m_1' - m_2'| = rp_1 + |R_1 2^w + m_1 - R_2 2^w - m_2| \quad \text{(See Eq. (7))}$$
$$= rp_1 + |(R_1 - R_2)2^w| = rp_1 + R'2^w$$

where r, R_1, R_2 and $R' = |R_1 - R_2|$ are all random numbers. In each such subtraction, all the adversary (including the authorized agent D_a who might be malicious) can get is a difference $d = rp_1 + R'2^w$, which is most likely not multiple of p_1 and thus the $\gcd(d, N) \neq p_1$. The case for d being multiple of p_1 is when the random numbers $R_1 = R_2$ and therefore $R' = R_1 - R_2 = 0$.

Now let's estimate how many homomorphically equal ciphers the adversary needs to collect so that they can recover the encryption key p_1. Let n be the number of homomorphically equal ciphers collected by the adversary. There are a total of $\binom{n}{2}$ pairs of homomorphically equal ciphers and thus we have $\binom{n}{2}$ possible d's. The odd for each such d being multiple of p_1 is $\frac{1}{2^z}$, where z is the number of random padding bits. If $z = 64$ and with n homomorphically equal ciphers, the odd to recover p_1 becomes $\binom{n}{2} \cdot \frac{1}{2^{64}}$. In order to have a good odd of about 50% chance to recover p_1, the adversary may need to know more than $n = 2^{32}$ homomorphically equal ciphers since It is unrealistic for the adversary to know $n = 2^{32}$ homomorphically equal ciphers.

5   Efficiency

Key size: PHE and PPHE algorithms have a modulus N which is similar to the one in the RSA algorithm. If PHE uses a modulus N with 2048 bits, the encryption key p_1 will be roughly 1024 bits. With this key length, PHE can support homomorphic additions and multiplications until the sum or product (plaintext value) exceeds $p_1 (\simeq 2^{1024})$. For the PPHE algorithm with 1024-bit p_1, if both w and z are 64 bits, it can support at least 7 homomorphic multiplications and about 900 homomorphic additions (if the product/sum does not exceed 2^{64}). With this key length setting, Gentry-like algorithms are not even possible to be built. If we increase the key length to megabytes comparable to the best implementation [9] of Gentry-like schemes, our PPHE algorithm can support 4,000 homomorphic multiplications even with both w and z set to 1024 bits.

Ciphertext size: The sharp growth of ciphertext after homomorphic operations is another obstacle preventing Gentry-like algorithms from being practical. This is not an issue in PHE/PPHE since all homomorphic additions and multiplications are mod N operations. The ciphertexts are always less than N.

Computation: Encryption, decryption, and homomorphic operations only need one/two modular operations in both PHE and PPHE. The PHE equality test needs one modular exponentiation, which is similar to an RSA operation.

6   Conclusion

In this paper, we proposed a secure and efficient homomorphic encryption algorithm PPHE over integers, which is a variant to an existing PHE algorithm. PPHE adds random padding technique to its encryption process to eliminate the security weakness presented in the PHE algorithm. Both the PHE and the PPHE algorithms are extremely efficient since all operations are just a few modular additions and/or multiplications. The proposed homomorphic encryption algorithms enable the cloud servers to perform arithmetic operations over encrypted data. With such encryption algorithms, the ciphers stored in clouds will never need to be decrypted and thus the privacy of these data can be preserved.

References

1. Top ten big data security and privacy challenges (2012). https://downloads. cloudsecurityalliance.org/initiatives/bdwg/Big_Data_Top_Ten_v1.pdf
2. The notorious nine - cloud computing top threats in 2013. https://downloads. cloudsecurityalliance.org/initiatives/top_threats/The_Notorious_Nine_Cloud_ Computing_Top_Threats_in_2013.pdf
3. Cunsolo, V.D., Distefano, S., Puliafito, A., Scarpa, M.L.: Achieving information security in network computing systems. In: The 8th IEEE International Conference on Dependable, Autonomic and Secure Computing (2009)
4. Linthicum, D.: Afraid of outside cloud attacks? You're missing the real threat (2010). http://www.infoworld.com/d/cloud-computing/ afraid-outside-cloud-attacks-youre-missing-real-threat-894

5. Gentry, C.: Fully homomorphic encryption using ideal lattices. In: The 41st ACM Symposium on Theory of Computing (STOC) (2009)
6. van Dijk, M., Gentry, C., Halevi, S., Vaikuntanathan, V.: Fully homomorphic encryption over the integers. International Association for Cryptologic Research (2009). http://eprint.iacr.org/2009/616
7. Smart, N., Vercauteren, F.: Fully homomorphic encryption with relatively small key and ciphertext sizes. http://www.info.unicaen.fr/M2-AMI/articles-2009-2010/smart.pdf
8. Gentry, C., Halevi, S.: Implementing gentry's fully homomorphic encryption scheme. In: EUROCRYPT 2011, pp. 129–148 (2011)
9. Lauter, K., Naehrig, M., Vaikuntanathan, V.: Can homomorphic encryption be practical? In: 3rd ACM Workshop on Cloud Computing Security, pp. 113–124 (2011)
10. Yeh, J.H.: A probabilistic homomorphic encryption algorithm over integers - protecting data privacy in clouds. In: 12th International Confenerce on Advanced and Trusted Computing (2015)
11. Rivest, R., Shamir, A., Adleman, L.: A method for obtaining digital signatures and public-key cryptosystems. Commun. ACM **21**(2), 120–126 (1978)
12. ElGamal, T.: A public-key cryptosystem and a signature scheme based on discrete logarithms. IEEE Trans. Inf. Theory **31**(4), 469–472 (1985)
13. Paillier, P.: Public-key cryptosystems based on composite degree residuosity classes. In: Stern, J. (ed.) EUROCRYPT 1999. LNCS, vol. 1592, pp. 223–238. Springer, Heidelberg (1999). doi:10.1007/3-540-48910-X_16
14. Okamoto, T., Uchiyama, S.: A new public-key cryptosystem as secure as factoring. In: Nyberg, K. (ed.) EUROCRYPT 1998. LNCS, vol. 1403, pp. 308–318. Springer, Heidelberg (1998). doi:10.1007/BFb0054135
15. Hayes, B.: Alice and Bob in cipherspace. Am. Sci. - Mag. Sigma Xi **100**, 362–367 (2012). Computing Science
16. Pomerance, C.: On the distribution of pseudoprimes. Math. Comput. **37**(156), 587–593 (1981)

Big Data Management the Mass Weather Logs

Hao Wu[✉]

Department of Computer Science,
JLUZH, Zhuhai College of Jilin University, Zhuhai, China
haowu_mouse@hotmail.com

Abstract. The log generated by the software system becomes bigger and bigger, which are including many important and valuable information, so it will have a very high commercial value in the future. Because the log information is complex, it is difficult to mine numerous relative data and receive valuable one by means of the traditional technology. The capability of the big data process on Hadoop is much faster than that of traditional modus. This research is primary to analyze the weather log apace, taking advantage of the parallel process on Hadoop. It forms the B/S structure system and utilizes the High charts plug-in to generate charts to display to the user, in the light of that people may forecast the temperature change and make a decision. By applying a test it shows a perfect application, a good graphical interface and an elevated operability and inter-activity, so it is more convenient and much faster, and much simpler to show the results of the analysis directly.

Keywords: Mapreduce · Log · Hadoop

1 Introduction

Because any system can produce a lot of logs in the operation process, Internet becomes one of the biggest driving the rapid growth of the big data progressively. The mass log data can obtain a lot of valuable information, so processing, analyzing and mining that become essential. On analyzing and processing the big data Hadoop is much more effective than the traditional means. Hadoop theory is applied to the finance, the education, social aspects and so on. The log format is various sorts and varieties. Sometimes that makes users be dazzled, but some regular patterns can be followed. In any systems the log is significant extremely, on account of the process recorded from starting to perishing, even the operational procedure of every functions and uses can be minute in detail. Owing to preserving the log to become mounting more and more, it is indispensable to propose a patulous system to deal with the log more efficaciously and more apace. Hadoop is a distributed storage system, which means to reserve the data on the different place. It can handle the terabytes of data parallel in high efficiency and saving time. It would be better commercial value in future, taking advantage of the Hadoop distributed system to discover the useful information on the log.

Taking advantage of the management mass weather logs system, it can get the maximum and the minimum annual temperature through analyzing the mass weather

© Springer International Publishing AG 2017
M. Qiu (Ed.): SmartCom 2016, LNCS 10135, pp. 122–132, 2017.
DOI: 10.1007/978-3-319-52015-5_13

logs, so as to help people to grasp the change of temperature and to predict the future tendency and make decisions. Because the type of log information is character, which also contains a lot of unnecessary information, it is much more difficult to mine numerous relative data and receive valuable one by means of the traditional technology. The capability of the big data process on Hadoop is much faster than that of traditional modus, and its coding is uncomplicated. The capability of the big data process on Hadoop is much faster than that of traditional modus, and its coding is uncomplicated. This research is primary to analyze the weather log apace, taking advantage of the parallel process on Hadoop. Hadoop consists of two core simulations, which is the distributed file system (HDFS) and the other is MapReduce mathematical framework, making use of that can implement the weather log. It forms the B/S structure system and utilizes the High charts plug-in to generate charts to display to the user, in the light of that people may forecast the temperature change and make a decision. By applying a test it shows a perfect application, a good graphical interface and an elevated operability and interactivity, so it is more convenient and much faster, and much simpler to show the results of the analysis directly.

2 Main Methodology

2.1 Linux Fedora

Fedora is a Linux distribution, which is a fast, stable, and powerful operating system. Fedora Project task is to serve as a cooperative community to lead the free and open source software and its spiritual progress. Three elements of the task are as follows.

- Fedora Project is always to strive to a leadership, rather than following.
- Fedora Project consistently tries to create, to improve, and to spread freedom and the spirit of free code.
- Fedora Project succeeds through collaboration and sharing of community members.

2.2 Hadoop

Hadoop Distributed File System (HDFS) has a high fault tolerance features, and is designed to be deployed in low-cost hardware, and it provides high through putto access the application data for those with large data sets applications. The core of the design framework Hadoop is HDFS and MapReduce. HDFS provides storage of vast amounts of data, and the MapReduce provides calculations for the vast amounts of data.

2.3 MapReduce

MapReduce is a programming model for large datasets (greater than 1 TB) of parallel computing. The concept of "Map" and "Reduce," is its main thoughts that are borrowed from the functional programming language, and borrowed from the vector programming language features. It is designated a Map function for a group of key-value pairs that are mapped to a new set of key-value pairs specified concurrent the Reduce function to ensure that all key mappings share the same key for each a group.

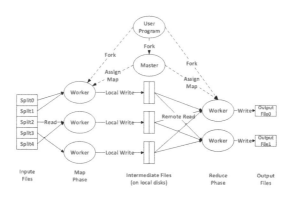

Fig. 1. MapReduce execution overview

As shown as Fig. 1, the starting is from the top of the user program that links MapReduce library to achieve the most basic functions of Map and Reduce functions. MapReduce library inputs user program file divided into m parts at first, and then uses the fork to copy user process to other machines in the cluster. A copy of the user program is called as a master, the rests called workers. The master is responsible for scheduling jobs assigned to idle workers. The worker assigned with Map job starts reading input the data corresponding slice. Map job extracts the key from the input data, for each key-value pairs are passed to map function as parameters, in the middle of which generate keys to be cached in memory. The intermediate key cached will be written to the local disk on a regular basis, and is divided into R zones that each zone corresponds to a Reduce job. Positions of these intermediate key-value pairs are communicated to the Master, which is responsible for forwarding information to the Reduce worker. The Master notifies the worker assigned with the Reduce job which is responsible for the district in the position. When the Reduce worker being responsible for all the intermediate key-value pairs are read, ordering them at first, the same key value pairs are get together. The Reduce worker passes the key and the value associated to a Reduce function whose output function generated will be added to the partition of the output file. When all of the Map and Reduce jobs are completed, the master wakes the genuine user program, MapReduce function call the user program code returned.

3   Deploying the Server Environment

It adopts the Hadoop distributed completely, and the user names are all Hadoop.

3.1 Installing the Fedora-14

The minimum requirements for processor speed depend on the end use, installation and specific hardware. Fedora's Anaconda installation tool using the default will start the graphical interface and ACPI support, so that when to install the required hardware can improve the compatibility.

3.2 Installing the Centralization

Environment directions. Hadoop cluster includes three nodes. One is Master, the other two are Workers. Nodes are connected by the local area network (LAN), and can ping each other.

These three nodes are the same operating system (Fedora-14), and have the same user on Hadoop. The Master is mainly on two roles of NameNode and JobTracker, to undertake the main distributed data and decomposition of the task execution. Two Workers deploy the role of DataNode and TaskTracker, to be responsible for the distributed data storage and the task execution.

In the cluster, Hadoop must be installed on each machine. To install and to configure Hadoop need the supreme authority identity of "root". To deploy the file of "hadoop-env.sh", the file of "core-site.xml" which attributes are HDFS address and the port number, the file of "hdfs-site.xml" which copies of the configuration is three, the file of "mapred-site.xml" which attributes are JobTracker address and the port number, the file of masters and to remove "localhost", then to join the Master node IP (192.168.1.67), and the file of Workers and to remove "localhost", then to join the Worker node IP (192.168.1.65 and 192.168.1.66).

Deploying the hosts file. The hosts file is used to configure the information on the DNS server host, and to record the connection of the host LAN corresponds to the host name and IP. When the user finds the file in the network link at first, it gets the configuration of the corresponding host name (or the domain name) to the IP address, such as Fig. 2:

```
::1       localhost       localhost6.localdomain6 localhost6
192.168.1.67 Master.Hadoop
192.168.1.65 Slave1.Hadoop
192.168.1.66 Slave3.Hadoop
```

Fig. 2. Host file configration

Deploying the verification of password-less on SSH. The operational process need distal the daemons on Hadoop. After starting Hadoop, the NameNode is controlled by SSH on the DataNode to start and stop. It requires password-less to log in to perform the corresponding commands between nodes, so as to configure the SSH password-less in the form of the public key authentication. The NameNode uses the SSH password-less to login and to start the Data Name process. As the same principle, the DataNode can also use the SSH password-less to login the NameNode on the Master.

It performs the command of "ssh-keygen-tras-P", which creates a password-less key pair, such as Fig. 3:

```
6c:50:b5:47:8c:ee:28:66:50:c6:6c:24:b7:21:fa:64
The key's randomart image is:
+--[ RSA 2048]----+
|    o++ ...o.    |
|    . +*+  .o.   |
|   . E+o  .. .   |
|    +.  o  ..    |
|      .. So      |
|        +.. .    |
|        o .      |
|                 |
|                 |
+-----------------+
```

Fig. 3. Key generated

It examines whether there is a folder of ".ssh" below "/home/hadoop/" and if there are two password-less keys to produce id_rsa and id_rsa.pub, such as Fig. 4:

```
-rw-------  1 hadoop hadoop 2178 10月 10 16:43 authorized_keys
-rw-------.  1 hadoop hadoop 1679  1月  7 09:46 id_rsa
-rw-r--r--.  1 hadoop hadoop  402  9月 13 10:55 id_rsa.pub
```

Fig. 4. File of key pairs

Then to input the command of "cat ~ /.ssh/id_rsa.pub ≫ ~ /.ssh/authorized_keys" on the primary node, it appends id_rsa.pub to the authorized key, so the Master node can login this machine in password-less.

Now the Master node in password-less can login the Worker node. It just need copy the public key to these two nodes machines and is appended to the authorized key.

In the same way, the Worker node in password-less login to the Master node. It just need copy the generated public key on the Worker node to the Master node and be appended to the authorized key.

4   The System Analysis

4.1   System Requirements

The requirement of the mass weather log management system on Hadoop is to combine small files into a large file and to process, that can analyze the annual maximum and the minimum temperature. The graphical interface is good and can improve the operability and interactivity, which is to reduce the burden of operators and the operation to be risen more convenient, faster and simpler. It can display the analysis results objectively and directly, and can be downloaded to view.

4.2   The System Function Analysis

The system function is consists of three parts, which are "Select a data range", "Upload the Log analysis" and "Display graphical data".

5   Overall Design

5.1   The System Structure

At the client the user uploads the data wanted to view on the Hadoop cluster, and the backend code analysis is to be stored in HBase. The background code reads the data from that database to return the client that is showed in the form of chart at the client. Such as Fig. 5.

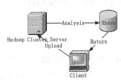

Fig. 5. System structure diagram

5.2   The Overall Functional Design

The overall goal of the system development is to use the Hadoop platform analysis and process the weather Log, to find out many years and the temperature corresponding to the maximum and the minimum temperature in annual and to display the data objectively. It realizes the systematic, standardized, scientific and automatic objective, so as to improve the temperature prediction. The function is comprehensive, and the interface is simple, easy to be operated.

5.3   Algorithm Thought

The data is cut according to certain rules, and then to take out the year and the temperature. The year is set as a key, and the temperature is set as a value, that is <year, temperature>key/value pair as the value of temperature. The value of the same year will be merged before the output is reduced, and it compares the maximum and the minimum value in Reduce at this time.

5.4   Uploading and Merging Data Idea

That HDFS provides API can be used to realize to upload local files to HDFS. When uploading, it can read the data of each small file, and then write to one file. So it can put the small file merged into a large file.

5.5   Database Design

HBase is a column database and need not set the data type. The line key in HBase can be sorted automatically, and so as MapReduce, that is shows that HBase can be combined with Hadoop very well, so the year value is to be the line key which will be sorted from small to large and these data is intuitive. The highest temperature is to be column name and the lowest one is to be value. That data is sent back to customers in a short time can reduce the network I/O request, thus to improve the transmission speed (Table 1).

Table 1. Datastructure

Linekey	Column cluster	
Year	The highest temperature	The lowest temperature

6   Detailed Design and Implementation

6.1   Data Preparation

To open the file to check the data format, the characteristics of the data can be found that "year-month-day hour-minute-second temperature", and it is to record a temperature in every two hours. Such as Fig. 12 (Table 2).

Table 2. Datacontents

Year-month-day	Hour-minute-second	Temperature
2001-01-02	00:00:00	15
2001-01-02	02:00:00	19
2001-01-02	04:00:00	20
2001-01-02	06:00:00	15
2001-01-02	08:00:00	16
....	......	......

The graphic design uses plug-in High charts generate mainly on the page references. Such as Fig. 6.

```
<script src="js/highcharts.js"></script>
<script src="js/exporting.js"></script>
```

Fig. 6.   References plugin highchair

6.2   Algorithm Design

According to the log format, this code will add the space into the inputting data value as "year-month-day hour-minute-second temperature" that is deposited into an array. The first part is to be the year cut up "-", so the year can be got. The third can get the temperature, then the year and the temperature composite the new key/value pair <key,value> will be passed to Reduce. For example, the inputting data is 2005-01-01 00:00:00 13, 2005-07-02 12:00:00 29, 2005-12-01 02:00:00 6, 2006-03-01 00:00:00 12,

```
public void map(Object key,Text value,Context context)
        throws IOException,InterruptedException{
    String[] str=value.toString().split(" ");
    if(str[0].matches("\\d{4}-\\d{1,2}-\\d{1,2}")){
        String st=str[0].split("-",2)[0];
        context.write(new Text(st),new Text(str[2]));
    }
}
```

Fig. 7. Map algorithm

2006-06-01 08:00:00 16, 2006-09-01 14:00:00 30 which is sent into Map will become <2005,13>, <2005,29>, <2005,6>, <2006,12>, <2005,16> and <2005,30> (Fig. 7).

These six groups of data sent to Reduce will be merged in to two sets of data of <2005,13 29 6> and <2006,12 16 30>, which is a mechanism of graphs framework. According to the same key, it is merged into the value automatically. The temperature in 2005 will not appear in 2006, so as to the same that which in 2006 doesn't appear in 2005. The Reduce will get the value from that two groups of data and compare their value, so as to determine the maximum and the minimum value. So the outputting data after Reduce will become <2005,629> and <2006,12 30>. At this time it need only put the value into the database one by one (Fig. 8).

```
public void reduce(Text key,Iterable<IntWritable> values,Context context)
        throws IOException, InterruptedException{
    int maxvalue=Integer.MIN_VALUE;
    int minvalue=Integer.MAX_VALUE;
    while(values.iterator().hasNext()){
        String s=values.iterator().next().toString();
        String str="";
        Pattern p = Pattern.compile("\\s*|\t|\r|\n");
        Matcher m = p.matcher(s);
        str = m.replaceAll("");
        int a=Integer.parseInt(str);
        if(maxvalue<a){
            maxvalue=a;
        }
        else if(minvalue>a)
        {
            minvalue=a;
        }
    }
}
```

Fig. 8. Reduce algorithm

6.3 Running Process and Results

The System will merge the content according to the data accepted. In the file of Log the user can select the date range of merger.

In Hbase "peak" is the highest temperature and "value" is the lowest one. The result analyzed in the database is as following (Fig. 9):

```
hbase(main):027:0> scan 'Temperature'
ROW                    COLUMN+CELL
 2001                  column=peak:28, timestamp=1389062900940, value=5
 2002                  column=peak:29, timestamp=1389062936416, value=3
 2003                  column=peak:31, timestamp=1389062948131, value=5
 2004                  column=peak:30, timestamp=1389063279300, value=4
 2005                  column=peak:29, timestamp=1389065299672, value=4
 2006                  column=peak:30, timestamp=1389065326058, value=3
 2007                  column=peak:32, timestamp=1389065353305, value=2
 2008                  column=peak:29, timestamp=1389065360671, value=3
 2009                  column=peak:30, timestamp=1389065381267, value=2
 2010                  column=peak:31, timestamp=1389065392624, value=1
10 row(s) in 0.0590 seconds
```

Fig. 9. Database analysis results

In the log folder the time interval files will be uploaded to HDFS and be handled with MapReduce, to analyze the annual maximum and the minimum temperature.

The data in HBase will read in the background, and be sent to the front desk. That result will display in the form of line chart, to let it objectively and readable (Fig. 10).

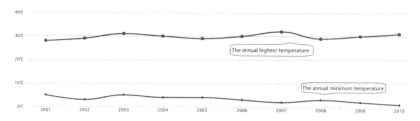

Fig. 10. Displaying the system data diagram

7 System Testing and Implementation

The purpose of the system performance testing is that the efficiency of processing and analyzing files in the same size and different copies. The design and result of the testing case is as follows (Table 3).

Table 3. Testing case designing

Cases	The size of files Uploaded	The copy of files Uploaded
1	5 M	5
2	25M	1

7.1 Testing Process and Methods

Accessing the system can view the status of the cluster, as well as the information detailed in each of MapReduce, including the running time, the size of the file, the utilization rate of CPU and the memory usage, etc. Using its own regulatory mechanism, it can be used to detect the efficiency of the system on the same file size and different copies.

- Usecase1: To upload 5 files of the weather log as 5 M, it is to check the processing time of running and opening the web page.
- Usecase2: To upload 5 files of the weather log as 25 M, it is to check the processing time of running and opening the web page.

To compare with these two operational time, it shows that the efficiency of the usecase2 is higher than the usecase1 obviously. The results are as follows (Table 4).

Table 4. Testing reslults

Usecases	Time/Second
1	2.8
2	1.2

After testing, it is found that the processing speed of the less copies of file is faster than these in the same file size and different copies. The conclusion is that Hadoop is not suitable for processing a large number of small files, but to process large files, which spends less time and the efficiency is higher. It is because that the Hadoop is chunking for data processing, which default is 64 M. If there are lots of small data files, such as a file of 2–3 M, which a small data file is far less than the size of a block of data has to process as a block of data. Storing a large number of small files occupies storage space, so its storage efficiency is not high and the retrieval speed is slower than a large file. Such small files consume the calculable capacity in MapReduce, because it is to allocate the Map tasks in a block.

8 Conclusion

In the process of the study, it uses Hadoop technology in the era of the cloud. After building a Hadoop platform, it adapts the core technology of HDFS and the operational framework of MapReduce to handle the analysis of the weather log. Combining with the programming techniques and ideas of Java Web, it forms a B/S structure and uses the plug-in High charts to generate charts.

After testing it finds that Hadoop is not good at handling small files, especially a large number of small files. The size of the file is bigger than HDFS block size, because of MapReduce computing in such a small file consumes the calculable capacity, which default is allocated on the Map tasks in a block. To let programmers modify them in purpose, it will make the system perfectly.

Acknowledgment. Thank Zhuhai Laboratory of Key Laboratory of Symbolic Computation and Knowledge Engineering of Ministry of Education.

References

1. Sharma, S., Mangat, V.: Technology and trends to handle big data: survey. In: 2015 Fifth International Conference on Advanced Computing & Communication Technologies, pp. 266–271 (2015)
2. Triguero, I., Peralta, D., Bacardit, J., Garca, S., Herrera, F.: MRPR: a MapReduce solution for prototype reduction in big data classification. Proc. Neurocomput. **150**(7), 331–345 (2015)
3. Cramer, J., O'Donnell, T.: Advancing big data and analytics capabilities. In: Proceedings - IBM Data Management Magazine, no. 11, 22 November 2013

4. Saito, K., Yoshimura, A., Lee, J.H.: Object oriented memory recollection and retrieval based on spatial log in the iSpace. In: 2012 IEEE/SICE International Symposium on System Integration (SII), pp. 271–276 (2012)
5. Lopez, J.C.B., Villaruz, H.M.: Low-cost weather monitoring system with online logging and data visualization. In: 2015 International Conference on Humanoid, Nanotechnology, Information Technology, Communication and Control, Environment and Management (HNICEM), pp. 1–6 (2015)

Application of a Parallel FSM Parsing Algorithm for Web Engines

Xin Ren[(⊠)], Jiong Zhang[(⊠)], Ying Li[(⊠)], and Jianwei Niu[(⊠)]

School of Computer Science and Engineering,
Beihang University, Beijing, China
{renxin, zhangjiong, liying}@buaa.edu.cn

Abstract. Finite State Machine (FSM) is widely applied to parsing the html pages in WebKit browser. In the traditional WebKit kernel, single thread is used to parse web pages, therefore cannot make full use of multi-core processors and page analysis in kernel is not granular down to html paragraphs, resulting in the browser working serially in each page. Our research is about realizing the page parallel parsing in WebKit, aiming at improving the browser's parsing speed. Made some optimizations on the method proposed in [13], we put forward a novel parsing model named PFBE (Parallel FSM Browser Engine) from the aspect of data parallelism, realizing the preliminary parser to load network resources and parsing string in parallel, thus using multi-core processors to improve browser's performance. PFBE carved up the input data into multiple segmentation which contained public characters for processing, through comparing the public characters to determine whether to merge the segmentation. PFBE utilized original serial FSM which has highly optimized, compatible with HTML5 standards and technology. We used Chromium web engine to present PFBE in detail and the result proved the improvement of PFBE comparing with serial FSM, and the page loading time has been reduced by 12.36%.

Keywords: Parallel computing · FSM · Web page parsing · HTML5

1 Introduction

Finite State Machine is aimed at studying the calculation of the limited memory and some kinds of abstract computational model [1]. In the process of loading page, the main task of web browser is to process html pages locally. On the one hand, if the speed of downloading is slower than processing web page, the network will become the bottleneck of browser performance. However, the network bandwidth has improved in recent years, such as some countries have already put 4G LTE mobile networks to use, which can provide tens of Mbps bandwidth [2]. In addition, the network cache technology, such as the classic HTTP cache buffer deposit and the new proposed web-based content caching scheme [3, 4]. On the other hand, with the enhancement of web standards, the browser needs to consume kinds of computing resources to parse,

© Springer International Publishing AG 2017
M. Qiu (Ed.): SmartCom 2016, LNCS 10135, pp. 133–143, 2017.
DOI: 10.1007/978-3-319-52015-5_14

lay out, and render the web page, and running the Sunspider JavaScript Benchmark [5]. The most widely used technology to improve the efficiency of JavaScript engine is instant compiled Just-In-Time technology [6]. Smart Caching project can avoid the redundant computing through reusing style sheets matching and the result of the page layout calculation to enhance performance [7]. Redundant CSS rules can be detected by CILLA [8]. Current browser can accelerate the page processing by taking advantage of the hardware of the PC and mobile devices, such as using the graphics processor to accelerate the page rendering process [9]. Furthermore, on the basis of practice, the web programmer summed up the rules of efficient HTML code [10, 11] and JavaScript code [12] to reduce the amount of computation of the client page processing.

According to different stages in page processing, researchers put forward many parallel algorithms, such as web analytic parallel algorithm [13], parallel script engine [14], parallel computing and laying out style sheets matching algorithm [15]. However, web language as HTML, CSS, JavaScript and so on, its internal data structures such as DOM, have not considered the needs for parallel processing at the beginning of design, so the current page parallel parsing algorithm needs to change the web standards, such as JavaScript was substituted by Flapjax language [14]; or restricting some web language, for example, only a subset of the CSS can be used [15]. DOHA accomplished a self-adapting web application runtime layer by using JavaScript [16], among which the parallel execution of multiple JavaScript tasks are realized through Web Workers [17]; OP and OP2 browser proposed task parallel method between modules through dividing the modules of the web browser [18]; Adrenaline [19] divided its structure at server according to the structure of html page, so that the client browser can load multiple sections of the page in parallel. These algorithms [16, 19] are only suitable for optimizing the processing efficiency of some web applications mostly, and the programs proposed in the paper need to modify the html page or web server, so it is difficult to be widely used. There are also some algorithms for parallel processing, such as parallel web analytic algorithm [13], our research is based on its thought to improve, without modifying the original FSM serial program but setting the pre-scanner in advance to download network resources.

The paper presents a new parallel web analytic algorithm, and its main innovation is parsing page from the perspective of data parallelism. The article focuses on the parallel web analytic algorithm and HTML5 technology, and our thought can be extended to other stages of processing the page, such as matching style sheets or rendering web pages. We named the new algorithm with Parallel FSM. Different with the existing parallel algorithms, Parallel FSM preserved the serial processing algorithm and achieved parallel processing on its basis. On the one hand, it inherited the optimization techniques of the serial processing algorithm; on the other hand, it has a good compatibility as serial-processing algorithm for web standards. The second part describes the model of the page and issues; the third part is to explain the parallel parsing algorithm steps and two detailed optimization techniques; fourth part is the experiment results; then we conclude final experiment and summary.

2 Web Analytic Model

In the mathematical model, the formal definition of Finite State Machine is a quintuple (Q, Σ, δ, q0, F), wherein: Q is a finite set, called the set of states; Σ is a finite set. It is called the alphabet; δ called the state transition function; q0 is the initial state; F is the set of accepting states.

FSM set has dozens of various types corresponding to the label, which is mentioned above Q. The input string, the output string and string these three events are the parser alphabet, which is mentioned above Σ. Parser in its current state, in response to an event, causes changes in the character state, which the state transition function is δ. The string in the initial state of parser is usually character "<", so it is the initial state of the parser. q0 is the start state, that has not been processed automatically (q0 ∈ Q). F is terminated state set, also known as acceptance set (that is F ⊆ Q). In the ideal case, the parser will always run, so it did not accept the state and the set F is empty.

From the above definition, FSM has three features:

(1) S is a finite number, so the transfer number of state is limited.
(2) F ⊆ S, that is at any time in the corresponding to F, only has one state.
(3) Σ and δ determine F, under certain premise, its transmission is from the current state to another state based on transferring function.

And Meyerovich [13], who summed up the fourth feature of FSM, which assumes an different initial state q0 strings, and in the same (S, δ, Σ) condition, while the accepting state set F may be different, but the status of the collection will gradually be stable.

FSM parsed the input data and expressed as a DOM tree in memory according to web standards. Figure 1 shows the basic flow of modern web browsers when it processes the web page.

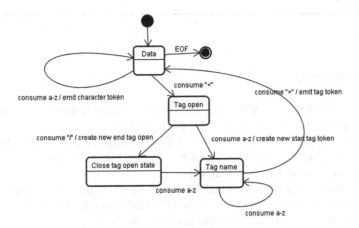

Fig. 1. Workflow of webpage processing

Our idea is to divide the html document into multiple fragments, each containing one or more of the html element, and then use traditional serial parsing algorithm while parsing a plurality of segments, the last the analytical results obtained for the fragment when put the final analysis results document together, pseudo-code description is as follows:

Algorithm.1 FSM Parse Html In Parallel Algorithm
Input: (1) An HTMLWeb w ; (2) the number of threads k
Output: the optimal lex parser in parallel
1: Initialize Partition HTML w to F_n
2: Initial $FSM_k \leftarrow$ k, keyword $\leftarrow$ overlap m-character
3: **for** $F_i \in F_n$ **do**
4:     parse F_n with FSM_n
4:     parse F_{n+1} with FSM_{n+1}
5:     **if** keyword in F_n = = keyword in F_{n+1}
6:       continue
7:     **else**
8:       keyword=keyword+1
9:       rescan F_n and F_{n+1}
10: **end for**
11: **while** (i + 1 < n) **do**
12:     merge F_i and F_{i+1}
13: **end while**

3   Current Design

This section introduces the parallel web analytic algorithm at first, and then introduces two important optimization techniques for the algorithm.

Given a html page w, in order to resolve it concurrently into a DOM tree, download the external data which was cited in w and then execute the JavaScript code blocks, the running algorithm is divided into four steps:

The pre-parse and pre-load is high priority loading mechanism, if browser kernel found resources on the page references, then the HTML-Resource-Pre-loader Class will request and execute script code. The pre-parse and pre-load mechanism is targeted at scanning img node; after that, html interpreter formally process web pages and run JavaScript code.

A Sub-algorithm 1 (Preload Network Resources): Pre-parse segment and scan the input character one by one in w, and find out all the "<" characters, if the script encountered img node or nodes, call *HTML-Resource-Pre-loader Class* to download pages, or JavaScript, etc.

All html tags, including the start tag, end tag, comments, etc., start with the character "<", so the parallel algorithm use "<" as the segment basis. It treats each "<" as a possible starting portion of the html tag. It is worth noting that the "<" character is not necessarily the starting html tags. For example, "<" may also be a "less than" comparison operator in JavaScript code. This error will be identified in the subsequent steps.

A Sub-algorithm 2 (Segment): Scanning w one by one and find all the "<" character. Split w into N segments F_1, F_2, F_3, ..., F_N, among them each F_N starts with "<".

The process of parsing a fragment is irrelevant with others, thus multiple html fragments can be resolved in parallel. Each html fragment may contain one or more html units and illegal html units, as previously described "<" operator as a start script. Illegal html units will make the FSM incorrect, which can be detected. In addition, if the parsed html tag cites the external content, then the citation will be downloaded.

A Sub-algorithm 3 (Parse in Parallel): Parallel parsing the fragments F_1, F_2, F_3, ..., F_N, which were parsed from sub-algorithm, and put the parsed html unit into global collection R (initially empty). Among them, the process of parsing a fragment is as follows:

(1) Create and initialize a blank Finite State Machine FSM_k.
(2) From the starting position SF_K of F_K, use FSM_k to parse the HTML text and join it into R.

It is worthy that, JavaScript code will not execute in the process of parallel parsing, thus ensuring a correct order to execute JavaScript code blocks.

Resolution process is terminated at the following cases:

It has been resolved to the last character of w; FSM_k happened wrongness; FSM_k finish parsing a html unit, and the position of the next character is greater or equal to the starting position SF_{k+1} of next html fragment F_{k+1}; matching sub-algorithm fragment obtained in F_1, F_2, F_3 ..., F_N, needs two adjacent matching, matching their total character m:

(1) *If the state of m characters in F_i and F_{i+1} are matched, then label F_i and F_{i+1} can be placed in R*
(2) *Otherwise, change the value of m and resegment F_i and F_{i+1} for matching;*

Merge is a process of combining and parsing the html unit in sequence to construct the DOM tree, and running the JavaScript block. In order to ensure correctness of the final results, the merge is a serial process.

A Sub-algorithm 4 (Merge): Sort the html unit in R, according to the order appeared in html page, then successively merge the html units to construct DOM tree, and run the html blocks. In the process, it will use a stack SU to maintain the nested relationship between DOM nodes, and use a marker CUR to maintain the end position of the last unit which was merged. For an html unit U_i, the merging process is as follows:

(1) If the start position Si ≤ CUR, then skip merging U_i;
(2) Otherwise, make the corresponding merge to U_i depending on the different types of U_i;

- If U_i is the start tag, then construct the node N according to the types, attributes, etc. of U_i. If SU is not empty, then push the SU stack node N_T, and put N as a sub-node of N_T, if U_i needs to have a corresponding end tag, then N will be pressed into SU;
- If the U_i is end tag, then pop up the top node of the SU;
- If the U_i is the text content, then construct a text node called N with the contents. Take the N_T of SU and make the N as the child node of the N_T. If N_T is script node, then run the script code of N.
- If U_i is other type of HTML element, such as annotations, then does anything.
- After completion, set the CUR to the end position E_i of the U_i.

4 Optimization Technology

The algorithm make all the "<" characters as the start of fragment. In order to realize the maximize of parallelism, which will cause each thread can only parse a smaller fragment every time, and it requires frequently switching analytic fragments, thus will increase the probability of data competition between multiple threads. It will bring a large additional overhead. In reality, if merge the html fragment according to the length of the html document and hardware conditions of the equipments, it will make the adjacent F_i, F_{i+1}, ..., F_m merged into a relatively large section F_k for parsing, balancing the additional overhead. When the CPU kernel is more, make full use of the parallel abilities of the equipment with smaller section; When the CPU cores are less, avoid too much additional overhead by using larger segments.

In Sect. 3, the merge of the algorithm happens after the completion of parallel parsing, and it needs a single merge thread, thus limits the parallelism of algorithm and also brings unnecessary computation and synchronization overhead. Situ merging can be used in the process of parallel parsing, completing the merge in the parallel parsing thread.

The algorithm is as follows:

A Sub-algorithm 5 (in Situ Merge): Maintain a single set R_i (initially empty) for each html fragment F_i, which is used to store the parsing html element of F_i. Put the parsing html element into R_i rather than global R; Maintain flag P_i for F_i and use it to identify whether the parse has been completed; preserve a global symbol WF (initialize to 0), which aimed at identifying the next html fragment to merge in the parsing process. If complete the interpretation F_i:

- if WF is not equal to i, don't do anything;
- Otherwise, from F_i, traverse F_i, F_{i+1}, ..., F_m, and merge the html element from the parsing of R_i, R_{i+1}, ..., R_m, until it has already reached the end of the html document;

If P_{m+1} is false, then F_{m+1} is not finished parsing yet. At this time, set the value of WF is $i + 1$.

5   Implementation and Results

This section introduces the algorithm implementation and experimental platform, and then analyze the results, finally discuss the difficulties and challenges in the process of achieving the algorithm.

5.1   Experiment Platform

Based on the widely used Chromium browser engine to achieve the parallel html parsing algorithm. In the experiment, invoke parallel algorithm Chromium engine or unmodified engine. The performance test focused on page load. When the page finished loading, Chromium engine evokes a load-finished signal. When the signal is received, Chromium automatically shut down. The experiment used 2 GB DDR2 memory in HuaWei tablet which has four QUALCOMM core processor (1.2 GHz), running Android 4.4.4 version of the operating system and installing all system patches.

5.2   Experiment Results

Firstly test the time of pre-parser completing downloading network resources by Chrome browser. The inspect remote monitor observed network resources by accessing google site.

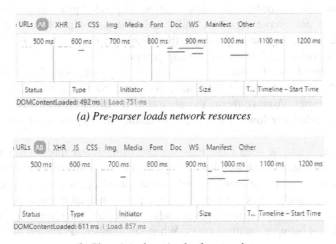

(a) Pre-parser loads network resources

(b) The original version loads network resources

Fig. 2. Inspect tool to test network resource loading time (Color figure online)

In Fig. 2(a), the blue vertical line represents the created time of building DOM tree, and the red vertical line records the time of loading network resources. The pre-parser loads network resources by 751 ms, while the unmodified version consumes 857 ms; the pre-parser loads network resources improving 12.36%. When building DOM tree, JavaScript will be in the barriers described in Chap. 2, and pre-loading of network resources can prevent from blocking DOM tree, which constructed DOM time relatively reduced 19.47%.

In order to measure the efficiency of the parallel algorithm in Sect. 3, the experiment measured the consumed time in web page parsing by using different number of analytical thread. Experiments used three popular web site home page, which is: MSN News, Amazon.com and Sina.com.cn. It shows the parsing speed of the page by using different number of analytical thread (1–8), in order to measure the efficiency of Parallel parsing algorithm accurately.

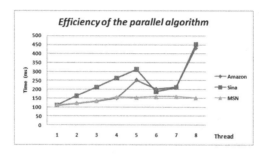

Fig. 3. Efficiency of the parallel algorithm

Figure 3 considered the time of parsing html page into a DOM tree and downloading the network. In order to maximize the use of parallel processing hardware, the experiment did not merge segments and enable merging algorithm. The data was normalized to the parsing speed in single thread. As can be seen from Fig. 3, parallel algorithm effectively accelerated the parsing process.

Using 8 parsed parallel threads can enhance the parsing speed of Amazon.com Website by 5.13 times; enhance the Sina.com.cn by 5.57 times; for MSN News can improve 2.01 times. Among them, the algorithm on MSN News, reasoning for its acceleration is relatively low on this page containing the large size of the JavaScript blocks.

In addition, it can be seen from Fig. 3, when using fewer threads to resolve, the parsing enhance speed of the web has nearly linear trend with the number of threads. However, when using five or more parallel parsing threads, page parsing speed showed a slight decline, mainly because of the hardware platform used in the experiment with four core processor. With multiple cores cache be shared, communicating with each other is via a bus between multiple cores. When the numbers of threads are more than four, these threads will be scheduled on a multi-CPU, thus increasing the cost of communications and data synchronization. In a nutshell, use four parallel threads can achieve the best analytic parallel efficiency.

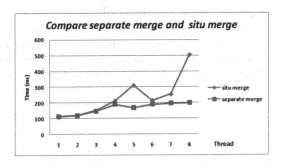

Fig. 4. Compare separate merge and situ merge (Amazon.com)

The situ merging algorithm can effectively improve the overall parallelism of the algorithm. Figure 4 showed loading Amazon.com alone or in situ merge and its comparison of page parsing speed. The experimental data used a separate thread to merge the parse and single thread was normalized according to the use, which demonstrated the situ merge algorithm efficiency. When merging using one thread, with the growth in the number of threads in parallel parsing, web analytic performance grows slowly. For example, use six parallel threads to resolve before they can reach twice the speed of parsing the page, while using up to eight threads parsing speed improved only 2.22 times; situ merge algorithm significantly improves parallel algorithm. The use of three parallel parsing thread can reach twice the speed of parsing the page, while using up to eight threads to enhance the speed of 5.13 times.

5.3   Difficulties and Challenges

In the beginning of design don't consider the demand of the parallelism, and original algorithm is designed and optimized for the serial processing, so many difficulties are faced in the process of design and realization of parallel algorithm. The main difficulties are listed below:

Although conceptually, web process can be divided into download, parse and JavaScript execution process, but in the concrete implementation, close coupling relationship is between the various modules. For example, in the process of web parsing, JavaScript block will be passed to the JavaScript engine, but the code can contain html code through the inner HTML attributes. Therefore it would trigger html parsing in the process of the execution of JavaScript code. Parallel algorithms must be synchronized across the invocation of the module properly.

6 Summary

This article focuses on the parallel processing for pages parsing algorithm to adapt to the current development trend of multi-core systems. Comparing with existing parallel algorithms, *parallel FSM* based on the data parallelism divided the rational html input data and processed in parallel, which can make full use of existing highly optimized serial FSM algorithm, compatible with web standards and technology, thus can be applied to parsing characters or web crawlers. Our follow-up studies will be based on highly structured data, such as DOM tree and render tree, realizing more targeted and efficient division, without dividing the input data entirely based on the "<" character.

References

1. Malik, N.A., Hanan, A.A., Omprakash, K.: FSM-F: finite state machine based framework for denial of service and intrusion detection in MANET. PLoS ONE **11**(6), e0156885 (2016)
2. He, C., Chen, Z., Huang, H., et al.: Summary of Web caching technology. Mini-Micro Syst. **25**(5), 836–842 (2004)
3. Rhea, S.C., Liang, K., Brewer, E.: Value-based Web caching. In: Proceedings of the 12th International Conference on World Wide Web, pp. 619–628 (2003)
4. Mickens, J.: Silo: exploiting JavaScript and DOM storage for faster page loads. In: Proceedings of the 2010 USENIX Conference on Web Application Development (2010)
5. Chromium. SunSpider JavaScript benchmark [EB/OL] (2010). http://www.Chromium.org/perf/sunspider/sunspider.html
6. Google. V8 JavaScript engine [EB/OL] (2008). http://code.google.com/p/v8/
7. Zhang, K., Wang, L., Pan, A., et al.: Smart caching for Web browsers. In: Proceedings of the 19th International Conference on World Wide Web, pp. 491–500 (2010)
8. Mesbah, A., Mirshokraie, S.: Automated analysis of CSS roles to support style maintenance. In: Proceedings of the 2012 International Conference on Software Engineering, pp. 408–418 (2012)
9. Weber, J.A.: Closer look at Internet explorer 9 hardware acceleration through flying images [EB/OL] (2010). http://blogs.msdn.com/b/ie/archive/2010/04/07/a-closer-look-at-interact-explorer-9-hardware-acceleration-through-flying.images.aspx
10. Yahoo. Best practices for speeding up your Web site [EB/OL] (2008). http://developer.yahoo.com/performance/roles.html
11. Waterson, C.: Notes on HTML reflow [EB/OL] (2004). http://www.mozilla.org/newlayout/doc/reflow.html
12. Wilton-Jones, M.: Efficient JavaScript [EB/OL] (2008). http://dev.opera.com/articles/view/efficient-javascript
13. Jones, C.G., Liu, R., Meyerovich, L.A., et al.: Parallelizing the Web browser. In: 1st USENIX Workshop on Hot Topics in Parallelism (2009)
14. Meyerovich, L.A., Guha, A., Baskin, J., et al.: Flapjax: a programming language for Ajax applications. In: Proceedings of the 24th ACM SIGPLAN Conference on Object Oriented Programming Systems Languages and Applications, pp. 1–20 (2009)
15. Meyerovich, L.A., Bodik, R.: Fast and parallel Webpage layout. In: Proceedings of the 19th International Conference on World Wide Web, pp. 711–720 (2010)

16. Erbad, A., Hutchinson, N.C., Krasic, C.: DOHA: scalable real-time Web applications through adaptive concurrent execution. In: Proceedings of the 21st International Conference on World Wide Web, pp. 161–170 (2012)

17. W3C. Web workers [EB/OL] (2012). http://www.w3.org/TR/workers/

18. Crier, C., Tang, S., King, S.T.: Designing and implementing the OP and OP2 Web browsers. ACM Trans. Web 5(2), article 11 (2011)

19. Mai, H., Tang, S., King, S.T., et al.: A case for parallelizing Web pages. In: Proceedings of the 4th USENIX Conference on Hot Topics in Parallelism (2012)

20. Badea, C., Haghighat, M.R., Nicolau, A., et al.: Towards parallelizing the layout engine of firefox. In: Proceedings of the 2nd USENIX Conference on Hot Topics in Parallelism (2010)

PTrack: A RFID-based Tracking Algorithm for Indoor Randomly Moving Targets

Gang Feng, Jian-qiang Li$^{(\boxtimes)}$, Chengwen Luo, and Zhong Ming

College of Computer Science and Software Engineering,
Shenzhen University, Shenzhen, China
`fenggang2@email.szu.edu.cn`, {`lijq,chengwen,mingz`}`@szu.edu.cn`

Abstract. RFID (Radio Frequency Identification) technology, with its multiple advantages, such as low power consumption, non-line-of-sight, non-contact, has been playing an important role in large-scale storage systems, underground parking systems, exhibition halls, supermarkets, construction sites and other scenarios. Many of those scenarios also require indoor positioning technologies, for example, warehouse goods positioning, item positioning in production assembly lines, worker positioning in construction sites. However, related researches about indoor positioning using RFID system has been having trouble in improving positioning accuracy, especially when tracking a randomly moving target. In this paper, we propose *PTrack*, a track prediction algorithm for tracking moving targets in indoor positioning systems which is based on RFID technology and the correspondences between the RSSI (Received Signal Strength Indicator) changes and the moving status of the target. Results show that the proposed algorithm effectively improves the positioning accuracy and achieves 1.7 m localization error in indoor environments, which makes a promising technology to support future pervasive RFID-based tracking applications.

Keywords: Indoor locationing · RFID · RSSI · PSO · Trajectory tracking

1 Introduction

Since the beginning of the 21st century, automatic positioning technology has gathering more and more attention of the public for its wild-rang benefits, and related technologies have developed increasingly rapidly. Among those technologies, indoor positioning technologies are urgently demanded in many fields of our daily lives. With the improvement of the performance of embedded chips, it is possible to process more complex signals on embedded devices while meeting the request of real time, which laid a foundation for the indoor positioning service [15].

Based on indoor positioning technologies, indoor tracking technologies have more and more applications. Tracking indoor moving targets in real time has

© Springer International Publishing AG 2017
M. Qiu (Ed.): SmartCom 2016, LNCS 10135, pp. 144–153, 2017.
DOI: 10.1007/978-3-319-52015-5_15

also showed a good future in logistics management, construction workers management, navigation in parking lots, etc. In these situations, outdoor positioning technologies like GPS and Compass which work well outdoors and in the wide-range positioning service, can't perform satisfyingly, because of complexity of indoor structures and significant signal attenuation. RFID is an identification and tracking technology using the radio signal propagation characteristics and tagging targets [6]. Because of its advantages such as low cost and fast response, it has become a focus of many research institutions and enterprises of indoor positioning research and a lot of researches have been conducted and different solutions have been proposed during the last decades [7,16]. But many current approaches are either too costly or not accurate enough, and low-cost and accurate ones can only work under limited or specific situations. There needs to be a inexpensive, accurate and generic solution to indoor positioning.

The rest of this paper is organized as follows. In Sect. 2, we discuss several former indoor positioning technologies. In Sect. 3, the technical background and main algorithms of *PTrack* are introduced. And We present a experimental system to evaluate *PTrack* in Sect. 4. We conclude the work in Sect. 5.

2   Related Works

Wireless location technologies can be divided into two categories based on different working scenarios: outdoor and indoor. After years of development, outdoor positioning systems have been able to meet the required accuracy, quality of real time and acceptability of cost for human activity. For example, the Global Positioning System (GPS), the Galileo satellite positioning system (Galileo), and the BeiDou Navigation Satellite System (BDS) are mature positioning systems [12]. However, due to the unacceptable signal attenuation of the those systems when in situations where targets are indoor, the positioning accuracy is greatly reduced. And the complexity of the indoor environment makes it necessary to adopt other technologies depending on certain circumstances. For now, there are already a number of indoor location solutions proposed.

- *Distance-based Positioning*
  Distance Based Positioning technologies are developed to figure out the distances between the target and multiple conference points whose positions are previously known. Based on those distances, the position of the target can be easily worked out. The following approaches are typical examples. Tagoram (RF hologram) is a complete RFID-based ind oor positioning system, designed by a Tsinghua University team [16]. Tagoram can deal with the interference from the diversity of RF tags by a method based on the phase angle of RF waves. However, due to the high computational complexity of its algorithm and the high error rate when locating a target with a random path (for example, construction workers places), its future remains limited.
- *Fingerprint-based Positioning*
  Position fingerprint based positioning technology, first proposed in Radar [2], takes a RSSI value as a position fingerprinting matching against a database

to determine the location of the target. In Horus [17], the author introduced a probabilistic fingerprint matching scheme and achieves a much higher localization accuracy. LANDMARC [7] introduces this technique to RFID localization with the help of multiple reference tag. Later on several other improvements over RSSI fingerprinting have been proposed [1,10,11], and extending to outdoor scenarios [11]. But RF fingerprinting based techniques require too much data fetching, database building and maintaining work. What's worse is when it comes to an environment, all those work will need a redo.

– *Vision-based Positioning*
Machine vision based positioning technologies require the help of real-time images capturing and recognizing. Heesung Chae and Kyuseo Han proposed an method for global localization incorporating signal detection from artificial landmark consisted of RFID tags, and for fine localization incorporating feature descriptor derived from a view of scene [3]. The system incorporates a RFID reader on a mobile robot checking the signal from RFID tags to localize the robot with respect to global position. After determining the global position of the robot, the feature matching can be used to checking the local position of it in a predetermined global position.

3   *PTrack* Algorithm

3.1   Signal Noise Reduction

In the experiments, we found that when repeatedly measuring the same signals with the same receiving device, the measurement values will wave in a certain range, but only few values which are considered as bad measurements are close to the bounds of that range. Here, we find and abandon those bad measurements by using minimum variance of the measurement values and take the average value of the remaining values.

First, assuming the raw data set gained from original signals during a time interval of Δt seconds as $X = \{x_1, K, x_n\}$, we define its average value as $\bar{x}$.

Then we define a new data set as $X_1 = \{x_2, K, x_n\}$ by removing the first element x_1 from X. In this way, we defined n new data sets to form a bigger data set $\{X_1, K, X_n\}$. For an element data set X_n in $\{X_1, K, X_n\}$, we define $\bar{x}_n$ as its average value, $n' = n - 1$ as its capacity and DX_n as its variance, and DX_n are calculated in the following way:

$$DX_n = \begin{cases} 0 & n = 1 \\ \frac{1}{n'} \sum_{i=1}^{n'} (x_i - \bar{x}_n)^2 & n \geq 2 \end{cases} \qquad (1)$$

An element x_k with a big DX_k value will be regarded as a bad measurement and removed from the data set X.

3.2   Mutating Probability Function

PSO (Particle Swarm Optimization) [5] Algorithm works by finding a best position *gbest* in a iteration for every particle to approach to in the next

Algorithm 1. PSO Algorithm Advanced With Mutating Probability Function

> **for** $i = 1$ to *size* **do**
>> $V[i] = randV()$;
>> $X[i] = randP()$;
>> $pbest[i] = X[i]$;
> **end for**
> $gbest = bestFitness(pbest)$;
> $\sigma^2 = convergence()$;
> **while** $\sigma^2 \geq \Delta$ && other loop conditions **do**
>> **for** $i = 1$ to *size* **do**
>>> $V[i] = V[i] + c_1 r_1 (pbest[i] - X[i]) + c - 2r_2(gbest - X[i])$;
>>> $X[i] = X[i] + V[i]$;
>>> **if** $fitness(pbest[i]) > fitness(X[i])$ **then**
>>>> $pbest[i] = X[i]$
>>> **end if**
>>> **if** $fitness(pbest[i]) < fitness(gbest)$ **then**
>>>> $gbest = pbest[i]$
>>> **end if**
>> **end for**
>> $\sigma^2 = convergence()$;
>> $P = 0$
>> **if** $\sigma^2 \geq \Delta \&\& fitness(gbest) \geq \hat{f}$ **then**
>>> $P = \frac{2}{1+e^{\omega x}} - 1$
>> **end if**
>> $[X, V] = mutate(X, V, P)$;
> **end while**

iteration [8,14]. Since many intelligent algorithms have well applied in many fields [4,9], we take advantage of PSO Algorithm into our approach to get better result. It is logically possible that the *gbest* is merely the best position in a certain range of its neighborhood but not the best position that is really the closest to the target, because the initial positions of all particles are randomly generated. That is how a local optimum occurs and it can cause a wrong result.

The fitness function plays an important role in PSO Algorithm. We define the fitness of a certain particle as $f_i, i \in \{1, 2, K, n\}$ (Given the swarm has n particles), and the average and the variance of all f_i as $\bar{f}$ and σ^2. σ^2 shows the degree of convergence of all particles. When σ^2 is close to 0, a optimum is generated, but it may not be the global optimum, which may lead us to a wrong result. To solve that problem, we need find out a way to make it possible to search other areas when it is stuck in a local search. That is where the mutating probability function functions. Here we define the probability controlling function P:

$$P = \begin{cases} \frac{2}{1+e^{\omega x}} - 1 & x \leq \phi \ \& \ f(gbest) \geq \hat{f} \\ 0 & others \end{cases} \quad (2)$$

In the formula above, x is the independent variable of P and stands for the variance σ^2, ω is an adjusting parameter, ϕ is the threshold for the variance

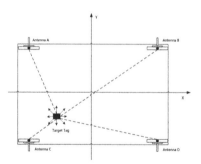

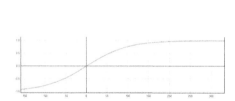

Fig. 1. Mutating probability function   **Fig. 2.** System overview 2D

which is much less than the maximum of the variance, and $\hat{f}$ is the theoretical optimal value of the fitness. When the value of the probability function P is not zero, P is a monotonically-increasing bipolar sigmoid function whose domain and range is $[-\infty, +\infty]$ and $[-1, 1]$. When $\omega = 0.02$, the function image is shown as the Fig. 1. As indicated by the function image, when the global fitness meets the conditions of mutating, the mutating probability P decreases as the variance σ^2 decreases. With the mutating function's joining, the improved PSO algorithm process is as follows:

3.3   Dynamic Trajectory Tracing Algorithm

In the paragraphs above, we've talked about how to locate a non-moving target, which can provide the initial position of the target as a basic reference information when tracking a moving target. Due to the data of signal read from devices may contain errors, we need to correcting the trajectory timely.

We use 4 antennas to gather the signal strength of the target tag, as showed in Fig. 2. According to the changes of the signal strength each antenna, we can figure out whether the target is approaching to or moving further from the antenna. That's how the moving direction of the target is determined. We assume

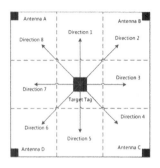

Antenna Direction	A	B	C	D
direction 1	+	+	-	-
direction 3	-	+	-	+
direction 5	-	-	+	+
direction 7	+	-	+	-
direction 2	+	+	-	+
direction 2	-	+	-	-
direction 4	-	+	+	+
direction 4	-	-	-	+
direction 6	+	+	+	+
direction 6	-	-	+	-
direction 8	+	+	+	-
direction 8	+	-	-	-

Fig. 3. Directions of tag   **Fig. 4.** Direction determining

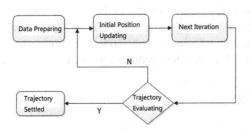

Fig. 5. Flowchart of the tracking algorithm

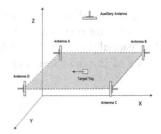

Fig. 6. System overview 3D

that the target only go in the 8 directions showed in Fig. 3. Determining the moving direction depends on all changes of the signal strength read from all 4 antennas, whose detail is shown in Fig. 4 where a plus sign "+" indicates an increase of signal strength read from a certain antenna, which means the target is approaching the antenna; whereas a minus sign "−" has the opponent meaning. When in 3-dimensional space, an auxiliary antenna is needed to measure the change of height which is planted as shown in Fig. 6. With the moving direction determined, the trajectory will be able to be tracked. Given that the reading rate of the reader is $F(Hz)$ and the moving speed of the target tag at direction i is defined as ν, the two factors should fit the following formula:

$$|\nu| = \frac{1}{\varepsilon F}, \quad i = 1, 2, 3, \ldots, 8 \tag{3}$$

where ε is a parameter which can be determined when in an actual scene. For the reading rate of readers current available in the market is about 2 to 3 times per second, ε ranges from 1 to 2.

Tracking algorithm flow is shown in Fig. 5. The details of the algorithm are described as follows: The original position and moving speed are defined as O and $\dot{\nu}$, then the location is updated in the following way:

$$O_{next} = O + \dot{\nu}, \quad i = 1, 2, 3, \ldots, 8 \tag{4}$$

After each update, we should calculate the fitness function which determines whether we should operate the update. The fitness function f is defined as formula (5) where d_{ji} is the distance between particle i and antenna j and d_{j0} is the distance between the target and antenna j.

$$f_i = \sqrt{\sum_{j=1}^{m} (d_{ji} - d_{j0})^2} \tag{5}$$

If m antennas read data for n times, we can get a matrix about information of distance:

$$D = \begin{bmatrix} D_{11} & K & D_{1n} \\ M & O & M \\ M_{m1} & L & D_{mn} \end{bmatrix} \tag{6}$$

D_{mn} is the distance between the target tag and the mth antenna during the nth reading process. In actual situations, the value of m usually is 4 in the 2-dimensional space, or 4 or 5 whereas in 3-dimensional space. To judge the accuracy of the detected trajectory, we define a deviation function Δ:

$$\Delta = \sum_{i=1}^{n} \sqrt{\sum_{j=1}^{m}(D_{ij} - \hat{D}_{ij})^2} \tag{7}$$

Where $\hat{D}_{ij}$ is the theoretical distance between each antenna and the target tag after each reading. We can set a threshold for Δ to determine whether the new position is invalid and a re-determining of the initial position is needed.

4   Implementation and Evaluation

4.1   Hardware and Software Design

Devices used in the experiment are active RFID tags and readers, as shown in Figs. 7 and 8, and the parameters are listed in Tables 1 and 2.

Due to the RSSI of RFID system reduced greatly when tags and reads are getting further from each other, we choose a $5\,\text{m} \times 4\,\text{m} \times 3\,\text{m}$ 3-dimensional

Table 1. Parameters of RFID tags

Type	Active, read-only
Operating frequency	391 MHz–464 MHz
Recognition range	100 m (depending on its antenna)
Sending rate	2 times/s
Modulating mode	MSK GDSK
Communication speed	500 Kbps
Data security	AES encryption and link layer encryption
Size	$87 \times 56 \times 5$

Fig. 7. Sample tag   **Fig. 8.** RFID reader

Table 2. Parameters of RFID tags

Type	Active, directional
Operating frequency	391–464 MHz, ISM, Microwave band
Antenna	Built-in, −112 dBm
Communicating interface	RS-232, TTL
Reading capacity	500 tags at the same time
Operating range	0–100 m

space and 5 readers for the experiment. 4 readers are placed in same horizontal plane and an auxiliary antenna is planted on the ceiling above the other 4, as shown in Fig. 6. The exact positions of all 5 antennas are $A(0, 0, 1.5)$, $B(5, 0, 1.5)$, $C(5, 4, 1.5)$, $D(0, 4, 1.5)$, $AUX(2.5, 2, 3)$. To test the tracking of moving targets, we move the target tag between certain specified positions whose coordinates are previously known: $M_0(0, 0, 1.5) \rightarrow M_1(1, 2, 1.5) \rightarrow M_2(4.5, 2, 1.5) \rightarrow M_3(4.5, 0.5, 1.5) \rightarrow M_4(0, 0.5, 1.5)$. The target tag follows a straight line when moving between every 2 of those 5 positions. For the reading rate of the devices used in the experiment is about 3 time per second, so the value of the moving speed $\hat{\nu}$ is 0.15 m/s, according to Formula (3).

4.2   Results of the Experiment

The result track and the planned route of the target tag in the experiment were drawn in the 3-dimensional space, as shown in Fig. 9. The real track is hardly as straight as the planned route, for we moved the tag by attaching it to a moving person who can not walk that straightly. We can see from the figure that the result track generally shows the movements of the person the tag was attached to. And when the result track went a little far from the real track, it could returned back to the real track gradually, owing to the functioning of the algorithm: when the parameter Δ is greater than a certain threshold, a re-determining of the initial position will emerge to fix the track.

After testing a variety of trajectories, such as straight-line routes, poly-line routes and back-and-forth routes, we found that the more complex the real trajectory was, the worse the result would be. When there are only straight lines and poly-lines, the results are better than when there are back-and-forth routes or standstills among the routes or when the moving speed is getting faster. The instability of the signal strength is main factor to be blamed to. Because when the tag moves back and forth, signal strength changes so obviously that it may be treated as noise when preprocessing the data. And when the tag moves too fast, readers may not be able to get enough data during the tag's trip. Solving the above problems requires a further optimization of the pre-processing algorithm and a hardware upgrading.

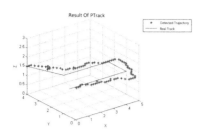

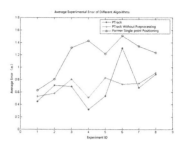

Fig. 9. The detected track and the real track

Fig. 10. Average experimental error (Color figure online)

We compared 3 kinds of positioning algorithms talked about in this paper in Fig. 10: *PTrack*, the algorithm similar to the first one but without data pre-processing, and the traditional tracking algorithm based on single point positioning. The blue line, representing *PTrack*, shows the positioning error generally waves around 1 m when tracking a random track. But the accuracy of former algorithms, such indoor door tracking algorithm using "dead reckoning" method [13], is 1.7 m at a 80% probability. We improves the positioning accuracy by 50%.

5   Discussion and Conclusion

In this paper, we discussed an indoor positioning system in which RFID technology was applied to detect the target, and the RSSI was chosen to measure the distances. In order to reduce system errors and increase positioning accuracy, we performed a pre-processing to the original data acquired from the devices and optimized PSO Algorithm for indoor positioning. And we took a further increase in positioning accuracy when it came to track a moving target. This paper may contribute to the field of indoor positioning and tracking systems using RFID technology. However, there are still unignorable shortcomings which also show us something for our future work. First, it doesn't work properly when tracking multiple targets because of the existence of electromagnetic interference between RFID tags. Second, the universality problem of RSSI technology that the signal strength of RFID is so easily affected by many factors of the environment and requires too much preparation before positioning. Those are also a meaningful future research directions.

Acknowledgements. The authors gratefully acknowledge the contribution of the National Science Foundation of China [61572330][61472258], the Natural Science foundation of Guangdong Province [2014A030313554], and the Technology Planning Project (Grant No. 2014B010 118005) from Guangdong Province.

References

1. Azizyan, M., Constandache, I., Roy Choudhury, R.: Surround-sense: mobile phone localization via ambience fingerprinting. In: International Conference on Mobile Computing and Networking, MOBICOM 2009, Beijing, China, pp. 261–272, September 2010
2. Bahl, P., Padmanabhan, V.N.: Radar: an in-building RF-based user location and tracking system. In: Institute of Electrical & Electronics Engineers Inc., vol. 2, pp. 775–784 (2000)
3. Chae, H., Han, K.: Combination of RFID and vision for mobile robot localization. In: International Conference on Intelligent Sensors, Sensor Networks and Information Processing Conference, pp. 75–80 (2005)
4. Gai, K., Qiu, M., Zhao, H.: Cost-aware multimedia data allocation for heterogeneous memory using genetic algorithm in cloud computing, p. 1 (2016)
5. Kennedy, J., Eberhart, R.: Particle swarm optimization. In: IEEE International Conference on Neural Networks, Proceedings, vol. 4, pp. 1942–1948 (1995)
6. Lockman, M.T., Selamat, A.: Multi-agent verification and validation for RFID system architecture. In: International Conference on Electronic Design, ICED 2008, pp. 1–5. IEEE (2008)
7. Ni, L.M., Liu, Y., Lau, Y.C., Patil, A.P.: Landmarc: indoor location sensing using active rfid. Wirel. Netw. **10**(6), 701–710 (2004)
8. Pinto, A.M., Moreira, A.P., Costa, P.G.: A localization method based on map-matching and particle swarm optimization. J. Intell. Robot. Syst. **77**(2), 313–326 (2015)
9. Qiu, M., Zhong, M., Li, J., Gai, K., Zong, Z.: Phase-change memory optimization for green cloud with genetic algorithm. IEEE Trans. Comput. **64**(12), 1 (2015)
10. Rai, A., Chintalapudi, K.K., Padmanabhan, V.N., Sen, R.: Zee: zero-effort crowd-sourcing for indoor localization, pp. 293–304 (2012)
11. Sen, S., Radunovic, B.R., Choudhury, R.R., Minka, T.: You are facing the Mona Lisa: spot localization using PHY layer information. In: International Conference on Mobile Systems, Applications, and Services, pp. 183–196 (2012)
12. Stoleru, R., He, T., Stankovic, J.A.: Walking GPS: a practical solution for localization in manually deployed wireless sensor networks. In: 29th Annual IEEE International Conference on Local Computer Networks, pp. 480–489. IEEE (2004)
13. Student, S.H., Connell, S., Milligan, I., Austin, D., Hayes, T.L., Chiang, P.: Indoor localization using pedestrian dead reckoning updated with RFID-based fiducials. In: 2011 Annual International Conference of the IEEE Engineering in Medicine and Biology Society, pp. 7598–7601, August 2011
14. Wang, H.: An Application of PSO algorithm in indoor positioning system. Master's thesis, Xidian University (2009)
15. Wang, Y., Jia, X., Lee, H., Li, G.: An indoors wireless positioning system based on wireless local area network infrastructure. In: 6th International Symposium on Satellite Navigation Technology Including Mobile Positioning & Location Services, no. 54 (2003)
16. Yang, L., Chen, Y., Li, X.Y., Xiao, C., Li, M., Liu, Y.: Tagoram: real-time tracking of mobile RFID tags to high precision using cots devices. In: Proceedings of the 20th Annual International Conference on Mobile Computing and Networking, pp. 237–248. ACM (2014)
17. Youssef, M., Agrawala, A.: The Horus wlan location determination system. In: International Conference on Mobile Systems, Applications, and Services, pp. 205–218 (2005)

Predicting the Change of Stock Market Index Based on Social Media Analysis

Rui Ma[1] and Honghao Zhao[2(✉)]

[1] Zhuhai Laboratory of Key Laboratory of Symbolic Computation and Knowledge Engineering of Ministry of Education, Department of Computer Science, Zhuhai College of Jilin University, Zhuhai, China
408779609@qq.com
[2] School of Business, Macau University of Science and Technology, Macau, China
hhzhao@must.edu.mo

Abstract. Stock market is definitely one of the common choices for investment throughout the world. Stock market index is well known as the composite of several representative stocks and reflects the trends of future stock market. Predicting the change of stock market index becomes crucial for individual investor, companies and stock holders. Recent research from behavioral finance implies that emotions of investors from social media can also influence stock market index in addition to the commonly used financial factors. Taking the advantages from the development of modern network and the age of big data, we are able to obtain information from different sources. More specifically, accessing user data from social media is no longer a challenge. Thus we apply data mining methods and propose to use the key words of emotions of market participates obtained from social media through text mining to predict the change of stock market index. Compared with traditional methods, we are able to utilize full information from the social network. We apply the propose approach in a dataset collected from Xueqiu forum, and the results show that linear discriminant analysis could give relatively good predictions of stock market index.

Keywords: Classification · Data mining · Emotions · Social Media Analysis

1   Introduction

Stock market is definitely one of the common choices for investment and attracts the attention from both individual investors, companies and stock holders. As we all know, the price of a stock is highly influenced by the financial status of the corresponding company, and the possible profits obtained by investors are determined by the difference between their buying price and the current price of a particular stock. Thus correct prediction of stock price is curial for purchasing stocks obviously. Normally, investors tends to invest several stocks simultaneously to reduce the risk of holding only one stock. Therefore, the trend

© Springer International Publishing AG 2017
M. Qiu (Ed.): SmartCom 2016, LNCS 10135, pp. 154–162, 2017.
DOI: 10.1007/978-3-319-52015-5_16

of stock market index which is a composite of representative stocks becomes more important than individual stock. Consequently, predicting the change of Stock Market Index occupies a vital position in investors' decision making.

Due to the nature of stock market index, it is reasonable to treat the stock market index as time series data. Thus time series analysis becomes one of the typical ways to predict the stock market index in the literature. As mentioned above, the stock market index is the intergraded information for a set of representative stocks, so the methods for predicting the price of stocks could be applied for the prediction of stock market index theoretically. Numbers of research studies have been done along this direction, examples can be found in Fama [1], Campell and Shiller [2], Chang et al. [3], Frino et al. [4] and Pai and Lin [5].

Although the above research results are appealing, there are two major drawbacks. The first drawback is due to the information used in the model. The above research use time and financial related factors as the predictor variables in their studies. Although financial factors definitely affect the value of stock market, other factors still have influence on it. From the prospective of behavioral finance, investors as human being tends to make emotion-based decisions [6] and affect the stock price consequently. With the development of internet, social media provide a wide platform for investors to express their emotions on the topic they are interested and offer a fast communication environment for them to exchange ideas. Consequently the potential impact of emotions on stock is enhanced.

The second drawback is related with the response variables in the model. Common research focus on predicting the true value of stock price or the stock market index. However, most of the time the trend of stock market index is more important, so predicting the change of stock is desirable. As we mentioned above, there are very few studies on predicting the change of stock market index by using classification methods based on emotions data from social media. Motivated by this research gap, we get started by investigating an empirical case. More specifically, we collect user generated contents from Xueqiu forum, and extract the emotion keywords by text mining. These original key words related to emotions are treated as the predictor variables for prediction purpose, and we apply popular classification methods to predict the change of stock market index, which fully utilize the information of emotions keywords compared with traditional methods.

The rest of this paper is organized as follows. In Sect. 2, five popular classification methods are discussed. In Sect. 3, an empirical analysis is given to illustrate the effectiveness of the above classification methods with a focus on predicting the SSE Composite Index in China. Furthermore, details about the entire case will be illustrated, including raw data, feature selection, sampling methods and the results. In Sect. 4, a concluding remark is given and limitation of this research is discussed.

2   Description of Methods

In this section, we will review 5 popular used classification methods which will be applied in our empirical analysis. We provide a very brief explanation for each of the method below, and one could find detailed information from the reference.

1. k-nearest neighbor(KNN) is a simple classification method. The predicted class of a new testing observation belongs to the majority group of the k nearest data point in the training set. "Nearest" is usually defined as the distance between the new observation and the training data point [7].
2. Logistic regression(LR) is a method based on regression. The classification results of LR are according to the probability that the new data point belongs to each class. Normally, LR applies to binary classification problem and its fundamental assumption is that the nature logarithm of the ratio on the probability between each class has a linear relationship with the predictor variables. Such idea could be written as the following equation.

$$log(\frac{q_i}{1 - q_i}) = \boldsymbol{X}_{\boldsymbol{i}}'\boldsymbol{\beta}, \tag{1}$$

where $q_i = P(y_i = 1|\boldsymbol{X_i})$ is the probability that $y_i = 1$ given $\boldsymbol{X}_i$.
3. Linear discriminant analysis(LDA) is a classification method derived from normal distribution and Bayes theorem. Normally, the model assumes that the distribution of predictor variables from different classes follows multivariate normal distribution with different means. Based on the prior probability distribution of different classes, one could derive the probability belongs to each class for a new data point [7].
4. Classification tree(TREE) is proposed according to the idea of splitting. The fundamental motivation is to split each variable so that the classification error is minimized for each split, which is known as the gain ratio idea. Thus Gini index or cross-entropy is applied to determine the size of a tree [8].
5. Support vector machines(SVM) is developed based upon the concept of a maximal margin hyperplane. A test observation is assigned a class depending on which side of the hyperplane it is located [9].

In our research, we are going to apply each of the 5 methods to the empirical case in the following section. To make it simple, we do not discuss the parameter selection in this case, rather we use the default settings in R program for implementation.

3   Empirical Analysis

In this section, we will illustrate the entire details on implementing and comparing the 5 classification methods based on a real example for predicting the stock market index.

3.1 Description of Data

As we all know, there are many different stock market index. In this research, we focus on the prediction for the SSE Composite Index in China. The ups and downs of the SSE Composite Index on time t becomes the response variable in our study. The predictor variables are the emotion information we obtained from social media on time $t-1$. In particular, we extract the contents generated by the potential stock investors through Xueqiu which is a popular forum in China with a particular focus on stocks and financial related products. Then we manually selected the keywords sets for emotions and count the number of occurrence for these keywords. These numbers becomes the predictor variables in our model. More specifically, The raw data are collected from December 2015 to February 2016 and the effective time duration is 50 days. The keywords we selected are manually collected from the first three days and the number of keywords is 57 finally. The curve of SEE Composite Index within the above time duration is shown in Fig. 1.

3.2 Feature Selection Method

As we mentioned before, we have 57 keywords, so there are 57 predictor variables in the model. To better fit the model and avoid over fitting, we need to select a subset of influential variables from the total. A intuitive idea is to select the variables which can separate the response variable effectively. So the commonly used approach is to order the important of each variables by a t-test value and choose the final variables according to their variance inflation factor (VIF) [10]. Suppose X_i is the ith variable in the model and $Y \in \{0,1\}$ is the corresponding response variable the t-test value is calculated below:

$$t_i = \frac{|\bar{X}_i^1 - \bar{X}_i^2|}{\sqrt{\frac{v_i^1}{n_2} + \frac{v_i^2}{n_2}}}, \tag{2}$$

where $\bar{X}_i^1 = mean(X_i|Y = 1)$ and $\bar{X}_i^2 = mean(X_i|Y = 0)$. Furthermore, $v_i^1 = var(X_i|Y = 1)$ and $v_i^2 = var(X_i|Y = 0)$ represent the variance of the corresponding group of data. Finally, n_1 and n_2 are the number of observations in each group.

After calculating the t value for each variable X, we sort the variables in descending order according to their t value. Then the first value is included in the model. After that we add the second variable in the model and calculate the corresponding VIF value. If the VIF is larger than 10, we omit this variable, otherwise we keep this variable in the final model. By repeating this process, we stop the feature selection process until we got the desired number of variables. In our study, the number of variables we chose is from 5 to 15.

3.3 Sampling Method

In order to compare the performance of different methods, we apply two schemes to divide the original data set. The first scheme is just to split the data set into two sets. One set is for training and the other is for testing. In particular, we select the first 20, 30 and 40 days as the training set respectively. The second scheme is cross validation. More specifically, we choose 5 folds cross validation and 10 folds cross validation in our case.

3.4   Results

The results of our research are presented based on the two sampling scheme. All results are measured by the classification error in the testing set. In particular,

Table 1. Classification errors for the 5 methods with different numbers of selected variables based on 20 training samples

Numbers of variables	LDA	SVM	TREE	KNN	LR
5	0.467	0.433	0.467	0.433	0.467
6	0.467	0.500	0.467	0.367	0.500
7	0.500	0.467	0.467	0.367	0.467
8	0.500	0.467	0.467	0.400	0.467
9	0.500	0.467	0.467	0.400	0.500
10	0.500	0.400	0.467	0.400	0.500
11	0.533	0.400	0.467	0.400	0.533
12	0.533	0.367	0.467	0.400	0.533
13	0.533	0.400	0.467	0.367	0.533
14	0.533	0.400	0.467	0.367	0.533
15	0.533	0.400	0.467	0.367	0.533

Table 2. Classification errors for the 5 methods with different numbers of selected variables based on 30 training samples

Numbers of variables	LDA	SVM	TREE	KNN	LR
5	0.250	0.400	0.300	0.400	0.250
6	0.350	0.400	0.400	0.500	0.350
7	0.450	0.500	0.400	0.500	0.450
8	0.450	0.550	0.400	0.500	0.450
9	0.500	0.500	0.400	0.450	0.550
10	0.450	0.550	0.400	0.500	0.450
11	0.450	0.550	0.400	0.450	0.450
12	0.450	0.500	0.500	0.450	0.450
13	0.450	0.500	0.500	0.350	0.500
14	0.400	0.500	0.500	0.350	0.550
15	0.400	0.450	0.500	0.450	0.550

Table 3. Classification errors for the 5 methods with different numbers of selected variables based on 40 training samples

Numbers of variables	LDA	SVM	TREE	KNN	LR
5	0.400	0.400	0.500	0.500	0.500
6	0.400	0.400	0.500	0.500	0.400
7	0.400	0.400	0.500	0.500	0.500
8	0.500	0.400	0.500	0.500	0.500
9	0.500	0.300	0.500	0.500	0.600
10	0.300	0.300	0.500	0.400	0.500
11	0.400	0.400	0.500	0.400	0.500
12	0.400	0.700	0.500	0.500	0.600
13	0.400	0.400	0.400	0.500	0.600
14	0.600	0.400	0.400	0.500	0.600
15	0.300	0.400	0.400	0.500	0.400

the results obtained by the first sampling scheme are shown in Tables 1, 2 and 3. In Table 1 the KNN method provides the smallest classification errors among the 5 methods. For all the 5 methods the classification errors are affected by the number of variables in the model, but there is no straight forward trend for it. However, if we increase the number of training samples from 20 to 30 and 40. We could discover that the LDA method tends to outperform the other methods in Tables 2 and 3.

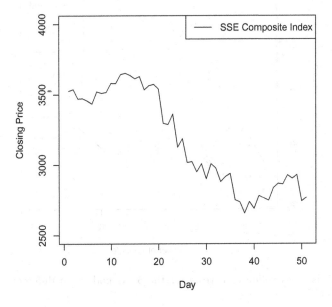

Fig. 1. SSE composite index between the interested time duration

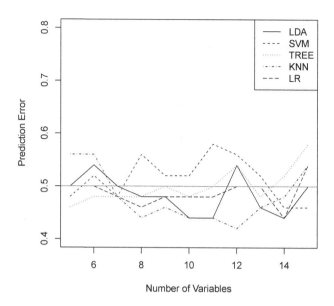

Fig. 2. 5 folds cross validation errors for the 5 methods with different numbers of selected variables

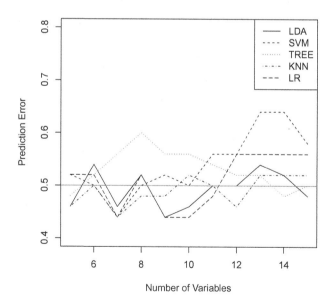

Fig. 3. 10 folds cross validation errors for the 5 methods with different numbers of selected variables

For the second sampling scheme, the results are presented in Figs. 2 and 3. We could still find that the LDA method provides relatively stable performance and give reasonable classification error in most scenarios. Consequently, the LDA method is generally preferable according to the current research results.

4 Concluding Remarks

Stock market is definitely one of the common choices for investment throughout the world. Predicting the change of stock market index becomes crucial for individual investor, companies and stock holders. Recent research from behavioral finance implies that emotions of investors from social media can also influence stock market index in addition to the commonly used financial factors. Thus we apply data mining methods and propose to use the key words of emotions of market participates obtained from social media through text mining to predict the change of stock market index. We apply the propose approach in a dataset collected from Xueqiu forum, and the results show that linear discriminant analysis could give relatively good predictions of stock market index.

In this paper, we only consider 5 popular used classification methods and it could be interesting to include more new methods to solve this problem. Furthermore, for each of the 5 methods we do not touch the issue for parameter selection to highlight a direct application of the empirical case and this is a limitation of our current research. We are going to extent this studies by discussing the parameter effects in the further research.

For the current study, we collect data from a time duration of 3 months. In the future, it is reasonable to collect more data to ensure the capabilities of each method and quantify the errors for each method. If one is interested in prediction with very limited number of observations, the current problem will lead to a high dimensional analysis problem which is worth future research as well.

Acknowledgement. The work described in this paper was supported by the grants from Zhuhai Laboratory of Key Laboratory of Symbolic Computation and Knowledge Engineering of Ministry of Education at the Department of Computer Science in Zhuhai College of Jilin University, Zhuhai, China.

References

1. Fama, E.F.: Random walks in stock market prices. Financ. Anal. J. **51**, 55–69 (1965)
2. Campell, J.Y., Shiller, R.J.: Stock prices, earnings, and expected dividends. J. Financ. **43**, 661–676 (1988)
3. Chang, P.C., Wang, D., Zhou, C.: A novel model by evolving partially connected neural network for stock price trend forecasting. Expert Syst. Appl. **39**, 611–620 (2012)
4. Frino, A., Walter, T., West, A.: The lead-lag relationship between equities and stock index futures markets around information releases. J. Futur. Market **20**, 467–487 (2000)

5. Pai, P.F., Lin, C.S.: A hybrid ARIMA and support vector machines model in stock price forecasting. Omega **33**, 497–505 (2005)
6. Danial, K., Hirshleifer, D., Teoh, S.H.: Investor psychology in capital markets: evidence and policy implications. J. Monet. Econ. **49**, 139–209 (2002)
7. Murphy, K.P.: Machine Learning: A Probabilistic Perspective. The MIT Press, Cambridge (2012)
8. Quinlan, J.R.: C4.5: Programs for Machine Learning. Morgan Kaufmann, San Francisco (1993)
9. Suykens, J.A., Vandewalle, J.: Least squares support vector machine classifiers. Neural Process. Lett. **9**, 293–300 (1999)
10. Zhou, L.G., Tam, K.P., Fujita, H.: Predicting the listing status of Chinese listed companies with multi-class classification models. Inf. Sci. **328**, 222–236 (2016)

Human Activity Recognition Based on Smart Phone's 3-Axis Acceleration Sensor

Shubin Cai[1,2(✉)], Zhiguang Shan[2,3], Tian Zeng[1,2], Jianfei Yin[1,2], and Zhong Ming[1,2]

[1] Shenzhen University, Shenzhen, Guangdong Province, China
{shubin,zengt,yjf,mingz}@szu.edu.cn
[2] Engineering Research Center of Mobile Internet Application Middleware,
Shenzhen, Guangdong Province, China
[3] State Information Center of China, Beijing, China
shanzg@cei.gov.cn

Abstract. With the rapid development of smartphone, human activity recognition based on acceleration sensors attracts much attention in the academic and industry recently. However, the recognition accuracy is not ideal due to the diversity of human activities and other environmental factors. A real-time user activities monitoring system is developed on android, and comparison of several feature extraction and classification algorithms is carried out. Based on the monitoring system, a feature called (TF4+FFT10) is proposed. Experiment result shows that the recognition accuracy rate of feature (TF4+FFT10) with the adopted KNN algorithm is 98.6 %.

Keywords: Acceleration sensor · Android · Human activity recognition · KNN

1 Introduction

Recently, the sensor technology has developed rapidly. 3-axis acceleration sensor becomes smaller, lower power consumption and remains quiet during operation. Almost every smart phone is quipped with 3-axis acceleration sensors. Since smart phone is widely used in modern life, human activity recognition based on smart phone's 3-axis acceleration sensors have a promising future. For example, if smart phones can recognize users' motions and activities correctly, they can interact with humans more intelligently and improve users experience.

Each person has different characteristics of acceleration while walking, running or jumping etc. Further experiment has shown that environmental factors, such as uneven road or rainy days can dramatically affect the characteristics of acceleration of the same person doing the same activity. Thus the recognition accuracy of human activity is rather low and needs to be improved.

© Springer International Publishing AG 2017
M. Qiu (Ed.): SmartCom 2016, LNCS 10135, pp. 163–172, 2017.
DOI: 10.1007/978-3-319-52015-5_17

2   Data Collection

2.1   Chosen of Data Collection Device

Collecting sample data from 3-axis acceleration sensor is the first step to recognize human activities. In most past research, researchers design some simple circuits with an acceleration sensor to realize a wearable device in the data collection section. Some of these data collection devices are in the size of a mobile phone, and some are as small as a coin. However, since 3-axis acceleration sensor has been embedded in almost every today's mobile phone, there is no need to design an additional data collection devices. Mobile phone's operating systems, such as Android and iOS, provide expedient way to access the data of acceleration sensors. Its convenient to carry out the human activity recognition research based on mobile phones acceleration sensors.

2.2   Device Placement and Coordinate Axis Adjustment

Trouser pocket is the most common place where cellphone is kept. In order to minimize the restrains, this paper selects trouser pocket as the place for data sampling.

In reality, cellphone is not always positioned in the same direction in trouser pocket. And it is impossible to regulate that all cellphones should be positioned in the same direction every time. Therefore, the coordinate axis must be adjusted and an absolute coordinate axis needs to be selected. The flip angles of cellphones can be determined with the built-in direction sensors of cellphones, so as to adjust the axis.

2.3   Sampling Frequency and Window Size

By concluding a large number of documents and physiological laws of human activities, this paper has adopted 50 Hz as the sampling frequency for the data sampling.

In order to facilitate the processing of data, the samples need to be cut, namely windowed. The size of the time window depends on the types of actions to be recognized. If the adopted time of the sliding window is too short, the window data may not have covered the information of a complete action. If the width of the sliding window is too long, it will not only make the data sophisticated and increase the amount of calculation, but also lead to delays, dispossessing the real-time character of the system. Through multiple tests and considering the coming calculations, the window selected is 2.56 s, which means every 128 points are considered as one data sample in the data analysis.

2.4   Sampling Procedure

The Flow Diagram of Sampling Procedures formulated according to the above settings is shown in the Flow Diagram Figure (see Fig. 1). In the sampling procedure, user start the application, select the activity, such as stand still, walking,

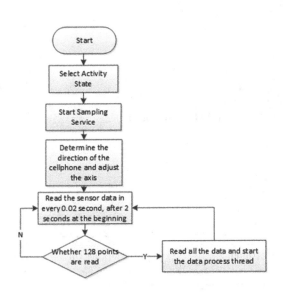

Fig. 1. Flow diagram of sampling procedure   **Fig. 2.** UI of sampling system

running etc., then perform the selected activity in the following a few seconds. When the sampling application is started, it will start the sampling service, determine the direction of the cell phone and adjust the coordinate axis, then read the acceleration sensor's data in every 0.02 s after a 2 s delay. The purpose of the 2 s duration is to allow user to put the cellphone in pocket and start activity. After every 2.56 s, which means 128 points of acceleration sensor's data are read, a data processing service is started. The UI Figure of the activity selection stage is shown in the UI of Sampling System Figure (see Fig. 2).

3   Data Feature Extraction

The data of acceleration sensor are not intuitive, the following figure (see Fig. 3) has listed the 3-axis time domain diagram of 8 actions. If the initial data are directly adopted to conduct the comparison, it needs to be guaranteed that the starting point of every datum sampled is the same. As we can see that the starting point of every action is different, the initial data cannot be directly categorized as the feature. Therefore, we must process the initial data and extract some sample characteristics which are not relevant to the time but can be adopted to distinguish these actions.

3.1   Recognition Accuracy of 4 Simple Time Domain Features

It can be clearly figured out from the curves in all axes that the curves of running is relatively dense, and its peak values in all axes are much larger than those of

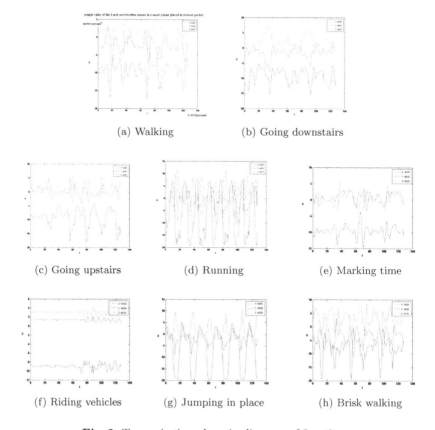

(a) Walking

(b) Going downstairs

(c) Going upstairs

(d) Running

(e) Marking time

(f) Riding vehicles

(g) Jumping in place

(h) Brisk walking

Fig. 3. Tree-axis time domain diagram of 8 actions

other actions. The most evident features are the average value and the peak-to-peak value. The standard deviation can well distinguish the dispersion degree of actions, and such distinguishment is even more significant on actions of strenuous exercise. Also, the abruptness of some curves has well reflected the variability of actions, and this feature has never been seen in other similar documents. The characteristic of this feature reflects the suddenness and intensity of actions. Therefore, the 4 features of time domain which are finally determined are:

(1) Average value;

(2) Standard deviation $\sigma = \sqrt{\frac{1}{N} \sum_{i=1}^{N} (X_i - \overline{X})^2}$ (in which N is an average value representing the number of samples)

(3) Peak-to-peak value: the difference between the peak and the trough;

(4) Abruptness: the maximum value of the intervals between all samples.

After the 10-fold cross-validation has been conducted with KNN, Bayes and decision tree respectively, the following table is obtained: The experiment has proved that these four time-domain features are relatively effective in

Table 1. Result of the time domain features

	Still	Walk	Run	Go down	Go up	Shake	Brisk walk	Jump	Brush teeth	Average
KNN	0.98	0.88	1	0.82	0.80	0.7	0.93	0.75	1	0.89
Bayes	0.97	0.93	1	0.86	0.83	0.79	0.94	0.2	1	0.89
DTree	0.96	0.88	1	0.86	0.87	0.75	0.87	0.7	0.94	0.92

distinguishing these actions. However, the accuracy has not achieved the desired result (Table 1).

3.2 Recognition Accuracy of 4 FFT Features

When the FFT transformation is carried out on the time-domain values of the samples, Fig. 4 represents the three-axis frequency domain feature of four actions. It can be known from the physical meaning of FFT that the coefficient of every dimension of FFT represents a trigonometric function. In this paper, the data of the first dimension of FFT is direct component, which represents the component of gravitational acceleration the accelerometer bears, therefore, this value can be eliminated. And it can be figured out from the features of FFT that the data obtained should be symmetric, which can also be seen from the figure. Therefore, it is feasible to take half of the FFT coefficient as the feature. The diagram drawn with fft results through matlab has shown that all movements vary greatly in triaxial fft phase, especially for the ones placed at the front

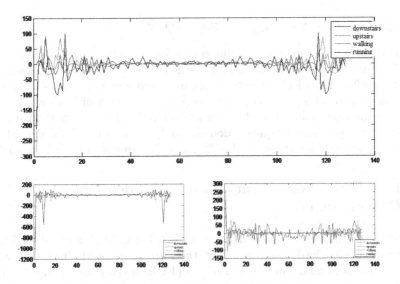

Fig. 4. Diagram of the frequency domain features

Table 2. Result of frequency domain features

	Dimensions	Still	Walk	Run	Go down	Go up	Shake	Brisk walk	Jump	Brush teeth	Average
KNN	5	.98	.95	.92	.92	.85	.92	.90	.64	1	.92
	10	.98	.99	.98	.96	.91	.94	1	.88	1	**.97**
	20	.98	.99	.98	.96	.93	.94	.97	.88	1	.96
	30	.98	.93	.97	.97	.89	.92	.97	.88	1	.94
	40	.98	.92	.98	.95	.91	.95	.93	.88	1	.94
Bayes	5	.97	.91	.89	.91	.89	.84	.90	0	.95	.84
	10	.97	.92	1	.96	.96	.89	1	.70	.95	**.91**
	20	.97	.94	1	.99	.96	.8	1	.6	.91	.90
	30	.95	.91	1	.98	.95	.84	1	.706	.91	.90
Dtree	5	.95	.91	.79	.90	.8	.88	.81	.47	.92	.86
	10	.93	.88	.82	.89	.90	.91	.87	.70	.88	**.89**
	20	.90	.91	.90	.93	.89	.91	.87	.5	.91	.89
	30	.95	.90	.83	.89	.89	.91	.90	.64	.87	.88

with greater difference. As the motion frequency of all human bodies is very low, the differences mainly arise at the front. 5 dimensions, 10 dimensions, 20 dimensions and 30 dimensions are selected respectively as features for another 10-fold verification.

In Table 2, the accuracy first increases progressively and then keeps decreasing. It reaches the peak around 10-dimension feature. Compared with time-domain features, all sorting algorithms of the frequency domain have gone up, and KNN algorithm is the best one, reaching 97%.

3.3   Recognition Accuracy of TF4+FFT10

Cross validation is performed again by combining previous time-domain features with features of the first ten dimensions in the paper. Various accuracies still rise somewhat after TF4+FFT10 features are used and new movements are added. The overall accuracies stay good, and the average accuracy of KNN remains the top and comes to 98.6%. And the accuracies have improved a lot than those of algorithms simply using frequency domain analysis methods. KNN algorithm is 1.6% and Bayesian algorithm reaches 2%. Decision tree algorithm is up to 5%.

4   Real-Time Forecasting Results Under Daily Circumstances

All the results above come from a series of modeling validation of data collected in advance. Under the daily environment, there are many interfering factors and a real-time prediction system needs to be completed by cellphone for the real-time detection. According to the methodology thought of feature extraction of data collection mentioned in the previous chapter, prediction parts are finished and

Table 3. TF4+FFT10 feature results

Human activity	KNN	Bayesian	Decision tree
Stay still	0.98	0.97	0.93
Walking	0.99	0.96	0.95
Running .	1	1	1
Go upstairs	1	0.99	0.95
Go downstairs	0.99	0.96	0.94
Shaking phones	0.97	0.94	0.95
Brisk walking	1	1	0.93
Jumping in place	0.89	0.7	0.66
Brushing teeth	1	0.95	0.91
Elevator up	0.92	0.7	0.88
Elevator down	0.93	0.67	0.8
Taking bus	0.98	0.69	0.95
Marking time	1	0.94	0.94
Squat jump	0.77	0.78	0.5
Average value	0.986	0.93	0.94

the flow chart is shown in Fig. 5. While current samples are collected, the previous sample is processed and predicated with a model which should be established beforehand. The design of testing movements is listed in the table (see Table 4). Training data come from the acquisition system. There are 73 still data, 130 walking data, 68 running data, 200 data of going upstairs, 165 data of going downstairs, 53 jogging data, 30 data of jumping in place, 35 data of boarding the elevator, 35 data of leaving the elevator and 62 data of taking a bus. All the data above are collected while cellphones are placed in the back pockets. During the test, movements are performed in line with the rules of the table after cellphones are put in the back pockets of pants, and results are obtained as below:

Table 5 shows that the accuracy is very high. For misclassifications of going upstairs and downstairs, it is understandable to misclassify them as walking because there is a certain length of gap. However, misclassification happens a lot while the test is performed in an elevator. People also always put cellphones in the front pockets of their pants, so the accuracy of testing front pockets is attempted through the training model for back pockets in the paper. The table below is drawn by repeating the movements in the Table 6:

Since the training model does not have any datum collected from the front pocket, the result is similar to that of back pockets regardless of relatively higher accuracy. However, the accuracy of handheld ones has sharply decreased, so for the software there is plenty of room for improvement.

Table 4. Testing movements standard

Human activity	Descriptions
Running	Keeps running in 30 s
Walking	Keeps walking in 60 s
Brisk walking	Keeps brisk walking in 60 s
Teeth brushing	Keeps teeth brushing in 30 s
Going upstairs	Keeps going upstairs in 30 s
Going downstairs	Keeps going downstairs in 30 s
Jumping in place	Keeps jumping in place in 30 s
Elevator up	Taking elevator from 1st floor to 14th floor
Elevator down	Taking elevator from 14th floor to 1st floor
Marking time	Keeps marking times in 30 s

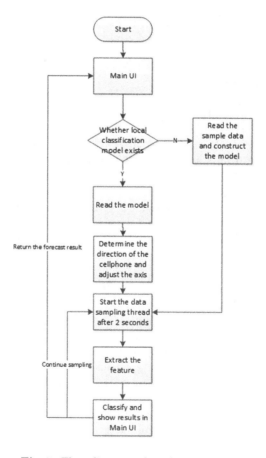

Fig. 5. Flow diagram of prediction system

Table 5. Accuracy of back pocket data

Human activity	Sampling data	Correct data	Failed data	Correct rate
Running	16	15	Brisk walking	93%
Walking	23	23	/	100%
Brisk walking	23	23	/	100%
Going upstairs	16	15	Walking	93%
Going downstairs	16	15	Walking	93%
Jumping in place	16	16	/	100%
Elevator up	10	8	Elevator down, stay still	90%
Elevator down	10	9	Elevator up	90%
Marking time	16	15	Walking	93%

Table 6. Accuracy of front pocket data

Human activity	Sampling data	Correct data	Failed data	Correct rate
Running	16	16		93%
Walking	20	18	Going upstairs and talking	86%
Brisk walking	23	20	Going upstairs	86%
Going upstairs	16	16	/	93%
Going downstairs	16	16	/	93%
Jumping in place	16	16	/	100%
Elevator up	10	9	Elevator down, stay still	90%
Elevator down	10	9	Stay still	90%
Marking time	16	14	Walking	93%

5 Conclusion

An analysis of body activity features collected by cellphone sensors and multiple verifications is made in the paper to prove that FT4+FFT10 feature extraction method which is put forward in the paper can effectively recognize 14 movements such as walking, running, going upstairs and downstairs, taking a bus and brushing teeth with an average accuracy over 98.6% under the testing environment. And a real-time predication system is realized on the Android system. The accuracy is relatively ideal while front or back pockets of pants are not analyzed separately. However, the accuracy of the handheld version is lower, so the system is expected for further improvement.

Acknowledgements. This work is supported by the National Natural Science Foundation of China under Grants NSFC 61672358.

References

1. Anguita, D., Ghio, A., Oneto, L., Parra, X., Reyes-Ortiz, J.L.: Human activity recognition on smartphones using a multiclass hardware-friendly support vector machine. In: Bravo, J., Hervás, R., Rodríguez, M. (eds.) IWAAL 2012. LNCS, vol. 7657, pp. 216–223. Springer, Heidelberg (2012). doi:10.1007/978-3-642-35395-6_30
2. Xue, Y.: Human activity recognition based on single acceleration sensor. South China University of Technology (2011)
3. Li, D.: Design of a human daily activity recognition system based on three-axis acceleration sensor. Instrum. Technol. (2013)
4. Mathie, M.J., Coster, A.C.F., Lovell, N.H., et al.: Accelerometry: providing anintegrated, practical method for long-term, ambulatory monitoring of human movement. Physiol. Meas. **25**(2), 1–20 (2005)
5. Kern, N., Antifakos, S., Schiele, B., et al.: A model of human interruptability: experimental evaluation and automatic estimation from wearable sensors. In: Proceedings of ISWC, Washington DC, USA, pp. 158–165 (2004)
6. Karantonis, D.M., Narayanan, M.R., et al.: Implementation of a real-time human movement classifier using a triaxial accelerometer for ambulatory monitoring. IEEE Trans. Inf. Technol. Biomed. **10**(1), 156–167 (2006). A Publication of the IEEE Engineering in Medicine and Biology Society
7. He, Z., Jin, L.: Activity recognition from acceleration data using AR model representation and SVM. In: IEEE International Conference on Machine Learning and Cybernetics, pp. 2245–2250 (2008)
8. Khan A.M., Lee Y.-K., Lee S.Y., et al.: Human activity recognition via an accelerometer-enabled-smartphone using kernel discriminant analysis. In: International Conference on Future Information Technology, pp. 1–6 (2010)
9. Jennifer, R.K., Gary, M.W., Samuel, A.M.: Activity recognition using cell phone accelerometers. In: SensorKDD 2010, 25 July, Washington, DC, USA, 9 pp (2010)
10. Li, N.: Research on wearable health monitoring system based on human activity recognition. Beijing University of Technology (2013)
11. Qiu, M., Zhong, M., Li, K., Gai, K., Zong, Z.: Phase-change memory optimization for green cloud with genetic algorithm. IEEE Trans. Comput. **64**(12), 3528–3540 (2015)
12. Gai, K., Qiu, M., Zhao, H.: Cost-aware multimedia data allocation for heterogeneous memory using genetic algorithm in cloud computing. IEEE Trans. Cloud Comput. (2016)

Bug Analysis of Android Applications Based on JPF

Libin Wu[1,2], Yahui Lu[1,2], Jing Qi[1,2], Shubin Cai[1,2(✉)],
Bo Deng[1,2], and Zhong Ming[1,2]

[1] Shenzhen University, Shenzhen City, Guangdong Province, China
{wulb,luyahui,qij,shubin,dengb,mingz}@szu.edu.cn
[2] Engineering Research Center of Mobile Internet Application Middleware,
Shenzhen, Guangdong Province, China

Abstract. Smart phones have occupied an irreplaceable place in our daily life. As software in mobile systems is a far cry from software in traditional computer operating systems, we can't directly use existing technologies to verify the correctness and reliability of mobile applications. JPF (Java Pathfinder) is a tool to make model detection of Java programs, but it doesn't support the detection of Android programs. This paper proposes a method which can make JPF support Android in bug detection, especially in the detection of no sleep bugs of energy leak. Using this tool, we analyzed ten open-source Android applications and successfully detected common bugs and no sleep bugs of energy leak, which means we have made progress in enhancing detection speed and in lowering down misjudgement rate.

Keywords: Java pathfinder · JPF · No-sleep bugs of energy leak · Android · Automated testing

1 Introduction

Smart phones have occupied an important place in our daily life and the power consumption of mobile phones has become a problem. Technology in batteries of mobile phones doesn't develop as fast as other hardware devices batteries' endurance is entering a straitened circumstance. In order to solve this problem so as to make use of limited battery resources in a more reasonable and effective way, we set out using a set of flexible power management strategies in mobile operation systems to manage the operation of functional devices. This set of default strategies tends to keep functional devices in a dormant state or in a state of low power consumption, unless an application actively applies to the mobile operating system for letting it remain in an operating state of high power consumption.

It is commonplace that a smart phone system will be added with a set of API during SDK development so that the developer of an application can apply this set of API to informing the system not to put a device into the sleep state

© Springer International Publishing AG 2017
M. Qiu (Ed.): SmartCom 2016, LNCS 10135, pp. 173–182, 2017.
DOI: 10.1007/978-3-319-52015-5_18

when the device is in operation, and also apply this set of API to informing the system to put the device into the sleep state after its tasks complete. As Wake Lock mechanism is adopted in Android systems to ban them from entering the sleep state [2], developers use specific wake locks to awake devices or put them into the sleep state. This programming mode, in which application programs are needed to manage the awaking and sleeping of devices, is referred to as Power Encumbered Programming [1]. Obviously, this mode gives higher requirements to developers. If a developer fails to put the devices into the sleep state in time after their tasks complete, it will lead to energy waste, which is called no sleep bug in general [4]. No sleep bug is defined as one of smart phones' bugs of energy leak, which is caused by developers' incorrect operation of API for power control and will lead to a great amount of electrical leak [5].

Thus a JPF (Java Pathfinder)-based bug detection method of Android application is proposed in this paper. This method is good at detecting no sleep bugs of energy leak and experiments shows that we have made progress in enhancing detection speed and in lowering down misjudgement rate.

2   Related Work

At present, research on detection of energy consumption and on tests of mobile platforms mainly concentrate on two aspects: One is static analysis technology, and the other is dynamic analysis technology.

Static analysis refers to a kind of theory and technology through which researchers analyze behavior of a program under the circumstance that the program is not running. At the very beginning, static analysis was applied to verifying and analyzing Java programs. But with the requirements of quality and safety for Android applications getting raised gradually, checking whether an Android application has no bug or won't throw exceptions by using appropriate approaches has become particularly important. For this reason, some researchers apply static analysis technology to Android to adapt to different requirements of testing and validation. Abhinav Pathak et al. from Purdue University carried out a detailed analysis on the types of no sleep bugs and applied reaching definitions dataflow analysis a classic algorithm, to detecting no sleep bugs in applications. They analyzed 500 Android applications, where 42 no sleep bugs were found, 13 bugs were misjudged, and 31 no sleep bugs failed to be detected [4]. Gottschalk et al. applied GReQL (Graph Repository Query Language) to checking codes of energy leaks and reconstructed the codes with reconstructing tools [6]. Applying static analysis approach may lead to misjudgment and overlook of bugs of energy leak. There were two reasons caused misjudgments. One was that the method to awake and release a wake lock was packaged by other "helper" methods or the mobile devices were operated through variables, and the other was that the device operation was taken over by a highest level of code logic [4]. Applying dynamic analysis technology is more advantageous in terms of precision of detection.

Dynamic analysis technology dynamically executes application codes by simulating device contexts and makes real-time analysis related to loopholes of energy consumption according to the results of operation. Yepang Liu et al. developed GreenDroid, which helps developers diagnose and locate codes that may cause energy leaks [7] by monitoring the situations of data being used under different circumstances. Lide Zhang et al. developed ADEL, which helps developers find places where energy leaks by using dynamic stain technique [8]. ADEL dilates through Dalvik virtual machine and TaintDroid [9]. which supports all-system stain trailing by virtue of sustainable storage. It will mark the data coming from the Internet and track the situations of data being used in the end. If the data is not used, it means that the request on the Internet is meaningless. Jindal et al. discovered a new energy loophole, known as "sleep conflicts". The loophole refers to mobile devices (such as sensors) being unable to enter the sleep state from the state of high power consumption. Jindal et al. analyzed the cause of this loophole and developed HYPNOS a running system to avoid "sleep conflicts" by analyzing the driving of hardware [10].

3   No Sleep Loophole

"No Sleep" loophole is the most prominent loophole among energy loopholes related to applications [5], and it is also the main content that this chapter needs studying on and solving. Loopholes of this type mostly result from developers' negligence of manually capturing anomalies that may appear or their failing to deal with situations which need judgement. These loopholes are likely to lead to the collapse of the applications or cause a lot of electrical waste although they may not affect the normal operation of applications. Therefore, detecting and repairing such loopholes is an important means to reduce anomalies and errors of applications and to lower down energy consumption.

3.1   No Sleep Code Path

The fundamental cause of most "no sleep" loopholes that have been found is that there exists a code path in the single threaded activity where a mobile component has been awoken. For example, a component was supposed to be awoken by a "wake lock", but the wake lock failed to be released afterwards. There are three causes that may lead to the existence of such a code path [4], where a component is started but can't enter its sleep state. The first reason: the programmer forgot to release the wake lock in the code, or the leasing statements were set in an "if" conditional judgment branch, which failed to run in actual operation. The second reason, the programmer released the wake lock in the code (such as Fig. 1), but before the release code was executed, an anomaly occurred to the program which caused the release to fail to be executed, and the programmer forgot to release the wake lock in anomaly capturing statements.

Then Fig. 1 caused a no sleep loophole due to its escape from anomaly capturing. The third reason, a higher level of conditional statements (such as deadlocks for applications) stopped the execution statements for releasing the wake lock.

```
1 try{
2     wl.acquire();
3     net_sync();
4     wl.release();
5 } catch(Exception e) {
6     System.out.println(e);
7 } finally{
8 } //End try-catch block
```

Fig. 1. Example of a no sleep loophole

3.2 No Sleep Race Condition

It is a problem of race condition caused by the second type of no sleep loopholes in multi-threaded applications. In a multi-threaded program, one threaded part turns on a component and the other threaded part is responsible for turning off the component under a certain condition. In extreme cases, if the two threaded parts have a problem of race condition in operating the wake lock, the threaded part responsible for turning on the component is likely to operate following the other part responsible for turning off the component, which will then cause a no sleep loophole. Such as the no sleep loophole caused by Fig. 2 due to race condition.

```
1 public void Main_Thread(){
2     mKill = false;
3     wl.acquire();
4     start(worker_thread);
5     //....
6     mKill = true;
7     stop(worker_thread);
8     wl.release();
9 }
10
11 public void Worker_Thread(){
12     while(true){
13         if(mKill) break;
14         net_sync();
15         wl.release();
16         sleep(180000);
17         wl.acquire();
18     }
19 }
```

Fig. 2. Example of a no sleep loophole caused by race condition

3.3 No Sleep Dilation

There is a circumstance existing in the running of an application, where the wake lock is released after a long waiting time rather than within expected time. For example, the "net sync ()" in Fig. 2 is supposed to complete within several seconds in general, but its actual operating time is far beyond expected. During this time, a great amount of power is consumed. When a critical task is being executed, no matter how long the operating time is, no matter whether it is the programmer's intention or not, literature [4]. categorizes such problems into the third type of no sleep loopholes, namely, no sleep dilation.

Table 1. Callback locations for releasing the wake locks

Android components	Callback function of exit point
Activity	onPause
Service (Bound)	onUnbind
Service (Started)	onStartCommand
IntentService	onHandleIntent
BroadcastReceiver	onReceive
Runnable	run

4   Means for Detecting No Sleep Loopholes

The four components in Android system, which are Activity, Services, Broad-castReceivers and ContentProviders, will cause electrical waste or extra energy consumption caused by no sleep loopholes if their wake locks fail to be released in time. Literature [13] holds that the components must release the wake locks at the several exit points of the applications, otherwise they should be considered as causing a no sleep loophole.

Behavior of Android components can be seen as a series of callback functions, including onCreate(), onStart(), onPause(), onResume(), onStop() and onDestory(). These callback functions decide the time of the components being set up, activated and turned off. Classifications of components in this paper are shown in Table 1. The method of exit point callback is based on the completion of a component's operation. For example, "Services" has completed its tasks at onUnbind, and "Activity" exits from OnPause. So, it can be considered that this type of exit point callback function is the final position of the wake lock which must be released by the assembly. The strategy of this paper can be summarized as: Dynamically analyze Android application through JPF, so that exit point function of each assembly can be detected. If the exit point function still holds a wake lock, then it is regarded that a nondormant bug is found.

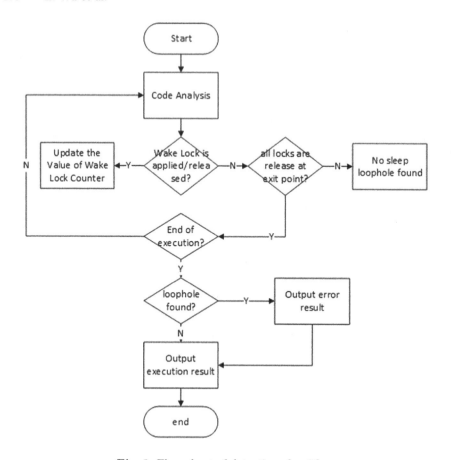

Fig. 3. Flow chart of detection algorithm

Detection algorithm of the process is shown in Fig. 3, and the entire algorithm mainly includes two stages: Firstly, reload the instructionExecuted of instruction calling method in JPF listening class; monitor execution code; conduct update to lock count value when the execution reaches releasing the lock or applying the lock; determine whether the exit point releases all wake locks. Secondly, print and output the result of nondormant bug detection.

JPF (java pathfinder) is a java model detection tool, which does not support Android program itself. The jpf-android plug-in developed by Heila and others [14] is used to support JPF to detect Android application program.

Figure 4 describes the general bug detection process. First of all, the tester needs to write a script file describing the UI event according to the UI in the program be tested, and the format of the script file is inherited from the jpf-awt plug-in. Among them, the variable name begins with the symbol @, and the variables can be used to express most of the Intent objects as shown in Fig. 5. In addition, the test script also supports the fields REPEAT and ANY

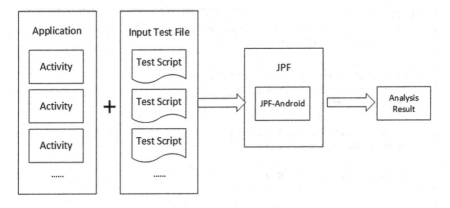

Fig. 4. Jpf-android testing process

in JPF-AWT. REPEAT can construct a set of repeated events sequences, and ANTY uses the ChoiceGenerator in JPF as well as the status and backtracking mechanism of JPF to construct a set of non-deterministic events sequences. After the completion of writing detection script, JPF can be used to test the application program, and detect the resource competition of whole program, deadlock, abnormalities and other bugs in run-time.

5   Experiment Results and Future Work

This paper has downloaded 10 open resource Android applications from Google Code website and Github website, and the tests are conducted according to the detection scheme put forward in this paper. Table 2 gives the codes number of application program, the number of abnormality found in run-time, the number of nondormant bugs and the time consumption of nondormant bugs detection.

The detection method of this paper is assessed from the following three aspects:

(1) False Positive. In the experiment, regarding abnormality detected in run-time, after one by one investigation and repair, then detect nondormant bugs again. The detected nondormant bugs are indeed due to that they are not released at the location of exit point, with error rate of 0. Compared with the static detection method in literature [4], great progress has been made. In the manual written test script, totally 11 nondormant bugs are detected, among which, 2 fail to be released in onPause method, 4 fail to be released in onReceive method, 2 fail to be released in onHandleIntent method and 3 fail to be released in onUnbind method.
(2) False Negative. In the experiment, although it is difficult to accurately detect the existing number of nondormant bugs or abnormal situation in applications, false negative situation does exist in the scheme put forward by this paper. It is because that the written test script inevitably cannot cover

```
SECTION default {
  @intent1.setComponent("com.example.com.ActivityTest")
  startActivity(@intent1)
}

SECTION com.example.com.ActivityTest {
  $buttonPrintHallo.onClick()
  $buttonResult.onClick()

  $buttonPrintHallo.onClick()
  $buttonResult.onClick()

  $backButton.onClick()

}

SECTION com.example.vdm.SampleProjectActivity {
  $button1.onClick()
  $button2.onClick()
  $backButton.onClick()
  $backButton.onClick()
}
```

Fig. 5. Detection script file

Table 2. Detection result (time is nondormant bug detection time)

App	LOC	Exception	Loopholes	Time (s)
Andoku	11059	1	0	25
DealDroid	2364	1	0	8
GuessTheNumber	2036	2	0	8
CMIS	10984	0	1	22
Opensudoku	9707	3	1	24
Rokon	17157	8	3	55
Skylight	3743	10	3	15
MonolithAndroid	7313	2	0	28
ConnectBot	52161	12	2	149
Delicious	2634	0	0	11
Scientific calculator	395	2	1	1
Summary	119553	41	11	346

all the detection path owing to negligence, thus making part of the events not triggered. Therefore, this part of codes will not be run, and it cannot guarantee if there are any existing bugs, which are not detected in the code.
(3) Detection time. It can be seen from Table 2 that with the increase in the number of lines of codes, the detection time increases correspondingly.

In addition, the detection time is also related to the complexity of the detection script. The more detailed the detection script is, the more coverage of the detection path is, and the longer corresponding detection time is Table 2 test script per thousand lines of detection time is 2.89 s (346*1000/119553).

Although the research of this paper has achieved preliminary results, there are still some problems to be solved, such as detection needs application source code to complete the detection and the whole testing framework is still combined together with jpf-android without forming an independent test tool yet. For these detects, the next step of the work mainly includes the following aspects:

(1) Release the testing tool of this paper in the form of JPF plug-in. Independent release of the plug-in is conducive to the use and dissemination of the tool.
(2) Support the detection of android setup program (.apk file). The detection in this paper still needs program source code, which cannot directly detect apk files, and in most cases, program source code cannot be obtained. In apk file, Dalvik byte code file is compressed, so the tool is needed to decompile Dalvik byte code into Java source code, such as the use of ded. and other tools.

References

1. Jindal, A., Pathak, A., Hu, Y.C., Midkiff, S.: On death, taxes, and sleep disorder bugs in smartphones. In: Proceedings of the Workshop on Power-Aware Computing and Systems, pp. 1–5. ACM (2013)
2. Qiu, M., Zhong, M., Li, J., Gai, K., Zong, Z.: Phase-change memory optimization for green cloud with genetic algorithm. IEEE Trans. Comput. **64**(12), 3528–3540 (2015)
3. Gai, K., Qiu, M., Zhao, H.: Cost-aware multimedia data allocation for heterogeneous memory using genetic algorithm in cloud computing. IEEE Trans. Cloud Comput. (2016)
4. Pathak, A., et al.: What is keeping my phone awake? characterizing and detecting no-sleep energy bugs in smartphone apps. In: Proceedings of the 10th International Conference on Mobile Systems, Applications, and Services. ACM (2012)
5. Pathak, A., Charlie Hu, Y., Zhang, M.: Bootstrapping energy debugging on smartphones: a first look at energy bugs in mobile devices. In: Proceedings of the 10th ACM Workshop on Hot Topics in Networks. ACM (2011)
6. Gottschalk, M., et al.: Removing energy code smells with reengineering services. In: Beitragsband der 42. Jahrestagung der Gesellschaft für Informatik e.v, pp. 441–455 (2012)
7. Liu, Y., Xu, C., Cheung, S.C.: Where has my battery gone? finding sensor related energy black holes in smartphone applications. In: proceedings of IEEE International Conference on Pervasive Computing and Communications (PerCom). IEEE (2013)
8. Zhang, L., et al.: Adel: an automatic detector of energy leaks for smartphone applications. In: Proceedings of the Eighth IEEE/ACM/IFIP International Conference on Hardware/Software Codesign and System Synthesis. ACM (2012)

9. Enck, W., et al.: TaintDroid: an information-flow tracking system for realtime privacy monitoring on smartphones. In: OSDI, vol. 10 (2010)

10. Jindal, A., et al.: Hypnos: understanding and treating sleep conflicts in smartphones. In: Proceedings of the 8th ACM European Conference on Computer Systems. ACM (2013)

11. Cai, S., Sun, H., Gu, S., Ming, Z.: Learning concept hierarchy from folksonomy. In: Proceedings of Eighth Web Information Systems, Applications Conference, pp. 47–51. IEEE Computer Society (2011)

12. Havelund, K., Pressburger, T.: Model checking java programs using java pathfinder. Int. J. Softw. Tools Technol. Transf. 2(4), 366–381 (2000)

13. Vekris, P., et al.: Towards verifying android apps for the absence of no-sleep energy bugs. In: Proceedings of the 2012 USENIX Conference on Power-Aware Computing and Systems. USENIX Association (2012)

14. van der Merwe, H., van der Merwe, B., Visser, W.: Verifying android applications using Java PathFinder. ACM SIGSOFT Softw. Eng. Notes 37(6), 1–5 (2012)

A Genetic-Ant-Colony Hybrid Algorithm
for Task Scheduling in Cloud System

Zhilong Wu[1,2], Sheng Xing[3], Shubin Cai[1,2], Zhijiao Xiao[1,2], and Zhong Ming[1,2(✉)]

[1] College of Computer Science and Software Engineering,
Shenzhen University, Shenzhen, China
{woozylong,shubin,cindyxzj,mingz}@szu.edu.cn
[2] Engineering Research Center of Mobile Internet Application Middleware,
Shenzhen, Guangdong Province, China
[3] College of Mathematics and Information Science,
HeBei University, Baoding, Hebei Province, China
cssxing@126.com

Abstract. As the task load of cloud system grows bigger, it becomes very important to design an efficiency task scheduling algorithm. This paper proposes a task scheduling algorithm based on genetic algorithm and ant colony optimization algorithm. The hybrid task scheduling algorithm can help the cloud system to complete users' tasks faster. Simulation experiment results in CloudSim show that, comparing with genetic algorithm and ant colony optimization algorithm alone, the hybrid algorithm has better performance in the aspects of load balancing and optimal time span.

Keywords: Cloud computing · Task scheduling · Ant colony algorithm · Genetic algorithm

1 Introduction

The applications in cloud environment have a great variety [1, 2]. The users' tasks and tasks' execution time keep changing. It's difficult yet valuable to tackle the task scheduling challenge in cloud environment. Currently, task scheduling in cloud computing mainly include first-come-first-service, load balancing, energy consumption minimization and so on. Along with the expanding of the scale of data center, it's extremely important to increase the efficiency of task scheduling.

However, task scheduling is an NP complete problem in the cloud computing environment [3]. Intelligent optimization algorithms such as simulated annealing algorithm, genetic algorithm, ant colony algorithm and particle swarm optimization are very suitable for solving NP problems [4]. This paper proposes a genetic-ant-colony hybrid algorithm to deal with the task scheduling problem. The hybrid algorithm has rapid convergence and optimization searching capacity features, which shortens the scheduling time of carrying out task scheduling and makes the load of virtual machine balanced as much as possible, greatly increases the efficiency of task scheduling in cloud computing.

© Springer International Publishing AG 2017
M. Qiu (Ed.): SmartCom 2016, LNCS 10135, pp. 183–193, 2017.
DOI: 10.1007/978-3-319-52015-5_19

2   Task Scheduling in Cloud System

The research on task scheduling can be divided into online mode and batch processing mode. Online mode is to produce a mapping relation of user task as soon as it arrives and carries out scheduling [5]; batch-processing mode is to carry out the mapping of distribution relation till the tasks are accumulated to a certain number. The research of this paper is conducted in the batch-processing mode, which means there are N tasks waiting for scheduling in the unit time.

In order to reduce the complexity of experiment and centralize the research on the algorithm itself, the experiment we conducted has following constraints on the relation between user task and virtual node resource:

1. There is no dependency relationship between tasks.
2. Node resources are unshared and independently occupied. Only when tasks finish (success or fail), other tasks can invoke the resource.

The problem of task scheduling can be described as: to assign m tasks Task $= \{T_1, T_2, \ldots T_m\}$ to n virtual node resources Vm $= \{vm_1, vm_2, \ldots vm_n\}$, m > n in the cloud environment. ETC (Expected time to completion) shows the expected executive time of user task and virtual node, and ETC_{ij} represents the handling time spent by task T_i to get exclusive access to virtual node resources vm_j, with the unit is millisecond; all relationships between ETC_{ij} form a ETC matrix. Matrix (1) can be used to represent the distribution relationship comprised by m user tasks and n virtual node resources [6]:

$$\begin{bmatrix} x_{00} & \cdots & x_{0n} \\ \vdots & \ddots & \vdots \\ x_{m0} & \cdots & x_{mn} \end{bmatrix} \tag{1}$$

Wherein, $x_{ij} \in (0, 1)$ represents the mapping relation between user task and virtual node resource and $\sum_{i=1}^{m} x_{ij} = 1$, $j \in \{1, 2, \ldots n\}$, if the user task j chooses virtual node resources i to run, then $x_{ij} = 1$, otherwise, $x_{ij} = 0$. ETC is the execution time that task run on the node which can be comprised by ETC matrix in Formula 2:

$$\begin{bmatrix} ETC_{00} & \cdots & ETC_{0n} \\ \vdots & \ddots & \vdots \\ ETC_{m0} & \cdots & ETC_{mn} \end{bmatrix} \tag{2}$$

The procedure of solving the scheduling problem by ant colony algorithm is to solve the set s $= \{x_1, x_2, \ldots, x_m\}$ and get decision variable X_i. M in the set indicates the user's task number; set $\{T, Vm\}$ can be worked out by the Cartesian product through m tasks $\{t_1, t_2, \ldots, t_m\}$ and n node resources $\{vm_1, vm_2, \ldots, vm_n\}$. The set $\{T, Vm\}$ means the specified virtual node map set that the assignment of user's task. T in each tuple of the set $\{T, Vm\}$ only has one chance to assign; the derived solution is

$\left\{ X_{t_1 1}, X_{t_2 2}, \ldots, X_{t_j j}, \ldots, X_{t_m n} \right\}, t_j \in \{1, 2, \ldots, m\}$, solution set should be as reasonable as possible and make the execution time span

$$\text{timespane} = \sum_{i=1}^{m} \sum_{j=1}^{n} X_{ij} \times ETC_{ij}$$

as short as possible.

3   The Genetic-Ant-Colony Hybrid Algorithm

Genetic algorithm has remarkable effects on global searching ability but often causes low efficiency and large quantities of redundancy iteration [7]. Ant colony algorithm converges to optimal solution quickly in later stage but don't work very well in earlier stage due to the lack of pheromone. This paper combines advantages of two algorithms, and puts forward task scheduling algorithm called genetic-ant-colony hybrid algorithm. The basic idea is using genetic algorithm at the earlier stage and ant colony algorithm at the later stage.

3.1   Genetic Algorithm at Earlier Stage

(1)   Rapid Global Search Based on Genetic Algorithm

In this algorithm, $T_i(0 \leq i \leq m-1)$ and $V_n(0 \leq j \leq n-1)$ respectively represents ID number of user tasks and ID number of virtual node resources. In initial time, the chromosome of each individual in the population will randomly create equivalent numbers of virtual node resources V_j. When the chromosome trough crossover, mutation and iteration, the task T_i will be distributed to any virtual node V_j. When many users are distributed to the same virtual node resources, the tasks will be carried out according to its sequence, while others will be ready to wake up in the waiting lines. The final goal is to get an optimal service quality solving scheme which is compliant with users' defines.

Let's assume that there are 15 user tasks to be distributed and 5 free virtual node resources in the cloud environment, so that the length of chromosome in the genetic algorithm model is 15, the value of single gene ranks from 1 to 5, shows the user tasks will be distributed to one of these 5 virtual nodes randomly. For example, {2 2 3 4 1 1 2 0 3 0 1 2 4 0 3} represents randomly generated code form of one chromosome, after decoding, it represents that the ID 1 user task reflects ID 2 virtual machine, while the second user task reflects ID 2 virtual machine too. The genes in chromosome are saved in the form of array.

(2)   Crossover operation of genetic algorithm

Crossover operation means the pairwise coupling process of chromosomes according to a certain rate. In this process, some chromosome's place changed and creating two new chromosomes.

Fig. 1. PMX diagram

The crossover algorithm this paper used is Partially Matching Crossover (PMX). And the crossover parts are limited to smaller than one fifth of the original length of the chromosome. A simple example is illustrated in Fig. 1.

(3) Mutation operation of genetic algorithm

Mutation operation means that genes in one or a number of places in chromosome changed and then generate a new chromosome. Mutation operation can help population create different individuals to keep diversity. It determines directly local search ability of genetic algorithm.

In this paper, the Genetic Algorithm's operation to mutation is based on Displacement Mutation (DM). As Fig. 2 shows, replacement variation firstly chooses one starting point in parent generation, then randomly chooses an ending point in the following gene sequence, and insert gene segment into that substring.

before mutation: 1 2 3 4 5 6 7 8

1 2 3 7 8

after mutation: 1 2 3 7 4 5 6 8

Fig. 2. Replacement mutation diagram

To avoid the chromosome variation causing the offspring inferior to the parent, this paper don't distribute one node to current user task randomly in the field of mutation operator design, but wipe out the original virtual node it occupied in the nodes set and choose the virtual node which makes current user task solved at the soonest. Choose virtual resource node which costs least time to finish current user task, then reflect this task to current virtual node, which can ensure that offsprings come from parents.

Crossover ratio function and mutation ratio function are:

$$P_c = \begin{cases} \dfrac{k_1 \left(f_{max} - f'\right)}{f_{max} - f_{avg}}, & f' \geq f_{avg} \\ k_2, & f' < f_{avg} \end{cases} \tag{3}$$

$$P_m = \begin{cases} \dfrac{k_3 \left(f_{max} - f\right)}{f_{max} - f_{avg}}, & f \geq f_{avg} \\ k_2, & f < f_{avg} \end{cases} \tag{4}$$

Wherein, f_{max} is the maximum colony evaluation value among the group, f_{avg} is the average colony evaluation value of each generation, f' is the bigger colony evaluation value of the two crossover individuals, f is the colony evaluation value of the individual to mutate. Calculate the crossover ratio and mutation ratio respectively and choose the bigger one as the final result P_c, P_m.

3.2   Ant Colony Algorithm at Later Stage

(1) Initialization of pheromone

Connect the user task i to virtual node resources j with pheromone cumulant τ_{ij}. Set the pheromone value at each side as a quite small positive constant τ_0.

(2) The rule of user task assigning virtual node resource

In each iteration of ACO, is every ant k, k = 1, …, m, (m is the quantity of ants) choose proper node resources, we need to calculate the dealing ability of pheromone density of other nodes' resources, then calculate the ratio they would be chosen by Formula (5). According to "roulette" selection to assign proper virtual node resources for next user task. After a node assigned, the node included this task will be added into Tabu and delete this node from Allowed list. Execute this process repeatedly until all the virtual nodes have been traversed and the quantity of element in Allowed list is 0. Then calculate the ETC matrix value according to Formula (6). Choosing the best mapping set between virtual node and user task at last, compare each ant's solution, then compare to the best fitness [8]. If this iteration fitness is smaller than the best fitness, this fitness should be seemed as the best one, and the chosen ordered mapping relation should be seemed as the best distribution relation in Tabu.

$$P_{ij}^k(t) = \begin{cases} \dfrac{\left[\tau_{ij}(t)\right]^\alpha * \left[\eta_{ij}\right]^\beta}{\sum_{s \in allowed_k} \left[\tau_{is}(t)\right]^\alpha * \left[\eta_{is}\right]^\beta}, & \text{if } j \in allowed_k \\ 0, & else \end{cases} \tag{5}$$

Wherein, τ_{ij} shows the pheromone concentration of user task i and mapping relation of virtual resource V_j; $allowed_k$ means the virtual node resources ant k can arrive in the next step and the virtual node resources records stored in Tabu of ant k; $\eta_{ij} = 1/d_{ij}$ is heuristic information, representing the visibility of ant t, thus it can be seen that the smaller d_{ij} is, the bigger η_{ij} is, i.e., the higher visibility of ant is; d_{ij} shows the execution time and transferring time of user task i on virtual node resources V_j which can be calculated by Formula (6), which TL_Task_i represents the total length of user task i submitted to virtual node resource V_j; $InputFileSize$ means the length of other information besides the execution part, VM_bw_j means the corresponding network bandwidth of virtual node resource j; EV_j means the calculating ability of virtual node resource V_j, its calculating method shows as Formula (7), pe_num_j means the CPU core quantity of virtual node resource j, pe_mips_j means the MIPS of per core CPU of virtual node resource V_j.

$$d_{ij} = \frac{TL\_Task_i}{EV_j} + \frac{InputFileSize}{VM\_bw_j} \tag{6}$$

$$EV_j = pe\_num_j \times pe\_mips_j \tag{7}$$

(3) Update of pheromone

The update of pheromone on virtual machine will vary with the change of disaggregation of ant colony search [9, 10]. If the pheromone density of virtual node resource V_j at t is $\tau_{ij}(t)$, then at t + 1, the pheromone cumulant of virtual node resource V_j needs to update, and the updating formula is as follows:

$$\tau_{ij}(t+1) = (1 - \rho)\tau_{ij}(t) + \Delta\tau_{ij}(t) \tag{8}$$

Wherein, $\rho \in [0, 1)$ is pheromone volatility coefficient, $1-\rho$ means residual coefficient of information, value of ρ is bigger, the sooner pheromone volatilized, then the influence of the searched path on choosing present path will be smaller, $\Delta\tau_{ij}(t) = \sum_{k=1}^{m} \Delta\tau_{ij}^k$ means the mapping (i, j) information quantity of the virtual node and the user i's task ant k allocated in this cycle.

The value of ρ is adjusted as follows:

$$\rho(t) = \begin{cases} 0.95 \times \rho(t - 1), & \rho_{min} \leq 0.95 \times \rho(t - 1) \\ \rho_{min}, & else \end{cases} \tag{9}$$

Wherein, ρ_{min} is the minimum value of ρ, in order to avoid algorithm stagnation when it is too small, this method can not only improve the rate of convergence of the algorithm, but also guarantee the ACO won't get locally optimal solution because of premature convergence, which is helpful to find better solutions and helpful to the dimension of the problem.

After ant k finishing a journey, leaving the iteration value of pheromone on each line (i, j) the ant went by user task and virtual node resource mapping. As the formula (10) shows:

$$\Delta\tau_{ij}^k(t) = \begin{cases} \dfrac{Q}{L_k(t)}, & if(i,j) \in T^k(t) \\ 0, & else \end{cases} \tag{10}$$

Wherein, $T^k(t)$ represents the collection of virtual nodes ant k visited, $L_k(t)$ is the summary of fitness value, and Q is a control parameter.

3.3 Links Genetic Algorithm to Ant Colony Algorithm

Set the initial parameter τ_s of pheromone as $\tau_s = \tau_C + \tau_G$, τ_C is an assigned pheromone constant, τ_{min} can be seen as in MMAS algorithm, its value is up to the degree of problem, and τ_G is the pheromone density transformed from the result of genetic algorithm.

The proposed hybrid task scheduling algorithm:

(1) Set the maximum and minimum value of the circulation times of genetic algorithm.
(2) According to the process of searching optimal solution of genetic algorithm, record the best fitness value of the present circulation and replace with the present best solution when better solution occurred in iteration process.
(3) In assigned algebra, if the fitness values of continuous generations are greater than the present fitness value, it can be considered that the execution efficiency of genetic algorithm starts to decrease, meaning this algorithm can be ended and AOC can start.

4 Implementation of the Hybrid Algorithm

The basic execution steps of task scheduling algorithm based on AOC and genetic algorithm are as follows:

Step 1 Define the goal function of task scheduling of cloud computing.

Step 2 Set related parameters and end conditions of this scheduling algorithm.

Step 3 Code the population of genetic algorithm.

Step 4 Calculate the fitness value of each chromosome and choose the best individual among the chromosomes.

Step 5 Run the crossover and variation calculation on chosen individual.

Step 6 When arriving assigned end condition, end the calculation of genetic algorithm and save this scheduling result and redirect to Step 7. Otherwise, redirect to Step 4.

Step 7 According to the scheduling result of genetic algorithm, initialize the pheromone of AOC with pheromone transform formula.

Step 8 M ants are placed on n virtual machines. According to the state transition probability formula (5), the suitable virtual node resource is selected for the next user task.

Step 9 Every task selected is placed in the taboo table. When the table is filled up, it indicates that the ants have finished one generation's search. According to the formulae, the positions of the virtual machines that the ants pass in the pheromone matrix are updated.

Step 10 When the scheduling of all the tasks is completed, according to the task scheduling list, the evaluation value of the scheduling result is calculated, and according to formulae (8), (9) and (10), the global pheromone is updated.

Step 11 The iterative algebra is auto-incremented 1, gen = gen +1; the optimal solution set is derived and stored in the list.

Step 12 If the iterative algebra satisfies the termination condition, namely, gen > MAX_GEN, then the algorithm is terminated and the optimal solution is output; otherwise, it hops to Step 8 to continue the implementation until the algorithm satisfies the termination condition.

The overall block diagram of the scheduling idea is as indicated in Fig. 3.

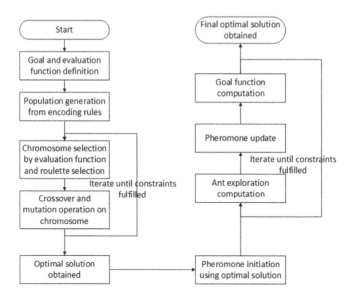

Fig. 3. Flow chart of the genetic-ant-colony hybrid algorithm

5 Simulation Experiment and Result Analysis

Cloudsim is the cloud calculation simulation emulator developed by the team led by Doctor Rajkumar Buyya of the University of Melbourne, Australia. It not only schedules the service model, but also the applications that change dynamically. Cloudsim makes a large-scale comparison and quantification of the allocation strategy of the resource integrity, and realizes the objective of controlling the cloud computation resources.

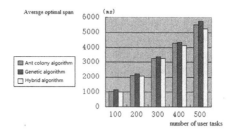

Fig. 4. The average optimal span

The realize of experiment is to simulate one data center under the Cloudsim simulation platform, including 10 virtual machines and 100 to 1000 different user tasks. The length of a user task varies from 300 MI (Million Instructions) to 20,000 MI, the values of parameters that directly or indirectly influence the algorithm α, β, ρ, t_{max}, the ant number m and the value of Parameter Q, and the values of the default parameters are $\partial = 1$, $\beta = 1$, $\rho = 0.5$, $Q = 100$, $t_{max} = 150$ and m = 20. The parameter values for the solutions to the crossover probability and mutation probability in the genetic algorithm are k_1, k_2, k_3, k_4, the values are 0.39, 0.85, 0.096, 0.056 respectively.

The follow-up experiment in this paper compares the average optimal span of the task scheduling under different user tasks. In the Cloudsim environment, we carry out 50 simulation experiments on the ant colony algorithm and the genetic algorithm in the task scheduling, and the ant colony genetic hybrid algorithm respectively. The mean values of the optimal span obtained are shown in Table 1 and Fig. 4:

Table 1. Execution times of different algorithms under the same task number (Unit: ms)

Algorithm	Tasks	$Time_{min}$	$Time_{max}$	$Time_{avg}$
Ant colony algorithm	100	936.3	1681.7	1030.86
	200	1712.9	2684.3	2115.99
	300	2863.5	3515.8	3264.35
	400	3923.6	4725.2	4283.07
	500	4983.9	5913.4	5520.07
Genetic algorithm	100	1003.2	1417.5	1161.42
	200	1915.8	2794.1	2213.84
	300	2860.4	3771.9	3365.25
	400	4108.7	4651.3	4337.24
	500	5157.3	6108.1	5758.93
Hybrid algorithm	100	941.5	1321.6	991.29
	200	1623.5	2438.2	2070.32
	300	2481.7	3121.8	3056.17
	400	3653.9	4584.3	4035.49
	500	4347.2	5506.8	5130.35

From the figure, it can be seen that with the increase of the user task number, the optimal span continuously increases as well. This is because in the situation of a certain virtual node number, as the nodes are exclusive and the user tasks are relatively independent, so greater task number leads to more tasks that line up for the virtual nodes to deal with, resulting in the increase in time. It can be seen in the figure that under the same user task, the ant colony genetic hybrid algorithm designed in this paper is better than the ant colony algorithm, whose execution is better than that of the genetic algorithm. This also reasonably accounts for the characteristic that the solutions of the colony algorithm are better than those of the genetic algorithm in the case of a specified algebra, and further demonstrates the plausibility of the algorithm in this paper.

DLB (I) is used to evaluate the load balancing degree of the task allocation plan "I". We set the following formula for calculation:

$$DLB(I) = \frac{\sum_{j=1}^{M} vmTime(VM_j)}{M \times completeTime(I)} \tag{11}$$

wherein $vmTime(VM_j) = \sum_{i=1}^{n} Time(T_i, VM_j)$, n is the number of sub-tasks allocated to the

virtual machine VM_j, $Time(T_i, VM_j)$ indicates the time needed for the completion of the ith sub-task T_i allocated to the virtual machine VM. The computation for $Time(T_i, VM_j)$ is shown in the Formula (6). Due to the parallelism that cloud computation deals with the sub-task sequence, I indicates the solution provided for the user task by genetic algorithm and ant colony, $completeTime(I)$ is the maximum value of $vmTime$ in the above calculations, and expressed by the formula $completeTime(I) = \max(vmTime(VM_j), j \in [1, m])$.

Bigger load balancing factor indicates higher utilization rate of all the virtual machines in the system. The more balanced the system load is, the higher the balancing degree of the load is. The average load factor distributions of the genetic algorithm and the hybrid algorithm are shown in the following figure when the number of user tasks is between 100 and 500.

It can be seen from the figure that the load balancing factors of the ant colony hybrid algorithm proposed in this paper are all lower than those of task scheduling algorithm. From the analysis of Formula (11), it is known that the load balancing degree of DLB gets poorer and poorer with the increase of user task number. Taken together with Fig. 5, it is concluded that under the same task, the less the time of the scheduling execution, the greater the load balancing degree is, which fully demonstrates that the system's load is more balanced and the utilization rate of the system is higher. Through the above analysis, it is concluded that from the perspective of time and load balancing degree, the ant colony hybrid scheduling algorithm has more excellent performance than the ant colony algorithm and the genetic algorithm under the cloud environment.

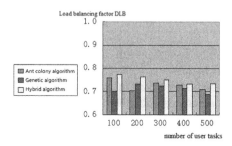

Fig. 5. Load balancing factor

6   Conclusion

This paper proposes a cloud computation hybrid genetic algorithm and ant colony based on the genetic algorithm and the ant colony optimization algorithm to solve the issue of task scheduling in the cloud environment. Through the experiment simulation, this paper demonstrates that this algorithm has good effect in terms of load balancing degree and optimal time span and is an effective scheduling algorithm. To verify the validity of the experiment algorithm, this paper conducts the simulation experiment. Although the platform can simulate the task scheduling scene of cloud computation to a maximum degree, it can be problematic in the actual cloud environment. Therefore, performance test of this algorithm in the actual cloud environment in the follow-up study can more accurately verify the accuracy, practicality and effectiveness of the algorithm.

Acknowledgement. This work is supported by the National Natural Science Foundation of China under Grants NSFC 61672358.

References

1. Armbrust, M., Fox, A., Griffith, R., et al.: Above the Clouds: A Berkeley View of Cloud Computing, mimeo, UC Berkeley, RAD Laboratory (2009)
2. Gai, K., Qiu, M., Zhao, H., Tao, L., Zong, Z.: Dynamic energy-aware cloudlet-based mobile cloud computing model for green computing. J. Netw. Comput. Appl. **59**, 46–54 (2015)
3. Gai, K., Du, Z., Qiu, M., Zhao, H.: Efficiency-Aware workload optimizations of heterogeneous cloud computing for capacity planning in financial industry. In: Proceedings of IEEE 2nd International Conference on Cyber Security and Cloud Computing, pp. 1–6 (2015)
4. Qiu, M., Zhong, M., Li, J., Gai, K., Zong, Z.: Phase-Change memory optimization for green cloud with genetic algorithm. IEEE Trans. Comput. **64**(12), 1–13 (2015)
5. Gai, K., Qiu, M., Zhao, H.: Cost-Aware multimedia data allocation for heterogeneous memory using genetic algorithm in cloud computing. IEEE Trans. Cloud Comput. 1 (2016)
6. Stutzle, T., Dorigo, M.: ACO algorithms for the traveling salesman problem. In: Evolutionary Algorithms in Engineering and Computer Science, pp. 163–183 (1999)
7. Li, K., Xu, G.: Cloud task scheduling based on load balancing ant colony optimization. In: 2011 Sixth Annual Chinagrid Conference (ChinaGrid), pp. 3–9 (2011). doi:10.1109/ChinaGrid.2011.17
8. Ding, J., Chen, Z., Yuan, Z.: On the combination of genetic algorithm and ant algorithm. J. Comput. Res. Dev. **9**(40), 1351–1356 (2003)
9. Buyya, R., Ranjan, R., Calheiros, R.N.: Modeling and simulation of scalable Cloud computing environments and the CloudSim toolkit: challenges and opportunities. In: International Conference on High Performance Computing & Simulation, HPCS 2009, pp. 1–11. IEEE (2009)
10. Wickremasinghe, B., Calheiros, R.N., Buyya, R.: CloudAnalyst: a CloudSim-based visual modeller for analysing cloud computing environments and applications. In: Proceedings of the 24th IEEE International Conference on Advanced Information Networking and Applications, pp. 446–452 (2010)

A Hybrid Algorithm of Extreme Learning Machine and Sparse Auto-Encoder

Yu Lin[1,2], Yanchun Liang[1,3], Shinichi Yoshida[4], Xiaoyue Feng[1(✉)], and Renchu Guan[1,3]

[1] Key Laboratory of Symbolic Computation and Knowledge Engineering of Ministry of Education, College of Computer Science and Technology, Jilin University, Changchun 130012, China
fengxy@jlu.edu.cn
[2] China Unicom Cloud Data Company Limited, China United Network Communications Corporation Liaoning Branch, Shenyang 110002, China
[3] Zhuhai Laboratory of Key Laboratory of Symbolic Computation and Knowledge Engineering of Ministry of Education, Zhuhai College of Jilin University, Zhuhai 519041, China
[4] School of Information, Kochi University of Technology, Kochi 782-8502, Japan

Abstract. This paper proposes two deep learning modes which combine sparse auto-encoder with extreme learning machine (ELM) and kernel extreme learning machine (KELM), namely as Stacked Sparse Auto-encoder-Extreme Learning Machine and Stacked Sparse Auto-encoder-Kernel Extreme Learning Machine. The proposed models are applied to the image recognition task. To learn features from the original input data, SSAE with deep architecture is employed. Then, to construct a unified neural network learning model, ELM and KELM are selected to classify the extracted features. To evaluate the performance of the two proposed models, we carry out experiments on three different image data sets respectively. The results show that the proposed models' performance can be not only superior to the shallow architecture models of Support Vector Machine, ELM and KELM but also better than the deep architecture models, such as Stacked Auto-encoder, Deep Belief Network and Stacked Denoising Auto-encoder.

Keywords: Deep learning · Extreme learning machine · Sparse Auto-encoder image recognition · Stack denoising Auto-encoder

1 Introduction

Recent rapid development of Internet-related technologies has facilitated the significant growth of novel research fields, such as cloud computing [1] and deep learning. In May of 2015, LeCun et al. pointed out that deep learning has dramatically improved the state-of-the-art models in speech recognition, object detection, genomics etc. [2]. Because of the discovery of novel approaches demonstrated successful at learning parameters, the neural networks which are deep architectures with good generalization capability have attracted more and more attentions [3–5]. For example, Hinton, et al.

© Springer International Publishing AG 2017
M. Qiu (Ed.): SmartCom 2016, LNCS 10135, pp. 194–204, 2017.
DOI: 10.1007/978-3-319-52015-5_20

has showed that restricted boltzmann machine (RBM) and auto-encoders well performed feature engineering [4]; Stacked auto-encoder (SAE) [6] and stacked denoising auto-encoder (SDAE) [7] also achieve good generalization performance on classification problems.

A stacked sparse auto-encoder is a deep neural network, which consists of multiple layers of sparse auto-encoders. A sparse auto-encoder is an artificial neural network. A sparsity constraint on the hidden units is imposed, and then the data's interesting structure will be discovered in spite of the large number of the hidden units [4, 8]. Extreme learning machine is proposed by Huang et al. [9, 10], which is an efficient learning algorithm of single hidden layer feed-forward neural networks (SLFNs) with extremely fast learning speed. And kernel extreme learning machine is a further improved version proposed by Huang et al. [11] combining the kernel method with ELM, which can achieve fast learning speed and good generalization capability.

In this paper, we firstly proposed a unified deep learning model which is applied to the image recognition problem, namely Stacked Sparse Auto-encoder Extreme Learning Machine (SSAE-ELM). The extracted features are learned by multiple-layer sparse auto-encoders and used by ELM to perform the classification. Moreover, to remove the impact of the number of hidden neurons and achieve a better performance based on kernel learning, we propose a Stacked Sparse Auto-encoder Kernel Extreme Learning Machine (SSAE-KELM). The paper is organized as follows: the second section introduces the model of ELM, KELM and sparse auto-encoder. Section 3 shows the proposed algorithms of SSAE-ELM and SSAE-KELM in details. The fourth section represents the experiment results and the conclusions. Future works are given in the last section.

2   Brief of Relevant Algorithms

2.1   Extreme Learning Machine

Extreme learning machine (ELM) is proposed by Huang et al. in [9]. It shows that its hidden nodes can be randomly generated. The input data is mapped to L dimensional random feature space and the network output is given by:

$$f_L(x) = \sum_{i=1}^{L} \beta_i h_i(x) = h(x)\beta \tag{1}$$

where $\beta = [\beta_i, \cdots, \beta_L]^T$ is the output weight matrix between hidden and output nodes, $h(x) = [g_1(x), \cdots g_L(x)]$ are the outputs of hidden nodes. With randomly assigning the weights for input x, $g_i(x)$ is the output of the i-th hidden node. Given N training samples $\{(x_1, t_1),\ldots, (x_N, t_N)\}$, the ELM equation can be shown by:

$$H\beta = T \tag{2}$$

where $T = [t_1, \ldots t_N]^T$ are the target labels of input data and $H = [h^T(x_1), \cdots, h^T(x_N)]^T$. In order to achieve better generalization, ELM minimizes the training errors as well as the norm of the output weights according to:

$$\text{Minimize} : \sum_{i=1}^{N} \|h(x_i)\beta - t_i\| \tag{3}$$

The output weights can be calculated by:

$$\widehat{\beta} = \mathbf{H}^{\dagger}\mathbf{T} \tag{4}$$

Where $\mathbf{H}^{\dagger}$ is the Moore-Penrose generalized inverse of matrix $\mathbf{H}$.

2.2   Kernel Extreme Learning Machine

In order to make the output of ELM classifier $h(x)\beta$ be close to the class labels, a constrained-optimization-based ELM can be formulated as [11]:

$$\text{Minimize} : L_{P_{ELM}} = \frac{1}{2}\|\beta\|^2 + C\frac{1}{2}\sum_{i=1}^{N}\xi_i^2$$

$$\text{subject to} : h(x_i)\beta = t_i^T - \xi_i^T, i = 1, \ldots, N \tag{5}$$

where C is the regularization constant, $t_i = [t_{i1}, t_{i2}, \ldots, t_{im}]^T$ and $\xi_i = [\xi_{i1}, \ldots \xi_{im}]^T$ is the training result of the m output nodes with respect to the training sample x_i. Based on the Karush–Kuhn–Tucker (KKT) conditions, ELM training is equivalent to solving the following dual optimization problem:

$$L_{D_{ELM}} = \frac{1}{2}\|\beta\|^2 + C\frac{1}{2}\sum_{i=1}^{N}\xi_i^2 - \sum_{i=1}^{N}\sum_{j=1}^{m}\alpha_{ij}\big(h(x_i)\beta_j - t_{ij} + \xi_{ij}\big) \tag{6}$$

where $\alpha_i = [\alpha_{i1}, \ldots, \alpha_{im}]^T$ is the vector of Lagrange multipliers and β_j is the vector of the weights of hidden layer to the jth output node. The output function of ELM classifier can be shown based on KKT corresponding optimality conditions:

$$f_L(x) = h(x)\beta = h(x)H^T\left(\frac{I}{C} + HH^T\right)^{-1}T \tag{7}$$

If a feature mapping $h(x)$ is unknown to users, we can apply Mercer's conditions on ELM. The kernel matrix Ω_{ELM} can be constructed and the output function of ELM classifier can be shown as:

$$f_L(x) = \begin{bmatrix} K(x,x_1) \\ \vdots \\ K(x,x_N) \end{bmatrix}^T \left(\frac{I}{C} + \Omega_{ELM}\right)^{-1} T \tag{8}$$

2.3 Sparse Auto-Encoder

Sparse auto-encoder is a feed-forward neural network with unsupervised learning using back propagation and batch gradient descent algorithm. It tries to learn an approximation to one identity function between the input and output data [4, 8]. A sparse auto-encoder neural network is constructed with one hidden layer in this paper.

Suppose we have m fixed training examples $\{(x^{(1)},y^{(1)}),...,(x^{(m)},y^{(m)})\}$, the overall cost function of sparse auto-encoder can be defined by:

$$J_{sparse}(W,b) = \left[\frac{1}{m}\left(\frac{1}{2}\left\|h_{w,b}\left(x^{(i)}\right) - y^{(i)}\right\|^2\right)\right] + \frac{\lambda}{2}\sum_{l=1}^{n_l-1}\sum_{i=1}^{s_l}\sum_{j=1}^{s_{l+1}}\left(w_{ji}^{(l)}\right)^2 + \varphi\sum_{j=1}^{s_l}KL(\rho\,\|\,\hat{\rho}_j) \tag{9}$$

where $W_{ji}^{(l)}$ denotes the parameter of the connection between unit j in layer $l + 1$ and unit i in layer l; $b_i^{(l)}$ is the bias of the units i in layer $l + 1$; $h_{w,b}(x)$ is the network output; n_l is the number of layers and s_l is the number of units in layer l. λ is the weight decay parameter. φ controls the weight of the sparsity penalty term; ρ is a sparsity parameter and $\hat{\rho}_j$ is the average activation of hidden unit j, which can be calculated by:

$$\hat{\rho}_j = \frac{1}{m}\sum_{i=1}^{m}\left[a_j^{(2)}\left(x^{(i)}\right)\right] \tag{10}$$

where $a_j^{(2)}$ denotes the activation of unit j in the network. Sparsity penalty term is used to approximate $\hat{\rho}_j$ to ρ which is a small value close to zero. It is based on the concept of Kullback-Leibler divergence which measures the difference between two probability distributions.

Our goal is to minimize $J_{sparse}(W,b)$ as a function of W and b. For each training example (x,y), a feed-forward round is firstly performed to compute all activations. Then, for each node i in layer l, we compute an "error term" $\delta_i^{(l)}$ that measures how much that node was "responsible" for the errors in the output [9]. For each output unit i in layer n, the following equation can be given:

$$\delta_i^{(n_l)} = \frac{\partial}{\partial z_i^{(n_l)}}\frac{1}{2}\left\|y - h_{W,b}(x)\right\|^2 = -\left(y_i - a_i^{(n_l)}\right)\cdot f'\left(z_i^{(n_l)}\right) \tag{11}$$

where $z_i^{(l)}$ denotes the sum of inputs weights to unit i in layer l. $f(\cdot)$ is the activation function of each layer. The error term of each node i in hidden layer can be expressed after taking sparsity into account:

$$\delta_i^{(l)} = \left(\sum_{j=1}^{s_{l+1}} W_{ji}^{(l)} \delta_j^{(l+1)} + \varphi \left(-\frac{\rho}{\hat{\rho}_i} + \frac{1-\rho}{1-\hat{\rho}_i} \right) \right) \cdot f'\left(z_i^{(l)} \right) \tag{12}$$

Then, the desired partial derivatives for each layer can be calculated by:

$$\frac{\partial}{\partial W_{ij}^{(l)}} J_{sparse}(W, b; x, y) = a_j^{(l)} \delta_i^{(l+1)} + \lambda W_{ij}^{(l)} \tag{13}$$

$$\frac{\partial}{\partial b_i^{(l)}} J_{sparse}(W, b; x, y) = \delta_i^{(l+1)} \tag{14}$$

where the weight decay term is applied to W but not b, $J_{sparse} = (W, b; x, y)$ is the overall cost function with respect to a single training example (x, y), then by referring to Eq. (9), we can use gradient descent to update parameters W and b.

3 Proposed Approach

For our new proposed model (SSAE-ELM), stacked sparse auto-encoder (SSAE) is employed to initialize the model and learn useful features from input data. The stacking procedure of unsupervised training part is used to train SSAE. The hidden layer units' outputs of the first sparse auto-encoder are used as the inputs of the second one. Instead of adding a logistic regression on the top of the encoders and using fine-tuning strategy for all parameters at last in [8], we introduce the extreme learning machine (ELM) as the classifier. The inputs of the ELM are the extracted features from hidden layer units of the sparse auto-encoder and its outputs are the labels of the target pattern. The features extraction effectiveness of SSAE-ELM can be easily observed from the result of ELM. The training procedure of SSAE-ELM is illustrated in Fig. 1, where H_n denotes the n-th hidden layer units, W_n and $W_{n'}$ are the n-th sparse auto-encoder weights, x is the input, Rec is the reconstructed input data of this layer's sparse auto-encoder and y is the output of the proposed deep network.

From the Fig. 1, it can be seen that a sparse auto-encoder is firstly initialized. Then, an ELM is added on the encoder layer to train the whole network. After obtaining SSAE-ELM with one layer sparse auto-encoder, we start to train SSAE-ELM with two layers sparse auto-encoder with the same training strategy. Figure 1(c) and (d), we add another one layer sparse auto-encoder based on the previous training procedure. Then, ELM is added on the encoder layer of the second sparse auto-encoder. The second sparse auto-encoder and ELM are trained at the same time. But the weights of the first layer sparse auto-encoder keep fixed while we train SSAE-ELM for the second layer sparse auto-encoder. Then, to take the advantage of KELM's good generalization capability, SSAE-KELM model is constructed. The saved parameters learned from SSAE-ELM are used as the weights of SSAE to SSAE-KELM, which is illustrated by Fig. 2, where W_n is the n-th sparse auto-encoder weights learned by SSAE-ELM and y is the output target label.

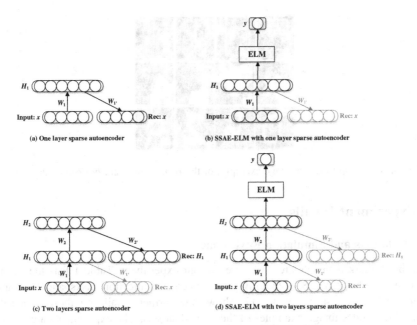

Fig. 1. The training procedures of SSAE-ELM

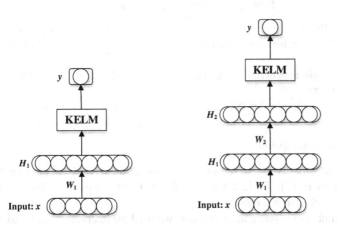

(a) SSAE-KELM with one layer sparse autoencoder   (b) SSAE-KELM with two layers sparse autoencoder

Fig. 2. The framework of SSAE-KELM

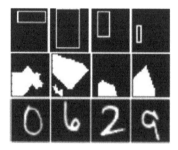

Fig. 3. From top to down image examples of the *rect*, *convex* and *basicmnist* data sets

4   Experiment Details

4.1   Data Sets and Simulation Environment

Three benchmark data sets are considered in our experiment. Table 1 lists the details about them. They are 28×28 gray-scale image classification tasks and the input dimensionality is 784. Reference [7] shows the further details on these data sets. Figure 3 illustrates image examples of the three data sets. The experiments are carried out in MATLAB 2013a environment running in Intel(R) Xeon(R) 3.07-GHz CPU with 70-GB RAM.

Table 1. Description of data sets

Name	Description	Classes	Train-Valid-Test
rect	Discriminate between white tall and wide rectangles	2	1000-200-50000
convex	Discriminate between convex and non-convex	2	6000-2000-50000
basicmnist	Smaller subset of MNIST	10	10000-2000-50000

4.2   Experiment Description

All the three datasets are divided into training set, validation set and testing set with the same standard used in [7]. Table 2 lists the SSAE-ELM parameters that would be chosen based on grid search method.

The epoch numbers in sparse auto-encoder, we tried 50 intervals from 50 to 1500. For each epoch sweeping through the training set, *rect* was randomly divided into 50 mini-batches, *convex* was randomly divided into 120 mini-batches and *basicmnist* was randomly divided into 200 mini-batches to train SSAE of the network using the batch gradient descent algorithm. To reduce the searching space, we keep the same number of hidden units, sparsity parameter, sparsity penalty term control weight, and learning rate for all hidden layers of SSAE. Sigmoid activation function is used for ELM to train SSAE-ELM model. The hidden layer unit numbers of ELM, we tried 500 intervals from 500 to 5000. For each specific experiment, we search the optimal parameter from

Table 2. Parameters for SSAE-ELM

Parameters	Description	Considered value
SSAE	Number of hidden layers	{1,2}
	Number of units in each hidden layer	{1000,2000,3000}
	Learning rate	{0.005,0.05,0.1}
	Number of epochs (passages through the whole training set)	{50,100,...,1500}
	Weight decay parameter	{0.0001,0.01}
	Sparsity parameter	{0.01,0.05}
	Weight of the sparsity penalty term	{2}
ELM	Number of hidden layers	{500,1000,...,5000}

the following sets: {0.001, 0.01, 0.1, 1, 5, 10, 100, 1000, 10000, 100000, 1000000} is used for the selection of regularization parameter and {0.0001, 0.001, 0.01, 0.1, 0.5, 1, 5, 10, 100, 1000, 10000} for kernel parameter. The RBF and linear kernel is used for KELM of SSAE-KELM. For each dataset, we also simulate the individual ELM and KELM for comparison. In our experiments, the test data set classification performance is reported with a 95% confidence interval.

4.3 Performance Evaluation

The comparison results are presented in Tables 3, 4, 5 and 6 based on the testing error rates (in percentages). The best results are in bold. RBF kernel and linear kernel based KELM are listed in Tables 4 and 5, respectively. Figures 4 and 5 show the sensitivity on the number of hidden layer nodes in ELM. Error bars show 95% confidence interval. In this paper, $SSAE_n$-ELM and $SSAE_n$-KELM denote SSAE-ELM and SSAE-KELM model with n layers SSAE, respectively. As shown in Table 3, Figs. 4 and 5, the results of SSAE-ELM improve gradually with the increase of the SSAE layer. Clearly, the performance of SSAE-ELM can benefit significantly from the features learned by SSAE. The testing error rates of SSAE-ELM decrease, respectively, by 52.68% and 45.97% on *convex* and *basicmnist* data set. For *rect* data set, SSAE-ELM's testing error rate can get the most significant reduction of 89.63% compared with other two cases.

From Tables 4 and 5, it can be concluded that SSAE-KELM outperforms KELM on each data set. It shows that using kernel KELM, the testing error rates decrease 66.1%, 22.06% and 18.80% on *rect*, *convex* and *basicmnist* data set, respectively. And the testing error of the models using linear kernel KELM decrease 96.97%, 38.70% and 82.18% on *rect*, *convex* and *basicmnist* data set, respectively.

Table 3. Comparison results of SSAE-ELM

Data sets	ELM	$SSAE_1$-ELM	$SSAE_2$-ELM
rect	17.36 ± 0.33	2.74 ± 0.14	**1.80 ± 0.12**
convex	36.05 ± 0.42	25.08 ± 0.38	**17.06 ± 0.33**
basicmnist	5.83 ± 0.21	3.87 ± 0.17	**3.15 ± 0.15**

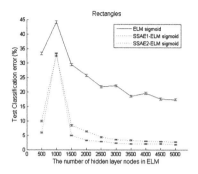

Fig. 4. Experiment results of *rect* data set

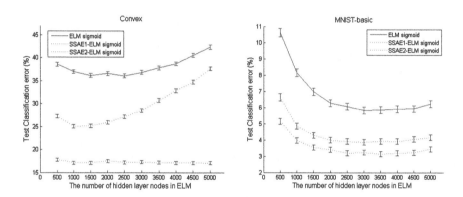

Fig. 5. Experiment results of *convex* and *basicmnist* data set

Table 4. Comparison results of SSAE-KELM (RBF kernel)

Data sets	KELM	SSAE₁-KELM	SSAE₂-KELM
rect	2.92 ± 0.15	**0.99 ± 0.09**	1.01 ± 0.09
convex	22.84 ± 0.37	18.76 ± 0.38	**17.78 ± 0.34**
basicmnist	2.81 ± 0.14	2.39 ± 0.13	**2.31 ± 0.13**

Table 5. Comparison results of SSAE-KELM (Linear kernel)

Data sets	KELM	SSAE₁-KELM	SSAE₂-KELM
rect	29.99 ± 0.40	1.06 ± 0.09	**0.91 ± 0.08**
convex	46.84 ± 0.44	38.54 ± 0.43	**28.71 ± 0.40**
basicmnist	16.61 ± 0.33	3.51 ± 0.16	**2.96 ± 0.14**

Table 6. Comparison results of different algorithms

Data sets	SVM	SAE	DBN	SDAE	SSAE-ELM	SSAE-KELM
rec	2.15 ± 0.13	2.41 ± 0.13	2.60 ± 0.14	1.99 ± 0.12	1.80 ± 0.12	**0.91** ± 0.09
con	19.13 ± 0.34	18.41 ± 0.34	18.63 ± 0.34	19.06 ± 0.34	**17.06** ± 0.33	17.80 ± 0.34
basicmnist	3.03 ± 0.15	3.46 ± 0.16	3.11 ± 0.15	2.48 ± 0.14	3.15 ± 0.15	**2.31** ± 0.13

In Table 6, we use the two proposed deep network algorithms in this paper to compare with Support Vector Machine (SVM), Stacked Auto-encoder (SAE), Deep Brief Network (DBN) and Stacked Denoise Auto-encoder (SDAE). As observed from Table 6, the experimental results show that the new proposed models' performance can not only be superior to classical ELM, KELM and SVM but also better than deep learning algorithms, such as SAE, DBN and SDAE.

5   Conclusions and Future Work

In this paper, we propose SSAE-ELM and SSAE-KELM, which are the combination of a deep network and ELM. The experiments on image recognition tasks show that they achieve competitive classification accuracies. While this paper shows the performance using only partial SSAE-ELM and SSAE-KELM parameter cases, we are going to try more different activation functions of ELM and more different kernel functions of KELM in our proposed model in the future.

Acknowledgement. The authors are grateful to the support of the National Natural Science Foundation of China (61602207, 61572228, 61272207, 61472158, 61373067, and 61373050), the National Key Basic Research Program of China (2015CB453000), and Science Technology Development Project from Jilin Province (20140520070JH).

References

1. Gai, K., Qiu, M., Zhao, H.: Cost-aware multimedia data allocation for heterogeneous memory using genetic algorithm in cloud computing. IEEE Trans. Cloud Comput. **99**, 1–2 (2016)
2. LeCun, Y., Bengio, Y., Hinton, G.: Deep learning. Nature **521**, 436–444 (2015)
3. Hinton, G.E., Osindero, S., Teh, Y.-W.: A fast learning algorithm for deep belief nets. Neural Comput. **18**, 1527–1554 (2006)
4. Hinton, G.E.: Reducing the dimensionality of data with neural networks. Science **313**, 504–507 (2006)
5. Bengio, Y., LeCun, Y.: Scaling learning algorithms towards AI. Presented at the MIT Press (2007)
6. Larochelle, H., Erhan, D., Courville, A., Bergstra, J., Bengio, Y.: An empirical evaluation of deep architectures on problems with many factors of variation. In: Proceedings of the 24th International Conference on Machine Learning, pp. 473–480. ACM, New York (2007)

7. Vincent, P., Larochelle, H., Lajoie, I., Bengio, Y., Manzagol, P.-A.: Stacked denoising autoencoders: learning useful representations in a deep network with a local denoising criterion. J. Mach. Learn. Res. **11**, 3371–3408 (2010)
8. UFLDL Tutorial. UFLDL. http://deeplearning.stanford.edu/wiki/index.php/UFLDL_Tutorial
9. Huang, G.-B., Zhu, Q.-Y., Siew, C.-K.: Extreme learning machine: theory and applications. Neurocomputing **70**, 489–501 (2006)
10. Huang, G.-B., Zhou, H.-M., Ding, X.-J., Zhang, R.: Extreme learning machine for regression and multiclass classification. IEEE Trans. Syst. Man Cybern. Part B Cybern. **42**, 513–529 (2012)
11. Huang, G.-B.: An insight into extreme learning machines: random neurons. Random Features Kernels Cogn. Comput. **6**, 376–390 (2014)

Bank Card and ID Card Number Recognition in Android Financial APP

Shubin Cai[1,2(✉)], Jinchun Wen[1,2], Honglong Xu[3,4], Siming Chen[1], and Zhong Ming[1,2,3]

[1] Shenzhen University, Shenzhen, Guangdong Province, China
{shubin,mingz}@szu.edu.cn, wenion@qq.com, 631941319@qq.com
[2] Engineering Research Center of Mobile Internet Application Middleware, Shenzhen, Guangdong Province, China
[3] Guangdong Province Key Laboratory of Popular High Performance Computers, Shenzhen, Guangdong Province, China
longer597@163.com
[4] School of Mathematics and Big Data, Foshan University, Foshan 528000, Guangdong, China

Abstract. In almost every financial management related Android application, users should input bank card and ID card number before transferring money between their financial accounts. In order to reduce user-input and improve user experience, a bank card and ID card number recognition method is proposed. The method consists of image preprocessing, numeral segmentation and numeral recognition. All the procedures are performed based on OpenCV and run on Android platform. Test results show that the correctness rate is 80% and its useful in practice.

Keywords: Numeral recognition · OpenCV · Convenient financial APP

1 Introduction

There's never a better time than now to start thinking about how you manage your money. Today, mobile banking is a service provided by a bank or other financial institution that allow us to conduct a range of financial transactions remotely using a mobile device. There are many stock market app provided us more convenient commercial business. Most financial app would request you to fill in your card number and ID number for a sequence of follow steps. However, because of the irregularity and much of numerals, some people may make a mistake in the first input and were required to input once more. The process of registering of a new account is not easy job. Here we want to use the camera of smartphone and image processing technology to simplify the process.

Currently, most approaches can segment and recognize the numeral [1]. Optical character recognition (OCR) [2] is one of the most successful applications of automatic pattern recognition in this field. Although the vast number of papers published on OCR every year, research on the application of numeral recognition to financial Apps is rare.

Despite little attention and the challenge involved, there are still some difficulties of this work. Below are a few points about the challenges:

© Springer International Publishing AG 2017
M. Qiu (Ed.): SmartCom 2016, LNCS 10135, pp. 205–213, 2017.
DOI: 10.1007/978-3-319-52015-5_21

1. There are two printed categories of card, one's numerals are printed and the other are embossed, we need to design a recognition algorithm suited for these two kinds, especially in pre-processing segmentation;
2. Due to the principles image formation of camera, the intensive and direction of light has an important influence on the image effect. Such as image pixel value of two characters after the reduction in character location failure and so on [3]; and
3. Bank card's background is a good way to show their feature, so they designed to attractive picture to the background. Some bank card background color and character color is very similar, they will also influence the segmentation and numeral recognition.

The focus on this paper is an approach practiced in financial and image processing to provide an efficient way. We propose a flexible numerals recognizer which will be applied to automatic numeral recognition of both card and ID card. The result shows that this method is a flexility of controlling the trade-off between accuracy and computational complexity.

2   The Overall Processing Procedure

ID card is designed to uniform background and explicit symbol of information while the bank cards are more complicated because of different bank. As an example, Fig. 1 is a pattern of credit card. It is noteworthy that the character form of id numeral is OCR-B, the embossed numeral of card is Farrington-7B-Qiqi and printed numeral is Arial [4].

Fig. 1. Credit card sample

The typically consists of the following processing steps and shown in Fig. 2.

1. Data collection. The images of card are collected by smartphone, which is data source. In this step, the images are required to take photographs parallel;
2. Image preprocessing. This step includes operations of image gray processing, binarization, tilt correction, image denoising, imaged dilation. This step is an important step in achieving good performance;
3. Numeral location, which can be used to enhance the recognition accuracy in accordance with some specified conditions;

4. Segmentation. Image segmentation is the task of finding groups of pixels that "go together". In this paper, the numeral image segmentation algorithm based on method of outline extraction is recommended [5]; and

5. Numeral recognition. Before this step, the individual images need to be normalization to keep its invariance. Because of different type of numeral and getting best performance, we take advantage of Artificial Neural Network to train dataset and recognition. Through ANNs model-based method [6], only the matching similarity up to 80% will be considered to this numeral.

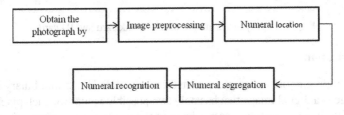

Fig. 2. The processing

3   Image Preprocessing

A card contains a lot of information, including background, name, number, validity and so on. How to extract the key is our concern. The results of image preprocessing are raw materials to follow steps.

3.1   Image Graying and Tilt Correction

With different weight of R, G, B three kinds of components, pictures show varieties of color. However, it is not practical to use RGB representation features. While color images can be treated as arbitrary vector-valued functions or collections of independent bands, it usually contains more information and needs to have a larger memory space [7]. Instead, the depth of the gray-scale image is only 8 bits, which will improve the working efficiency of the whole recognition process.

This step using the common method of image preprocessing, transforms the colorized image to gray and then makes it only two values by a critical point threshold. Figure 3 is the result handled by image graying.

Sometimes, the images already are a little skew, but people probably don not realize it. The skew image will affect the positioning of the late segmentation and recognition. So it is required to judge whether the image is tilted. Typically, we follow this process: Canny edge detection, Hough transformation, find the longest line, calculate tilt and rotation image according to the tilt.

Fig. 3. A gray-scale images each of bank card and ID card

3.2 Binarization

In order to highlight the text area, we transform grayscale image into binary image. A binary image is a digital image that has only two possible values for each pixel. It often arise in digital image processing as masks or as the result of certain operations such as segmentation, thresholding, and dithering.

We set a threshold, where the pixel value is 1 or 0 depends on that it greater than or less than threshold.

$$G(x,y) = \left\{ \begin{array}{ll} 0, & f(x,y) < threshold \\ 1, & f(x,y) \geq threshold \end{array} \right\} \tag{1}$$

Otsu's method [8], which named after its inventor Nobuyuki Otsu, is one of many binarization algorithms. Otsu's method involves iterating through all the possible threshold values and calculating a measure of spread for the pixel levels each side of the threshold, i.e. the pixels that either fall in foreground or background. The aim is to find the threshold value where the sum of foreground and background spreads is at its minimum.

Suppose we have N pixels in image, t is the threshold. There is different number of foreground N0 and background N1. The average gray value μ as follow:

$$\mu = \omega_0 \times \mu_0 + \omega_1 \times \mu_1 \tag{2}$$

The next step is to calculate the 'Within-Class Variance'. This is simply the sum of the two variances multiplied by their associated weights.

$$\sigma^2(k) = \omega_0 \times \left(\mu_0 - \mu \right)^2 + \omega_1 \times \left(\mu_1 - \mu \right)^2 \tag{3}$$

This same calculation needs to be performed for all the possible threshold t values 0 to N. To sum up, we choose the Otsu's method as approach of binarization step.

4   Numeral Location

Although we got a binary image and it looks like easy to search the text area, there are still noises. There is a way to improve the quality of image by denoising algorithm. In addition, we use inflation algorithm to make the characters connection. So we can more easy to locate the text area and then to segregation.

4.1   Image Denoising

The most commonly used type of neighborhood operator is a linear filter, in which an output pixel's value is determined as a weighted sum of input pixel values. A better filter to use in this case is the median filter [9], which selects the median value from each pixel's neighborhood. Since the shot noise value usually lies well outside the true values in the neighborhood, the median filter is able to filter away such bad pixels.

4.2   Image Dilation

After the binarization image processed by smoothing operation, the numeral is not continuous, but independent. It is difficulties that locate the text area position by methods of finding outline directly. The numeral should be dilated so that they link together for segregation.

Figure 4 shows that dilation repeats 10 times for ID card and 13 times for bank card make best result.

Fig. 4. Images from binarization, denoising and dilation

4.3   Numeral Area Location

One primary reason why dilate image is that locate the candidate numeral area. We can use function findContours() to obtain its enclosed outline position's information and then cut it according to its position and size. For more precisely, the outline must be:

- The width of area is bigger than 70% of image's;
- The ratio of width to height is more than 10; and
- The y of area is between 40% to 80% of image (bank card) and upon 70% of image (ID card).

5 Numeral Segmentation

The numeral should be divided into individuals before the recognition process. There are a lot of methods to segregate the image [10]. These involve a lot of related algorithms which are based on projection, active contours and level sets, normalized cuts.

5.1 Improved Active Contours Method

In order to avoid losing the pixel detail and incomplete image, we need to dilate the image twice with the kernel 2×1. After that we should segregate the numeral by method of outline extraction [11]. If there is more than one numeral in separate image, we should determine whether it is meet the condition and do once again until each separate image have single numeral.

The first segmentation

We invoke function findContours() to segmentation with condition that the area's width is longer than 10% and the height of area is larger than original's 80%. However, some numerals still have adhesion together. For more precisely, the unqualified images need to second segmentation.

The second segmentation

Our ID number is 18 digits while the bank card number is uniform, likely from 16 to 19 digits. There is reasonable assume to 20 digits. With this condition, if the length of images divided by d is greater than 2, it should be segregated once again and as follow:

$$length/d > 2\sigma^2(k) = \omega_0 \times \left(\mu_0 - \mu\right)^2 + \omega_1 \times \left(\mu_1 - \mu\right)^2 \tag{4}$$

The steps are as follow:

- use the length of individuals divide by standard width and the result will be rounding;
- segregate the image averagely according to the precious result; and
- locate the position for the new coordinate of image.

The result of improved method was showed in Fig. 5.

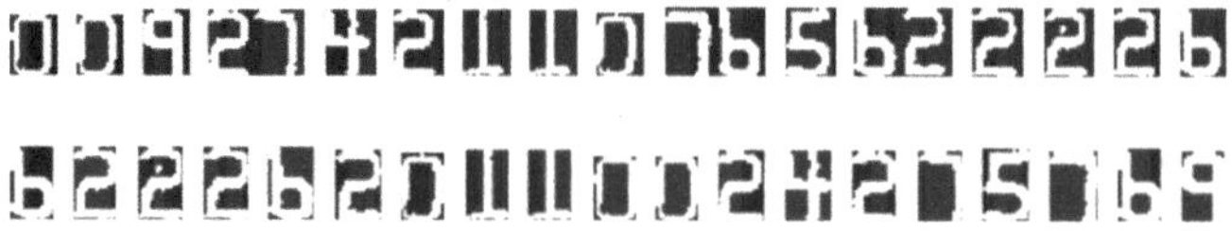

Fig. 5. The result of improved active contours method

6   Numeral Recognition

Numeral recognition is the last step of the process and very crucial. In this part, there are series of process, which consisted of normalization, recognition. The method of data recognition is based on ANN, which includes feature detection, obtaining datasets, training dataset and numeral recognition [12].

6.1   Image Normalization

The numeral images need to be normalization before recognition because of the uniform style will be benefit to recognition. The normalization may change the size of images, the blank pixel should be filled by interpolation. In this paper we invoke bilinear interpolation function to calculate the blank pixel value [13].

The bilinear interpolation is an extension of linear interpolation for interpolating functions of two variables on a rectilinear 2D grid. The formula is as follow:

$$f(i + u, j + v) = (1 - u)(1 - v)f(i,j) + (1 - u)vf(i, j + 1) + u(1 - v)f(i + 1, j) + uvf(i + 1, j + 1) \qquad (5)$$

where i and j are the Integer part of floating point coordinate, u and v are Decimal part of floating point coordinate, f is the value of this pixel. There is a function resize() provided by OpenCV.

6.2   Features Extracted

For each numeral, the pixel of it is different features. A common method utilizes the histogram and low resolution sampling image from vertical and horizontal axes of image as numeral features [14].

During the normalization processing, each image are set to 28×41 pixels. The histogram data comes from the number of nonzero pixel value from each column or each row. On the other hand, we use low resolution image to sample in order to improve the result.

6.3   Data Training

The network has 11 output nodes, one for each of the ten digit classes "0"–"9" and letter "x" that the network tries to recognize. Three intermediate layers were used. Each node in a layer has connections from a number of nodes in the previous layer and during the training phase, connection weights are learned [15]. The output at a node is a function of the weighted sum of the connected nodes at the previous layer. One can think of a feedforward neural network as constructing decision boundaries in a feature space and as the number of layers and nodes increases, the flexibility of the classifier increases by allowing more and more complex decision boundaries. However, it has been shown that this flexibility must be restricted to obtain good recognition performance.

Based on completed histogram and feature matrix of low resolution, we invokes the function predict() to make a classification of feature matrix, the best of similarity is the predicted numeral.

7   Result

This paper has shown the result of numerals (Table 1). Here we collected 10 samples of embossed bank card, 10 samples of printing bank card and 20 ID card. The initial procedure of pre-processing and feature extraction is done. This is used for classification and recognition of numerals.

Table 1. The result of the improved active contours method

Class	Samples	Average running time	Recognition correctly	Recog rate
Bank card	20	3.21 s	14	70%
ID	20	3.72 s	18	90%
Total	40	3.465 s	32	80%

From the table, because of complex background, preprocessing stage is not able to work well. In the other hand, the light is another factor, the reflective would influence the photo.

8   Conclusion

From our work we presents that the method used for the pre-processing as well as segmentation of numeral characters are more correct. Compared to the other method, the ANN allows more flexible application. In future we are going to improve robustness of algorithm to adapt more bank card with complex background and speed of operation.

Acknowledgements. This work is supported by the National Natural Science Foundation of China under Grants NSFC 61672358.

References

1. Oliveira, L.S., Sabourin, R., Bortolozzi, F., Suen, C.Y.: Automatic recognition of handwritten numerical strings: a recognition and verification strategy. IEEE Trans. Pattern Anal. Mach. Intell. **24**(11), 1438–1454 (2002)
2. Trier, Ø.D., Jain, A.K., Taxt, T.: Feature extraction methods for character recognition-a survey. Pattern Recogn. **29**(4), 641–662 (1996)
3. Jain, A.K., Topchy, A., Law, M.H.C., Buhmann, J.M.: Landscape of clustering algorithms. vol. 1, pp. 260–263 (2004)
4. Pujol, O., Escalera, S., Radeva, P.: An incremental node embedding technique for error correcting output codes. Pattern Recogn. **41**(2), 713–725 (2008)

5. Kim, K.K., Suen, C.Y., Jin, H.K.: Recognition of unconstrained handwritten numeral strings by composite segmentation method. In: International Conference on Pattern Recognition, Proceedings, vol. 2, p. 2594 (2000)
6. Zhang, G., Patuwo, B.E., Hu, M.Y.: Forecasting with artificial neural networks: the state of the art. Int. J. Forecast. **14**(1), 35–62 (1998)
7. Hsu, R.L., Abdelmottaleb, M., Jain, A.K.: Face detection in color images. IEEE Trans. Pattern Anal. Mach. Intell. **1**(5), 696–706 (2008)
8. Ohtsu, N.: A threshold selection method from gray-level histograms. IEEE Trans. Syst. Man Cybern. **9**(1), 62–66 (1979)
9. Wang, Z., Zhang, D.: Progressive switching median filter for the removal of impulse noise from highly corrupted images. IEEE Trans. Circ. Syst. II Analog Digit. Sig. Process. **46**(1), 78–80 (1999)
10. Seo, W., Cho, B.: Efficient segmentation path generation for unconstrained handwritten hangul character. In: Bussler, C., Fensel, D. (eds.) AIMSA 2004. LNCS (LNAI), vol. 3192, pp. 438–446. Springer, Heidelberg (2004). doi:10.1007/978-3-540-30106-6_45
11. Li, N., Gao, X., Jin, L.: Curved segmentation path generation for unconstrained handwritten Chinese text lines. In: IEEE Asia Pacific Conference on Circuits and Systems, APCCAS 2008, pp. 501–505 (2008)
12. Armano, G., Chira, C., Hatami, N.: Ensemble of binary learners for reliable text categorization with a reject option. In: Corchado, E., Snášel, V., Abraham, A., Woźniak, M., Graña, M., Cho, S.-B. (eds.) HAIS 2012. LNCS (LNAI), vol. 7208, pp. 137–146. Springer, Heidelberg (2012). doi:10.1007/978-3-642-28942-2_13
13. Hussain, F., Cowell, J.: Character recognition of arabic and latin scripts, pp. 51–56 (2000)
14. Naz, S., Hayat, K., Razzak, M.I., Anwar, M.W., Akbar, H.: Arabic script based language character recognition: Nasta'liq vs Naskh analysis. In: Computer and Information Technology, pp. 1–7 (2013)
15. Lin, Y., Lv, F., Zhu, S., Yang, M., Cour, T., Yu, K., et al.: Large-scale image classification: fast feature extraction and SVM training. In: IEEE Computer Society Conference on Computer Vision & Pattern Recognition, IEEE Computer Society Conference on Cvpr, vol. 1, pp. 1689–1696 (2011)

Mining Association Rules from a Dynamic Probabilistic Numerical Dataset Using Estimated-Frequent Uncertain-Itemsets

Bin Pei$^{(\boxtimes)}$, Fenmei Wang, and Xiuzhen Wang

Computer Research and Teaching Section,
New Star Research Institute of Applied Technology, Hefei, China
pei_ice@aliyun.com, wangfenmei205@126.com,
18149394@qq.com

Abstract. In recent years, many new applications, such as location-based services, sensor monitoring systems, and data integration, have shown a growing amount of importance of uncertain data mining. In addition, due to instrument errors, imprecise of sensor monitoring systems, and so on, real-world data tend to be numerical data with inherent uncertainty. Thus, mining association rules from an uncertain, especially probabilistic numerical dataset has been studied recently. However, a probabilistic numerical dataset often grows as new data append. Thus, developing a mining algorithm that can incrementally maintain discovered information is quite important. In this paper, we have designed an efficient, incremental mining algorithm to mine association rules from a probabilistic numeric dataset using estimated-frequent uncertain-itemsets. By using a user-specified support threshold, estimated-frequent uncertain-itemsets could act as a gap to avoid small itemsets becoming large in the updated dataset when new transactions are inserted. As a result, the algorithm has execution time faster than that of previous methods. An illustrated example is given to demonstrate the procedures of the algorithm.

Keywords: Association rule · Dynamic probabilistic numerical dataset · Estimated-frequent uncertain-itemsets · Data mining · Uncertain data

1 Introduction

Association rule mining (ARM) is a popular and well-researched method for discovering interesting relations between attribute values of objects (data records) especially in transaction datasets. Thus, it becomes one of the most important research fields in data mining. Traditionally, ARM is applied to data that are certain and precise. Existing ARM algorithms, such as the well-known Apriori algorithm [1] and other variants, were designed for mining "certain" data.

However, uncertainty is prevalent in many application domains, since imprecision usually exists in collecting, transmitting, storing and understanding data. Several algorithms of discovering association rules from uncertain, especially probabilistic database have been proposed by several authors [2–5]. This problem has recently become a hot topic in database and data mining domains [6–8].

© Springer International Publishing AG 2017
M. Qiu (Ed.): SmartCom 2016, LNCS 10135, pp. 214–223, 2017.
DOI: 10.1007/978-3-319-52015-5_22

As described in [5], all above algorithms have been focused on handling proba-bilistic database with asymmetric binary attribute values. These kinds of algorithms can be labeled as Boolean Association Rules from Probabilistic Database. Yet, this is not always the case. In lots of real-world applications, the existence of numerical attribute values in a transaction is best captured by a likelihood measure or a probability. Considering the following example.

Example. Table 1 shows a probabilistic dataset with values of attributes *coal con-sumption, transport density* and *SO_2 concentration* in the air of a city reported by sensor readings. In fact, the sensor readings are inherently uncertain due to the factory accuracy, the environmental erosion, its reduced power consumption, et al. To model this phenomenon, we put these readings in a dataset under the existential uncertain data model, in which each numerical attribute value is associated with an existential probability that indicates the likelihood of its presence in a transaction. For instance, the value of attribute *Coal Consumption* in transaction 1 is represented as "(0.6) 0.8". This value describes that the sensor reading value is 0.6, and the probability that the real value in that place at that time is true is 0.8. Here, the true value is unknown because of the uncertainty of the sensor which measures and collects data in an imprecise way.

Table 1. A sample probabilistic numerical dataset

TID	Coal consumption	Transport density	SO_2 concentration
1	(0.6)0.8	(3)0.7	(0.02)0.8
2	(0.7)0.9	(15)0.8	(0.06)0.7
3	(2.5)0.6	(9)0.3	(0.09)0.7
4	(1.2)0.6	(10)0.6	(0.02)0.8
5	(2.1)0.7	(40)0.8	(0.11)0.7
6	(1.7)0.8	(10)0.3	(0.03)0.7

The problem of mining fuzzy association rules from a probabilistic numerical dataset (PND) was first studied in [9]. By transforming the original PND to a proba-bilistic dataset with fuzzy sets, reference [9] introduces new definitions of support and confidence suitable for a dataset with both fuzziness and randomness. Based on these new measures, the paper then develops an Apriori-based algorithm, called FARP, to mine fuzzy association rules (FARs) from a PND.

In [9], the rules discovered from a PND only reflect the current state of the data-base. In fact, a PND is frequently updated than static. When new transactions are appended frequently, association rules discovered in the previous PND possibly no longer valid and interesting rules in the updated dataset. One possible approach is to re-run FARP algorithm again in order to find FARs in the total updated PND. This method has obvious disadvantages. All the information of old frequent FARs given by initial running FARP are wasted, and all frequent itemsets have to be computed from the very beginning. This leads to the inefficiency of the processing time.

In this paper, we propose a new incremental association rule discovery algorithm, called FARP_D (Fuzzy Association Rules mining from a Dynamic PND), for the dynamic PND. The algorithm is capable of dynamically discovering new association rules when a number of new records have been added to a PND. In our approach, we propose the notion of estimated-frequent itemsets that are capable of being frequent itemsets after a number of new records have been added to a PND. Our algorithm could reduce the number of times to scan the original database. As a result, the algorithm has execution time faster than that of previous methods.

2 Preliminaries

In this section, we give the PND model used in this paper. We then briefly introduce the framework of algorithm FARP described in [9], which is used to mine fuzzy association rules from a static PND.

The attribute is called a probabilistic numerical attribute because the value type of an attribute is numerical. A *probabilistic numerical item (p-item)* is defined as (v, p), where v is the numerical value of item and p is probability associated with this value $(0 \leq p \leq 1)$. A *probabilistic numerical dataset (PND)* in a collection of p-items $D = \{t_1, \ldots, t_l, \ldots, t_n\}$, where $t_l = \{(v_{11}, p_{11}), \ldots, (v_{jl}, p_{jl}), \ldots, (v_{ml}, p_{ml})\}$, and (v_{jl}, p_{jl}) $(1 \leq j \leq m, 1 \leq l \leq n)$ is a p-item.

An *uncertain item (u-item)* is denoted as $(f/F)p$, where *p* indicates the probability of fuzzy event (f/F), and *f* represents degree of fuzzy set F. So a u-item aggregates fuzzy uncertainty with random uncertainty.

Another kind of uncertain item is *rule uncertain item (r-item)* defined as $(I_j : F_{jk})(1 \leq j \leq m, 1 \leq k \leq K)$, where F_{jk} is one of the fuzzy sets related to attribute I_j. A *rule uncertain itemset (r-itemset)* is denoted as $UI = \{(I_j : F_{jk})\}(1 \leq j \leq m, 1 \leq k \leq K)$, where $(I_j : F_{jk})$ is a r-item.

Problem Statement. Given an original PND DB with its frequent r-itemsets and fuzzy association rules discovered already, and an incremental PND db, which is newly added to DB, find all FARs having support $\geq$ minsup and confidence $\geq$ minconf in DB+=DB$\cup$db. This problem is called *mining FARs from a dynamic PND*.

PARP, a modified version of the Apriori algorithm, was presented in [9] as a baseline algorithm to solve the problem of mining association rules from a static PND. FARP first converts a PND to a probabilistic database with fuzzy sets (i.e., mapping numerical data to its corresponding fuzzy sets). Next, new support and confidence measures for the probabilistic database with fuzzy sets are introduced. FARP then gives an Apriori-like algorithm to mine FARs from a PND. One can refer to reference [9] for more details (Table 2).

Table 2. Notations used in the paper

Symbol	Definition and description
$|D|$	The number of transactions in dataset D
L_k^{DB}	The set of FRIs from DB
L_k^{db}	The set of FRIs from db
L_k^{DB+}	The set of FRIs from DB+
E_k^{DB}	The set of ERIs from DB
E_k^{db}	The set of ERIs from DB
E_k^{DB+}	The set of ERIs from DB+

3 Maintaining Frequent and Estimated-Frequent r-Itemsets

According to Tasi et al. [10], an itemset X is categorized into four cases (shown in Table 3).

Table 3. Four cases of a r-itemset

	Frequent r-itemset	Infrequent r-itemset
Frequent r-itemset	Case 1	Case 2
Infrequent r-itemset	Case 3	Case 4

Case 1, 2: The support of r-itemsets in case 1 and 2 in DB+ can be easily computed, because we just need to scan db to find $Sup^{db}(X)$, while $Sup^{DB}(X)$ is already computed initially. So for r-itemset X in case 1 and 2, $Sup^{DB+}(X) = Sup^{db}(X) + Sup^{DB}(X)$.

Case 3: The support of r-itemset X in case 3 is the hardest work because it needs to re-execute FARM_S algorithm and scan DB $\cup$ db to get $Sup^{DB+}(X)$.

Case 4: The r-itemset X in case 4 cannot become frequent r-itemset in DB+ and generate new FARs, because X is infrequent in both DB and db.

Therefore, how to efficiently discover the support of r-itemsets in case 3 is the most important task in the incremental algorithm.

In order to make full use of the old information and avoid scanning original dataset, we hope less r-itemsets in case 3 are to be processed. Thus, we propose a support for estimated-frequent r-itemsets and maintain estimated-frequent r-itemsets, which have promise to be frequent when an updated probabilistic dataset are inserted into original PND. Obviously, the more estimated-frequent r-itemsets stored in the original dataset processing procedure, the less time we will re-scan the original dataset, thus greatly reducing the cost time spent on mining FARs from the total dataset. In this way, discovering FARs in the total dataset is efficient and costs less processing time because it can use the information from frequent and estimated-frequent r-itemsets in the original dataset.

Given a lower support threshold σ_m ($\sigma_m < \sigma_s$), a r-itemset UI is said to be estimated-frequent if $\sigma_m \leq \mathrm{Sup}^{DB}(UI) < \sigma_s$. We call UI as *estimated-frequent r-itemset* (ERI).

A lower support threshold σ_m for ERI can be defined as $\sigma_m = (1+c)\sigma_s - c$, where σ_s is the support threshold for frequent r-itemset (FRI), and $c = |db|/|DB|$. The support threshold for ERI defied here grantees support degree of r-item below σ_m cannot become ERI or FRI.

Theorem: Let UI be a r-itemset such that $\mathrm{Sup}^{DB}(UI) \leq (1+c)\sigma_s - c$, then UI $\notin L^{DB+}$, where σ_s is the support threshold for FRI, and $c = |db|/|DB|$.

Proof:

$$\mathrm{Sup}^{DB+}(UI) = \frac{\mathrm{Sup}^{DB}(UI) * |DB| + \mathrm{Sup}^{db}(UI) * |db|}{|db| + |DB|}$$

$$\leq \frac{\left(\left(1 + \frac{|db|}{|DB|}\right)\sigma_s - \frac{|db|}{|DB|}\right) * |DB| + 1 * |db|}{|db| + |DB|}$$

$$= \frac{|DB| * \sigma_s + |db| * \sigma_s - |db| + |db|}{|db| + |DB|} = \sigma_s.$$

This theorem proves that any infrequent r-itemset in the original dataset will not become frequent in the updated dataset. Also, the theorem shows that FRIs and ERIs probably become FRIs in the updated dataset. Because these two kinds of r-itemsets were computed and stored in the initial mining algorithm, the incremental algorithm uses the existing information and avoids the additional re-scanning the original dataset, which will reduce the processing time of the total updated dataset.

Lemma: Let UI be a r-itemset such that UI $\notin (L^{DB} \cup E^{DB})$, then UI $\in L^{DB+} \cup E^{DB+}$ only if UI $\in (L^{db} \cup E^{db})$.

Proof: Since UI is not in $L^{DB} \cup E^{DB}$, $\mathrm{Sup}^{DB}(UI) \leq \sigma_m * |DB|$. If $\mathrm{Sup}^{db}(UI) \leq \sigma_m * |db|$, then $\mathrm{Sup}^{DB+}(UI) \leq \sigma_m * |DB| + \sigma_m * |db| = \sigma_m * (|DB| + |db|)$. That is, UI cannot be FRI or ERI. Thus, we have the lemma.

4   Mining FARs from a Dynamic PND

In this section, we design FARP_D algorithm to discover interesting FARs from a dynamic PND.

4.1   FARP_D Algorithm

Similar to FARP in [9], the main algorithm of FARP_D consists of three sub-functions: Dataset Mapping, FRIs and ERIs Generation, and FARs Generation. The first and third

---

Function: *Update_FreqRuleItemsets-gen.* Discover estimated-frequent r-itemsets and frequent r-itemsets in the updated dataset DB+.

Input: DB; db; membership fuctions FS; σ_s; σ_m; L^{DB}; E^{DB}.

Output: L^{DB+}, E^{DB+}.

Method:

Scan db from D^T transformed from db using function Dataset-Transform Function

{ find L^{db} and E^{db};

//get all supports of those r-itemsets in $L^{DB} \cup E^{DB}$ in incremental dataset db

for each r-itemset UI $\in (L^{DB} \cup E^{DB})$

      compute $Sup^{db}(UI)$;

}

k=1;

repeat {

// get all supports of those r-itemsets in $(L_k^{DB} \cup E_k^{DB}) \cap C_k^{DB+}$ in updated dataset DB+

for each r-itemset UI $\in (L_k^{DB} \cup E_k^{DB}) \cap C_k^{DB+}$

{

$Sup^{DB+}(UI) = (Sup^{DB}(UI)*|DB| + Sup^{db}(UI)*|db|)/(|DB|+|db|)$;

    if $Sup^{DB+}(UI) \geqslant \sigma_s$ then

        $L_k^{DB+} = L_k^{DB+} \cup UI$;

    if $\sigma_s > Sup^{DB+}(UI) \geqslant \sigma_m$ then

        $E_k^{DB+} = E_k^{DB+} \cup UI$;

}

// get all supports of those r-itemsets that are not in $(L_k^{DB} \cup E_k^{DB}) \cap C_k^{DB+}$ but in $(L_k^{db} \cup E_k^{db}) \cap C_k^{DB+}$ in updated dataset DB+

for each r-itemset UI $\notin L_k^{DB} \cup E_k^{DB}$

    for each r-itemset UI $\in (L_k^{db} \cup E_k^{db}) \cap C_k^{DB+}$

rescan DB to compute $Sup^{DB}(UI)$;

        {

$Sup^{DB+}(UI) = (Sup^{DB}(UI)*|DB| + Sup^{db}(UI)*|db|)/(|DB|+|db|)$;

        if $Sup^{DB+}(UI) \geqslant \sigma_s$ then

            $L_k^{DB+} = L_k^{DB+} \cup UI$;

        if $\sigma_s > Sup^{DB+}(UI) \geqslant \sigma_m$ then

            $E_k^{DB+} = E_k^{DB+} \cup UI$;

}

// generate C_{k+1}^{DB+} from all frequent and estimated-frequent k r-itemsets $(L_k^{DB+} \cup E_k^{DB+})$ using σ_m

    $C_{k+1}^{DB+} = $ apriori-gen$(L_k^{DB+} \cup E_k^{DB+}, \sigma_m)$;

    k=k+1;

}**until** $(E_k^{DB+}=\emptyset$ and $L_k^{DB+}=\emptyset)$;

return$L^{DB+} = \cup_k L_k^{DB+}$, $E^{DB+} = \cup_k E_k^{DB+}$;

Fig. 1. Update_FreqRuleItemsets-gen function

procedures are similar to corresponding steps in FARP. Figure 1 shows FRIs and ERIs Generation procedure in detail.

First, FRIs and ERIs are computed in incremental dataset db using FARP-like algorithm. Of course, in order to get ERIs in db, FARP should be modified a bit to compute and store ERIs and their supports values in addition to FRIs. Simultaneously, we also count the support values for those r-itemsets in $L^{DB} \bigcup E^{DB}$ in db.

For a $UI \in L^{DB} \bigcup E^{DB}$, we can easily get its support count in DB+, because we get the support count of UI in db in the previous step, and we have the support count of UI in DB. Then if the support of UI is greater than minimum support in DB+, it is added to L^{DB+}; similarly, if the support of UI is greater than lower support threshold for ERIs in DB+, it is then added to E^{DB+}. On the other hand, there could be some new r-itemsets which become FRIs and ERIs in the updated dataset DB+. Let t be a r-itemset that gets added to FRIs and ERIs in DB+. By lemma, we know that t has to be $L^{db} \cup E^{db}$. In order to get support count of t in DB+, we should know the support of t in DB and db. Therefore, the original dataset DB should be re-scanned to get its support in DB. The same procedure is repeated until no FRIs and ERIs are found.

Based on the above discussions, we give Update_FreqRuleItemsets-gen procedure (see Fig. 1) to discover FRIs in the updated dataset efficiently.

4.2   Algorithm Analysis

From the algorithm aforementioned, we find ERIs can act as a pool to avoid infrequent r-itemsets becoming frequent in the updated dataset when the incremental dataset is added to the original dataset. For all r-itemsets in $L^{DB} \bigcup E^{DB}$, we just scan db and get FRIs in the updated dataset DB+. For those r-itemsets that not in $L^{DB} \bigcup E^{DB}$ but become frequent in DB+, we grarantee that these r-itemsets must be in $L^{db} \cup E^{db}$, and scan them on DB. By using ERIs that are expected to become frequent when new transactions are inserted to the original dataset, less r-itemsets are scanned on DB, thus avoiding re-scanning the original dataset only when required and reducing the processing time dramatically.

5   An Illustrated Example

An illustrated example is given to show how FARP_D algorithm is used to generate FRIs when the incremental dataset is added to the original dataset. When all FRIs in DB+ have been discovered, fuzzy association rules are easily mined using the procedures described in [9]. Assume the original PND DB is shown in Table 1. Suppose an incremental PND db containing one transaction shown in Table 4 is inserted after the original PND is processed.

Table 4. An incremental PND *db*

TID	Coal consumption	Transport density	SO$_2$ concentration
7	(0.6)0.7	(10)0.6	(0.12)0.7

According to the definition of σ_m, we get $\sigma_m = (1+c)\sigma_s - c = \left(1+\frac{1}{6}\right)*0.25 - \frac{1}{6} = 0.125$. The proposed FARP_D algorithm proceeds as follows.

Step 1: Finding all FRIs and ERIs in db. The method of finding FRIs and ERIs is similar to step 1 and 2 of FARP algorithm using support values σ_s and σ_m. Tables 5, 6, 7, 8, 9 show the procedures and results.

Table 5. Transformed dataset after mapping

TID	Coal consumption	Transport density	SO₂ concentration
7	(0.33/EL)0.7 + (0.67/L)0.7	(1/EL)0.6	(1/EH)0.7

Table 6. The temporary dataset with SL value of each u-item

TID	Coal consumption	Traffic density	SO2 concentration
7	(EL, 0.271) + (L, 0.378)	(EL, 0.654)	(EH, 0.696)

Table 7. L_1^{db} itemsets and their supports in db

L_1^{db}	(CC : EL)	(CC : L)	(TD : EL)	(SC : EH)
Support in db	0.271	0.378	0.654	0.696

E_1 itemsets and their supports in db: null

Table 8. L_2^{db} itemsets and their supports in db

L_2^{db}	(CC : EL) (TD : EL)	(CC : EL) (SC : EH)	(CC : L) (TD : EL)	(CC : L) (SC : EH)	(TD : EL) (SC : EH)
Support in db	0.271	0.271	0.378	0.378	0.654

E_2 itemsets and their supports in db: null

Table 9. L_3^{db} itemsets and their supports in db

L_3^{db}	(CC : EL) (TD : EL) (SC : EH)	(CC : L) (TD : EL) (SC : EH)
Support in db	0.271	0.378

Table 10. L_1^{DB} itemsets and their supports (≥ 0.25) in DB+

L_1^{DB}	(CC : EH)	(TD : EL)	(SC : EL)
Support in DB+	0.237	0.561	0.262

Step 2: Getting all supports of those r-itemsets in $(L_k^{DB} \cup E_k^{DB}) \cap C_k^{DB+}$ *in updated dataset DB+.* Since all support degree of r-itemsets in $L_k^{DB} \cup E_k^{DB}$ have already been computed and stored when DB has been processed, we just need to scan db once to get support degree of these r-itemsets in db. Finally, we get support degree of these r-itemsets. Tables 10, 11, 12, 13 show the procedure and results.

Table 11. E_1^{DB} itemsets and their supports (≥ 0.125) in DB+

E_1^{DB}	(CC : L)	(SC : EH)
Support in DB+	0.237	0.245

Table 12. L_2^{DB} itemsets and their supports in DB+

L_2^{DB}	(TD : EL, SC : EL)
Support in DB+	0.231

Table 13. E_2^{DB} itemsets and their supports in DB+

E_2^{DB}	(CC : EH, SC : EH)	(CC : L, TD : EL)
Support in DB+	0.146	0.171

Table 14. Itemsetsthat are in L_1^{db} but not in L_1^{DB} and their supports in db

L_1^{db}	(CC : L)
Support in db	0.237

Itemsets that are in E_1^{db} but not in E_1^{DB} and their supports in db: null

Table 15. Itemsetsthat are in L_2^{db} but not in L_2^{DB} and their supports in db

L_2^{db}	(CC : EL) (TD : EL)	(CC : EL) (SC : EH)	(CC : L) (TD : EL)	(CC : L) (SC : EH)	(TD : EL) (SC : EH)
Support in db	0.086	0.039	0.171	0.054	0.131

Itemsets that are in E_2^{db} but not in E_2^{DB} and their supports in db: null

Table 16. Itemsetsthat are in L_3^{db} but not in L_3^{DB} and their supports in db

L_3^{db}	(CC : EL) (TD : EL) (SC : EH)	(CC : L) (TD : EL) (SC : EH)
Support in db	0.039	0.054

Table 17. All frequent r-itemsets in DB+

L_1^{DB+}	(TD : EL)	(SC : EL)
Support in DB+	0.561	0.262

Step 3:Getting all supports of those r-itemsets that are not in $\left(L_k^{DB} \cup E_k^{DB}\right) \cap$ C_k^{DB+} *but in* $\left(L_k^{db} \cup E_k^{db}\right) \cap C_k^{DB+}$ *in updated dataset DB+.* For those r-itemsets that are not in $\left(L_k^{DB} \cup E_k^{DB}\right) \cap C_k^{DB+}$ but in $\left(L_k^{db} \cup E_k^{db}\right) \cap C_k^{DB+}$, we need to re-scan DB to get support degree of these r-itemsets in DB. Finally, we get support degree of these r-itemsets. Tables 14, 15, 16 and 17 show the procedures and results.

6 Conclusion

In this paper, we have proposed the concept of estimated frequent itemsets, and designed an efficient, incremental mining algorithm, called FARP_D, to mine association rules from a dynamic probabilistic numeric dataset. Using a user-specified support threshold, estimated-frequent r-itemsets could act as a gap to avoid small itemsets becoming large in the updated dataset when transactions are inserted. In the future, further researches and experiments on the proposed algorithm will be presented.

Acknowledgement. This research is supported by Anhui Provincial Natural Science Foundation (1408085MF117).

References

1. Agrawal, R., Srikant, R.: Fast algorithms for mining association rules. In: Proceedings of the 20th Conference on VLDB, Santiago, Chile, pp. 487–499 (1994)
2. Aggarwal, C., Li, Y., Wang, J., Wang, J.: Frequent pattern mining with uncertain data. In: KDD (2009)
3. Chui, C.K., Kao, B., Hung, E.: Mining frequent itemsets from uncertain data. In: PAKDD (2007)
4. Zhang, Q., Li, F., Yi, K.: Finding frequent items inprobabilistic data. In: SIGMOD (2008)
5. Sun, L., Cheng, R., Cheung, D.W., Cheng, J.: Mining uncertain data with probabilistic guarantees. In: KDD (2010)
6. Carvalho, J.V., Ruiz, D.D.: Discovering frequent itemsets on uncertain data: a systematic review. In: Perner, P. (ed.) MLDM 2013. LNCS (LNAI), vol. 7988, pp. 390–404. Springer, Heidelberg (2013). doi:10.1007/978-3-642-39712-7_30
7. Wang, Y., Li, X., Li, X., et al.: A survey of queries over uncertain data. Knowl. Inf. Syst. **37**(3), 485–530 (2013)
8. Aggarwal, C.C., Philip, S.Y.: A survey of uncertain data algorithms and applications. IEEE Trans. Knowl. Data Eng. **21**(5), 609–623 (2009)
9. Pei, B., Zhao, S., Chen, H., et al.: FARP: mining fuzzy association rules from a probabilistic quantitative database. Inf. Sci. **237**, 242–260 (2013)
10. Tsai, P.S.M., Lee, C.-C., Chen, A.L.P.: An efficient approach for incremental association rule mining. In: Zhong, N., Zhou, L. (eds.) PAKDD 1999. LNCS (LNAI), vol. 1574, pp. 74–83. Springer, Heidelberg (1999). doi:10.1007/3-540-48912-6_10

An Optimized Scheme of Mel Frequency Cepstral Coefficient for Multi-sensor Sign Language Recognition

Nana Wang$^{(\boxtimes)}$, Zhiyuan Ma, Yichen Tang, Yi Liu,
Ying Li, and Jianwei Niu

Beijing, China
1264693253@qq.com

Abstract. This paper proposes an optimized scheme of Mel Frequency Cepstral Coefficient (MFCC) and Deep-learning for sign language recognition using gyroscopes, accelerometers and surface electromyography, explores the possibility of building a sign language translation system by using these algorithms. Meanwhile, MFCC was compared with the traditional feature extraction methods, Linear Predictive Coding (LPC), Linear Predictive Cepstral Coefficient (LPCC), wavelet transform etc., in order to verify the effectiveness and superiority of MFCC for extracting characteristics of the surface of muscle. With continuous acquisition of surface electromyogram signals, an optimized scheme of MFCC was proposed to extract feature points. Then we cluster and classify the gesture features to match with the sign language words. The sign language words which can be matched successfully adjusts the language model training to be more consistent with the communication habits. This method provides a feasible way to realize a sign language recognition system which translates signs performed by deaf people into text/sound. The experimental results show that the accuracy of the feature extraction gesture recognition based on MFCC is about 90%, which is at least 4% higher than that of LPC, LPCC and wavelet transform. For translation of a complete statement, Bilingual Evaluation Understudy (BLEU) scores increases 5% after the adoption of the language model adjustment.

Keywords: Sign language recognition · MFCC · Sensor fusion · Deep-learning · Feature extraction gesture recognition

1 Introduction

Sign language is a language based on simultaneously combing hand shapes, orientation and movement of the hands, arms or body, and facial expressions to express a speaker's thoughts, as opposed to acoustically conveyed sound patterns [1]. Wherever communities of hearing impaired (i.e., a deaf, hereinafter referred to as hearing impaired people) exits, sign language have developed, and are 'the important assistant

J. Niu—Senior Member, IEEE.

© Springer International Publishing AG 2017
M. Qiu (Ed.): SmartCom 2016, LNCS 10135, pp. 224–235, 2017.
DOI: 10.1007/978-3-319-52015-5_23

tool of sound language' for mutual communication and exchange of ideas of hearing impaired people.

Sign language recognition is one depth and wide application of hand gesture recognition. With the development of hardware and technology, human computer interaction plays a more and more important role in our daily life [2]. Hand gesture is a more natural and convenient interactive mode comparing with traditional mouse and keyboard devices. Hand gesture recognition has been successfully applied to various types of consumer electronics.

Nowadays the human body motion recognition can mainly be divided into three categories: (1) The identification technology based on the motion sensor parameters. It uses all kinds of acceleration sensors to get the users' various related physical index when performing any actions. The acceleration sensor generally is a three-dimensional acceleration sensor and can feel objects in three dimensions. This technology allows the users and the sensors be together, so a number of wearable devices emerge as the times require. The representative of this technology used in gesture recognition is smart gloves which can capture a lot of the finger gestures' details. The smart gloves usually have very high precision of recognition, but they are more cumbersome and limit the freedom of users' fingers. (2) The identification technology based on computer vision. It uses images or videos to capture human body movements, to obtain the expression and gestures of the graphic information. In theory, computer vision technology can obtain the amount of information as much as human eyes. So it has been widely applied in detection, tracking and orientation analysis of human body and facial recognition etc. Kinect and Leap Motion are excellent products which use computer vision to get a good interactive function. The complexity of the algorithm and the portability and cost of the products are the challenges in the future. (3) The multi-sensor fusion technology based on electromyography transducer and other biological information. The human body action is much more complex with multiple joints. It has many hybrid patterns, high degree of freedom and continuity so on. The exploring the fusion of multi sensors and introducing of biological information has become one of the breakthrough in the field of human body motion recognition. In this paper, the surface electromyography (sEMG) is mainly discussed and investigated. sEMG [3–6] signal is weak biological electrical signal of neural muscle system relating with action information which can be obtained by surface electromyography electrodes recorded from the human skin. Different actions need different groups of muscle to participate in. Different body movements correspond to different muscle contraction modes and different sEMG. So it can be inferred that actions which users performed by recognized different patterns of sEMG. Portable sensor of sEMG acquisition and processing can provide a better user experience of wearable devices under the premise of ensuring the accuracy of human body motion recognition.

In recent years, domestic and overseas scholars have taken large-scale research on using sEMG for the gesture identification. The researchers have done a lot of work and obtained tremendous analyze results on this filed. Many algorithms and methods have been proposed and discussed mainly including hidden Markov model method (HMM), fuzzy pattern recognition method, Support Vector Machines [7] (SVM) and so on. Two vision-based real-time ASL recognition systems which are both tested for 40 signs and achieve 92% and 98% accuracy are studied for continuous American Sign Language at sentence level using Hidden Markov Model (HMM) [8]. In the first system, the camera is mounted on the desk while in the second system, the camera is mounted on a cap which is worn by the user. Another vision-based SLR system is proposed for Chinese Sign Language in vocabulary level [9]. Also one system is studied for German sign language using an accelerometer and one channel sEMG, which achieve a 99.82% accuracy for seven sign words [10]. In 2009, sign language recognition using intrinsic-mode sample entropy on sEMG and accelerometer data with an accuracy of 93% in recognizing 60 Greek sign words is proposed. Three years later, a sign-component-based framework for Chinese sign language recognition using an accelerometer and 4-channel sEMG is proposed. The main differences between our work and the previous works are as follows: (1) An optimized strategy of MFCC and Deep-learning for sign language recognition. (2) Using different modalities (6-channel sEMG, 3-D accelerometer, 3-D gyroscope) to get more feature information.

The contribution of this paper is a part of the design and implementation of Chinese sign language real-time translation device. This paper mainly studies the application of MFCC and neural network algorithm in gesture recognition. The rest of this paper is structured as follows: Firstly, Sect. 2 describes each step of our method. Then, Sect. 3 discusses the results of our experiments with more details given. Finally, conclusions and more discussions are presented.

2   Proposed Method

This section focuses on the extraction of surface EMG signal, the real-time filtering and processing and the classification and recognition algorithms.

A. *Signal extraction*

Surface electromyography signal is a kind of non-stationary and time-varying signal depending on the states of motion units. When people do not perform any action, the surface EMG signal will remain in a relatively stable and within a small range. Once the sign language action is performed, the surface EMG will have a larger amplitude signal increase.

Figure 2 shows a 6-channel Bluetooth-enabled physiological signal acquisition system.

Fig. 1. The acquisition device

Generally, sEMG signals are in the frequency range of 0 Hz–500 Hz depending on the space between electrodes and muscle type [11]. In our system, the sample data from SEMG, 3-D ACC and 3-D GYRO is sent to a PC via Bluetooth.

B. Preprocessing

Surface EMG signal intensity is usually weak, so the signal always need be amplified and noise filtered, then the useful signal is extracted. The energy mostly distributes in frequency band between 10 and 500 Hz. The following picture is the time-filed picture of surface EMG signal in the case that acquisition frequency is 200 Hz (Figs. 3 and 4).

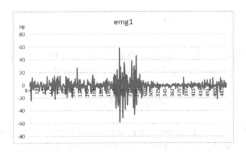

Fig. 2. Raising the arm

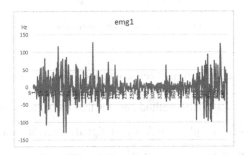

Fig. 3. The wrist turning inward

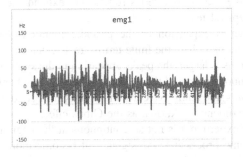

Fig. 4. The wrist turning outward

- One-dimensional Gauss filter

Gaussian filter is usually used for processing image signals, removing the noise signal in the images. The disturbance noise is always existence in data collection process, which would be confused with the Surface EMG signal and the analysis and the research. So we firstly processed the signals using Gaussian Filter.

We used 1D masks in this paper, the processing procedure is shown below (Fig. 5):

Fig. 5. The processing procedure

Three convolution kernels have been contained. Then a number will be inserted before the first number, and those two have to be equal to each other. The last element is changed in the same way. This method ensures that the signal stays the same length.

- Pre-emphasis, sub-frame, plus windows on signals after filtered

The purpose of pre-emphasis is to highlight high frequency signals and to avoid the signal to noise ratio is too low in high frequency band. The equation to calculate phase values is as follows:

$$H(z) = 1 - \mu * z \tag{1}$$

During the 10–30 ms, the surface EMG signal can be considered as a stationary signal, therefore, the formula for calculating frame length is

$$Length = fs * ts \tag{2}$$

Some of the symbols is as follows: Fs is sampling frequency, ts is sampling time. In this experiment, fs = 200 Hz, ts = 0.03 s, so frame length is six, that means one frame includes six sampling points.

After the surface EMG signal is divided into frames, then each frame as a smooth signal is processed. We use Fourier series function to expand each coefficients to get the Mel frequency cepstral features.

As the Fourier series expansion of the periodic functions (such as rectangular pulses) with discontinuous points, the finite terms are selected for synthesis. When the number of selected items is larger, the peaks appearing in the synthesized waveform are closer to the discontinuous points of the original. When the number of selected items is enough large, the peak value tends to a constant, approximately equal to 9% of the total jump value. The phenomenon is known as the Gibbs effect.

The frame in our experiment is not so continuous at the beginning and the end of the time. So after frame processing, the signal will become more and more deviated from original signal. To avoid this, we need to increase the continuity of both ends of the frame through the dispose adding window. After adding windows, there will be some characteristics of periodic function in the surface EMG signals. The most commonly used method is every single frame multiplied by Hamming window.

Assuming per frame as $S(n), n = 1, 2, 3 \ldots n$, n is the frame length. Then the processed signals can be described as

$$S'(n) = S(n) * W(n, a) \tag{3}$$

$$W(n, a) = (1 - a) - a * \cos\left(\frac{2\pi n}{N - 1}\right)$$
$$a = 0.46, 0 \leq n \leq N \tag{4}$$

- Dispersed Fourier Transform

The frequency signal is transformed from time-domain signal by dispersed Fourier transformation. The formula is as follows:

$$X(k) = DFT[x(n)] = \sum_{n=0}^{N-1} x(n) \, e^{-jkn*\frac{2\pi}{N}}, 0 \leq k \leq N - 1 \tag{5}$$

In the formula, $x(n)$ is the input sEMG signal, N is the number of points in Fourier transform.

C. Feature Extraction

- Triangular Bandpass Filters

On one hand, the first purpose of the Triangular Bandpass Filters is the continuous processing of spectrum and the removal of harmonics. In this way, the formant frequency of the surface EMG signal is emphasized. On the other hand, the Triangular Bandpass is performed to reduce the data dimension in domain.

Gesture is mainly produced by the movement of the arm, wrist and finger. According to the experiment's data, the power spectrum of the arm movement is far larger than that of the finger movement, and the power of the wrist movement is somewhere in between.

In our experiment, it is concluded that the frequency components of arm movement is concentrated in the low frequency band (0–50 Hz), the frequency spectrum of wrist movement is distributed in the middle frequency band (50 Hz–100 Hz), and the frequency spectrum of finger motion is occurs at the high frequency (100 Hz–500 Hz). Since the frequency components of sEMG beyond the range of 5 Hz–450 Hz are negligible [12], the low frequency noise in sEMG will be removed by a 5 Hz IIR high pass filter.

Our experiment assume that from the perspective of the amount of information that is expressed, the finger is more important than the arm, and the arm is more significant than the wrist.

The formula of filter frequency response is defined as

$$
H_m(k) = \begin{cases} 0, & k < f(m-1) \\ \frac{2(k-f(m-1))}{(f(m+1)-f(m-1))(f(m)-f(m-1))}, & f(m-1) \le k \le f(m) \\ \frac{2(f(m+1)-k)}{(f(m+1)-f(m-1))(f(m)-f(m-1))}, & f(m) \le k \le f(m+1) \\ 0, & k \ge f(m+1) \end{cases}
\tag{6}
$$

The symbols is as follows: $f(m)$ is center frequency, $m = 1, 2, 3 \ldots M$. M is the number of filter in the filter bank.

Then we need calculate the logarithmic energy of each filter:

$$
S(m) = \ln\left(\sum_{k=0}^{N-1} |x_a(k)|^2 H_m(k)\right), 0 \le m \le M
\tag{7}
$$

- Dispersed Fourier Transform

The logarithmic energy above is brought into the discrete Fourier transform to obtain the L-order Mel-scale Cepstrum parameters.

$$
C(n) = \sum_{m=0}^{N-1} s(m) \cos\left(\frac{\pi n(m - 0.5)}{M}\right), n = 1, 2, \ldots, L
\tag{8}
$$

- Output Feature Vector

The standard Mel-scale Cepstrum [13] parameters can only reflect the static characteristics of sEMG signal. Experiments show that the dynamic and static features can effectively improve the recognition performance. So we use the difference spectrometer of the static features to replace the dynamic features. The calculation of difference parameters can be calculated by the following formula:

$$
d_t = \begin{cases} C_{t+1} - C_t, & t < K \\ \frac{\sum_{K=1}^{K} K(C_{t+k} - C_{t-k})}{\sqrt{2\sum_{K=1}^{K} k^2}} \\ C_t - C_{t-1}, & t \ge Q - K \end{cases}
\tag{9}
$$

In the formula, d_t stands the first rank difference. C_t is the Mel-scale Cepstrum parameter. Q is the order of Mel-scale Cepstrum parameter. K is the time difference of first derivative, and here it can equal 1 or 0. Using this formula, we can get the second-order difference parameter.

MFCC is a commonly method of feature extraction in speech recognition. This method combines the hearing characteristics of human ears and sets up a group f Mel filter banks, which are characterized by the closely spaced low frequency and high frequency sparse. For sEMG signal, we cannot determine the distribution characteristics of the main frequency at first, so we set equal-phase interval filters in the

5 Hz–500 Hz frequency region. In the training process, the training various deep-learning models will be compared through adjusting the filter number. Finally, this paper will give a model which is most suitable for Chinese Sign Language recognition according to our samples.

D. Classification

In order to obtain a higher recognition rate, this paper makes simple classification of sign language gestures. There are single-hand gestures and double-hand gestures. For the single-hand gesture, they can be divided into gestures based thumb, gestures based forefinger, gestures based middle finger and so on. Also there are several gestures mixed different fingers. This decision tree [14] is based primarily on the signals of 3-D ACC and 3-D GYRO (Fig. 6).

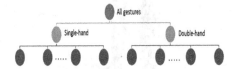

Fig. 6. The decision tree

- Logistic Regression

Logistic regression [15] is one of the most widely used algorithms of deep-learning. This paper uses it in the sub-segment after the decision tree. The logistic regression model is described by the weight matrix and bias vector, and the probability that the sample X belongs to the category Y is

$$P(Y = i|x, W, b) = softmax_i(Wx + b)$$
$$= \frac{e^{W_i x + b_i}}{\sum_j e^{W_i x + b_i}} \tag{10}$$

The equipment trains the model by maximizing the likelihood function. That means the opposite of minimized logarithm likelihood function:

$$\mathcal{L}(\theta = \{W, b\}, \mathcal{D}) = \sum_{i=0}^{|\mathcal{D}|} \log\left(P\left(Y = y^{(i)}|x^{(i)}\right), W, b\right) \tag{11}$$
$$\ell(\theta = \{W, b\}, \mathcal{D}) = -\mathcal{L}(\theta = \{W, b\}, \mathcal{D})$$

The classification model can be built after the training of a large amount of gestures samples.

3 Experiments

A. *Data Collection*

Data collection uses the device that made by our research group. The device is referenced to Fig. 1 above. The device placements are referenced to the researches of one group of Computer Science and Engineering, Texas A&M University [16] (Fig. 7).

The Chinese sign-language can involve not only fingers, but also the wrists and the arms. And there are some gesture using two hands. In this paper, we chose the right hand movements as study subject of Chinese sign-language. If two hands can be studied in our experiment, the recognition will be increased. About this, we will continue to explore in the future work.

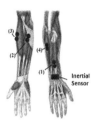

Fig. 7. The placement of devices

This paper selects five Chinese sign-language gestures [17] that are used widely in common life. 250 samples are collected from four individuals who are familiar with sign-language, so each gesture has 1000 samples. The gestures and their corresponding Chinese are showed in the below figure (Fig. 8).

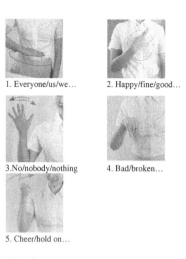

1. Everyone/us/we...   2. Happy/fine/good...

3.No/nobody/nothing   4. Bad/broken...

5. Cheer/hold on...

Fig. 8. Gestures and their meanings

In our experiment, the sample time of each gesture is 3 s. The acquisition frequency is 1000 Hz. So the sum of each gesture's sample point is 3000. In the data processing, we take the frame length as 300 sample points, and the overlapping area contains 100 sample points, so each sample has 14 frames. The order of MFCC coefficient is 5. Then adding the first order and the second order difference coefficient, we can get a 15-dimension MFCC coefficient. Finally, each sample can be represented by a matrix of 14 * 15. We expand the matrix into 210 dimensional vector for using the principle components analysis (PCA).

After the original data is extracted by the feature, this paper divides it into two parts which are the training set and the test set in the percentage 7:3. That is, the training set contains 5 * 700 samples, and the test set contains 5 * 300 samples.

B. Experimental Results

Table 1 gives out the recognition results of 5 Chinese sign-language. In this table, P1, P2, P3 and P4 stands for four subjects. The column "LPC" are the results using the LPC algorithm. The column "LPCC" are the results using the LPCC algorithm. The column "WT" are the results using the wavelet transform algorithm. The last column are the results using optimized strategy MFCC which is presented in this paper.

From the result of experimental data, the correct rate of the feature extraction gesture recognition based on MFCC is up to 91.2%, which is at least 4% higher than that of LPC, LPCC and wavelet transform. For translation of a complete statement, Bilingual Evaluation Understudy (BLEU) scores rose 5% after the adoption of the language model adjustment.

Table 1. Recognition results of 5 Chinese sign-language

	LPC	LPCC	WT	MFCC
P1	0.820	0.842	0.869	0.912
P2	0.816	0.837	0.861	0.903
P3	0.793	0.827	0.848	0.906
P4	0.833	0.844	0.864	0.925
Aver	0.816	0.838	0.860	0.912
Stdev	0.017	0.007	0.009	0.009

Also this paper performs twice data processing. The first has the decision tree while the second use logisitc regression directly without the decision tree. Changes in recognition results are compared between the two groups based on the samples. The results proves that the decision tree is a new classification method applied to uncertain gesture data and shows good performance in improve the recognition accuracy. The experiment result is given in the bar chart below (Fig. 9).

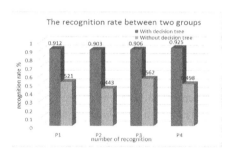

Fig. 9. Recognition rate between twice processing

4 Conclusion

In this paper, we propose a sign-language recognition algorithm based on the surface electromyography, acceleration and angular velocity. An optimized strategy of MFCC and Deep-learning for sign language recognition using gyroscopes, accelerometers and surface electromyography is proposed in our paper. The experimental results validate our proposed algorithm and show it promising to realize the high accuracy of wearable system for Chinese sign-language recognition. Regarding the future work, we will work on the basis of this paper to develop a translation equipment for hearing impaired people to help them communicate with other people conveniently in daily life.

References

1. Stokoe, W.C.: Sign language structure: an outline of the visual communication systems of the American deaf. J. Deaf Stud. Deaf Educ. **10**(1), 3–37 (2005)
2. Barberis, D., Garazzino, N., Prinetto, P., Tiotto, G., Savino, A., Shoaib, U., Ahmad, N.: Language resources for computer assisted translation from Italian to Italian sign language of deaf people. In: Proceedings of Accessibility Reaching Everywhere AEGIS Workshop and International Conference, Brussels, Belgium, November 2011
3. Liu, K., Chen, C., Jafari, R., et al.: Multi-HMM classification for hand gesture recognition using two differing modality sensors. In: 2014 IEEE Dallas Circuits and Systems Conference (DCAS), pp. 1–4. IEEE (2014)
4. Mebarkia, K., Bekka, R.E., Reffad, A., et al.: Fuzzy MUAP recognition in HSR-EMG detection basing on morphological features. J. Electromyogr. Kinesiol. **24**(4), 473–487 (2014)
5. Mane, S.M., Kambli, R.A., Kazi, F.S., et al.: Hand motion recognition from single channel surface EMG using wavelet & artificial neural network. Procedia Comput. Sci. **49**, 58–65 (2015)
6. Dhage, S.S., Hegde, S.S., Manikantan, K., et al.: DWT-based feature extraction and radon transform based contrast enhancement for improved iris recognition. Procedia Comput. Sci. **45**, 256–265 (2015)
7. Hong, J., Wang, L., Wang, C., et al.: Based on the artificial fish swarm algorithm to optimize the SVM of the hand movement sEMG recognition. Transducer Microsyst. Technol. **35**(2), 23–25 (2016)

8. Starner, T., Weaver, J., Pentland, A.: Real-time American sign language recognition using desk and wearable computer based video. IEEE Trans. Pattern Anal. Mach. Intell. **20**(12), 1371–1375 (1998)
9. Zhang, L.-G., Chen, Y., Fang, G., Chen, X., Gao, W.: A vision-based sign language recognition system using tied-mixture density HMM. In: Proceedings of the 6th International Conference on Multimodal Interfaces, pp. 198–204. ACM (2004)
10. Kim, J., Wagner, J., Rehm, M., André, E.: Bi-channel sensor fusion for automatic sign language recognition. In: 8th IEEE International Conference on Automatic Face and Gesture Recognition, FG 2008, pp. 1– 6. IEEE (2008)
11. De Luca, C.J., Gilmore, L.D., Kuznetsov, M., et al.: Filtering the surface EMG signal: movement artifact and baseline noise contamination. J. Biomech. **43**(8), 1573–1579 (2010)
12. Merletti, R., Di Torino, P.: Standards for reporting EMG data. J. Electromyogr. Kinesiol. **9**(1), 3–4 (1999)
13. Imai, S.: Cepstral analysis synthesis on the mel frequency scale. In: IEEE International Conference on Acoustics, Speech, and Signal Processing, ICASSP 1983, vol. 8. IEEE (1983)
14. Nowozin, S., Rother, C., Bagon, S., et al.: Decision tree fields. In: 2011 International Conference on Computer Vision, pp. 1668–1675. IEEE (2011)
15. Peduzzi, P., Concato, J., Kemper, E., et al.: A simulation study of the number of events per variable in logistic regression analysis. J. Clin. Epidemiol. **49**(12), 1373–1379 (1996)
16. Wu, J., Sun, L., Jafari, R.: A wearable system for recognizing American sign language in real-time using IMU and surface EMG sensors. IEEE J. Biomed. Health Inform. **20**(5), 1281–1290 (2016)
17. Yinlin, D.: Chinese Sign Daily Conversation Crash. Publishing House of Electronics Industry, Beijing (2016)

Security and Privacy Issues:
A Survey on FinTech

Keke Gai[1], Meikang Qiu[1(✉)], Xiaotong Sun[1], and Hui Zhao[2]

[1] Department of Computer Science, Pace University, New York City, NY 10038, USA
{kg71231w,mqiu,xs43599n}@pace.edu
[2] Software School, Henan University, Kaifeng 475000, Henan, China
zhh@henu.edu.cn

Abstract. As a new term in the financial industry, FinTech has become a popular term that describes novel technologies adopted by the financial service institutions. This term also covers aspects in security and privacy issues, such as threats, malicious behaviors, attacks, and adversaries, as well as the existing or potential solutions. This work aims to produce a survey of FinTech by collecting and reviewing contemporary achievements in security and privacy issues of the financial industry. The findings of this work can be used for forming the theoretical framework of FinTech in the security and privacy dimension, which will a fundamental support for establishing a solid security mechanism in FinTech.

Keywords: FinTech · Security · Privacy · Data mining

1 Introduction

As an emerging technical term, *Financial Technology* (FinTech) has been considered a distinguishing taxonomy that mainly describes the financial technology sectors in a wide range of operations for enterprises or organizations, which mainly addresses the improvement of the service quality by using *Information Technology* (IT) applications. Recent development of FinTech [1] has been powered by the advances of multiple technologies, such as mobile networks [2], big data [3], mobile embedded systems [4,5], cloud computing [6], and data analytic techniques [7]. However, security and privacy issues have been raised along with the consistent emergence of novel FinTech in recent years.

A quantity of recent investigations have emphasized the significance of security and privacy development in FinTech domains. Gartner's statistical reports [8] present that the investment of cybersecurity is expected to turn into $170 billion by 2020 globally. Moreover, only 35 % companies that highly rely on the usage of technologies for their businesses are confident of their security, according to the statistics done by Silicon Valley Bank [9]. The reality is that most modern *Financial Service Institutions* (FSIs) are applying IT-related techniques to support the delivery of financial services. Cyber threats often exist at multiple layers throughout the technical process in the financial industry, from private

© Springer International Publishing AG 2017
M. Qiu (Ed.): SmartCom 2016, LNCS 10135, pp. 236–247, 2017.
DOI: 10.1007/978-3-319-52015-5_24

to public sectors. An accurate awareness of FinTech, as well as its main issues and solutions, is an urgent demand for most current enterprises that are using finance-oriented technologies.

This paper concentrates on understanding the main security and privacy issues of FinTech from a comprehensive survey on recent updated researches. Two major aspects addressed by this survey are discerning primary concerns of security and privacy in FinTech and the emerging solutions or protection techniques. Throughout this survey, the principal research targets are appearing objects in FinTech, including both new cyber challenges and novel security techniques, even though traditional cyber risks are still seriously threatening the financial industry, such as hack attacks and malicious actions.

The main contributions of this work include:

1. This work has a concise focus that researches on security and privacy issues in FinTech and provides a solid survey. The major concentrations of this survey covers both issues and solutions deriving from recent researches.
2. The findings of this work highlight the foremost factors and concerns of cyber security in FinTech, which can be used as a reference for future researches.

The rest of this paper is organized by the following order. Section 2 synthensizes security and privacy issues in FinTech domain. Next, Sect. 3 reviews recent updated security solutions in FinTech. Furthermore, Sect. 4 represents a discussion as well as main findings of our work. Finally, we give our conclusions in Sect. 5.

2 Security and Privacy Issues in FinTech

2.1 Risks in Business Operations

Cyber concerns in the financial industry used to be a business operation issue at the early era of using electronic transactions and networking techniques [10]. A variety of surveys had been done for forming solid IT security strategies [11,12]. One of the concerns for most financial firms was that the firms concerned about business operations using updated techniques, since the return of the investment on security was difficult to predicate and govern. To address this concern, a recent investigation [13] showed that the return of the security investment had a positive relationship with the level of the corresponding investment. Another study [14] further proved that perceiving privacy concerns and building up a trust mechanism were two critical tasks for securing transactions conducted electronically.

Meanwhile, some researches have addressed the threat sources in financial business operations. Roumani et al. [15] completed a study on examining whether the records of the financial organizations are associated with the security vulnerabilities. This investigation considered the potential impacts caused by the assessed aspects, which included business scope, performance, markets, and sales. More work [16] have proved that the level of the IT transparency could impact on multiple business operations in financial advisory encounters from the perspective of trustworthiness. On the whole, most concerns of business operations mainly derive from unknown technical details, masked implementation process, and IT strategy-making.

2.2   Threats in Cloud Computing

As a popular Web-based service model, cloud computing has been widely accepted by numerous industries, including the financial industry [17,18]. For example, *Bank of America* (BoA) has recently announced that the financial firm is collaborating with Microsoft enterprise to improve financial transactions by developing *Blockchain* technologies [19]. The benefits of using cloud-based solutions are enabling financial businesses to closely connect to the target markets by increasing system performances [20,21], but this paradigm also introduces new threats due to outsourcing workloads. We converge major concerns of using cloud-based solutions in the next paragraph.

First, the masked complexity leaves FSIs a great concern due to lack of data controls in clouds [22], which makes private clouds a mainstream in the financial industry. A typical example is that FSIs may not know the physical server locations when using public clouds. Next, the data stored in remote cloud servers are still facing various threats, since the services settled on the complicated networks and intercrossed service participantships bring a lot of vulnerability opportunities. It is difficult for system designers and cloud vendors to fully predict or prevent the happenings of cyber risks in a cloud-based operating environment. Finally, the complicated and unanticipated communications between *Virtual Machines* (VM) can result in unpredictable vulnerabilities [23]. Ni et al. [24] pointed out that cloud data could be tampered due to the vulnerabilities of active protocols, even though the solution had been explored by the prior research. In summary, main threats of using cloud computing in FinTech derive from the complexity of the Web-based systems, lack of data controls, and uncertainty of technical details.

2.3   Emerging Fin-Privacy Issues

First, privacy protection is generally considered one of the most significant aspects in the financial security domain and preserving data carrying privacy is a critical task in producing a privacy protection strategy. A recent study evaluated the trade-off between the data usage and privacy protection by implementing a machine learning-based method [25]. This work used a K-means clustering algorithm to discern data carrying privacy out of the multi-party clustering scenario. Additionally, location-based services usually carry users' movement privacy, such that the applications or devices have become common attack targets [1].

Furthermore, new financial services bring new concerns in security and privacy. For example, implementing financial insurance is impacting on financial service organizations in making IT-related decisions, such as cybersecurity insurance [26]. Understanding cyber risks and discerning the relationships between the cyber incidents and the insurance covered items are challenging issues for many companies that are tightly attached to the Web-based applications. In addition, using electronic approaches for financial frauds [27] is another emerging issue in Fin-Tech. Traditional fraud detection methods usually relied on statistical methods, which were insufficient to find out the continuous real-time deception happenings.

3 Solution Synthesis

Many scholars have accomplished a great amount of researches in securing data. Some work exactly focused on the financial industry and some other studies addressed the universal solutions to security and privacy problems. In this section, we aim to synthesize the updated achievements that are either FinTech-oriented solutions or protection techniques that can be applied in the financial industry.

3.1 Risk Detection Explorations

The implementations of cloud computing have powered the cyber risk management by providing a flexible service deployment, either centralized or decentralized manners. For instance, one of the recent researches [28] has proved that cloud-based cyber risk management system could assist in classifying cyber-related information in the financial industry. This approach classifies information releases that can cause potential privacy leakage by using supervised learning techniques, which are combined with taxonomy. The work provides a feasible approach for classifying cyber incidents and align them with business items by using semantic techniques.

Next, financial frauds are frustrating FSIs as well. Glancy et al. [29] developed a model for detecting financial frauds. The model used quantitative analysis focusing on the frauds conducted by textual data. However, the quantitative approach could not ensure a stable detection accuracy, which meant the model could be only used as a supporting tool for distinguishing suspected transactions. An additional research [30] suggested an method that used multiple criterion decision making to select clustering algorithms for financial risk analysis. Moreover, analyzing correlation coefficients is an alternative approach for detecting abnormal operation behaviors, which has been proved by the prior work [31,32]. Improper activities can be detected when the correlation coefficient values deriving from multiple elements are different from the coefficients obtained from the clean dataset. In summary, most current work in financial risk detections intend to apply state analysis techniques.

3.2 Authentication and Access Control Mechanisms

Many FSIs conduct authentications and access controls by using cryptography-based approaches. Besides the data encryptions for protecting financial information, a few novel security mechanisms are also alternations for securing financial privacy. One solution direction was exploring the mechanism of strengthening access controls throughout multimedia [33]. The semantic accesses are associated with the service requestors' features identified by ontology techniques. Financial service acquisitions can be reached by creating semantic-based access controls that are supported by multimedia. Another ontology-based solution [34] was proposed to power up cybersecurity ecosystem by generating the knowledge graph that showed the interrelations between cyber risks and their causes. However, ontology-based approaches usually has a limitation that the accuracy of access controls can hardly reach a perfect when the amount of ontologies in the system is large.

Moreover, some previous work addressed developing multiple constraints for data accesses. One approach [35] was proposed to address strengthening biometric authentication systems by considering three dimensional constraints in creating protection strategies, which included security, privacy, and trust. The researchers formulated the trade-offs from these three aspects in order to increase the efficiency of the protection when biometric technologies are applied. However, this approach is restricted in many application scenarios, since three dimensions in criterion, including security, privacy, and trust, have vague boundaries in practice. Adjusting the protection mechanism is also a challenging issue in that forming an implementation strategy takes a long time and frequently switching the mechanism of biometric authentication systems is not applicable for the demands of most financial services [36].

In addition, to improve the performance of the data encryptions, some scholars have developed a few approaches for dynamically determining the strategy of data protections. For example, one approach [37] was developed to selectively encrypt data based on the privacy classifications with privacy weights under the fixed timing constraint. This approach could produce optimal solutions to maximize the total privacy weight value. In summary, the main trade-off of access security in FinTech is attached to the conflict between security and service performance. Most current FSIs intend to maximize the level of security to protect all transactions as well as financial customers' privacy.

3.3   Data Usage Cycle Protections

Financial service institutions generally emphasize the importance of data usage, since its performance has a direct relationship with the service quality. The data governance may become dramatically complicated when the size the system increases or new functions are added [38]. A research direction addresses the critical concern of financial data protections, which is to prevent data from malicious behaviors launched by the unexpected third party during the data usage cycles. Many previous studies have explored a variety of methods to reduce the risk level when data are shared, transmitted, or exchanged through different parties. Chang et al. [39] proposed a mechanism that used *Business Process Modeling Notation* (BPMN) to discern the processes and implementations of the data usage. The outcome of this approach can point at the period or data range that requires additional security protection operations based on the data usage simulations. Another research [40] also considered the research perspective of the business process but focused on payment systems. This work emphasized the vulnerabilities caused by poor business process management in e-commerce, such as improper control flows.

Furthermore, Xiao et al. [41] proposed a novel security system for delivering location-aware services, which did not rely on techniques of key sharing. This approach utilized the proximity-based authentications and temporal location tags for establishing the session key. However, this work did not reach a non-error rate goal because of the radio propagation properties. Another research attempt [42] focused on developing a privacy protection strategy of grouping mobile users. By using this method, the data carrying location information are not needed to frequently share or transmit, since mobile users are grouped based on locations. Nevertheless, this method was hardly to be implemented because the inputs and geographical locations are generally dynamic for most location-oriented applications [43]. Therefore, hazards in data usage cycle mainly come from a few dimensions, which include participations of the unexpected third party, unclear business processes, and large-range distributed usage.

3.4   Secure Data Storage and Processing

Improving the security mechanism in data storage and processing is also vital research direction in FinTech and many researches in this field have been done in recent years. Gai et al. [44] proposed a solution to securing data by applying a distributed data storage in cloud systems. This approach considers two potential adversaries that include both internal and external attackers. The data carrying sensitive data are divided into two parts before the data are sent out on the network, so that the privacy can be protected even though the transmissions are monitored by adversaries. This approach can be further strengthened by adding a judgement process by which the data requiring distributive storage are determined [45]. The improvement is an optimization of reducing workload when the volume of data is large.

Next, some other scholars looked into privacy-assured searchable data storage in cloud computing. One of the recent developed methods [46] used symmetric-key encryption primitives and covered three functions, which included ranking results, identifying similarities, and searching structured data. This method allows users to share the encrypted data in the distributed service deployment. Moreover, prior privacy-related researches not only explored text-based file storage but also investigated the image maintenance in cloud computing. For example, one research [47] proposed a scheme that used *Content-Based Image Retrieval* (CBIR) technique to encrypt images before they are stored in cloud servers.

In summary, FSIs intend to obtain a higher level data protection no matter what technologies are selected. Centralized data storage and processing can reduce the risks of the privacy release at server side, but it has a limited impact on improving security level during the data transmissions. Meanwhile, a decentralized data storage/processing is also facing challenges from various dimensions, such as monitoring communications and database abuses. Therefore, securing financial data is associated with protecting data within a network system and all threats existing in a web should be addressed in this field.

3.5   Risk Reduction and Prevention

In general, there are two basic types of methods for risk reductions and preventions, which include physical and application-based methods. The physical method refers to the approach of securing data by conducting operations on the physical infrastructure, such as preventing a jamming attack or avoiding a network damage. An application-based method refers to the security solution achieved by cryptographic methods, such as creating a secure protocol or configuring access controls [48]. For example, using logical authorization language is an alternative approach for forming an access control rule in social networks [49]. Besides these two aspects, a number of new research directions in reducing financial cyber risks have been proposed along with the development of new technologies.

First, developing a proactive protection approach is an alternative research direction for reducing risks. One scheme was proposed to protect financial customers' privacy by using attribute-based access controls [50]. This approach only allows those third parties that are configured as the trustable parties to decrypt their data either fully or partially. The approach was improved by introducing semantic web such that the relations between the data owners and unknown third parities can be clarified [51]. However, this approach needs configurations for defining trusted parties done by data owners, which means it may bring dramatical extra workloads when financial customers have the authentications for establishing their own trust parties.

Next, the financial production selection can be influenced by various elements when there exist a few alternative and available service choices. To address a proper decision-making on determining financial productions, some recent researches have tried to develop applicable solutions under certain constraints. For example, a research weighted all available financial services and measured coefficients in order to obtain the optimized solution to choosing services [52]. This method simply used an ordered sequence such that the weight values were not well addressed. Another research had similar research focus on weighting service items, which developed a method of classifying cyber risks [53]. This research had stronger contribution to cyber risk classifications by using semantic techniques to create knowledge representation graphs. The work was further improved [54] by applying *Monte Carlo* (MC) simulations for efficient data analytics in forming security framework.

Moreover, many prior researches concentrated on reducing the risks related to physical locations. Ma et al. [55] proposed a method of increasing the security level for *Radio-Frequency IDentification* (RFID)-based applications. This approach considered the geographical information-related data the constraints of the data accesses; thus, a financial transaction will be denied when an abnormal location-related movement is detected. The limitations of this approach are that the mechanism highly relies on RFID techniques and the accuracy of the adversaries detections is under debate. Another work [56] developed an authentication protocol of RFID that used a hash function to reach mutual authentication between the data and databases. Using a hash function for the purpose of creating secure RFID protocols had been also used in other work [57]. The crucial part of using RFID techniques in securing financial data/information is to successfully protect hash tags.

In summary, it is difficult for FSIs and system administrators to perceive potential cyber hazards from emerging technologies or applications until the adversaries attack. Typical solution is to understand the active system and reduce the rate of attacks by discerning technical details and business processes.

4   Findings and Discussions

Our work primarily focused on security and privacy issues in FinTech and completed a comprehensive survey on the target field. Traditional threats still exist throughout all security dimensions in the financial industry. Moreover, FSIs were always primary attack targets in a networking environment so that new cyber risks kept coming out along with the emergence of new technologies. We summarized our main findings addressing major appearing challenges and solutions in this section, as given in the following statement.

1. An explicit perceptibility of financial business processes was a critical task for formulating secure data flows and predicating user scopes. Cyber hazards could be caused by the participations of unknown, untrusted, or unexpected data users.
2. Using multimedia was a trend to strengthen the access controls in that multiple constraints could be applied in validations. The logical restrictions defining role authentications could increase the protections in network-based financial systems.
3. Mobile financial services highly relied on the utilizations of mobile devices and networks, such that cyber risks attached to distributed networking systems should be addressed, such as monitoring communications and hacking cloud storage.
4. New technologies could bring unanticipated cyber risks although the technologies were desired to introduce benefits. The challenges were generally attached to technical vulnerabilities of new systems, uncertain business process designs, and high complexity of governance.

5   Conclusions

This work accomplished a survey on major security and privacy issues in Fin-Tech. The literature review not only covered the traditional threats but also looked into new challenges in the field. We categorized the main issues into three dimensions, including business operations, threats in cloud computing, and emerging Fin-privacy issues. The solution synthesis was made in three aspects, which included risk detections, authentication and access controls, data usage cycles, and data storage/processing. The outcomes of this work provide a solid theoretical support for future Fin-Tech strength in security and privacy domains.

References

1. Li, Y., Dai, W., Ming, Z., Qiu, M.: Privacy protection for preventing data over-collection in smart city. IEEE Trans. Comput. **65**, 1339–1350 (2015)
2. Gai, K., Qiu, M., Tao, L., Zhu, Y.: Intrusion detection techniques for mobile cloud computing in heterogeneous 5G. Secur. Commun. Netw. **9**, 1–10 (2015)
3. Yin, H., Gai, K.: An empirical study on preprocessing high-dimensional class-imbalanced data for classification. In: The IEEE International Symposium on Big Data Security on Cloud, New York, USA, pp. 1314–1319 (2015)
4. Gai, K., Qiu, M., Chen, M., Zhao, H.: SA-EAST: security-aware efficient data transmission for ITS in mobile heterogeneous cloud computing. ACM Trans. Embed. Comput. Syst. **PP**, 1 (2016)
5. Gai, K., Du, Z., Qiu, M., Zhao, H.: Efficiency-aware workload optimizations of heterogenous cloud computing for capacity planning in financial industry. In: The 2nd IEEE International Conference on Cyber Security and Cloud Computing, New York, USA, pp. 1–6. IEEE (2015)
6. Qiu, M., Zhong, M., Li, J., Gai, K., Zong, Z.: Phase-change memory optimization for green cloud with genetic algorithm. IEEE Trans. Comput. **64**(12), 3528–3540 (2015)
7. Qiu, M., Cao, D., Su, H., Gai, K.: Data transfer minimization for financial derivative pricing using Monte Carlo simulation with GPU in 5G. Int. J. Commun. Syst. **29**(16), 2364–2374 (2016)
8. Morgan, S.: Cybersecurity market reaches \$75 billion in 2015; expected to reach \$170 billion by 2020 (2015)
9. SVB: Cybersecurity report 2015 (2015). Accessed https://www.svb.com/uploadedFiles/Content/Trends_and_Insights/Reports/Cybersecurity_Report/cybersecurity-report-2015.pdf
10. Gai, K., Qiu, M., Elnagdy, S.: A novel secure big data cyber incident analytics framework for cloud-based cybersecurity insurance. In: The 2nd IEEE International Conference on Big Data Security on Cloud, New York, USA, pp. 171–176 (2016)
11. Gai, K., Steenkamp, A.: A feasibility study of Platform-as-a-Service using cloud computing for a global service organization. J. Inf. Syst. Appl. Res. **7**, 28–42 (2014)
12. Gai, K., Steenkamp, A.: Feasibility of a Platform-as-a-Service implementation using cloud computing for a global service organization. In: Proceedings of the Conference for Information Systems Applied Research ISSN, vol. 2167, p. 1508 (2013)
13. Chai, S., Kim, M., Rao, H.: Firms' information security investment decisions: stock market evidence of investors' behavior. Decis. Support Syst. **50**(4), 651–661 (2011)
14. Liao, C., Liu, C., Chen, K.: Examining the impact of privacy, trust and risk perceptions beyond monetary transactions: an integrated model. ECRA **10**(6), 702–715 (2011)
15. Roumani, Y., Nwankpa, J., Roumani, Y.: Examining the relationship between firm's financial records and security vulnerabilities. IJIM **36**(6), 987–994 (2016)
16. Nussbaumer, P., Matter, I., Schwabe, G.: "enforced" vs. "casual" transparency-findings from IT-supported financial advisory encounters. ACM Trans. Manag. Inf. Syst. **3**(2), 11 (2012)
17. Gai, K., Li, S.: Towards cloud computing: a literature review on cloud computing and its development trends. In: 2012 Fourth International Conference on Multimedia Information Networking and Security, Nanjing, China, pp. 142–146 (2012)

18. Gai, K., Qiu, M., Zhao, H.: Cost-aware multimedia data allocation for heterogeneous memory using genetic algorithm in cloud computing. IEEE Trans. Cloud Comput. **PP**(99), 1–11 (2016)
19. Hernandez, P.: Microsoft, Bank of America announce Blockchain collaboration (2016). Accessed http://www.eweek.com/cloud/microsoft-bank-of-america-announce-blockchain-collaboration.html
20. Gai, K., Qiu, M., Zhao, H., Tao, L., Zong, Z.: Dynamic energy-aware cloudlet-based mobile cloud computing model for green computing. JNCA **59**, 46–54 (2015)
21. Gai, K., Qiu, M., Zhao, H., Liu, M.: Energy-aware optimal task assignment for mobile heterogeneous embedded systems in cloud computing. In: 2016 IEEE 3rd International Conference on Cyber Security and Cloud Computing (CSCloud), Beijing, China, pp. 198–203. IEEE (2016)
22. Gai, K.: A review of leveraging private cloud computing in financial service institutions: value propositions and current performances. Int. J. Comput. Appl. **95**(3), 40–44 (2014)
23. Gai, K., Qiu, M., Zhao, H., Dai, W.: Privacy-preserving adaptive multi-channel communications under timing constraints. In: The IEEE International Conference on Smart Cloud 2016, New York, USA, p. 1. IEEE (2016)
24. Ni, J., Yu, Y., Mu, Y., Xia, Q.: On the security of an efficient dynamic auditing protocol in cloud storage. IEEE Trans. Parallel Distrib. Syst. **25**(10), 2760–2761 (2014)
25. Banu, R., Nagaveni, N.: Evaluation of a perturbation-based technique for privacy preservation in a multi-party clustering scenario. Inf. Sci. **232**, 437–448 (2013)
26. Elnagdy, S., Qiu, M., Gai, K.: Understanding taxonomy of cyber risks for cybersecurity insurance of financial industry in cloud computing. In: The 2nd IEEE International Conference of Scalable and Smart Cloud, pp. 295–300. IEEE (2016)
27. Sharma, A., Panigrahi, P.: A review of financial accounting fraud detection based on data mining techniques. arXiv preprint. arXiv:1309.3944 (2013)
28. Gai, K., Qiu, M., Elnagdy, S.: Security-aware information classifications using supervised learning for cloud-based cyber risk management in financial big data. In: The 2nd IEEE International Conference on Big Data Security on Cloud, New York, USA, pp. 197–202 (2016)
29. Glancy, F., Yadav, S.: A computational model for financial reporting fraud detection. Decis. Support Syst. **50**(3), 595–601 (2011)
30. Kou, G., Peng, Y., Wang, G.: Evaluation of clustering algorithms for financial risk analysis using MCDM methods. Inf. Sci. **275**, 1–12 (2014)
31. Gai, K., Qiu, M., Zhao, H., Dai, W.: Anti-counterfeit schema using Monte Carlo simulation for e-commerce in cloud systems. In: The 2nd IEEE International Conference on Cyber Security and Cloud Computing, New York, USA, pp. 74–79. IEEE (2015)
32. Löhr, S., Mursajew, O., Rösch, D., Scheule, H.: Dynamic implied correlation modeling and forecasting in structured finance. J. Futures Markets **33**(11), 994–1023 (2013)
33. Li, Y., Gai, K., Ming, Z., Zhao, H., Qiu, M.: Intercrossed access control for secure financial services on multimedia big data in cloud systems. ACM Trans. Multimedia Comput. Commun. Appl. **12**(4s), 67 (2016)
34. Asamoah, C., Tao, L., Gai, K., Jiang, N.: Powering filtration process of cyber security ecosystem using knowledge graph. In: The 2nd IEEE International Conference of Scalable and Smart Cloud, pp. 240–246. IEEE (2016)

35. Kanak, A., Sogukpinar, I.: BioPSTM: a formal model for privacy, security, and trust in template-protecting biometric authentication. Secur. Commun. Netw. **7**(1), 123–138 (2014)
36. Davis, M., Kumiega, A., Van, V.: Ethics, finance, and automation: a preliminary survey of problems in high frequency trading. Sci. Eng. Ethics **19**(3), 851–874 (2013)
37. Gai, K., Qiu, M., Zhao, H., Xiong, J.: Privacy-aware adaptive data encryption strategy of big data in cloud computing. In: The 2nd IEEE International Conference of Scalable and Smart Cloud (SSC 2016), Beijing, China, pp. 273–278. IEEE (2016)
38. DeStefano, R., Tao, L., Gai, K.: Improving data governance in large organizations through ontology and linked data. In: The 2nd IEEE International Conference of SSC, pp. 279–284. IEEE (2016)
39. Chang, V., Ramachandran, M.: Towards achieving data security with the cloud computing adoption framework. IEEE Trans. Serv. Comput. **9**(1), 138–151 (2016)
40. Yu, W., Yan, C., Ding, Z., Jiang, C., Zhou, M.: Modeling and validating e-commerce business process based on petri nets. IEEE Trans. SMCS **44**(3), 327–341 (2014)
41. Xiao, L., Yan, Q., Lou, W., Chen, G., Hou, Y.: Proximity-based security techniques for mobile users in wireless networks. IEEE Trans. IFS **8**(12), 2089–2100 (2013)
42. Bilogrevic, I., Jadliwala, M., Joneja, V., Kalkan, K., Hubaux, J., Aad, I.: Privacy-preserving optimal meeting location determination on mobile devices. IEEE Trans. IFS **9**(7), 1141–1156 (2014)
43. Wang, Y., Liu, J., Chen, Y., Gruteser, M., Yang, J., Liu, H.: E-eyes: device-free location-oriented activity identification using fine-grained WiFi signatures. In: Proceedings of the 20th Annual International Conference on MCN, Maui, Hawaii, USA, pp. 617–628. ACM (2014)
44. Gai, K., Qiu, M., Zhao, H.: Security-aware efficient mass distributed storage approach for cloud systems in big data. In: The 2nd IEEE International Conference on Big Data Security on Cloud, New York, USA, pp. 140–145 (2016)
45. Li, Y., Gai, K., Qiu, L., Qiu, M., Zhao, H.: Intelligent cryptography approach for secure distributed big data storage in cloud computing. Inf. Sci. **PP**(99), 1 (2016)
46. Li, M., Yu, S., Ren, K., Lou, W., Hou, Y.: Toward privacy-assured and searchable cloud data storage services. IEEE Netw. **27**(4), 56–62 (2013)
47. Xia, Z., Wang, X., Zhang, L., Qin, Z., Sun, X., Ren, K.: A privacy-preserving and copy-deterrence content-based image retrieval scheme in cloud computing. IEEE Trans. IFS **11**(11), 2594–2608 (2016)
48. Ma, L., Tao, L., Zhong, Y., Gai, K.: RuleSN: research and application of social network access control model. In: IEEE International Conference on IDS, New York, USA, pp. 418–423 (2016)
49. Ma, L., Tao, L., Gai, K., Zhong, Y.: A novel social network access control model using logical authorization language in cloud computing. CCPE **PP**(99), 1 (2016)
50. Gai, K., Qiu, M., Thuraisingham, B., Tao, L.: Proactive attribute-based secure data schema for mobile cloud in financial industry. In: The IEEE International Symposium on BigDataSecurity, New York, USA, pp. 1332–1337 (2015)
51. Qiu, M., Gai, K., Thuraisingham, B., Tao, L., Zhao, H.: Proactive user-centric secure data scheme using attribute-based semantic access controls for mobile clouds in financial industry. Future Gener. Comput. Syst. **PP**, 1 (2016)
52. Merigó, J., Gil-Lafuente, A.: New decision-making techniques and their application in the selection of financial products. Inf. Sci. **180**(11), 2085–2094 (2010)

53. Elnagdy, S., Qiu, M., Gai, K.: Cyber incident classifications using ontology-based knowledge representation for cybersecurity insurance in financial industry. In: The 2nd IEEE International Conference of Scalable and Smart Cloud, pp. 301–306. IEEE (2016)
54. Gai, K., Qiu, M., Hassan, H.: Secure cyber incident analytics framework using Monte Carlo simulations for financial cybersecurity insurance in cloud computing. CCPE **PP**(99), 1 (2016)
55. Ma, D., Saxena, N., Xiang, T., Zhu, Y.: Location-aware and safer cards: enhancing RFID security and privacy via location sensing. IEEE TDSC **10**(2), 57–69 (2013)
56. Wei, C., Hwang, M., Chin, A.: A mutual authentication protocol for RFID. IT Prof. Mag. **13**(2), 20 (2011)
57. Sun, D., Zhong, J.: A hash-based RFID security protocol for strong privacy protection. IEEE Trans. Consum. Electron. **58**(4), 1246–1252 (2012)

Multi-sensor System Calibration Approach Based on Forward-Model and Inverse-Model

Xiaojun Tang[1(⊠)], Feng Zhang[1], Hailin Zhang[1], Junhua Liu[1], and Yuntao Liang[2]

[1] State Key Laboratory of Electrical Insulation & Power Equipment,
Xi'an Jiaotong University, Xianning West RD 28#,
Xi'an City, Shaanxi Province 710049, China
xiaojun_tang@mail.xjtu.edu.cn
[2] State Key Laboratory of Coal Safety, Shenyang Branch of
China Coal Research Institute, Shenyang 110016, China

Abstract. Aimed at the calibration problem of multi-sensor system because of cross-sensitivity, a novel approach based on forward-model and inverse-model was proposed. For forward-model, measurants were taken as inputs, sensor outputs were taken as outputs. For inverse-model, its inputs and outputs were just the opposite. The forward-model was built with calibration samples by interpolation at first. And with this model, "additional samples" were produced. Then, inverse model was built with calibration samples and "additional samples" by training neural networks for every component of analyte. Finally, spectral analysis of light alkane gas mixture was taken as an example to test the calibration approach. The results showed that almost same accuracy, when the approach was applied, could be obtained with only 200 sets of samples as that obtained with 7000 sets of samples, which meaned the proposed approach could reduce the number of calibration sample. And calibration cost could be reduced.

Keywords: Multi-sensor system · Cross-sensitivity · Calibration approach · Neural network · Additional samples

1 Introduction

Most sensors are sensitive to not only their measurants. In other words, cross-sensitivity exists in many measurement systems. For instance, most methane sensor made with semiconductor is also sensitive to ethane and carbon monoxide beside methane [1, 2]. If one measure the concentration of ethane using optical sensor at well-head, he may doesn't know the output change of the sensor is caused by ethane or propane because the absorption spectrum of ethane overlap with that of propane [3]. In order to reduce cross-sensitivity and obtain accurate measurement result, in this case, multi-sensor system has to be designed and calibrated [4, 5].

In the past twenty years, many approaches such as Least Mean Square(LMS), Neural Network(NN), Support Vector Machine(SVM), and so on, have ever been used to build the static model of the multi-sensor system with calibration data [5–7]. However, every approach has its shortcomings. For LMS, it is hard to build a complex

M. Qiu (Ed.): SmartCom 2016, LNCS 10135, pp. 248–255, 2017.
DOI: 10.1007/978-3-319-52015-5_25

model because of its architecture. It is said that NN can fit any function with any accuracy. But this object is based on such an assumption that the number of calibration samples is enough large and even infinite. In practice, it is impractical to obtain so many calibration samples in some cases. Then, either under-fitting or over-fitting always bewilders the user. SVM is also used for multi-sensor system calibration in recent years. It is said that SVM can overcome the problem of over-fitting when calibration sample is not enough. But a new matter comes forth. There isn't a reliable rule for choosing the nonlinear function of the machine.

In this paper, a novel approach was presented to improve calibration accuracy of multi-sensor system with limited number of training samples based on our previous works [3, 4]. For this approach, the structure of the whole multi-sensor measurement system was shown in Fig. 1. The models, $y_j = f_j(x_1, x_2, \cdots, x_n)$, shown in Fig. 1 were called forward-model and were built through interpolation at first. Where, $i = 1, 2, \cdots, n$, $j = 1, 2, \cdots, m$, $m \geq n$. For the forward-models, measurants, $x_i (i = 1, 2, \cdots, n)$, were taken as inputs of the models and the outputs of sensors were looked as outputs. Then, forward-model was used to produce "additional samples". Finally, NN was used to build the measurement model called as inverse-model, $x_i' = g_i(y_1, y_2, \cdots, y_m)$, with both "additional samples" and calibration samples.

In the end of this paper, the presented approach was applied in spectral analysis of multi-component of gas mixture. In the measured gas mixture, methane, ethane, propane, iso-butane and n-butane were taken as analytes, iso-pentane and n-pentane were taken as interferents, that have ever been analyzed with Fourier transform infrared spectrometer (FT-IRS) and gas chromatography (GC) in our previous work [3].

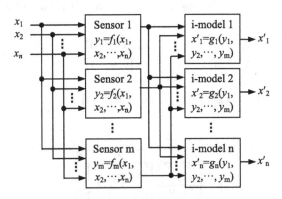

Fig. 1. Structure of multi-sensor measurement system

2 Methodology

Because measurants are normally uncorrelated to each other, all the measurants and one of the outputs of the multi-sensor system can be looked as a curve surface in an orthogonal coordinate system. It is well known, a curve surface function is easy to be fitted in such an orthogonal coordinate system rather in a nonorthogonal coordinate system. Furthermore, the output of every sensor is normally monotone increasing or decreasing to every measurant. And high fitting accuracy may be obtained with the forward-model built with calibration samples even using local linearization, spline fitting, trigonometry or other interpolation approach [8]. Then, enough "additional samples" can be produced with these forward-models. NN can be used to build the measurement model called as inverse-model, $x_i' = g_i(y_1, y_2, \cdots, y_m)$, with both "additional samples" and calibration samples. Because the samples used to train the inverse-model can be "produced" infinitely with forward-models, enough "samples" can be gotten, and over-fitting will not happen.

However, conventional interpolation is only used for curve fitting in two-dimensional plane or curve surface in three-dimensional space. The number of measurants may be higher than 2. In order to extend the application of interpolation to high-dimensional space, in this paper, an approach called as patulous-cross-section-interpolation (PCSN) was proposed for getting forward-models, the transfer function $f_i(\cdot)$, shown in Fig. 1. The idea of this approach was introduced though density calculation of a cube shown in Fig. 2. In this figure, a cube had been divided equally into 27 son-cubes. For the cube, the density wasn't well-proportioned, but the density of every vertex of every son-cube was known. For example, the density of dot $A(x_A, y_A, z_A)$ was ρ_A, and the density of dot $B(x_B, y_B, z_B)$ was ρ_B, where $x_A, y_A, z_A, x_B, y_B, z_B, \rho_A$ and ρ_B were known. If the coordinate of a random dot $W(x_W, y_W, z_W)$ in the object was given, where x_W, y_W, z_W were known, the density on dot W could be calculated with all the specimen dots or some of the specimen dots through following process.

From Fig. 2, it was obvious that dot W was on line JM in plane HISV, line JM was perpendicular to axis x (line OR) and axis y (line OC), so the coordinates of dot J, K, L and M could be written as (x_W, y_W, z_J), (x_W, y_W, z_K), (x_W, y_W, z_L) and (x_W, y_W, z_M)

$$
\begin{array}{lll}
X_0 \rightarrow M_0 & x_{0,r} & r \in [0, M_0 - 1] \\
X_1 \rightarrow M_1 & x_{1,s} & s \in [0, M_1 - 1] \\
\vdots & \vdots & \vdots \\
X_{N-1} \rightarrow M_{N-1} & x_{N-1,t} & t \in [0, M_{N-1} - 1]
\end{array}
$$

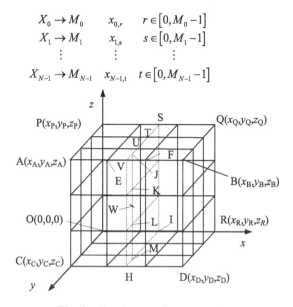

Fig. 2. Sketch map of space splitting

respectively. If the densities of these four dots, ρ_J, ρ_K, ρ_L and ρ_M were also known, the density of dot W could be calculated with the coordinates and densities of dots J, K, L and M through curve interpolation such as 3-order spline. Similarly, in plane ABQP, shown in Fig. 2, if the densities of dot V, U, T and S were known, the density of dot J could also be calculated through curve interpolation. In Fig. 2, four dots, A, E, F and B, were all vertex of son-cube, their coordinates and densities were known, so the density of dot V could be calculated with the coordinates and densities of these four dots through curve interpolation. In the same way, the densities of U, T and S could also be calculated in plane ABQP. Then, set the densities of V, U, T and S as known quantities, the density of dot J could be calculated through curve interpolation again. By the same way of calculating the density of dot J, the densities of K, L and M were calculated. Then, the density of dot W could be calculated with the densities of J, K, L and M by curve interpolation. In this way, a curve surface in four dimensional space was fitted by curve interpolation through splitting curve surface fitting in four dimensional space into curve surface fitting in three dimensional space, and then splitting curve surface fitting in three dimensional space into curve fitting in two dimensional space. The rest might be deduced by analogy, curve surface fitting in higher dimensional space could be performed with this reducing dimension method. Above idea induced, following theorem was concluded:

In a N + 1 dimensional space, there is a curve surface function: $y = f(x_0, x_1, \cdots, x_{N-1})$. For the curve surface, There are $\prod\limits_{l=0}^{N-1} M_l$ known discrete dots whose coordinates are known. The lengths of X_0, X_1, $\cdots$, X_{N-1} are respectively. Then, Y, $x_{0,i}$, $x_{1,j}$, $\cdots$, X_l, $\cdots$, x_{N-1}, l are assumed to be known in the N + 1 dimensional space. Where Y is $\{Y_l(x_{0,i}, x_{1,j}, \cdots, X_l, \cdots, x_{N-1,k}), 0 \leq l \leq N-1\}$, vector Xl is made up of $x_{l,h}(h \in [0, M_l - 1])$, and vector Y_l is made up of $y_h(x_{0,i}, x_{1,j}, \cdots, X_l, \cdots, x_{N-1,k})$, X_l is in normal to vector r $Y_l(x_{0,i}, x_{1,j}, \cdots, X_l, \cdots, x_{N-1,k})$ in the N + 1 dimensional space. The function for curve interpolation in plane $x_{0,i}$, $x_{1,j}$, $\cdots$, X_l, $\cdots$, $x_{N-1,k}$, Y_l is written as

$$y(x_{0,i}, x_{1,j}, \cdots, x_{l,m}, \cdots, x_{N-1,k}) = G_{x_{0,i}, x_{1,j}, \cdots, x_l, \cdots, x_{N-1,k}}[X_l, x_{l,m}, Y_n(x_{0,i}, x_{1,j}, \cdots, X_l, \cdots, x_{N-1,k})] \quad (1)$$

where $x_{l,m}$ is the interpolated point on the direction of vector x_l, x_l is the input vector for curve interpolation, and Y_l is the output vector for curve interpolation. Then, if a random point whose coordinate is $x_{0,a}$, $x_{1,b}$, $\cdots$, $x_{N-1,c}$, $\bar{y}$ is given in the N + 1 dimensional space, where $x_{0,a}$, $x_{1,b}$, $\cdots$, $x_{N-1,c}$ are known, $\bar{y}$ can be calculated by Eq. (2)

$$\bar{y}(x_{0,a}, x_{1,b}, \cdots, x_{N-1,c}) = G_{x_{0,a}, x_{1,b}, \cdots, x_{N-2,p}, x_{N-1,c}}[X_{N-1}, x_{N-1,c}, Y_{N-1}(x_{0,a}, x_{1,b}, \cdots, X_{N-1})] \quad (2)$$

where X_{N-1} is $\{x_{N-1,p}, p \in [0, M_{N-1}]\}$, $x_{N-1,c}$ is the interpolation point, and vector Y_{N-1} $(x_{0,a}, x_{1,b}, \cdots, X_{N-1})$ is $\{y_{N-1,0}, y_{N-1,1}, \cdots, y_{N-1,MN-1}, 0 \leq p \leq M_{N-1}\}$. For given p,

$$y(x_{0,i}, x_{1,j}, \cdots, x_{l,m}, \cdots, x_{N-1,k}) = G_{x_{0,i}, x_{1,j}, \cdots, x_l, \cdots, x_{N-1}, k}[X_l, x_{l,m}, Y_n(x_{0,i}, x_{1,j}, \cdots, X_l, \cdots, x_{N-1,k})] \quad (3)$$

where $x_{N-2,q}$ is the interpolation on the direction of vector X_{N-2}, and $x_{N-2,q} \in X_{N-2}$; vector $Y_{N-2}(x_{0,a}, x_{1,b}, \cdots, X_{N-2}, x_{N-1,p})$ can be calculated by the same method.

3   Application Results and Discussion

3.1   Application

In this section, the presented approach in Sect. 2 was applied in spectral analysis of multi-component gas mixture. Every absorbance was looked as a common sensor. In the measured gas mixture, there were five components of analytes, methane, ethane, propane, iso-butane and n-butane were included, and two components of interferents, iso-pentane and n-pentane were included. The spectrum resolution was 4 cm^{-1}. Because their molecular structures are close to each other, their absorption spectra overlap with each other extensively, see Fig. 3. In other words, there may be high cross-sensitivity.

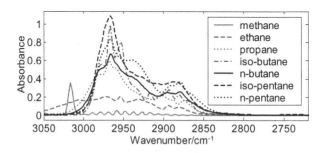

Fig. 3. Absorption spectra of seven components of light alkane gas

In our previous work [3], seven thousands of samples had ever prepared. In order to test the performance of this approach presented in this paper, two hundred of samples have been chosen from them and used to build analysis model by the method introduced above. Firstly, "additional sample" was produced with these samples. Then, baseline correction, feature extraction and model building were performed by the same approaches used in our previous work. Finally, testing samples collected at well-head of oil exploration were analyzed with the inverse models built in this paper[1].

3.2   Results and Discussion

The analysis results of every component of analyte were shown in Fig. 4. In this figure, curves marked with "FTIR0", "FTIR1" and "GC" indicated analysis results in our previous work [3], that obtained by the method introduced in this paper, and that gotten with GC, respectively. GC have been widely used for a long time in the field of gas well logging, its gas well log could be taken as the reference for testing the instrument based on FT-IRS [3]. Here, "GC" was still taken as a reference.

From every sub-figure of Fig. 4, one could found that "FTIR1" almost overlapped with "FTIR0", especially for methane and propane. For ethane, from the 100th minute,

---

[1] All the samples can be downloaded from my network disk: http://pan.baidu.com/s/1hqrKdJe.

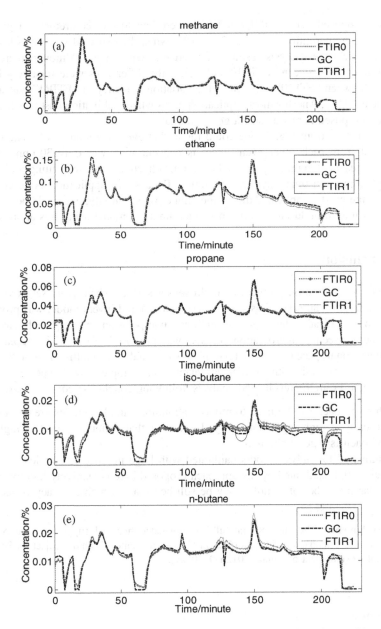

Fig. 4. Analysis results of light alkane mixture. (a) methane, (b) ethane, (c) propane, (d) iso-butane, (e) n-butane

"FTIR1" began to be less than "FTIR0", the difference increased as time went on. But for iso-butane and n-butane, the condition was almost the opposite. The maximum relative error appeared in Fig. 4(d). At the 144th minute, denoted with an ellipse, the analysis result of GC was 88×10^{-6} (88 ppm) while that of FT-IRS calibrated by the approach proposed here was 107×10^{-6}. According to the requirement of gas well logging [3], the permitted error was 30×10^{-6} when the expected value was less than 100×10^{-6}. So the analysis result still fitted the requirement. Although there were some differences between "FTIR1" and "GC", the differences were relatively higher than that between "FTIR0" and "GC", the variation tendency of "FTIR1" was always the same of "GC". And for most application cases where calibration cost was high, the approach proposed here was still a good alternative.

Additionally, from Fig. 4, one could found that every component of gas could be recognized from each other even when its concentration was only 30 ppm. So the approach presented in this paper could overcome effectively the over-fitting problem of NN applied in multi-sensor system calibration. Because the application example might be one of the most complicated due to the high cross-sensitivity, one could believe that it was possible to build any multi-sensor measurement model using this approach.

4   Conclusion

To overcome calibration problem of multi-sensor system where cross-sensitivity was high, a calibration approach based on forward-model and invers-model was proposed. Aimed to the disadvantage of conventional interpolation, an approach called as patulous-cross-section-interpolation (PCSN) was proposed to extend interpolation to high-dimensional space for getting forward-model. Finally, the multi-component of light alkane mixture analysis with FT-IRS was taken as an example to test the performance of the approach. From the results and analysis, following conclusion could be drawn:

(1) This approach can be used to reduce calibration sample number greatly, even in the case of high cross-sensitivity. In case of that calibration cost is high, this approach is a good alternative.
(2) Compared to analysis model calibrated with large number of samples, the accuracy of that calibrated by the approach proposed here is relatively less for a little. So large number of samples may be still necessary if high accuracy is needed.

Acknowledgments. The authors gratefully acknowledge financial support by the National Natural Science Foundation of China (51277144) and the National Major Project on Scientific Instrument Development of China (2012YQ240127).

References

1. Mitra, P., Mukhopadhyay, A.K.: ZnO thin film as methane sensor. B. Pol. Acad. Sci. Tech. **55**, 281–285 (2007)
2. Shoyama, M., Hashimoto, N.: CO sensing properties of ZnO\SnO2 composite thin films prepared by the sol gel method. J. Ceram. Soc. JPN (special Issue) **112**, s559–s561 (2004)
3. Tang, X., Li, Y., Zhu, L., Zhao, A., Liu, J.: On-line multi-component alkane mixture quantitative analysis using Fourier transform infrared spectrometer. Chemometr. Intell. Lab. **146**, 371–377 (2015)

4. Xiaojun, T., Liu, J.: Research on dynamic characteristics of multi-sensor system in the case of cross-sensitivity. Sci. China Ser.E Eng. Mater. Sci. **48**, 1–22 (2005)

5. Zhang, Y., Liu, J., Zhang, Y., Tang, X.: Cross sensitivity reduction of gas sensors using genetic algorithm neural network. Opt. Eng. **41**, 615–625 (2002)

6. Liao, R.J., Bian, J.P., Yang, L.J., Grzybowski, S.: Forecasting dissolved gases content in power transformer oil based on weakening buffer operator and least square support vector machine–Markov. IET Gener. Transm. Dis. **6**, 142–151 (2012)

7. Giorgi, M.G.D., Campilongo, S., Ficarella, A., Congedo, P.M.: Comparison between wind power prediction models based on wavelet decomposition with Least-Squares Support Vector Machine (LS-SVM) and Artificial Neural Network (ANN). Energies **7**, 5251–5272 (2014)

8. Sirakov, N.M., Granado, I., Muge, F.H.: Interpolation approach for 3D smooth reconstruction of subsurface objects. Comput. Geocis. UK **28**, 877–885 (2002)

A Virtual Communication Strategy for Smart Photovoltaic Generation Systems

Jing Yang[1], Hui Liao[2], Jing Chen[3], Yang Yu[2], and Yi Wang[3(✉)]

[1] Experimental and Innovation Practice Center,
Harbin Institute of Technology, Shenzhen 518055, China
[2] State Grid Electric Power Research Institute,
Shenzhen Branch, Shenzhen 518054, China
[3] College of Computer Science and Software Engineering,
Shenzhen University, Shenzhen 518060, China
yiwang@szu.edu.cn

Abstract. Renewable energy sources such as photovoltaic power generation is becoming a promising technology for modern computing and communication systems. Current power stations integrate multiple smart devices. Each of them uses a communication front-end computer to facilitate the data collection and transmission to the back-end servers. Compared to conventional power stations that use fossil fuel, the number of communication front-end computers in photovoltaic power station is nearly doubled, which causes maintenance and pricing issues. This paper presents a novel virtual communication strategy for smart photovoltaic power generation systems. We use serial port servers to replace communication front-end computers and establish a virtual communication channel to transparently manage communication. We implement the proposed communication strategy on a commercial photovoltaic power generation system. The experimental results show that the proposed strategy can reduce the costs by over 60% and achieve reliable performance.

Keywords: Communication strategy · Photovoltaic generation · Solar energy · Virtual port · Relay protection

1 Introduction

The burning of fossil fuels, such as coal and oil, will release greenhouse gases and other pollutants into the atmosphere. Solar energy is a typical renewable energy, which becomes a clean alternative to fossil fuels [1,2]. In modern photovoltaic systems, many smart devices are integrated into the power generation system. Most of current photovoltaic systems follow the distributed communication infrastructure in conventional fossil fuels powered system. In such systems, every inverter chamber is configured with a communication front-end computer. All the collected information will be transformed into standard format and sent to the back-end server. Photovoltaic systems necessarily allocate much more

M. Qiu (Ed.): SmartCom 2016, LNCS 10135, pp. 256–266, 2017.
DOI: 10.1007/978-3-319-52015-5_26

communication front-end computers. The large number of communication front-end computers causes high production costs and incurs difficulty in hardware maintenance and software update.

In order to solve this problem, this paper presents a novel virtual port management strategy for photovoltaic systems. We utilize the properties of ethernet communication protocols and Modbus protocol, to replace distributive communication front-end computers with serial port server. The serial port server builds up a virtual communication channel on top of smart devices. Using this transparent communication, it simplifies the connection of communications. Since serial port server is relatively cheaper and the maintenance cost is much easier compared to the conventional distributive front-end computer, the proposed strategy can effectively reduce the system cost and provide reliable performance. We have implemented the proposed strategy in a commercial photovoltaic generation system. Experimental results proved our design concept. The prototype can demonstrate reliable and efficient communication performance.

The remainder of this paper is organized as follows. Section 2 shows the background and the motivation. Section 3 presents our proposed virtual data communication strategy for photovoltaic generation systems. Section 4 shows the experimental results. Finally, we conclude the paper and discuss the future work in Sect. 5.

2   Background and Motivation

Serial port is a kind of communication process by sending data sequentially one bit at a time. It normally transits through a communication channel or computer bus. In modern relay protection systems, RS-485 and RS-232 are the typical standards for serial binary data transfer. These compatible serial interfaces define connector pin-outs, cabling, signal levels and parity checking information. Controllers can utilize the serial interface to recognize and use any types of networks without realizing which they are actually communicating on.

Modbus protocol, which is designed by Modicon Inc., provides a protocol with the predefined content structure. This open protocol can be used in Modicon controllers to communicate using a master-slave fashion, in which only one master (i.e., a device) can initiate transactions. Other slaves have to respond by taking the action requested in the query, or providing the demanded data to the master.

During data communications, Modbus protocol can determine the way to locate each device address to the associated controller. It will also determine the consecutive actions to be taken and how to extract any data or other information contained in the message. It builds up a common format for the content and the layout of the message field.

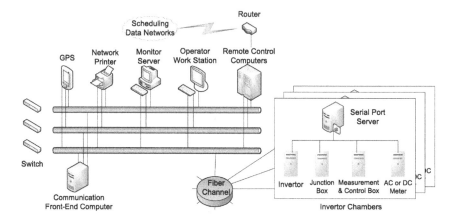

Fig. 1. The proposed network architecture of the photovoltaic monitoring system.

3   Design of Virtual Data Communication Strategy For Photovoltaic Generation Systems

3.1   System Architecture

This section presents the system architecture of the proposed virtual data communication strategy for photovoltaic generation systems. The system has the multiple-level, distribute, and open network infrastructure. As shown in Fig. 1, it consists of two functional layer, i.e., station control layer and site control layer. Among them, station control layer is a centralized management layer, and it adopts double-star ethernet infrastructure. All devices in the station, including back-end computers, remote-control computers, and front-end computers, are attached to station control networks. Site control layer handles network access. It uses RS-485 bus to collect all the operation states of devices, and it communicates with station control layer through optical fiber ethernet.

In the photovoltaic system, all inverter cabinets compose of optical fiber ring network. Serial port devices, including inverters, junction boxes, measurement and control boxes, AC and DC voltmeters are connected to the ethernet via serial port server in the chamber. Serial port server lets real serial ports map to virtual serial ports in communication front-end machine through TCP/UDP protocols. This step provides transparent translation between serial ports and network ports. In this way, communication front-end machine can directly access smart devices in inverter chamber, thereby reducing the number of communication front-end machines and improving the debugging efficiency in communication.

3.2   System Analysis

In our design, the serial ports in the server are mapped to virtual serial port channel in ethernet. The communication delay T_{delay} in ethernet mainly consists of queuing delay T_{queue}, send delay T_{send}, and transmit delay T_{trans}.

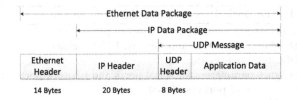

Fig. 2. The format of the UDP packet.

$$T_{delay} = T_{queue} + T_{send} + T_{trans} \qquad (1)$$

Among them, queue delay is due to the concurrent transition from multiple nodes. This conflict will cause the resend of data, and it is related to the number of conflicts. Modbus protocol is a query-based protocol. The query mechanism is controlled by communication front-end machine and the message flow is stable. Our experimental results also prove that this transition and scheduling mechanism will not cause queue delay, when the network budget is lower than the communication bandwidth of the ethernet.

Send delay is determined by the frame length L and the transition rate V of the ethernet. Then the send delay can be obtained by the following equation.

$$T_{send} = \frac{L}{V} \qquad (2)$$

For a 100 Mbit/s ethernet, suppose L is 256 bytes (including the leading code with 8 bytes), the send delay is 26.4 μs.

Transmit delay is affected by transmission rate of the signal, the number of repeaters, and the physical distance between the source and the destination nodes. Suppose a 200 MW large-scale photovoltaic generation station has 200 inverter chambers and 200 serial port servers, and every serial port server has 4 RS-485 ports. Serial port transmit rate is 9600 bit/s; data bits are 8 bits; starting bit is 1 bit; stop bit is 1 bit; and there is no correcting bit. Modbus sends 8 bytes of data for each query, and the response from the device takes 100 bytes. The utilization ratio of serial port channel is 90%. Therefore, the data flow of a single serial port takes $[9600/(8 + 1 + 1 + 0)] \times 90\% = 864$ bytes/s, which corresponds to $[864/(8 + 100)] \times 2 = 16$ packages/s.

Figure 2 illustrates the frame format of an UDP packet. Based on this format, every UDP package includes application data and the 42 bytes data header. Then the corresponding data flow for virtual serial port channel is $[(42 + 8) + (42 + 100)]/2 \times 16 = 1536$ bytes/s. The communication bandwidth is 12 kbit/s.

Based on the analysis, for a large-scale photovoltaic generation system with capacity of 200 MW under very high serial port utilization ratio, the total communication bandwidth is only 9.6 Mbit/s, which is less than 10% of the ethernet's 100 Mbit/s bandwidth. Therefore, the bandwidth for communication front-end computer is well under control. Using middle-end or even low-end centralized communication front-end computer can satisfy the requirement.

Algorithm 3.1. The Basic Process for Serial Port Server

Require: A number of S_{port} serial ports, network ports.
Ensure: Issue the data transaction and forward UDP package.
 1: System initialization and bootup.
 2: Read and configure parameters for serial ports and network ports.
 3: Monitor the condition of ports 9001-9008.
 4: **while** Ports 9001-9008 receive UDP package x **do**
 5:    $S_x \leftarrow$ network port number N_x- 9000.
 6:    Forward the application data in UDP package x to serial port S_x.
 7:    **for** each serial port S_y among S_{port} **do**
 8:       **if** Serial port S_y received data **then**
 9:          **if** the received data is over-length or overtime **then**
10:             Network port number $N_y \leftarrow$ serial port number S_y + 9000.
11:             Forward UDP package to network port N_y.
12:          **end if**
13:       **end if**
14:    **end for**
15:    Issue the data transaction.
16: **end while**

3.3   Design of Serial Port Server

Serial port server does not interpret specific application-level protocols. It only implements the transparent translation between serial port data and network data. This unique functionality makes serial port server feasible for various photovoltaic generation systems. Serial port server only needs to configure several parameters to facilitate the data transmission between serial port and network. Since the communication overhead should be carefully designed [3,4], this can significantly minimize the workload for the system maintenance and debugging, thereby reducing the communication error rate.

Our design for the serial port server has been implemented in the photovoltaic generation system. The basic process for serial port server is shown in Algorithm 3.1. Serial port server will monitor network ports 9001 to 9008. In our design, we allocate these eight network ports to serve the serial ports. It is a general strategy that can be extended to the case with multiple network ports. Serial port server will extract application data from the received UDP package, and forward the package to the corresponding serial port (1 to 8).

Serial port server will also monitor the eight serial ports to ensure the completion of data transition. The received data will be stored in the corresponding data buffer. When the data exceeds the maximum buffer length or the data encounters time out, serial port server will send the package in data buffer as UDP application data to the corresponding network port. In our prototype, the timeout is set as 3 ms.

3.4   Design of Communication Front-End Machine

Communication front-end machine creates a protocol thread for each virtual serial port. It manages each single RS-485 bus, and multiple threads can be executed in parallel. For one thread, only one protocol (e.g., Modbus protocol) can be interpreted. Communication front-end machine uses IP address and network port number to identify virtual serial port. Each virtual port will create a socket, which can be used to send or receive specific IP address and the UDP package of the network port number.

For the basic process for communication front-end machine, communication front-end machine checks the status of smart device with the same virtual serial port based on the round-robin manner. It sends UDP query through sockets. Communication front-end machine simultaneously monitors whether the socket received the UDP data package. If the socket received the data, it will extract application data and store it into data buffer. When the message in data buffer has been successfully assembled, protocol module will call analysis functions and handle the data. A new round of query will be immediately started after this step. If the serial port has been interrupted or the internet has encountered package loss, this case will be treated as failure after a certain time of overtime. Communication front-end machine will reset polling time and restart the previous query.

4   Evaluation

4.1   Experimental Setup

We have implemented the prototype of our design in a commercial photovoltaic generation system. We build and operate a solar power station with an initial capacity of 20 MW in Xinjiang in the dry north-west of China, whose abundant sunshine and deserts make it China's prime location for solar power generation. Figure 3 illustrates the prototype of the system. This photovoltaic power station contains twenty inverter chamber. Each inverter chamber has two inverters, twelve junction boxes, one measurement and control box, two AC voltmeters, and two DC voltmeters. A total of nineteen smart devices are attached to four RS-485 ports, and each RS-485 port handles about 4–5 devices. The query time for each RS-485 bus is 200 ms, which can ensure that data can be updated within one second. This setup is sufficient to satisfy the data refresh requirement for modern photovoltaic power station.

We configured one communication front-end machine, and its IP address is 198.121.0.203. We also configured twenty serial port servers. The IP addresses range from 198.121.0.101 to 198.121.0.120. Communication front-end machine consists of eighty Modbus protocol modules. These eighty modules are communicated with virtual serial ports 9002 to 9005 of each serial port server. Communication front-end machine can parallel process every protocol module, and each machine can simultaneously query eighty virtual serial port channels.

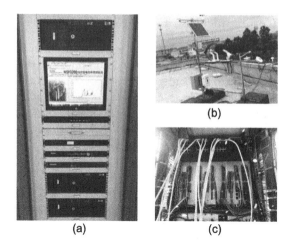

Fig. 3. The implementation of the prototype. (a) the photovoltaic monitoring system that incorporates multiple smart devices. (b) the photovoltaic system that monitors the radiation strength. (c) the back review of the photovoltaic system.

Table 1. Performance for virtual communication strategy for smart photovoltaic generation systems

Performance metrics	Experimental results
Average network traffic	151 kbit/s
Peak network traffic	1.5 Mbit/s
Effective message length	28 bytes

4.2   Experimental Results

Table 1 illustrates the performance of virtual communication strategy for smart photovoltaic generation systems. We test the average network traffic, and the result is 151 kbit/s. In terms of effective message length, which denotes the transmission of effective Modbus query and response messages, the average length is 28 bytes. Based on these results, even for the large-scale photovoltaic generation system with initial capacity of 200 MW, the actual network traffic is only 1.5 Mbit/s. This result is much lower than the theoretical bound presented in the analysis in Sect. 3. This result shows that our implementation can satisfy the design objective of the photovoltaic communication system.

In terms of communication costs, we have tested a set of experiments with different test cases. The experimental results are shown in Tables 2 and 3. For each serial port server with specific IP address (e.g., 198.121.0.103), its network port has a corresponding RS-485 port. Communication front-end machine will check the smart devices one after another based on the round-robin manner. After the response from the device, it will send the next query message. The communication cost denotes that the time between the query from communication

Table 2. Message interaction time and the total message length for virtual communication strategy

Test cases	Message interaction time (ms)	The total message length (bytes)
1	22.123	17
2	25.619	21
3	22.206	17
4	21.621	17
5	20.243	15
6	30.068	23
7	23.953	19
8	20.618	15
9	68.297	61
10	83.555	75
11	22.085	17

front-end machine and the request being serviced. The average communication cost is 4.638 ms. Among them, the configuration for message timeout of serial server is 3 ms. The rest of the costs belong to network communication delay and device response cost, and this time is 1.638 ms. Based on the analysis in Sect. 3, the network communication delay is less than $100\,\mu s$. Therefore, the majority of other time costs is caused by the response of the smart device itself.

In our implementation of photovoltaic generation system, we use only one centralized communication front-end machine and twenty serial port server to replace twenty distributed communication front-end machines. The price for

Table 3. Time costs for virtual communication

Test cases	Serial communication time (ms)	Other time costs (ms)
1	17.708	4.415
2	21.875	3.744
3	17.708	4.498
4	17.708	3.913
5	15.625	4.618
6	23.958	6.110
7	19.792	4.161
8	15.625	4.993
9	63.542	4.755
10	78.125	5.430
11	17.708	4.377

centralized communication front-end machine is close to that for distributed one, while the price for serial port server is only one-third of communication front-end machine. As a result, the total budget of the system reduces about two-thirds. Our experimental results show that the requested communication bandwidth is far below the theoretical bound. Therefore, our virtual communication strategy presents a feasible and reliable solution.

5   Conclusion

This paper presents a virtual communication strategy for photovoltaic generation system. We jointly consider the unique characteristics of communications in photovoltaic system, and the property of TCP/UDP, RS-485 and Modbus protocols. Different from the conventional design that uses the distributed communication infrastructure, our solution uses a centralized communication infrastructure, and it can significantly reduce the device expenses while achieving similar system performance. Our communication strategy has been implemented in a commercial 20 MV photovoltaic generation system. Experimental results show that our solution can achieve reliable and satisfactory performance.

Advanced computing and communication systems adopt emerging technologies to reduce the memory usage and boost the system performance. It is possible to integrate parallel and hybrid storage systems [5–8,21], flash memory [9–14], phase-change memory (PCM) [15–17], spin-transfer torque RAM (STT-RAM) [18], and 3D memory hierarchy [19,20] for better scalability, which may require different architectural support and communication optimization strategies.

Acknowledgements. The work described in this article is partially supported by the grants from National Natural Science Foundation of China (61502309), Guangdong Natural Science Foundation (2014A030313553, 2013B090500055, 2014A0303 10269 and 2016A030313045), Shenzhen Science and Technology Foundation (JCYJ20150529164656096 and JCYJ20150525 092941059), and Natural Science Foundation of SZU (701-000360055905, 701-00037134, and 827-000073).

References

1. Wang, Y., Chen, R., Shao, Z., Li, T.: SolarTune: real-time scheduling with load tuning for solar energy powered multicore systems. In: 2013 IEEE 19th International Conference on Embedded and Real-Time Computing Systems and Applications, pp. 101–110 (2013)
2. Li, C., Zhang, W., Cho, C.B., Li, T.: SolarCore: solar energy driven multi-core architecture power management. In: 2011 IEEE 17th International Symposium on High Performance Computer Architecture, pp. 205–216 (2011)
3. Wang, Y., Liu, D., Qin, Z., Shao, Z.: Optimally removing intercore communication overhead for streaming applications on MPSoCs. IEEE Trans. Comput. **62**(2), 336–350 (2013)

4. Wang, Y., Shao, Z., Chan, H.C.B., Liu, D., Guan, Y.: Memory-aware task scheduling with communication overhead minimization for streaming applications on bus-based multiprocessor system-on-chips. IEEE Trans. Parallel Distrib. Syst. **25**(7), 1797–1807 (2014)
5. Nijim, M., Qin, X., Qiu, M., Li, K.: An adaptive energy-conserving strategy for parallel disk systems. Future Gener. Comput. Syst. **29**(1), 196 (2013)
6. Qiu, M., Chen, L., Zhu, Y., Hu, J., Qin, X.: Online data allocation for hybrid memories on embedded tele-health systems. In: 2014 IEEE International Conference on High Performance Computing and Communications (HPCC 2014), pp. 574–579 (2014)
7. Wang, J., Qiu, M., Guo, B.: High reliable real-time bandwidth scheduling for virtual machines with hidden markov predicting in telehealth platform. Future Gener. Comput. Syst. **49**(C), 68 (2015)
8. Li, X., Xiao, L., Qiu, M., Dong, B., Ruan, L.: Enabling dynamic file I/O path selection at runtime for parallel file system. J. Supercomput. **68**(2), 996 (2014)
9. Guan, Y., Wang, G., Wang, Y., Chen, R., Shao, Z.: BLog: block-level log-block management for NAND flash memorystorage systems. SIGPLAN Not. **48**(5), 111–120 (2013)
10. Wang, Y., Liu, D., Qin, Z., Shao, Z.: An endurance-enhanced flash translation layer via reuse for NAND flash memory storage systems. In: 2011 Design, Automation Test in Europe, pp. 14–19 (2011)
11. Huang, M., Liu, Z., Qiao, L., Wang, Y., Shao, Z.: An endurance-aware metadata allocation strategy for MLC NAND flash memory storage systems. IEEE Trans. Comput. Aided Des. Integr. Circ. Syst. **35**(4), 691–694 (2015)
12. Zhang, Q., Li, X., Wang, L., Zhang, T., Wang, Y., Shao, Z.: Lazy-RTGC: a real-time lazy garbage collection mechanism with jointly optimizing average and worst performance for NAND flash memory storage systems. ACM Trans. Des. Autom. Electron. Syst. **20**(3), 43:1 (2015)
13. Chen, R., Qin, Z., Wang, Y., Liu, D., Shao, Z., Guan, Y.: On-demand block-level address mapping in large-scale NAND flash storage systems. IEEE Trans. Comput. **64**(6), 1729–1741 (2015)
14. Chen, R., Wang, Y., Hu, J., Liu, D., Shao, Z., Guan, Y.: Unified non-volatile memory and NAND flash memory architecture in smartphones. In: 2015 20th Asia and South Pacific Design Automation Conference (ASP-DAC 2015), pp. 340–345 (2015)
15. Qiu, M., Ming, Z., Li, J., Gai, K., Zong, Z.: Phase-change memory optimization for green cloud with genetic algorithm. IEEE Trans. Comput. **64**(12), 3528 (2015)
16. Zhao, M., Jiang, L., Shi, L., Zhang, Y., Xue, C.: Wear relief for high-density phase change memory through cell morphing considering process variation. IEEE Trans. Comput. Aided Des. Integr. Circ. Syst. **34**(2), 227 (2015)
17. Pan, C., Xie, M., Hu, J., Qiu, M., Zhuge, Q.: Wear-leveling for PCM main memory on embedded system via page management and process scheduling. In: 2014 IEEE 20th International Conference on Embedded and Real-Time Computing Systems and Applications (RTCSA), pp. 1–9 (2014)
18. Li, J., Shi, L., Li, Q., Xue, C.J., Chen, Y., Xu, Y.: Cache coherence enabled adaptive refresh for volatile STT-RAM. In: Proceedings of Design, Automation Test in Europe Conference Exhibition (DATE 2013), pp. 1247–1250 (2013)
19. Wang, Y., Shao, Z., Chan, H., Bathen, L., Dutt, N.: A reliability enhanced address mapping strategy for three-dimensional (3-D) NAND flash memory. IEEE Trans. Very Large Scale Integr. (VLSI) Syst. **22**(11), 2402 (2014)

20. Wang, Y., Zhang, M., Dong, L., Yang, X.: A thermal-aware physical space allocation strategy for 3D flash memory storage systems. In: Proceedings of the 2016 International Symposium on Low Power Electronics and Design, ISLPED 2016, pp. 290–295. ACM, New York (2016)
21. Liu, M., Liu, D., Wang, Y., Wang, M., Shao, Z.: On improving real-time interrupt latencies of hybrid operating systems with two-level hardware interrupts. IEEE Trans. Comput. **60**(7), 978–991 (2011)

Hybrid One-Class Collaborative Filtering for Job Recommendation

Miao Liu[1], Zijie Zeng[1], Weike Pan[1], Xiaogang Peng[1(✉)], Zhiguang Shan[2], and Zhong Ming[1(✉)]

[1] College of Computer Science and Software Engineering, Shenzhen University, Shenzhen, China
mloud.vip@gmail.com, zengzijie1991@gmail.com,
{panweike,pengxg,mingz}@szu.edu.cn
[2] Information Research Department, State Information Center, Beijing, China
shanzhiguang@263.com

Abstract. Intelligent recommendation has been a crucial component in various real-world applications. In job recommendation area, developing an effective and personalized recommendation approach will be very helpful for the job seekers. In order to deliver more accurate job recommendations, we propose a system by incorporating users' interactions and impressions as the data source and design a hybrid strategy by taking the advantages of three existing one-class collaborative filtering (OCCF) algorithms. The proposed solution combines multi-threading techniques in the traditional item-oriented OCCF (IOCCF) and user-oriented OCCF (UOCCF) algorithms, and also applies an approximation of the sigmoid function in Bayesian personalized ranking (BPR) to improve the efficiency of the overall performance. Based on the experiment results, the proposed system shows the effectiveness of using users' interactions and impressions by an improvement of 23.78%, 15.61% and 11.50%, and the effectiveness of using the hybrid strategy by a further improvement of 34.55%, 16.20% and 20.90%, over IOCCF, UOCCF and BPR, respectively.

Keywords: One-class collaborative filtering · Hybrid recommendation · Job recommendation

1 Introduction

Intelligent recommendation has been an important solution to address various information overload challenges in industry, including product recommendation in e-commerce such as Amazon and TMall, and advertisement recommendation in search engines like Google and Baidu. Recently, job recommendation in professional networks, such as LinkedIn, becomes another important research topic that interests many researchers. For example, the ACM RecSys Challenge 2016 focuses on the task of job recommendation in XING[1], which is a typical social

---

[1] https://www.xing.com/.

© Springer International Publishing AG 2017
M. Qiu (Ed.): SmartCom 2016, LNCS 10135, pp. 267–276, 2017.
DOI: 10.1007/978-3-319-52015-5_27

website for businesses and individuals for job opportunities. Specifically, there are about 15 million users and 1 million job postings on the XING platform. The challenge is to find a number of suitable job postings that a XING user will probably interact with (e.g., click, bookmark and/or reply).

It is very complicated to develop a single super-model for this year's RecSys Challenge considering the extremely sparse dataset under such a big data volume. Generally speaking, in order to improve the recommendation performance, all useful data regarding users' interactions, temporal context and contents need to be combined as training data. However, with the limited computing resources, we decide to mainly concentrate on the public data of users' interactions and impressions, and formally define the recommendation task as a one-class collaborative filtering (OCCF) [4] problem.

In order to model the data properly, we adopt several state-of-the-art OCCF algorithms. Firstly, we implement user-oriented and item-oriented one-class collaborative filtering [1] (denoted as UOCCF and IOCCF, respectively), which achieves satisfying results as reported in [1]. Secondly, we choose Bayesian personalized ranking (BPR) [9] algorithm as the machine learning part due to its effectiveness as compared with [2,7]. Finally, in order to further boost the recommendation performance, we design a hybrid strategy, i.e., HyOCCF, to take the advantages of UOCCF, IOCCF and BPR. Moreover, we implement multi-threading techniques on traditional IOCCF and UOCCF, and utilize a portion of high quality open source code to optimize the efficiency of BPR model training.

2   Related Work

One-class collaborative filtering (OCCF) algorithms are widely known in the recommendation community. Most of them are used to handle problems of finding the most attractive items for different individuals accurately. The mainstream collaborative methods are divided into two branches: memory-based approaches (e.g., item-oriented OCCF [1]) and model-based approaches (e.g., Bayesian personalized ranking [9], factored item similarity models [2]). Both of these two kinds of methods rely on the input data that a user interacts with items positively such as clicks and browses.

For memory-based approaches, the most significant point is the measurement of similarity. An appropriate formula of neighbors' similarity can improve the recommendation result significantly [1]. Jaccard index and Cosine similarity are the most representative and easy-to-implement ones [10]. Some extensions of the measures above include: Cosine similarity with inverse user frequency [1], which captures an extra feature about the frequency of common users between two different items; and normalized similarity measurement [3] that makes the similarity of different items more comparable. In order to achieve a good tradeoff between the runtime and accuracy of similarity computation, we simply adopt the classical Jaccard index and Cosine similarity in our empirical studies.

On the other hand, for model-based approaches, plenty of machine learning based methods have been proposed. This kind of methods design some objective functions to model the data. Bayesian personalized ranking [9] and its extensions have been well studied and applied to the OCCF problem. The pairwise learning technique can leverage a mass of unobserved (user, item) pairs besides the observed data of users' behaviors. Moreover, the generalization of the BPR framework is also worth considering. For example, Pan et al. [5] studied the group preference based BPR for OCCF, where the group preference is introduced to relax the independence assumptions. And Zhao et al. [11] leveraged social connections to improve personalized ranking. What's more, the original authors of BPR pointed out that BPR is likely to suffer from the slow convergence problem and thus proposed an oversampling method to improve the pairwise learning of BPR based on the stochastic gradient descent (SGD) optimization framework [8]. This oversampling strategy leads to a faster convergence as well as a better prediction quality.

3 Background

3.1 Problem Definition

We believe that users' interactions and impressions are one of the most important parts of the data in a job recommendation system because they are very closely and directly related to users' preferences. In order to fully exploit the interactions and impressions in the form of (user, item) pairs, we convert the job recommendation task to a one-class collaborative filtering (OCCF) [4] problem, which is shown in Fig. 1. Specifically, the input of OCCF is abstracted as a set of (user, item) pairs, and the output is a personalized ranked list of items for each user.

We put some notations in Table 1.

Table 1. Some notations and explanations.

$\mathcal{U}$	the whole user set
$\mathcal{I}$	the whole item set
$u, v \in \mathcal{U}$	user ID
$i, j, k \in \mathcal{I}$	item ID
$\mathcal{R} = \{(u, i)\}$	(user, item) pairs in training data
$\hat{r}_{ui}$	predicted preference of user u on item i
$y_{ui} \in \{0, 1\}$	indicator variable, $y_{ui} = 1$ if $(u, i) \in \mathcal{R}$
$\mathcal{U}_i, \mathcal{U}_k$	users that prefer item i and item k, respectively
$\mathcal{I}_u, \mathcal{I}_v$	items that are preferred by user u and user v, respectively
$\mathcal{N}_j, \mathcal{N}_u$	K nearest neighbors of item j and user u, respectively

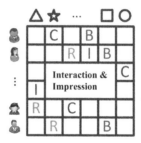

Fig. 1. An illustration of one-class collaborative filtering (OCCF) for job recommendation with both interaction data (i.e., "C" for click, "B" for bookmark and "R" for reply) and impression data (i.e., "I" for impression).

3.2 Popularity-Based Ranking

Popularity-based Ranking (PopRank) is a kind of basic method that estimates the popularity of each item (e.g., item i) as follows,

$$p_i = \sum_{u \in \mathcal{U}} y_{ui}/|\mathcal{U}|, \tag{1}$$

which is then used to generate the same set of most popular items for each user. Notice that PopRank is a very basic and non-personalized recommendation method.

3.3 Item-Oriented One-Class Collaborative Filtering

Item-oriented one-class collaborative filtering (IOCCF) [1] is a kind of well-known and classic collaborative filtering algorithm that has been widely used by many industry giants. The central theme can be summarized as calculating the similarity between each two items. For example, we may use Jaccard index or Cosine similarity, where the former turned out to be much better for the job recommendation task in our empirical studies. Specifically, the Jaccard index between item k and item j is as follows,

$$s_{kj} = \frac{|\mathcal{U}_k \cap \mathcal{U}_j|}{|\mathcal{U}_k \cup \mathcal{U}_j|}. \tag{2}$$

With the calculated similarity, we select the K most similar items of item i, i.e., $\mathcal{N}_i$, and then predict the preference of user u on item i as an aggregation of the similarity [1],

$$\hat{r}_{ui} = \sum_{k \in \mathcal{I}_u \cap \mathcal{N}_i} s_{ki}, \tag{3}$$

which can then be used to generate a personalized list of items for each user.

3.4 User-Oriented One-Class Collaborative Filtering

Similarly, user-oriented one-class collaborative filtering (UOCCF) captures the feature of users' similarity with a similar theme to IOCCF. Specifically, the predicted preference of user u on item i is as follows,

$$\hat{r}_{ui} = \sum_{v \in \mathcal{U}_i \cap \mathcal{N}_u} s_{vu} \tag{4}$$

where $s_{vu} = \frac{|\mathcal{I}_v \cap \mathcal{I}_u|}{|\mathcal{I}_v \cup \mathcal{I}_u|}$ is the Jaccard index between user v and user u, and $\mathcal{N}_u$ is a set of K nearest neighbors of user u.

In the experiments, we have implemented both IOCCF and UOCCF with multi-threading techniques in order to speed up the procedure of similarity calculation.

3.5 Bayesian Personalized Ranking

Bayesian personalized ranking (BPR) is a state-of-the-art machine learning based algorithm concentrating on the pattern of pairwise learning. In terms of accuracy and efficiency, BPR is one of the best models for the large-scale OCCF problem. The basic unit of BPR is the preference difference of a user u on two items $i \in \mathcal{I}_u$ and $j \in \mathcal{I} \backslash \mathcal{I}_u$ [9],

$$\hat{r}_{uij} = \hat{r}_{ui} - \hat{r}_{uj} = (U_u. V_i.^T + b_i) - (U_u. V_j.^T + b_j), \tag{5}$$

where $b_i \in \mathbb{R}$ is the item bias, and $U_u., V_i. \in \mathbb{R}^{1 \times d}$ are user u's and item i's latent features, respectively. The preference difference $\hat{r}_{uij}$ is then fed to a minimization problem [9],

$$\min_{\Theta} \sum_{u \in \mathcal{U}} \sum_{i \in \mathcal{I}_u} \sum_{j \in \mathcal{I} \backslash \mathcal{I}_u} (-\ln \sigma(\hat{r}_{uij}) + regularization), \tag{6}$$

where $\sigma(z) = 1/(1 + \exp^{-z})$ is the sigmoid function. Notice that $\Theta = \{U_u., V_i., b_i | u \in \mathcal{U}, i \in \mathcal{I}\}$ denotes the set of parameters to be learned. And by the optimization method of stochastic gradient descent (SGD), BPR will reach a satisfying convergence in an acceptable time for the parameter learning.

Besides, we find that the sigmoid function $\sigma(z)$ implemented by the default exponential function in C++ library takes a great proportion of time in every learning iteration of the BPR algorithm. One possible reason is that the default exponential function is implemented by Taylor formula, which takes polynomial time when it is invoked. In order to reduce the runtime in BPR, we utilize a section of source code from Google word2vec[2]. Finally, it only takes a constant runtime after optimization.

---

[2] https://code.google.com/p/word2vec/.

4   Hybrid One-Class Collaborative Filtering

In this section, we aim to integrate the aforementioned three OCCF algorithms, i.e., IOCCF, UOCCF and BPR, in one single framework. The reason we use hybridization is that we believe that the prediction power of two memory-based methods and one model-based method may be further boosted when they are put together, which is also often observed in previous studies on other recommendation problems. Specifically, for each user, we generate the final recommendation as follows:

1. Firstly, we obtain three lists of top-30 jobs, generated by IOCCF, UOCCF and BPR separately;
2. Secondly, we merge the two lists of IOCCF and BPR by putting the common items in the front in a random order, and then include items from each list one-by-one from top to bottom until we have 30 jobs;
3. Thirdly, we merge the temporary list from the above second step and the list of UOCCF in the same way and get the final list of jobs for recommendation.

Notice that we choose to merge IOCCF and BPR first because they perform worse than UOCCF, which is expected to achieve a better final result. This merging strategy can also be adapted to a scenario where more than three base algorithms are combined, i.e., merging the results in a bottom-up manner with worst-performing algorithms first.

We illustrate the merging procedure of two lists in Fig. 2.

Fig. 2. An illustration of merging two intermediate recommendation lists of jobs in HyOCCF. Notice that the symbols indicate jobs that the lists share in common.

5   Experimental Results

5.1   Dataset and Evaluation Metric

Due to the lack of public job recommendation data, we use the released data of ACM RecSys Challenge 2016. Specifically, there are 1.4 M users in the user profile file "users.csv" and 1.4 M items in the item description file "items.csv", among which 150 K users appear in the test user set (i.e., the number of users listed in a separate user profile file "target_users.csv") and 327 K items appear in the test item set (i.e., the number of items marked as "active_during_test = 1" in "items.csv").

For the above 1.4 M users and 1.4 M items, there is some interaction data (i.e., click, bookmark, reply and delete) for each (user, item) pair and some impression data for each user. We keep the positive interaction data only (i.e., click, bookmark and reply) for the users and items listed in "users.csv" and "items.csv", and get 7.7 M (user, item) interaction pairs. For the impression data, we only keep the very recent data (i.e., week-45), and those appear for only 1, ≥ 7, and ≥ 40 successive times within a certain user's impression record, and get 5.5 M, 2.8 M and 0.25 M (user, item) impression pairs, respectively.

We list the statistics of the data in Table 2. Notice that we have removed the repeated copies of (user, item) pairs.

Table 2. Description of the (user, item) interaction pairs and impression pairs used in the experiments.

Dataset (abbreviation)	# Pairs
Interaction (Int)	7.7 M
Impression_1, randomly take 1/7 (Imp_1)	0.56 M
Impression_7 (Imp_7)	2.8 M
Impression_40 (Imp_40)	0.25 M

As for the evaluation metric, we take the official evaluation method[3]. Specifically, for each user u from the test user set $\mathcal{U}_{te}$, we compare the recommended ranked list of 30 items and the items that are preferred by the user u (i.e., with positive interaction), and calculate the performance: Performance(u) = $20 \times$ (Prec@2 + Prec@4 + recall + 1-call) + $10 \times$ (Prec@6 + Prec@20). Finally, we get the overall performance Performance(u) $\times |\mathcal{U}_{te}|$. Notice that there are 150000 users in the test set.

5.2   Baselines and Parameter Settings

In order to validate the effectiveness of the proposed hybrid strategy on heterogeneous data, we include all those three base recommenders, i.e., IOCCF, UOCCF and BPR, on interaction data only, and on both interaction data and impression data. We will report the detailed configuration of the model parameters in the subsequent sections.

Notice that the experiments[4] are conducted on the machines installed with the Intel Xeon E5-2640 v3 32 cores and 64 GB RAM.

---

[3] https://github.com/recsyschallenge/2016/blob/master/EvaluationMeasure.md.
[4] The implementations of PopRank, IOCCF, UOCCF, BPR and HyOCCF can be downloaded at https://sites.google.com/site/weikep/HyOCCF.zip.

274   M. Liu et al.

Table 3. Recommendation performance using 7.7 M (user, item) interaction pairs.

Method	Performance
PopRank	67768.13
IOCCF	238293.14
UOCCF	295429.91
BPR	294404.37

5.3 Results with Interaction Data

In the beginning, we run those three base recommendation algorithms and
PopRank with interaction data only. Obviously, there is no need to adjust the
parameter in PopRank except selecting top-30 most popular jobs. For memory-
based methods, more specifically, IOCCF and UOCCF, we fix the number of
nearest neighbors as 200 and utilize Jaccard index as the similarity measurement in this stage. As for BPR, we fix the value of tradeoff parameter as 0.01,
the learning rate as 0.01, the number of latent dimensions as 150, and the number
of iterations as 1500, in the SGD optimization framework.

From the reported results in Table 3, we can see that the performance ordering is UOCCF $\approx$ BPR > IOCCF $\gg$ PopRank, which shows the effectiveness of
the memory-based and model-based methods.

5.4 Results with Heterogeneous Data

Evidently, there is not enough information of users' behaviors (i.e., interactions)
for more accurate recommendation. We thus must leverage some impression
records, i.e., the problem then becomes one-class collaborative filtering with
heterogeneous data. Based on extensive preliminary studies, we decided to use
impression data where the same job consecutively appears more than or equal
to 7 times within one single record, and convert the selected data to the format
of (user, item) pairs as additional *virtual* interactions in IOCCF and UOCCF.
For BPR, the criterion is more critical, i.e., the time of the same job appearance
within one single record must be larger than or equal to 40. We also randomly
pick up 1/7 of the selected impression data where the same job appears only once
within a single record and add them to the training data of BPR in the same
format. We think that if one user positively interacts with these job-postings as
soon as he/she had received the postings, the recommendation system by XING
will not post these jobs again later.

As has been noted, the time threshold is an important factor. We believe that
the latest impressions will help improve the result significantly. As for the reason
why we do not set the time threshold for interactions is that the interaction data
is rare and more reliable. We finally only use the impression data logged in week-
45 and drop the rest impressions.

Furthermore, we fix a set of parameters more critically as follows: for IOCCF
and UOCCF, we use Jaccard index and 250 nearest neighbors; and for BPR, we

Table 4. Recommendation performance using heterogeneous data of interaction pairs and impression pairs.

Method	Heterogeneous data	Performance
IOCCF	Int, Imp_7	294957.54
UOCCF	Int, Imp_7	341540.78
BPR	Int, Imp_1, Imp_40	328254.63

fix the value of the tradeoff parameters as 0.01, the learning rate as 0.01, the number of latent dimension as 200, and the number of iterations as 2000.

We report the results in Table 4. We can see that using heterogeneous data, i.e., both interactions and impressions, can significantly improve the recommendation performance, which can be easily observed from the contrast between Table 3 and Table 4. Specifically, we obtain an improvement of 23.78%, 15.61% and 11.50% over the interaction-only version of IOCCF, UOCCF and BPR, respectively.

5.5   Results of Hybrid Recommendation

Finally, we mix all those three recommendation lists by our hybrid strategy discussed before, and report the results in Table 5. We can see that the recommendation performance can be further improved significantly, i.e., a further improvement of 34.55%, 16.20% and 20.90%, over IOCCF, UOCCF and BPR, respectively, which clearly shows the effectiveness of our proposed hybrid recommendation approach.

Notice that, for a real application, there is no need to set the parameters to such critical values. Using 50 nearest neighbors in memory-based collaborative filtering will take only 8 h to run with our machine, and using 50 latent dimensions and 500 iterations in BPR will take less than 3 h to converge with such a large-scale dataset. All of the settings above will receive around 350 K scores as well after executing the hybrid strategy. Hence, the time cost for a real deployment should be acceptable for our solution.

6   Conclusions and Future Work

In this paper, we have studied an industrial recommendation problem, i.e., job recommendation, with heterogeneous data, including users' interactions and

Table 5. Recommendation performance of the proposed hybrid strategy. Notice that the base recommender IOCCF, UOCCF and BPR are trained using the heterogeneous data in the same way with that of Table 4.

Method	Performance
HyOCCF	**396857.40**

impressions. Specifically, we formulate it as a one-class collaborative filtering (OCCF) problem and design a hybrid approach called HyOCCF. For empirical studies, we have studied the recommendation performance of three state-of-the-art OCCF algorithms using homogeneous data (i.e., interaction data only) and heterogeneous data (i.e., both interaction data and impression data), and that of our HyOCCF. We find that heterogeneous data and hybrid strategy can both help improve the recommendation performance significantly.

For future works, we are interested in extending our hybrid recommendation strategy to model the interaction data in a finer granularity, e.g., we may treat click, bookmark, reply and delete actions differently [6].

Acknowledgements. We thank the support of National Natural Science Foundation of China No. 61502307 and No. 61672358, and Natural Science Foundation of Guangdong Province No. 2014A030310268 and No. 2016A030313038.

References

1. Deshpande, M., Karypis, G.: Karypis.: Item-based top-n recommendation algorithms. ACM Trans. Inf. Syst. **22**(1), 143–177 (2004)
2. Kabbur, S., Ning, X., Karypis, G.: Fism: factored item similarity models for top-n recommender systems. In: Proceedings of the 19th ACM SIGKDD International Conference on Knowledge Discovery and Data Mining, KDD 2013, pp. 659–667 (2013)
3. Karypis, G.: Evaluation of item-based top-n recommendation algorithms. In: Proceedings of the 10th International Conference on Information and Knowledge Management, CIKM 2001, pp. 247–254 (2001)
4. Pan, R., Zhou, Y., Cao, B., Liu, N.N., Lukose, R., Scholz, M., Yang, Q.: One-class collaborative filtering. In: Proceedings of the 8th IEEE International Conference on Data Mining, ICDM 2008, pp. 502–511 (2008)
5. Pan, W., Chen, L.: GBPR: group preference based bayesian personalized ranking for one-class collaborative filtering. In: Proceedings of the 23rd International Joint Conference on Artificial Intelligence, IJCAI 2013, pp. 2691–2697 (2013)
6. Pan, W., Zhong, H., Xu, C., Ming, Z.: Adaptive Bayesian personalized ranking for heterogeneous implicit feedbacks. Knowl. Based Syst. **73**, 173–180 (2015)
7. Rendle, S.: Factorization machines with libfm. ACM Trans. Intell. Syst. Technol. (ACM TIST) **3**(3), 57 (2012)
8. Rendle, S., Freudenthaler, C.: Improving pairwise learning for item recommendation from implicit feedback. In: Proceedings of the 7th ACM International Conference on Web Search and Data Mining, WSDM 2014, pp. 273–282 (2014)
9. Rendle, S., Freudenthaler, C., Gantner, Z., Schmidt-Thieme, L.: BPR: bayesian personalized ranking from implicit feedback. In: Proceedings of the 25th Conference on Uncertainty in Artificial Intelligence, UAI 2009, pp. 452–461 (2009)
10. Ricci, F., Rokach, L., Shapira, B.: Recommender Systems Handbook, 2nd edn. Springer, Heidelberg (2015)
11. Zhao, T., McAuley, J.J., King, I.: Leveraging social connections to improve personalized ranking for collaborative filtering. In: Proceedings of the 23rd ACM International Conference on Conference on Information and Knowledge Management, CIKM 2014, pp. 261–270 (2014)

Learning Quality Evaluation of MOOC Based on Big Data Analysis

Zihao Zhao$^{(\boxtimes)}$, Qiangqiang Wu, Haopeng Chen$^{(\boxtimes)}$, and Chengcheng Wan

Shanghai Jiao Tong University, Shanghai, China
zhaozihaoalbert@foxmail.com , chen-hp@sjtu.edu.cn

Abstract. The popularity of Massive Open Online Courses has been rapidly growing recently. However, the completion rates of MOOC appear to be quite low. Moreover, the learning quality is quite doubtful for administrators of Universities since there is no suitable tools to evaluate it. Benefitting from the online environment, MOOC platforms can collect and store a huge amount of data related to learning processes. We use Storm as the parallel computing tool to accomplish the data analysis of MOOC. Our research focuses on three types of learning quality evaluation: relationship between students' forum participation and their academic performance, relationship between students' forum emotion and their academic performance, relationship between students' video seeking operation and their academic performance.

Keywords: MOOC · Learning quality evaluation · Big data analysis

1 Introduction

The popularity of Massive Open Online Courses (MOOC) has been rapidly growing since 2012, when Stanford, MIT, Harvard, Columbia and many other prestigious universities started opening their courses to the society. Today, Coursera, edX, and Udacity have been the most well-known world-wide platforms, and thousands of users have studied variety of courses on them [1]. However, the completion rates of MOOC appear to be quite low, some research reported it is lower than 7% [2]. Moreover, the learning quality is quite doubtful for administrators of Universities since there is no suitable tools to evaluate it. This problem has stunted the credit acquisition from MOOC platforms even though the popularization of free study on MOOC is continuously growing.

Actually, benefitting from the online environment, MOOC platforms can collect and store a huge amount of data related to learning processes in which each individual action, such as assignments submitting and video viewing, will leave a digital fingerprint and provide data points to MOOC platforms. By analyzing these data, we can objectively evaluate the learning quality of MOOC. The results of such evaluation could provide personalized learning advices to students.

Accounting for the huge amount of data, we need an excellent parallel computing tool to accomplish the data analysis of MOOC. Without a doubt, Storm

© Springer International Publishing AG 2017
M. Qiu (Ed.): SmartCom 2016, LNCS 10135, pp. 277–286, 2017.
DOI: 10.1007/978-3-319-52015-5_28

is an ideal option for this task. Our research focuses on three types of learning quality evaluation: relationship between students' forum participation and their academic performance, relationship between students' forum emotion and their academic performance, relationship between students' video seeking operation and their academic performance.

The structure of the paper is as follows. Section 2 presents the related work of the paper. Section 3 describes system design and data architecture. Section 4 presents the analysis models. Section 5 describes data analysis and results. Section 6 is about summary and future work.

2   Related Work

Actually, there is much existing research focusing on big data analysis of MOOC. They can be classified as following types:

2.1   Forum Related Analysis

On the one hand, forum related analysis is helpful for teachers to get students' feedbacks about the course to improve the course itself. Through sentiment analysis of students' post texts and opinions on forums, students' trending opinions towards the course and major course tools can be monitored [3]. On the other hand, some interesting learning behaviors can be observed through forum related analysis. For example, relationship between students' academic performance and their participation in the course forum through combined analysis of forum and students' course grade. Analysis the semantics of both teachers and TAs' posts in the course forum is a good way to get their teaching behaviors [4].

2.2   Video Watching Related Analysis

VisMOOC designed by Hong Kong University of Science and Technology, is a visual analytic system to help analysts to extract user behavior patterns based on the original overview of clickstream clusters and intuitive grouping. At the same time, clickstream data such as "seek" actions contain specific time sequence information which can reveal more information on student learning behavior [5].

Another way of video watching analysis is based on events by extracting recurring subsequences of student behavior, which contain fundamental characteristics such as reflecting (i.e., repeatedly playing and pausing) and revising (i.e., plays and skip backs). It is found that some of these behaviors are significantly associated with whether a user will be Correct on First Attempt (CFA) or not in answering quiz questions [6].

2.3   Homework Related Analysis

Through analysis of homework deadline, behavior of students' submitting homework which is one of the most important learning behaviors can be observed [1].

Furthermore, combined homework deadline analysis of different courses, different sections and different types of students can result more interesting learning behaviors. This is one of our prior works.

2.4 Hybrid Analysis

Most hybrid analyses are based on behavioral data from distributed as training data and consisted of discussion forum data (in SQL) and clickstream data (in JSON format), in order to predict whether the student will cease to actively participate after that week [7–9]. Scott Crossley et al. combine click-stream data and natural language processing (NLP) approaches to examine if students' online activity and the language they produce in the online discussion forum is predictive of successful class completion [10]. Shu-Fen Tseng et al. intend to classify learning behaviors to examine learning outcomes in MOOCs by different types of learners, and in their study, the Ward's hierarchical and k-means non-hierarchical clustering methods were employed to classify types of learners' behavior while they engaged in learning activities on the MOOC platform [11].

Devendra Singh Chaplot et al. propose an algorithm based on artificial neural network for predicting student attrition in MOOCs using sentiment analysis and show the significance of student sentiments in this task [12]. Zhiyun Ren et al. develop a personalized linear multiple regression (PLMR) model to predict the grade for a student, prior to attempting the assessment activity and their developed model is real-time and tracks the participation of a student within a MOOC (via click-stream server logs) and predicts the performance of a student on the next assessment within the course offering [13]. Girish Balakrishnan et al. build a multidimensional, continuous-valued feature matrix for students across the time slices of the course, and quantize this feature space, using either k-means clustering or cross-product discretization, into a discrete number of observable states that are integral to a Discrete Single Stream HMM and then apply the Baum-Welch algorithm to train our HMM on a chosen number of hidden states [14]. Marius Kloft et al. present an approach that works on click-stream data which is able to predict dropout significantly better than baseline methods. Among other features, the machine learning algorithm takes the weekly history of student data into account and thus is able to notice changes in student behavior over time [15].

Although the existing research has presented much useful outcome, there isn't any research on the learning and teaching quality yet. So our research is based on Storm.

3 System Design and Data Architecture

3.1 System Design

Tools for big data analysis of MOOC built on the computing platform is based on Storm. The computing platform consists of three tiers: data storage layer, data

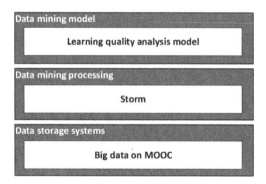

Fig. 1. System architecture

mining processing layer and data mining model layer. The system architecture is shown in Fig. 1.

Data storage layer: Due to terabyte data's heterogeneity, this layer set up a hybrid storage system with both relationship database and NoSQL database to store data from MOOC. In addition, some cloud technologies, such as storage virtualization, dynamic scheduling and distributed storage are utilized to ensure the high reliability and high scalability of the system.

Data mining processing layer: To achieve high performance, we leverage Storm to do the data mining processing.

Data mining model layer: To deal with big data, this layer builds a series of data mining models to meet analyzing requirements and goals. These models abstract valuable info out of massive data via common DM methods, such as clustering, classification, association rule, time series, prediction and so on. On the other hand, these models emphasize inter-data based analysis, dynamic analysis, business requirements oriented analysis and multi-dimensional analysis.

3.2 Data Source

All source data is from Coursera. There are three kinds of main entities involved in MOOC platforms: courses, students and teachers.

The data sample includes two main types: static data and dynamic data. For both static data and dynamic data, there are two dimensions: time and granularity. Time dimension represents generated timing of data while granularity dimension identifies if the data are related to a particular individual, a group of individuals or all individuals.

Analysis on static data: On time dimension, historical route of students' learning behaviors will be observed to research learning quality. On granularity dimension, differences between individuals, differences between one individual to average sample of group and entire individuals will be observed to research learning quality.

Analysis on dynamic data: On time dimension, the change of learning quality will be found during the entire course open period. For example, it is possible to warn students who have poor academic performance. On granularity dimension, learning quality of individuals and clusters will be observed dynamically.

3.3   Data Classification

MOOC data can be divided into three coarse-grained class (shown in Fig. 2): structured data, process data and result data.

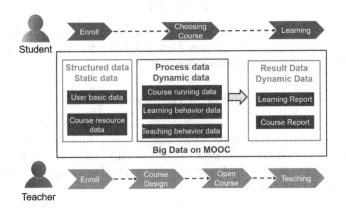

Fig. 2. Classification of MOOC Data

Structured Data, which is static data on MOOC platform which will not change with course running. There are two kinds of structured data: user basic data and course resource data.

Process Data, which is dynamic data during a course open period. There are three kinds of process data: course running data, learning behavior data.

Result Data, which is dynamic result of learning. When a course is closed, result data will be converted into archived static data. There are two kinds of process data: learning report and course report.

4   Analysis Models

The learning quality can be refined as two qualities, which are learning quality of running courses and overall learning quality of historical courses. We need to build two models for them in our research.

4.1 Learning Quality of Running Courses

A series of learning behavior data and score related courses running data are generated during the course running when the student interacts with learning related course running data. Some learning related course running data are quoted from static course resources data, such as homework, test, exam and project, and the others are running data, such as video, courseware and forum.

The learning quality will be derived from the dynamic data and expressed as a two-dimensional vector, including learning behavior vector (B) and score vector (S):

$$Q = \{B, S\}$$

It illustrates that learning quality is a complex vector which is made up of many factors, rather than a single one. Thus, it enables us to analyze the learning quality within multidimensional approaches.

4.2 Overall Learning Quality of Historical Courses

When a course is closed, all learning data generated from it will be archived as static structure data, which allow us to analyze individual student or a group's growth trajectory over time. So we also need to build an overall learning quality evaluation model to draw a whole picture of individual or group's learning quality across all the learned courses.

5 Data Analysis and Results

Students' academic performance and the distribution of students' scores is shown in Fig. 3.

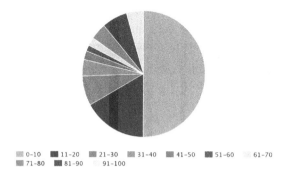

Fig. 3. The distribution of students' scores.

From Fig. 3, we can see that most students do not complete the course since their final scores are below the pass mark. This indicates the low completion rate of courses on Coursera.

Considering the low completion rate which may affect the results of data analysis, analyses in our research uses average value to evaluate the relationship between students' learning behaviors and their academic performance.

Our research focuses on three types of learning quality evaluation: relationship between students' forum participation and their academic performance, relationship between students' forum emotion and their academic performance, relationship between students' video seeking operation and their academic performance.

5.1 Relationship Between Students' Forum Participation and Their Academic Performance

The distribution of average forum posts of students with different final scores is shown in Fig. 4. The x-axis shows students' final scores in their course. The y-axis shows the average value of forum participation (both posts and replies) of students who get different scores.

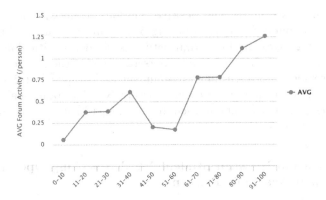

Fig. 4. The distribution of average forum posts of students with different final scores.

From Fig. 4, we can see that students who get higher scores spent more time in forum participation. But some thing interesting is that students who get final scores in 40 to 60 have low forum participations, it indicates that this group of students have poor academic performance.

5.2 Relationship Between Students' Forum Emotion and Their Academic Performance

The distribution of average forum posts with positive emotion and negative emotion of students with different final scores is shown in Fig. 5.

The x-axis shows the average value of forum participation (both posts and replies) which contain a certain emotion (positive emotion or negative emotion). The y-axis shows students' final scores in their course. The blue histogram

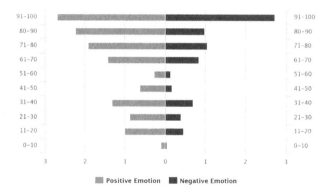

Fig. 5. The distribution of average forum posts with positive emotion and negative emotion of students with different final scores.

represents the positive emotions while the black histogram represents negative emotions.

In all forum posts and replies, positive emotion accounts for 66% while negative emotion accounts for 34%. From Fig. 5, we can see that for both positive emotion and negative emotion forum activities, students with different score level trend the same. It indicates that Coursera as one of the open online course platforms provides a free learning environment and a open learning mode, in this way students can share their opinions freely. Not like traditional classroom, students may fear to talk about their real opinions in forum because it may affect their final course score.

5.3 Relationship Between Students' Video Seeking Operation and Their Academic Performance

The distribution of average video seeking operations of students with different final scores is shown in Fig. 6.

The x-axis shows the average value of video seeking operations (including seeking forward and seeking backward). The y-axis shows students' final scores in their course. The blue histogram represents the seeking forward operation while the black histogram represents seeking backward operation.

In all video seeking operations, seeking forward operation accounts for 93% while seeking backward operation accounts for 7%, it indicates that the course is relatively easy to learn, so that most students seek forward during course video watching.

From Fig. 6, we can also see that students who get final scores in 40 to 60 have low video seeking operations, it also indicates that this group of students have poor academic performance. In the perspective of teachers, this group of students should be paid more attention on because they have poor academic performance and they do not study hard, if they work harder, they will pass the course.

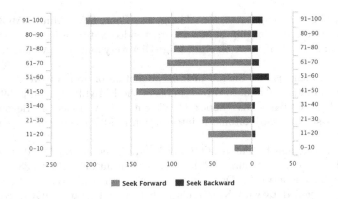

Fig. 6. The distribution of average video seeking operations of students with different final scores.

6   Conclusion

In our research, Storm is used as the parallel computing tool to accomplish the data analysis of MOOC. Three types of learning quality evaluation are taken into consideration: relationship between students' forum participation and their academic performance, relationship between students' forum emotion and their academic performance, relationship between students' video seeking operation and their academic performance.

Big data analysis of MOOC is more helpful for teachers to evaluate students' learning quality than simple statistical analysis on Coursera. Through big data analysis of MOOC, more interesting results can be found and learning behaviors can be reflected effectively.

Specific model of big data analysis of MOOC in our research is not overall. To build a more refined model is our future work.

Acknowledgment. This paper is supported by Ministry of Education - China Mobile Research Fund of China under Granted No. MCM20150605.

References

1. Chen, D., Feng, Y., Zhao, Z., Jiang, J., Yu, J.: Does MOOC really work effectively. In: 2014 IEEE International Conference on MOOC, Innovation and Technology in Education (MITE), pp. 272–277. IEEE (2014)
2. Parr, C.: MOOC completion rates below 7%. Times Higher Education (2013)
3. Wen, M., Yang, D., Rose, C.: Sentiment analysis in MOOC discussion forums: What does it tell us? In: Educational Data Mining 2014 (2014)
4. Chaturvedi, S., Goldwasser, D., Daumé III, H.: Predicting instructor's intervention in MOOC forums. In: ACL, vol. 1, pp. 1501–1511 (2014)
5. Shi, C., Siwei, F., Chen, Q., Huamin, Q.: VisMOOC: Visualizing video clickstream data from massive open online courses. In: 2015 IEEE Pacific Visualization Symposium (PacificVis), pp. 159–166. IEEE (2015)

6. Brinton, C.G., Buccapatnam, S., Chiang, M., Poor, H.V.: Mining MOOC click-streams: On the relationship between learner video-watching behavior and performance. In: ACD Conference on Knowledge Dicrovery and Data Mining (SIGKDD) (2015)
7. Rosé, C.P., Siemens, G.: Shared task on prediction of dropout over time in massively open online courses. In: Proceedings of the EMNLP, vol. 14, p. 39 (2014)
8. Whitehill, J., Williams, J.J., Lopez, G., Coleman, C.A., Reich, J.: Beyond prediction: First steps toward automatic intervention in MOOC student stopout. SSRN 2611750 (2015)
9. Taylor, C., Veeramachaneni, K., O'Reilly, U.-M.: Likely to stop? predicting stopout in massive open online courses (2014). arXiv preprint arXiv:1408.3382
10. Crossley, S., Paquette, L., Dascalu, M., McNamara, D.S., Baker, R.S.: Combining click-stream data with NLP tools to better understand MOOC completion. In: Proceedings of the Sixth International Conference on Learning Analytics & Knowledge, pp. 6–14. ACM (2016)
11. Tseng, S.-F., Tsao, Y.-W., Yu, L.-C., Chan, C.-L., Lai, K.R.: Who will pass? analyzing learner behaviors in MOOCs. Res. Pract. Technol. Enhanced Learn. **11**(1), 1 (2016)
12. Chaplot, D.S., Rhim, E., Kim, J.: Predicting student attrition in MOOCs using sentiment analysis and neural networks. In: Proceedings of AIED 2015 Fourth Workshop on Intelligent Support for Learning in Groups (2015)
13. Ren, Z., Rangwala, H., Johri, A.: Predicting performance on MOOC assessments using multi-regression models (2016). arXiv preprint arXiv:1605.02269
14. Balakrishnan, G., Coetzee, D.: Predicting student retention in massive open online courses using hidden markov models. Electrical Engineering and Computer Sciences University of California at Berkeley (2013)
15. Kloft, M., Stiehler, F., Zheng, Z., Pinkwart, N.: Predicting MOOC dropout over weeks using machine learning methods. In: Proceedings of the EMNLP 2014 Workshop on Analysis of Large Scale Social Interaction in MOOCs, pp. 60–65 (2014)

QoS-Driven Frequency Scaling for Energy Efficiency and Reliability of Static Web Servers in Software-Defined Data Centers

Lihang Gong$^{(\boxtimes)}$, Zhenhua Wang, Haopeng Chen, and Delin Liu

School of Software, Shanghai Jiao Tong University, Shanghai 200240, China
{lindaman,zjwzh1,chen-hp,sheng2010ren}@sjtu.edu.cn

Abstract. Conventional dynamic voltage and frequency scaling (DVFS) techniques use high CPU utilization as a predictor for user dissatisfaction, to which they react by increasing CPU frequency. However, QoS requirements are not linearly related to CPU utilization since they can vary from one business scenario to another. In this paper, we propose QoS-driven frequency scaling for energy-saving and reliability enhancement of static web servers in software-defined data centers (QoS-FS), an adaptive QoS requirement aware dynamic CPU frequency scaling technique. QoS-FS is implemented on a Linux-based static web server. Compared with the default Linux CPU frequency controller, QoS-FS reduces the measured system-wide power consumption of the static web server by about 5% while meeting user's QoS requirements. Besides, under some heavy workload, QoS-FS is 17% more reliable than the default Linux CPU frequency controller in the static web server.

Keywords: CPU frequency scaling · Energy saving · QoS-driven study · Static web servers · Software-defined data center

1 Introduction

Static websites are popular because they are efficient, extremely fast and usually free to host [1]. Many hot web applications, such as Blogs, resumes, marketing websites, landing pages, and documentation are all good candidates for static websites. Since static websites are common, how to reduce power consumption of static web servers while guaranteeing their quality of service (QoS) has become an essential topic for webmasters. Power consumption is a primary concern because the tremendous increase in computer performance has come with an even greater increase in power usage. According to Eric Schmidt, CEO of Google, what matters most to Google "is power – low power, because data centers can consume as much electricity as a city" [2]. What's more, Many prior researches have addressed various dimensions of the cloud system developments, such as energy-saving [3]. The recent work in Paul et al. [4] considered electricity cost reduction and increase in renewable energy integration through GLB

© Springer International Publishing AG 2017
M. Qiu (Ed.): SmartCom 2016, LNCS 10135, pp. 287–296, 2017.
DOI: 10.1007/978-3-319-52015-5_29

and Xu Zhou et al. exploit intelligent green power scheduling policies to provide efficiency-aware power management [5]. Barroso et al. [6] develop the claim that, in a datacenter, 42 percent of the total energy is consumed in CPUs. Therefore, CPU power management approaches are very important for saving energy. Dynamic frequency and voltage scaling (DVFS) [7] is one of the most commonly used energy reduction techniques for CPUs of Linux-based servers in data centers. Furthermore, application-level feedback has been widely used to determine which CPU frequency is used [1,8–11]. But feedback is not ideal enough because of its latency and inflexibility. As far as the data center is concerned, a new paradigm in data center, software defined data center (SDDC), advocates building the masking of server resources (including the number and identity of individual physical servers, processors, and operating systems) from server users. [12] In business, system performance requirements are defined by quality of service (QoS) requirements, which are driven by business needs [13]. Performance and reliability requirements are two typical QoS requirements. In this paper, we conduct a research on these two QoS requirements in the context of low-level CPU power management (CPU DVFS). There are also many prior works focused on reliability Enhancement. Hu et al. [14] and Hsu et al. [15] investigate how to analyze software reliability. Comparing with these works, we find that coarse-grained CPU power management approaches may lead to reliability problems in the context of QoS, which could often be neglected. To enhance reliability, we propose a fine-grained management approach, QoS-FS.

The contributions of this paper are summarized as follows: (**1**) We demonstrate that for static websites, traditional CPU power management approach (Linux CPU frequency controller) can lose sight of user's service-level QoS requirements. And we show there exist *surplus time* which, as Fig. 1a shown, is the difference between the QoS required time and the actual one and *excessive CPU throttling* which means that reducing power consumption through CPU throttling may cause performance not meet the requirements. (**2**) We propose an adaptive QoS requirement aware dynamic CPU frequency scaling technique (QoS-FS) for conventional Linux-based static web servers. It exploits QoS requirements to save energy and enhance reliability. With the proposed technique, the *surplus time* can be reduced and *excessive CPU throttling* can be avoided. Our proposed technique is able to automatically scaling the frequency based on Qos requirements, which spares the users from having to understand and manage complicated details of servers and makes the data center become a software-defined data center. (**3**) We implement two scheduling algorithms in the proposed technique for energy-saving and reliability enhancement in a traditional Linux-based static web server. (**4**) We conduct several experiments with various types of workload to demonstrate the effectiveness of the proposed technique and scheduling algorithms.

The rest of this paper is organized as follows. Section 2 introduces the motivation of this paper. Section 3 describes the design and implementation of the proposed technique. Section 4 presents the results of experiments to demonstrate the effectiveness of the technique. And Sect. 5 concludes the paper with a discussion.

2   Motivation

In this section, we first show conventional DVFS can cause surplus time and excessive CPU throttling in a Linux-based static web server under QoS constraints. Then, we demonstrate that surplus time can be reduced and excessive CPU throttling can be avoided by scaling the CPU frequency more reasonably. Finally, we measure the energy consumption (Joules) of the ENTIRE system (including other energy consuming components such as memory, graphic card, and hard disk) and calculate the reliability of the server before and after optimization respectively for comparison. **Note:** the experimental settings in this section are similar to those described in Sect. 4.1. Httperf [16], a web benchmark tool is used to generate workload for the server.

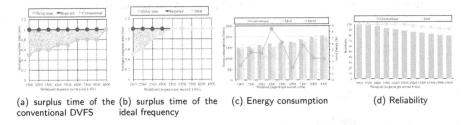

(a) surplus time of the   (b) surplus time of the   (c) Energy consumption   (d) Reliability
conventional DVFS   ideal frequency

Fig. 1. Comparison of conventional DVFS and ideal frequency

First, we focus on investigating feasibility of reducing surplus time by CPU frequency scaling in case of moderate workload. To start, we assume the QoS required average response time is less than or equal to 1ms. Then, for each sample workload, we manually try all available frequencies to execute it and record the values of surplus time. The frequency with the minimal value of surplus time is selected as the ideal frequency. The experimental results of surplus time for each workload at the ideal frequency are displayed in Fig. 1b. Results of the default Linux CPU controller (conventional DVFS) are also displayed in Fig. 1a. Comparing Fig. 1b with Fig. 1a, surplus time of the ideal frequency is obviously shorter than that of conventional DVFS. We also record the system-wide energy consumption of the server in the two cases respectively. Figure 1c displays the result, showing that energy consumption of the ideal frequency is 2% to 8% less than that of conventional DVFS. Next, we turn to reliability in case of heavy workload. We use the percentage of number of replies in number of requests to represent the functional reliability of the server. We assume QoS required timeout is 0.02 s. For each sample workload, we manually try all available frequencies to execute it and record the values of reliability. The frequency with the maximum value of reliability is selected as the ideal frequency. The experimental results of reliability for each workload at the ideal frequency are displayed in Fig. 1d. For comparison, the results of conventional DVFS are also displayed in this figure. The results show that reliability can be enhanced by scaling CPU frequency. For moderate workload, it strives to reduce surplus time to save energy while for heavy workload, it strives to maximum the functional reliability of the server.

3   Design and Implementation of QoS-FS

3.1   Workload Representation

Since for static web service, any access is read-only, the differences between requests are only resulted from differences between request file sizes. Thus, to send a file to a client, its data should be copied to a send buffer at first. Considering the send buffer size is limited, data of a large file may need to be copied to the buffer for several times before being completely sent. Because of this, the larger the file is, the more times the loop is executed. It means that the CPU workload of processing the file increases with the increase of the file size. Therefore, we use the times the loop is executed to represent the CPU workload of processing the file. Assume that there are n request(s) per second, f_i represents each request file size ($1 \leq i \leq n$). The workload, TC is calculated as:

$$TC = \sum_{i=1}^{n} \lceil \frac{f_i}{buffer_s} + 1 \rceil (1 \leq i \leq n) \tag{1}$$

3.2   Scheduling Algorithms

3.2.1   ART-Aware Scheduling

As discussed in Sect. 2, surplus time of conventional DVFS can be reduced to save energy in case of moderate workload. ART-aware scheduling is designed to address this issue. It selects the CPU frequency which minimizes the surplus time. To achieve this goal, a performance prediction model is used to predict the corresponding average response time of each available CPU frequency when executing input workload. Then, the algorithm selects the frequency whose predicted corresponding average response time is closest to the QoS required as its return result. By this way, the surplus time can be minimized. The performance prediction model is the key to ART-aware scheduling. Since the experimental observed average response time increases in an approximately polynomial way with increase of workload, we use the method of polynomial fitting to fit the performance data collected in the experiment. More specifically, we firstly fix the frequency of the CPU in the server. Then, we run different workloads represented by TC (See Sect. 3.1) on the server and collect the average response time data. Finally, we use the method of polynomial fitting to fit the collected data to obtain a curve for average response time prediction at this frequency. For each available frequency of the CPU, we repeat the above steps and fit a curve for it. The prediction model is evaluated in the experiment.

3.2.2   Timeout-Aware Scheduling

To implement timeout-aware scheduling, we evaluate all the available CPU frequencies under the QoS constraint. For instance, we firstly define the timeout is 0.02 s (QoS required) here. Then, we use httperf to generate a series of workloads represented by TC (See Sect. 3.1) and the parameter of httperf, timeout is set 0.02 s. We run the generated workloads in the static web server and collect the reported performance data by httperf to calculate the reliability. For each frequency, when the reliability is 100%, we define the heaviest workload as its workload capacity. For example, when timeout is 0.02 s (QoS required), the observed workload capacity of 1.6 GHz is 4352 because when workload exceeds 4352, the reliability is not 100%. The details of the implementation of timeout-aware scheduling is displayed in Algorithm 1. In Algorithm 1, workload capacity of each available CPU frequency is compared with the input workload, tc. If workload capacity of the frequency is greater than or equal to tc, the reliability of executing the workload at the frequency is 100%. Thus, the frequency is added to the frequencySet. In the end, the algorithm selects the minimum frequency in frequencySet as the return result for energy saving. However, if frequencySet is null, it means that the input workload is so heavy that the reliability cannot be 100% at any frequency. In this case, the algorithm returns the maximum available frequency to ensure the highest reliability of the server.

Algorithm 1. Timeout-aware scheduling. In this algorithm, frequencySet is the set of all available frequencies of the CPU. The function **GetWorkloadCapacity** is used to get the workload capacity of a frequency. And tc is the input workload.

```
1   Input: tc;
2   resultSet = ∅;
3   frequencySet = GetAllAvailableFrequencies(CPU);
4   for each f in frequencySet
5     workloadCapacity = GetWorkloadCapacity(f);
6     if workloadCapacity >= tc
7       resultSet = resultSet U f;
8     end if
9   end for
10  if resultSet == ∅
11    return max(frequencySet);
12  else
13    return min(resultSet);
14  end if
```

3.2.3   Architecture of QoS-FS

Finally, we display the architecture of QoS-FS in Fig. 2. QoS-FS is implemented in the scheduling controller. It takes the monitored workload of the web server as its input parameter. For multicore CPUs, it also collects the CPU utilization information through the agent monitor to calculate the workload distribution to its cores while for single-core CPU, the information is not necessary. It sends its result as command line (CMD) to be executed by the agent scheduler. The agent interacts with the operation system through CPU API provided by the operating system. Agent monitor collects the CPU utilization information and agent scheduler scales the CPU frequency by the provided API. Unlike the Linux CPU controller, QoS-FS resides in application level of the server.

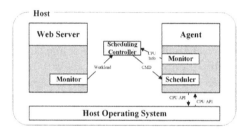

Fig. 2. Architecture of QoS-FS

Table 1. Hardware information

Basic parameters	Description
Model	HP Pro 6300 MT (C0Q80PA)
CPU series	Intel Core i5 3 generation family
CPU type	Intel Core i5 3470
Available CPU frequencies	1.6 GHz, 1.7 GHz, 1.8 GHz, 1.9 GHz, 2.1 GHz, 2.2 GHz, 2.3 GHz, 2.4 GHz, 2.5 GHz, 2.6 GHz, 2.7 GHz, 2.9 GHz, 3 GHz, 3.1 GHz, 3.2 GHz, 3.3 GHz
Memory	4 GB, DDR3 1600 MHz
Disk	500 GB, 7200
NIC	1000 Mbps Ethernet card

4   Experimental Evaluations

4.1   Experimental Settings

The important hardware information is displayed in Table 1. In the experiment, we use httperf [16] to generate various HTTP workloads and measure server performance. To measure the energy consumption of the ENTIRE server (including other energy consuming components such as memory, graphic card, and hard disk), we use an electric meter to probe how many joules the server consume to execute workload. And the operating system of the server is Ubuntu 12.04 LTS 64bit with 3.8.0 version of Linux kernel. Apache 2.2 is deployed in the Ubuntu operating system as the HTTP server. To make the results clear and straightforward, we use the workload description methods (TC) presented in Sect. 3.1 to represent workload in the experiment. We use two different types of workload in the following experiments. The first workload group, named PEAK, is consist of relative heavy workload hence the server is error-prone when executing workload in this group. In practice, most workloads during peak hours belong to this type. We choose workload that can lead to timeout errors in the server employing conventional DVFS as representations of this kind of workload. The other workload group is named MODERATION. Workload in this group is lighter than that

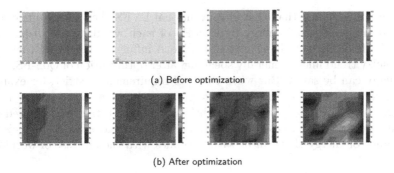

(a) Before optimization

(b) After optimization

Fig. 3. Percentage of actual time in requited time (results of 0.9 ms, 1.1 ms, 1.3 ms and 1.5 ms are displayed from left to right respectively) (Color figure online)

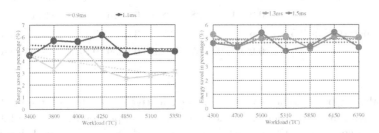

Fig. 4. Percentage of energy saved by QoS-FS compared with conventional DVFS

belongs to PEAK. When executing workload in this group, the server employing conventional DVFS is free of errors and its surplus time can be obvious.

4.2 ART-Aware Scheduling Evaluation

We first evaluate the effectiveness of ART-aware scheduling of QoS-FS in reducing surplus time and energy saving with the workload group named MODERATION. We use four different QoS required average response times (0.9 ms, 1.1 ms, 1.3 ms and 1.5 ms) and we run the workload on the server for different durations (10 s, 20 s, 30 s, 40 s, 50 s and 60 s). To demonstrate that how QoS-FS reduces surplus time clearly, we calculate the percentage of actual time of conventional DVFS and QoS-FS in QoS required time respectively and use different color to represent the value of the percentage. In Fig. 3, the y-scale represents the duration of the workload and x-scale is the workload (TC) used in the experiment. The color changes from blue to red, corresponding to the value of percentage from 0% to 100%.

The results of Fig. 3 demonstrate the effectiveness of QoS-FS in reducing surplus time under the condition of different QoS required times and workload durations. Compared with the conventional DVFS, the actual time after optimization of QoS-FS is closer to the QoS required time because the color after optimization is redder than that of conventional DVFS. And the results of average percentage of saved energy by QoS-FS of each workload are displayed in Fig. 4. To find out how the change of workload influences the saved energy, we also display trend lines. Generally speaking, after optimization of QoS-FS, about 5% energy can be saved. But when the QoS requirement is strict, for example, when required time is 0.9 ms in Fig. 5a, the percentage of saved energy ranges from 2.5% to 4.5%. The result is straightforward because the QoS required time is relatively strict in this case. And the trend of percentage of saved energy is that percentage of energy that can be saved declines with the increase of workload.

Table 2. Correlation coefficient (R^2) of fitted curves

Frequency (GHz)	1.6	1.7	1.8	1.9	2.1	2.2	2.3	2.4
R^2		0.9968	0.9884	0.9898	0.9834	0.9916	0.9943	0.971	0.966
Frequency (GHz)	2.5	2.6	2.7	2.9	3.0	3.1	3.2	3.3
R^2		0.9912	0.9861	0.9737	0.9701	0.9809	0.9739	0.9906	0.9794

Finally, we present the evaluation of the effectiveness of the polynomial fitting based performance prediction model used in the experiment, we calculate the correlation coefficient (R2) of each fitted curve and display them in Table 2. The results are consistent with what we observe.

4.3   Timeout-Aware Scheduling Evaluation

We next evaluate the effectiveness of timeout-aware scheduling in reliability enhancement with the workload group named PEAK. To evaluate how the QoS required timeout influence on the reliability of conventional DVFS and the performance of timeout-aware scheduling, we use four different timeout values in the experiment. They are 0.0002 s, 0.0008 s, 0.0014 s and 0.002 s respectively. We set timeout of httperf to those values while running the workload. And for the same reason mentioned in Sect. 4.2, we run the workload for various durations. They are 10 s, 20 s, 30 s, 40 s, 50 s and 60 s. In addition to reliability, we also measure the system-wide energy consumption before and after optimization respectively to evaluate the energy cost for reliability enhancement. To study reliability under the condition of various durations, we calculate the average reliability of the static web server under various durations for each workload. In addition, we also calculate the average energy cost of reliability enhancement in percentage. The results are displayed in Figs. 5 and 6.

Figure 5 displays the experimental results of the average reliability before and after optimization. And to evaluate energy cost of QoS-FS for reliability enhancement, we displayed the average increased energy consumption in percentage in Fig. 6. As the percentage displayed in the figure shown, QoS-FS consumes 3% to 8% more energy than conventional DVFS for reliability enhancement. Compared Fig. 6 with Fig. 5, we draw a conclusion that with the enhanced reliability increases, the increased energy consumption in percentage decreased.

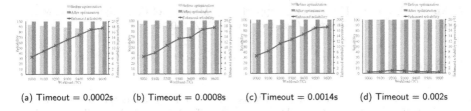

(a) Timeout = 0.0002s   (b) Timeout = 0.0008s   (c) Timeout = 0.0014s   (d) Timeout = 0.002s

Fig. 5. Reliability before and after optimization

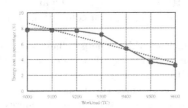

Fig. 6. Energy cost in percentage for reliability enhancement of QoS-FS compared with conventional DVFS

5 Conclusion

In this paper, we propose QoS-FS, an adaptive QoS requirement aware dynamic CPU frequency scaling technique, which helps the user automate the frequency scaling based on Qos which were previously manual. We implement two algorithms of it. Through the experimental results, we show that it is more energy-efficient and reliable than conventional DVFS technique.

To demonstrate the idea, we use a very widely used web application, static website to verify QoS-FS. With the proposed QoS-FS, energy can be saved about 5% while reliability can be increased by 17%. However, the application we selected for implementing QoS-FS does not stand for the most efficient energy optimization case of QoS-FS since static web application is more memory intensive than CPU intensive. For CPU intensive applications, the percentage of energy that can be saved by QoS-FS may increase. The cost of implementing QoS-FS is not high. All the data for building it can be collected and evaluated during web stress testing. Actually, the basic ideal of QoS-FS is utilizing some performance data obtained from web stress testing to decide how the CPU can be used better when taking QoS requirements into consideration. We believe the basic ideal of QoS-driven frequency scaling is scalable as there have been plenty researchers proposing numerous performance modeling methods for software. Thus, such kind of methods may be applied to implementing the basic idea of QoS-FS in different applications.

In the future work, we will discuss how to implement QoS-FS in various kinds of applications. Besides, we will incorporate the technique named request patching into QoS-FS to improve the performance of QoS-FS when workload is low.

Acknowledgment. This paper is supported by Ministry of Education - China Mobile Research Fund of China under Granted No. MCM20150605

References

1. What is a static website. http://nilclass.com/courses/what-is-a-static-website
2. Markoff, J., Lohr, S.: Intel's huge bet turns iffy. New York Times Technol. Sect. **3**(2) (2002)
3. Gai, K., Qiu, M., Zhao, H., Tao, L., Zong, Z.: Dynamic energy-aware cloudlet-based mobile cloud computing model for green computing. J. Netw. Comput. Appl. **59**, 46–54 (2016)
4. Paul, D., Zhong, W.-D., Bose, S.K.: Energy efficiency aware load distribution and electricity cost volatility control for cloud service providers. J. Netw. Comput. Appl. **59**, 185–197 (2016)
5. Zhou, X., Cai, H., Cao, Q., Jiang, H., Tian, L., Xie, C.: Greengear: leveraging and managing server heterogeneity for improving energy efficiency in green data centers. In: Proceedings of the 2016 International Conference on Supercomputing. ACM, p. 12 (2016)
6. Barroso, L.A., Clidaras, J., Hölzle, U.: The datacenter as a computer: an introduction to the design of warehouse-scale machines. Synth. Lect. Comput. Architect. **8**(3), 1–154 (2013)
7. Dhiman, G., Pusukuri, K.K., Rosing, T.: Analysis of dynamic voltage scaling for system level energy management. In: USENIX HotPower, vol. 8 (2008)
8. Alimonda, A., Carta, S., Acquaviva, A., Pisano, A., Benini, L.: A feedback-based approach to DVFS in data-flow applications. IEEE Trans. Comput. Aided Des. Integr. Circuits Syst. **28**(11), 1691–1704 (2009)
9. Poellabauer, C., Singleton, L., Schwan, K.: Feedback-based dynamic voltage and frequency scaling for memory-bound real-time applications. In: 11th IEEE Real Time and Embedded Technology and Applications Symposium, 2005, RTAS 2005, pp. 234–243. IEEE (2005)
10. Yang, L., Dick, R., Memik, G., Dinda, P.: Happe: human and application-driven frequency scaling for processor power efficiency. IEEE Trans. Mob. Comput. **12**(8), 1546–1557 (2013)
11. Lefurgy, C., Wang, X., Ware, M.: Server-level power control. In: Fourth International Conference on Autonomic Computing, 2007, ICAC 2007, p. 4. IEEE (2007)
12. Software-defined data center (sddc). http://searchsdn.techtarget.com/definition/software-defined-data-center-SDDC
13. Quality of service requirements. http://docs.oracle.com/cd/E19636-01/819-2326/gaxqg/index.html
14. Hu, H., Jiang, C.-H., Cai, K.-Y., Wong, W.E., Mathur, A.P.: Enhancing software reliability estimates using modified adaptive testing. Inf. Softw. Technol. **55**(2), 288–300 (2013)
15. Hsu, C.-J., Huang, C.-Y.: An adaptive reliability analysis using path testing for complex component-based software systems. IEEE Trans. Reliab. **60**(1), 158–170 (2011)
16. Mosberger, D., Jin, T.: httperf-a tool for measuring web server performance. ACM SIGMETRICS Perform. Eval. Rev. **26**(3), 31–37 (1998)

Making Cloud Storage Integrity Checking Protocols Economically Smarter

Fei Chen, Xinyu Xiong, Taoyi Zhang, Jianqiang Li, and Jianyong Chen[⊠]

College of Computer Science and Engineering, Shenzhen University, Shenzhen, China
{fchen,lijq,jychen}@szu.edu.cn, 1587564712@qq.com, 2547260515@qq.com

Abstract. In recent years, researchers are investigating plenty of efforts to design cloud storage integrity checking protocols to ensure the outsourced data's security. In such protocols, parameters are often chosen in an ad-hoc way. In this paper, we show that an optimal parameter selection exists in the economical sense when considering current cloud storage practices. We derive the optimal parameter selection by establishing a cost model for the previously already proposed cloud storage integrity checking protocols. The cost model takes account of storage cost, data transfer cost, and data access request cost as charged by current cloud storage practice. Besides the optimal parameter selection, we show that an optimal total cost also exists. Experimental analysis with one large Chinese cloud storage service provider, Ali OSS, further validates our model. Therefore, we believe that the optimal parameter selection suggested in this work helps the already proposed cloud storage integrity checking protocols more ready to be used in practice.

Keywords: Cloud storage · Integrity checking · Cost model · Optimal parameter · Economical analysis

1 Introduction

Cloud storage integrity checking has received plenty of research efforts in the past decade [1–15] with more and more users outsourcing their data to cloud storage service providers both in the private and public sectors [16]. The cloud storage integrity checking problem is then how a user can check the integrity of the outsourced data in the cloud with the challenge that the user no longer stores a local copy of the outsourced data. This problem is especially important for those who rely their business on data but have limited budget for in-house storage (e.g. new mobile app startups, etc.), and those who strictly adhere to law regulations. This is because the outsourced storage in the cloud is under risk being modified, damaged, or even deleted due to various financial incentives and management faults. To solve this problem, researchers are proposing various cloud storage integrity checking protocols [1–15] to ensure the integrity of the outsourced storage.

Among those proposed protocols, two approaches are employed. The first is the PoR (proof of retrievability) approach. This approach models the cloud

© Springer International Publishing AG 2017
M. Qiu (Ed.): SmartCom 2016, LNCS 10135, pp. 297–306, 2017.
DOI: 10.1007/978-3-319-52015-5_30

as a black box and only uses the open API from the cloud to conduct storage integrity checking. The other one is the PDP (proof of data possession) approach. This approach performs better in the communication cost than the PoR approach; however, this approach requires the cooperation of the cloud to perform computations on the outsourced storage. Both approaches leverage the sample-and-check idea, i.e. dividing the outsourced storage into blocks and then randomly checking the integrity of selected/challenged data blocks.

While the proposed protocols can verify the integrity of the outsourced storage, we note that the proposed protocols did not take account of practical economical cost of such protocols. Current cloud storage practice has a system of pricing strategies, which imply that the proposed protocols all have integrity checking costs. Thus, we aim at making cloud storage integrity checking protocols smarter economically in this work.

In this paper, we model the economical cost of cloud storage integrity checking protocols; we also propose a method to derive optimal parameters for the proposed protocols. This result can help to choose better parameters for the proposed cloud storage integrity checking protocols [1–15] when deployed into current cloud storage service providers.

We achieve this in two steps. First, we model the economical cost of cloud storage integrity checking protocols into three parts: storage cost, data transfer cost, and data access request/HTTP request cost. The three costs are computed according to current cloud storage pricing practices. Our model shows that the total cost depends on the storage size, block size, the successful detection rate, the challenge size during integrity checking, and the total number of integrity checkings. Second, once we have the total cost, we derive the optimal parameter in terms of the block size, given other parameters fixed.

Next, we simulate and analyze the cost for different parameter choices. The simulation results validate our cost model. Especially, an economically optimal block number exists for the cloud storage integrity checking protocols.

The rest of the paper proceeds as follows. Section 2 models the cloud storage integrity checking protocol and reviews current two solution approaches - the PoR approach and the PDP approach. Section 3 then constructs a cost model for such protocols, followed by simulations and analysis in Sect. 4. Section 5 reviews related work for cloud storage integrity checking. Finally, Sect. 6 concludes the paper.

2   Cloud Storage Integrity Checking Model

2.1   System Model

Figure 1 illustrates how a cloud storage integrity checking protocol works; this basic model underlies all proposed cloud storage integrity checking protocols [1–15]. Such a cloud storage integrity checking protocol involves two entities: user and cloud. The user could be an individual, an organization, an enterprise, etc.; the cloud could be any public cloud service provider, e.g. Amazon S3, Microsoft Azure, Google Cloud Storage, Ali OSS, etc. The user first processes the storage to be outsourced by embedding some secret information, then outsources its storage to the cloud. Later the user interacts with the cloud in a challenge-response manner to check the integrity of the outsourced data, e.g. whether the data is damaged, modified, or even lost. Specifically, the user sends an integrity checking (or audit) query to the cloud, requesting the cloud to prove that the outsourced data remains intact. After the cloud sending back the proof, the user verifies the integrity of the data by examining the proof.

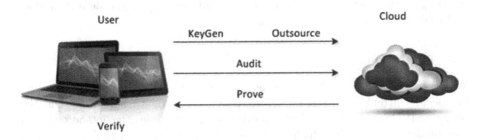

Fig. 1. Cloud storage integrity checking model

2.2 Modeling Proposed Protocols

We briefly model the previously already proposed cloud storage integrity checking protocols corresponding to the system model as in Fig. 1. The proposed protocols can be categorized into two approaches: one is the PoR (proof of retrievability) approach; the other is the PDP (proof of data possession) approach. Both approaches have similar syntax - they both model a cloud storage integrity checking (CSIC) protocol in five efficient algorithms CSIC = (KeyGen, Outsource, Audit, Prove, Verify) as follows:

- KeyGen(λ) $\rightarrow K$: On input a security parameter λ, the user runs this algorithm to generate a secret key K to enable audit and verification.
- Outsource($F; K$) $\rightarrow F'$: On input the data F to be outsourced, the user runs this algorithm to get a processed data F' using the secret key K. The processed data contains some authentication information of the outsourced data F and is then sent to the cloud.
- Audit(K) $\rightarrow q$: The user runs this algorithm to generate an audit query q which will be sent to the cloud.
- Prove(q, F') $\rightarrow \Gamma$: On input an audit query q, the cloud computes a proof Γ using the stored data F'.
- Verify($q, \Gamma; K$) $\rightarrow \delta$: On input an audit query q, the cloud's proof Γ, the user checks whether the cloud's proof is correct using the secret key K. The user outputs $\delta = 1$ if the proof is correct, else outputs $\delta = 0$.

We note that the PoR approach and the PDP approach differ substantially. The PoR approach takes the cloud as a black box which does not require the knowledge how the cloud stores and processes the data; in contrast, the PDP approach requires the cloud to conduct computation when proving the integrity of the outsourced data to the user. The PoR approach is more ready to be deployed while the PDP approach is more communication efficient.

Despite the difference, both the PoR and PDP approaches employ the same syntax and the sample-and-check idea, i.e. randomly selecting data blocks to check the data integrity in a probabilistic sense. This idea is contained in the algorithms CSIC.Audit, CSIC.Prove, and CSIC.Verify. We only require the understanding of the syntax as above to understand this paper; the details of PoR and PDP are vertical to this research.

2.3   Limitations

We note an important limitation of the proposed cloud storage integrity checking protocols. Current cloud storage service providers charge their services. But the proposed PoR-approach and PDP-approach protocols divide the outsourced data into blocks heuristically, which incurs different cost. There should be a way to parameterize the number of total blocks optimally in the economical sense, which we show in this work.

3   Cost Model for Cloud Storage Integrity Check Protocols

3.1   Current Cloud Storage Pricing Practice

We evaluate the cost of cloud storage integrity checking protocols by understanding how cloud charges. After reviewing some prominent cloud storage service providers from both international and Chinese enterprises, we find that the service providers charge on the storage size of the outsourced data, the volume of data transfer (or communication cost) when downloading the data from the cloud, and the times of data access requests in terms of HTTP requests. Table 1 lists an example for some large Chinese cloud storage service providers.

Table 1. Current cloud storage pricing practice

Cloud storage provider	Ali OSS	Baidu BOS	Tencent COS
Storage (/GB/Month)	0.158	0.15	0.156
Data transfer (/GB)	0.64	0.61	0.4
Data access request (/10000 requests)	0.01	0.01	0.01

We also find that different providers charge differently; but the difference only lies in the detailed numbers as exemplified in Table 1. We later unify these prices in variables in our cost model.

3.2   Constructing Cost Model

We now evaluate the total cost of a cloud storage integrity checking protocol. It comprises of three parts according to current cloud storage practices, i.e. storage cost, data transfer cost, and data request cost. We compute these costs respectively.

We give some notations for easy presentation. Let M denote the total cost of a cloud storage integrity checking protocol, m_1, m_2, m_3 denote the storage, data transfer, and data request cost with standard unit price p_1, p_2, p_3, respectively. Let s denote the outsourced storage size in bytes, n the total number of blocks, and s' the size of the authentication information of each data block. Let i denote the months for which the outsourced data is stored in the cloud, j the total number of integrity checking times, k the data damage rate, p the successful detection rate, and c the number of challenged data blocks in a integrity checking query. Let g be a constant denoting whether the cloud stores the authentication information for each data block; $g = 2$ if each data block is authenticated and $g = 1$ otherwise.

Storage cost. This cost consists of the outsourced data and the authentication data accompanying each data block if any. Note that the size of outsourced data is in bytes and the price charged by the cloud is in Gigabytes. We compute the storage cost in i months as

$$m_1 = \frac{s' \times n + s}{2^{30}} \times p_1 \times i.$$

Data transfer cost. This cost contains the data block and its authentication information in one challenge returned by the cloud. Compared with storage cost, this communication cost is much higher. This is because communication over the Internet incurs cost for the cloud service providers who buy bandwidth from internet service providers. Thus, when an integrity checking query consists of c challenges, the cost of j integrity checking queries is

$$m_2 = \frac{(s' + s/n) \times c}{2^{30}} \times p_2 \times j.$$

Data access request cost. This cost happens when the user gets/puts a data block from/to the cloud, typically in a HTTP request. Frequent reads/puts significantly influence the cloud's performance, thus it is a cost. When the user outsources the data to the cloud, the cost is $\frac{g \times n \times p_3}{10000}$. For c challenges in an integrity checking, the cost is $\frac{g \times c \times p_3}{10000}$. Thus for j integrity checkings, the cost is

$$m_3 = \frac{g \times n \times p_3}{10000} + \frac{g \times c \times p_3}{10000} \times j.$$

Total cost. Finally, we can have the total cost by adding the storage, data transfer, and data access request together. The total cost for a cloud storage integrity checking protocol is therefore as follows

$$M = \frac{s' \times n + s}{2^{30}} \times p_1 \times i + \frac{(s' + s/n) \times c}{2^{30}} \times p_2 \times j + \frac{g \times n \times p_3}{10000} + \frac{g \times c \times p_3}{10000} \times j. \quad (1)$$

3.3 Deriving Optimal Parameters

Optimal parameters for a cloud storage integrity checking protocol exist according to the total in Eq. (1). We simplify the total cost as follows

$$M = \alpha n + \frac{\beta}{n} + \gamma$$

where $\alpha = \frac{s'}{2^{30}} \times p_1 \times i + \frac{g \times p_3}{10000}, \beta = \frac{s \times c}{2^{30}} \times p_2 \times j, \gamma = \frac{s'}{2^{30}} \times p_2 \times j + \frac{g \times c \times p_3}{10000} \times j$. In practice, a successful detection rate p is first determined in advance using the assumed data damage rate k as follows

$$p = 1 - (1 - k)^c. \tag{2}$$

Then the total number of challenges in an integrity checking can be determined; normally it is constant, or only logarithmic in the variable n [1,2,12] depending on the assumption of k. Thus we have approximately α, β constant when fixing i and n being large.

Therefore, the minimal cost is $M \approx 2\sqrt{\alpha\beta} + \gamma$ which is achieved for $n \approx \sqrt{\frac{\beta}{\alpha}}$ roughly. It is worth noting that real cost also depends on how long the data is to be stored on the cloud and how many times integrity checking is to be conducted.

4 Analysis and Simulation

In this section, we analyze the cost of cloud storage integrity checking protocols by simulation; we aim to understand how to choose the total number of blocks, how the total cost grows. Our methodology is to study the influences of certain variable on the total cost and optimal block numbers while fixing other parameters unchanged. We employ the Ali OSS pricing practice [17] in our simulation. We use $p_1 = 0.0053, p_2 = 0.75, p_3 = 0.01$.

4.1 Challenge Size vs. Optimal Block Number

We study how challenge size influences the optimal block number. We fix $i = 1$, $j = 1, s = 100 \times 1024 \times 1024$. Assume that the data damage rate k is 1% and that the requirement of integrity checking accuracy is 99%, 95%, and 90% respectively. Then the corresponding value of challenge size c is 459, 299, and 230, respectively according to Eq. (2). Figure 2 depicts how the total cost M depends on n and c.

From Fig. 2, we find that the total cost generally grows with the block number and the challenge size. For different challenge sizes, the user needs to choose different optimal block numbers.

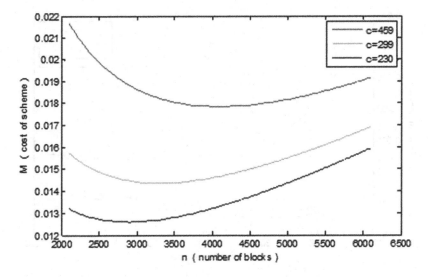

Fig. 2. Challenge size influences the optimal block number.

4.2 Storage Size vs. Optimal Block Number

We turn to investigate how outsourced storage size influences the optimal block number. We assume that the data damage rate k is 1% and the requirement of integrity checking accuracy is 90%, thus we take $c = 230$. We vary the value of s to be 100 MB, 200 MB, and 300 MB respectively. Figure 3 displays the result.

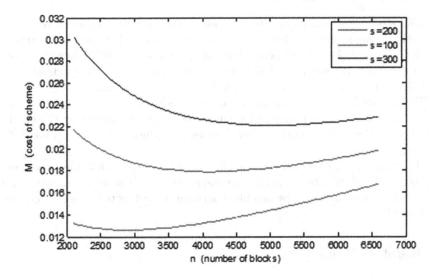

Fig. 3. Storage size influences the optimal block number.

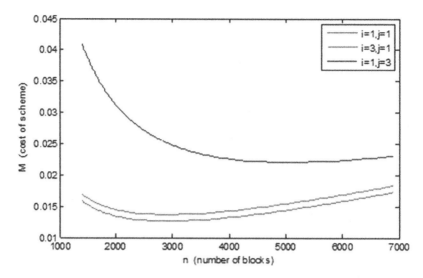

Fig. 4. Duration time and integrity checking frequency both influence the optimal block number.

From Fig. 3, we can observe that similar phenomenon with the challenge size happens here. The larger the file size is, the higher the total cost is. And the optimal number also differs when storage size changes.

4.3 Duration Time and Integrity Checking Frequency vs. Optimal Block number

We now study how duration time and integrity checking frequency influence the optimal block number. Same as above, we assume that the data damage rate k is 1%, and the requirement of checking accuracy is 90%. Then challenge size is $c = 230$. We simulate 3 different group values for i, j respectively, i.e. $i = 1, j = 1; i = 3, j = 1; i = 1, j = 3$. Figure 4 shows the dependency.

We can find from Fig. 4 that the total cost gets larger when we store the outsourced data in the cloud longer and when we have more frequent integrity checkings. The optimal block number also changes with duration time and integrity checking frequency.

The simulation results all together in this section show that the total cost, the optimal block number depend on various factors. In practice, a user could compute the total cost, the optimal block number according to the user's requirement and Eq. (1).

5   Related Work

We review most related work in this section. The problem of cloud storage integrity checking was first proposed in 2007 by Juels, A. and Kaliski Jr., B.S. [1], and Ateniese, G. et al. [2]. The former proposed the first PoR-approach solution while the later proposed the first PDP-approach solution. While the two solutions can solve the cloud storage integrity checking problem in theory, their functionalities are limited. Later works extended the two works by supporting data dynamics, third-party auditing [3–15]. Vertical to the detailed solutions, researchers also studied the relationship of cloud storage integrity checking with other problems, e.g. network coding [10], distributed string equality checking [12]. Researchers also made efforts to make the proposed solutions in use for current cloud storage practices [18–20].

Compared with previous work and to the authors' knowledge, this paper is the first to study how to make cloud storage integrity checking protocols more economically efficient, which may make cloud storage integrity checking more ready to practical deployment.

6   Conclusion

To make cloud storage integrity checking protocols more ready to be used for current cloud storage practices, in this paper, we establish a cost model for such protocols. The model integrates the storage cost, data transfer cost, and data request cost together to deduce the total cost. We show that an optimal cost and an optimal parameter selection exist for a cloud storage integrity checking protocol according to the total cost model. Our simulation results with one Chinese cloud storage provider Ali OSS further validate our cost model. Finally, we hope that the cost model in this paper may shed insights to other cloud security problems.

Acknowledgments. The authors are grateful for the anonymous reviewers' insightful comments. The work in this paper was supported in part by the National Natural Science Foundation of China (Nos. 61502314 61672358), the Science and Technology Plan Projects of Shenzhen (Nos. JCYJ20160307115030281 and GJHZ20160226202520268), the Tencent Rhinoceros Birds Scientific Research Foundation (2015), and the Technology Planning Project from Guangdong Province, China (Grant No. 2014B010118005).

References

1. Juels, A., Kaliski Jr., B.: Pors: proofs of retrievability for large files. In: Proceedings of ACM CCS (2007)
2. Ateniese, G., Burns, R., Curtmola, R., Herring, J., Kissner, L., Peterson, Z., Song, D.: Provable data possession at untrusted stores. In: Proceedings of ACM CCS (2007)
3. Shacham, H., Waters, B.: Compact proofs of retrievability. In: Pieprzyk, J. (ed.) ASIACRYPT 2008. LNCS, vol. 5350, pp. 90–107. Springer, Heidelberg (2008). doi:10.1007/978-3-540-89255-7_7
4. Erway, C., Küpçü, A., Papamanthou, C., Tamassia, R.: Dynamic provable data possession. In: Proceedings of ACM CCS (2009)

5. Wang, C., Ren, K., Lou, W., Li, J.: Toward publicly auditable secure cloud data storage services. IEEE Netw. Mag. **24**(4), 19–24 (2010)
6. Wang, C., Chow, S.S., Wang, Q., Ren, K., Lou, W.: Privacy-preserving public auditing for secure cloud storage. IEEE Trans. Comput. **62**(2), 362–375 (2013)
7. Yang, K., Jia, X.: An efficient and secure dynamic auditing protocol for data storage in cloud computing. IEEE Trans. Parallel Distrib. Syst. **24**(9), 1717–1726 (2013)
8. Shi, E., Stefanov, E., Papamanthou, C.: Practical dynamic proofs of retrievability. In: Proceedings of ACM CCS (2013)
9. Azraoui, M., Elkhiyaoui, K., Molva, R., Önen, M.: StealthGuard: proofs of retrievability with hidden watchdogs. In: Kutyłowski, M., Vaidya, J. (eds.) ESORICS 2014. LNCS, vol. 8712, pp. 239–256. Springer, Heidelberg (2014). doi:10.1007/978-3-319-11203-9_14
10. Chen, F., Xiang, T., Yang, Y., Chow, S.S.: Secure cloud storage meets with secure network coding. IEEE Trans. Comput. **65**(6), 1936–1948 (2016)
11. Armknecht, F., Bohli, J.-M., Karame, G.O., Liu, Z., Reuter, C.A.: Outsourced proofs of retrievability. In: Proceedings of the 2014 ACM SIGSAC Conference on Computer and Communications Security, pp. 831–843. ACM (2014)
12. Chen, F., Xiang, T., Yang, Y., Wang, C., Zhang, S.: Secure cloud storage hits distributed string equality checking: more efficient, conceptually simpler, and provably secure. In: Proceeding of INFOCOM (2015)
13. Li, J., Tan, X., Chen, X., Wong, D.S., Xhafa, F.: Opor: enabling proof of retrievability in cloud computing with resource-constrained devices. IEEE Trans. Cloud Comput. **3**(2), 195–205 (2015)
14. Yuan, J., Yu, S.: Pcpor: public and constant-cost proofs of retrievability in cloud1. J. Comput. Secur. **23**(3), 403–425 (2015)
15. Guan, C., Ren, K., Zhang, F., Kerschbaum, F., Yu, J.: Symmetric-key based proofs of retrievability supporting public verification. In: Pernul, G., Ryan, P.Y.A., Weippl, E. (eds.) ESORICS 2015. LNCS, vol. 9326, pp. 203–223. Springer, Heidelberg (2015). doi:10.1007/978-3-319-24174-6_11
16. Knapp, K.: Cloud storage adoption strong despite lock-in concerns (2015). http://searchcloudcomputing.techtarget.com/news/4500258090/Cloud-storage-adoption-strong-despite-lock-in-concerns
17. Alibaba: Ali OSS pricing (2016). https://www.aliyun.com/price/product-/oss/detail
18. Juels, A., Oprea, A.: New approaches to security and availability for cloud data. Commun. ACM **56**(2), 64–73 (2013). http://doi.acm.org/10.1145/2408776.2408793
19. Chen, F., Zhang, T., Chen, J., Xiang, T.: Cloud storage integrity checking: going from theory to practice. In: Proceedings of the 4th ACM International Workshop on Security in Cloud Computing, pp. 24–28. ACM (2016)
20. Juels, A., Oprea, A.M., Van Dijk, M.E., Stefanov, E.P.: Remote verification of file protections for cloud data storage, Jan 5 2016. US Patent 9,230,114

E^3: Efficient Error Estimation for Fingerprint-Based Indoor Localization System

Chengwen Luo[✉], Jian-qiang Li, and Zhong Ming

College of Computer Science and Software Engineering,
Shenzhen University, Shenzhen, China
{chengwen,lijq,mingz}@szu.edu.cn

Abstract. Wireless indoor localization has attracted extensive research recently due to its potential for large-scale deployment. However, the performances of different systems vary and it is difficult to compare these systems systematically in different indoor scenarios. In this work, we propose E^3, a Gaussian process based error estimation approach for fingerprint-based wireless indoor localization systems. With an efficient error estimation algorithm, E^3 is able to efficiently estimate the localization errors of the localization systems without requiring the expensive site evaluations. Our evaluation results show that the proposed approach efficiently estimates the performance of fingerprint-based indoor localization systems and can be used as an efficient tool to tune system parameters.

Keywords: Wireless fingerpring · Indoor localization · Error estimation · Gaussian process

1 Introduction

Accurate indoor localization is one of the fundamental building blocks for mobile applications. Due to the rising application requirements, recently indoor localization has attracted extensive research efforts [4,16,18,20,25,26].

Among different categories of wireless indoor localization approaches, *fingerprint-based indoor localization* [4] is one of the most popular due to the widespread availability of WiFi APs. State-of-the-art research on fingerprint-based indoor localization either focuses on improving the accuracy of the location estimation [13,19,26], or reducing the time and effort in constructing the fingerprint database [14,18,20,25]. While all these approaches have addressed various shortcomings in existing indoor localization systems, the configurations of these systems vary. Each of these localization systems is evaluated in settings with different physical layout and environmental effects, making it difficult to compare and evaluate them systematically. One of the key objectives of our work is to make systematic comparison feasible.

© Springer International Publishing AG 2017
M. Qiu (Ed.): SmartCom 2016, LNCS 10135, pp. 307–316, 2017.
DOI: 10.1007/978-3-319-52015-5_31

The main idea of this paper is as follow. Given a set of radio signal finger-prints collected, a Gaussian process (GP) [17] approach is used to model the signal distribution of access points that cover the area of interest. Using the signal distribution model derived, random sampling is performed to simulate the collection of fingerprint values collected at each location of interest during localization. Given a particular localization algorithm, the mapped location in the system can be determined. The average localization error of each location in the area of interest can now be estimated even though the original set of data collected as input may not be sufficient for localization purpose on its own.

To validate our approach, we implement E^3 and evaluate the system on two different indoor environments covering more than $300\,\mathrm{m}^2$ area. In both environments, point-level, region-level and floor-level error estimation are evaluated. For point level accuracy, evaluation result shows the difference between GP estimation and ground truth is small, demonstrating that accuracy awareness provides an accurate and practical assessment to fingerprint-based localization systems.

The rest of the paper is organized as follows. We discuss the technical background and Gaussian process in Sect. 2. Section 3 explains the concepts, algorithms for point-level, region-level and floor-level error estimations. The evaluation results are given in Sect. 4 and related works are discussed in Sect. 5. Finally, we conclude in Sect. 6.

2   Technical Background: Gaussian Process

To model the signal strength propagation continuously over the whole field, Gaussian process is used to capture the spatial correlation existed in signal strength distribution [7,8,24]. Gaussian Process (GP) [17] is a Bayesian non-parametric model that performs non-linear regression on the training data $\mathcal{D} = \{(\mathbf{x_i}, y_i)|i = 1, ..., n\}$. That is,

$$y_i = f(\mathbf{x_i}) + \varepsilon \tag{1}$$

where $\mathbf{x_i} \in \mathbb{R}^d$ is a d dimensional input value and y_i is the observation value, and ε is a zero-mean noise term with known covariance σ_n^2. Function $f \sim \mathcal{GP}(\mu(\mathbf{x}), k(\mathbf{x}, \mathbf{x}'))$ is a Gaussian process with mean function $\mu(\mathbf{x})$ and covariance function, or kernel, $k(\mathbf{x}, \mathbf{x}')$, where:

$$\mu(\mathbf{x}) = \mathbb{E}[f(\mathbf{x})] \tag{2}$$
$$k(\mathbf{x}, \mathbf{x}') = \mathbb{E}[(f(\mathbf{x}) - \mu(\mathbf{x}))(f(\mathbf{x}') - \mu(\mathbf{x}'))] \tag{3}$$

With GP priors and training data, prediction of the unobserved function value at any arbitrary location $\mathbf{x}^*$ can be made using the following equation [24]:

$$\mu_{\mathbf{x}^*|\mathcal{D}} = \mu_{\mathbf{x}^*} + \Sigma_{\mathbf{x}^*\mathcal{D}}\Sigma_{\mathcal{D}\mathcal{D}}^{-1}(y_\mathcal{D} - \mu_\mathcal{D}) \tag{4}$$

Here $\mu_{\mathbf{x}^*}$, $\mu_\mathcal{D}$ are the mean values of the data points and are specified by the GP prior $\mu(\mathbf{x})$. $\Sigma_{\mathbf{x}^*\mathcal{D}}$ is the $1 \times n$ vector of covariance between x^* and the n

training data $\mathcal{D}$, and $\Sigma_{\mathcal{DD}}^{-1}$ is the $n \times n$ covariance matrix. With this formulation, the observation value at any arbitrary location in the field can be predicted conditionally on the training data.

To model the signal strength distribution of access points covering certain area, input $\mathbf{x} = (x_h, x_v)$ is a two dimensional vector specifying the horizontal and vertical coordinate of the location. The observation value y_i is the signal strength received at the given location. Note that the input data $\mathcal{D}$ here can be obtained from the fingerprint database, or radio map, which are generally required and constructed by any fingerprint-based localization systems in the offline calibration phase in order to perform localization.

The radio map contains a sequence of records $(\mathbf{x}, \mathbf{fp})$ which associates wireless fingerprints $\mathbf{fp}$ to each location $\mathbf{x}$. Each fingerprint $\mathbf{fp} = (BSSID_i, r_i | i = 1, ..., k)$ consists of signal strength readings r of all k WiFi BSSIDs (MAC addresses of access points) observable. Hence for each BSSID in the system, the training data $\mathcal{D} = \{(\mathbf{x_i}, r_i) | i = 1, ..., n\}$ is available. With the availability of the training data, Gaussian processes can be applied to characterize the signal strength distribution of the whole area.

3 E^3: Efficient Error Estimation

In this section, we study the error estimation of fingerprint-based localization systems based on the GP signal strength model we presented in Sect. 2.

3.1   Point-Level Error Estimation

Point-level accuracy is commonly used in most localization system to measure performance. Such accuracy depends on the ability of fingerprints to uniquely identify a particular location. Hence, the fingerprints at different locations should display sufficient *location diversity*. In our work, signal strength models of access points derived from the training data $\mathcal{D}$ (radio map) provide the mean signal strength value $\mu_{\mathbf{x}^*|\mathcal{D}}$ and variance $\sigma^2_{\mathbf{x}^*|\mathcal{D}}$ of each access point at each location $\mathbf{x}^*$. We can then use these information to get the likelihood estimate for fingerprints and simulate fingerprint sampling at each location during localization to get the error estimate.

3.1.1   Error Estimation

Error of fingerprint-based localization comes from the fact that sampled signal strengths of access points fluctuate and can be different from the fingerprints in the radio map. By chance they will be mapped to different locations. To characterize the average localization error $E(\mathbf{x})$ at one location $\mathbf{x} = (x_h, x_v)$:

$$E(\mathbf{x}) = \sum_{\mathbf{x}'} p_{\mathcal{L}}(\mathbf{x}'|\mathbf{x}) \cdot d(\mathbf{x}, \mathbf{x}') \tag{5}$$

where $\mathbf{x}'$ is the reported location by the localization algorithm $\mathcal{L}$, and $p_{\mathcal{L}}(\mathbf{x}'|\mathbf{x})$ is the probability that the localization algorithm $\mathcal{L}$ reports $\mathbf{x}'$ when users are

---

Algorithm 1. Fingerprint Sampling Algorithm

---

1 **Input:** Location $\mathbf{x}$, mean $\mu_{\mathbf{x}|\mathcal{D}}$, variance $\sigma^2_{\mathbf{x}|\mathcal{D}}$, k

2 **Output:** Sampled fingerprint **fp**

3 **for** $i = 1{:}k$ **do**

4   If r_i hasn't been assigned, with probability p_i set $r_i = \mathrm{rand}(\mu_{\mathbf{x}|\mathcal{D}}, \sigma^2_{\mathbf{x}|\mathcal{D}})$, otherwise set $r_i = -100$;

5   For all $j > i$, set $r_j = r_i$ if $S_{ij} < \tau$;

6 **end**

---

actually in $\mathbf{x}$, and $d(\mathbf{x}, \mathbf{x}')$ is the Euclidean distance between two locations in the 2D plane in meters.

For evaluation purpose, the area of interest is discretized into a number of locations. By averaging over all possible locations, the expect error of each location can be obtained. $p_{\mathcal{L}}(\mathbf{x}'|\mathbf{x})$ is determined by the localization algorithm $\mathcal{L}$ and property of fingerprints **fp** collected at this locations:

$$p_{\mathcal{L}}(\mathbf{x}'|\mathbf{x}) = \sum_{\mathbf{fp}} p(\mathbf{fp}|\mathbf{x}) \cdot \delta_{fp} \tag{6}$$

Here $p(\mathbf{fp}|\mathbf{x})$ is the possibility that fingerprint **fp** can be sampled at location $\mathbf{x}$. $\delta_{fp} = 1$ if $\mathcal{L}(\mathbf{fp}) = \mathbf{x}'$, that is, the localization algorithm maps **fp** to location $\mathbf{x}'$ and $\delta_{fp} = 0$ otherwise. The mapping of $\mathcal{L}$ is deterministic once the fingerprint is given and localization algorithm is chosen. The localization error $E(\mathbf{x})$ hence depends on the fingerprint characteristics and the algorithms used. To get the error estimate from (5) and (6), fingerprints needs to be traversed. Though we already have the likelihood estimate for each fingerprint using Gaussian processes, however, consider a floor with k access points and each access point has q different signal strength readings, we have q^k different fingerprints.

3.1.2 Fingerprint Sampling

It is not feasible to traverse the fingerprint space in practice when k can easily exceeds 100 and $q = 71$ when signal strength ranges from $[-100, -30]$. Instead,

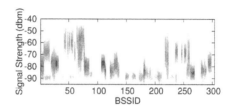

Fig. 1. Gaussian process sampling

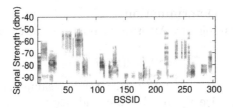

Fig. 2. Ground truth phone sampling

we use Monte Carlo sampling approach [10] to simulate fingerprint based localization and get the error estimate for each location.

To model the real fingerprint readings, generally each access point is considered to be independent [7–9]. This assumption is made based on the fact that access points are physically separated. However, modern access points allow multiple BSSID beacon setting, making access points able to broadcast itself to have multiple BSSID addresses [1]. Therefore, the different signal strength readings of different BSSIDs can belong to the same access point, resulting the readings of these BSSIDs to be mostly identical. These duplicated BSSIDs are recorded down by the sampling devices such as smartphones in the radio map in the calibration phase and are used to perform localization in the online phase as long as they can be received at the location. It is therefore not correct to assume the independence between these BSSIDs. We use the following metric to detect these duplicated BSSIDs:

$$S_{ij} = \frac{\sum\limits_{i=1}^{m} |r_i - r_j|}{m \cdot |r_{min}|} \tag{7}$$

where m is the total number of fingerprints collected in $\mathcal{D}$, r_i and r_j are the signal strength of two BSSIDs and is set to -100 if the BSSID is not detected in this fingerprint. $r_{min} = -100$ is the minimum signal strength observable. S_{ij} should be small if these two BSSIDs are broadcast by the same access point. In the sampling, BSSIDs with S_{ij} less than the threshold τ are set to have the same signal strength. In this work τ is set to 0.005. At each location, the probability p_i that one BSSID can be received can also be learned from the training data $\mathcal{D}$. As shown in Algorithm 1, for a fingerprint **fp** containing k BSSIDs, the signal strength r of each BSSID is sampled randomly from the mean $\mu_{\mathbf{x}*|\mathcal{D}}$ and variance $\sigma^2_{\mathbf{x}*|\mathcal{D}}$ learned from the Gaussian processes with probability p_i. Otherwise it is set to -100, indicating that the BSSID is not observed in this fingerprint.

Figure 1 shows the fingerprints sampled by the sampling algorithm for 304 BSSIDs on one floor in one randomly selected location. Figure 2 shows the ground truth fingerprints sampled by smartphones in the same location. We can see that the GP-based sampling algorithms follows actual fingerprint samples fairly well and provides a "smoother" distribution. With the sampling algorithm, we are now able to simulate fingerprinting at arbitrary locations in this floor to get the error estimate.

3.1.3   Sample Size Determination

Each fingerprint **fp** sampled by Algorithm 1 provides one error estimate $e(\mathbf{x})$ for the location $\mathbf{x}$:

$$e(\mathbf{x}) = d(\mathbf{x}, \mathcal{L}(\mathbf{fp})) \qquad (8)$$

To estimate the average error $E(\mathbf{x})$ with random sampling, we need to decide the minimum sample size n_e to achieve confidence interval α. From statistics theories [6]:

$$\frac{\overline{e(\mathbf{x})} - E(\mathbf{x})}{S/\sqrt{n_e}} \sim t(n_e - 1) \qquad (9)$$

where $\overline{e(\mathbf{x})}$ is the mean of all n_e estimates of $e(\mathbf{x})$ and S is the standard derivation of the n_e samples, and $t(n_e - 1)$ is the t-distribution with $(n_e - 1)$ degrees of freedom [6]. The confidence interval $[-\frac{S}{\sqrt{n_e}} \cdot t_{\alpha/2}(n_e-1), \frac{S}{\sqrt{n_e}} \cdot t_{\alpha/2}(n_e-1)]$ ensures the error estimation with α confidence. We set $\alpha = 99\%$ here.

To make the average error estimate less than ϵ:

$$2\frac{S}{\sqrt{n_e}} \cdot t_{\alpha/2}(n_e - 1) < \epsilon \qquad (10)$$

minimum sample size n_e can be calculated. ϵ is the maximum estimation error and is set to 0.1 m. Algorithm 2 shows the final algorithm for average localization error at each location $\mathbf{x}$. n_0 is the initial sample size and is set to 100. After that the sample size keeps increasing until it meets the constraint set by Eq. (10). The average error $\mathbf{E}(\mathbf{x})$ is then obtained from the sampling algorithm.

Algorithm 2. Error Estimation Algorithm

---

1 **Input:** Location $\mathbf{x}$, localization algorithm $\mathcal{L}$
2 **Output:** Average localization error $\mathbf{E}(\mathbf{x})$

3 **while** *Size of $e(\mathbf{x})$ list $< n_0$ or Eq. (10) not met* **do**
4 | Sample another **fp** using Algorithm 1;
5 | Calculate $e(\mathbf{x})$ using (8);
6 | Add $e(\mathbf{x})$ to the sampled error list;
7 **end**
8 Return the mean of all sampled $e(\mathbf{x})$ as $\mathbf{E}(\mathbf{x})$;

3.2   Region-Level Error Estimation

While the point-level accuracy provides the error characteristics of each location in the indoor environment viewing it at a coarse granularity gives a different perspective to the system behavior. In this section, we analyze the error characteristics at the region-level.

A region here consists of those nearby locations with similar localization errors. The region-level error summarizes the region error distribution and can help to identify blind spots for localization system. By identifying these regions, we have opportunities to improve these regions accordingly. For example, one possible way to improve the poor region performance in fingerprint-based indoor localization system is to place another access points in this region. Placing additional access points will add the "uniqueness" of the fingerprints to this region and hence reduce the localization error for the whole region.

With error estimation algorithm and region-level analysis, the error distribution of the indoor floor is visualized and the impact of each access point to the whole system also becomes easily observable. This capability is useful in both identifying poor performance regions, and deciding where to place new access points or which APs should be included in the fingerprint database.

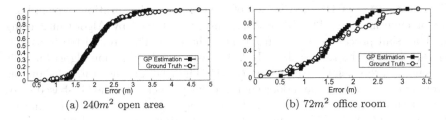

(a) $240m^2$ open area   (b) $72m^2$ office room

Fig. 3. CDF of point-level error

3.3 Floor-Level Error Estimation

The overall performance of localization system depends on many factors such as the localization algorithm $\mathcal{L}$ used, or the deployment of access points. The average error of the whole floor E_f is an important metric that is widely used in the literature to characterize the localization performance. Here

$$E_f = \sum_{i=1}^{n_l} E(\mathbf{x}_i)/n_l \tag{11}$$

is the average point-level error of all n_l locations in the same floor. We focus on the floor-level accuracy and study the factors that affect the overall accuracy in this section.

4 Evaluation

The accuracy awareness based on the Gaussian process provides direct assessment of different fingerprint-based localization systems. One key concern is that how well the error estimation results fit the ground truth and how useful the derived guideline information is. We discuss the evaluation results in this section.

4.1 Data

To evaluate the accuracy-awareness algorithms proposed, we collect data over two weeks period from a big $20 \times 12\,\text{m}^2$ indoor open area and inside a smaller $8 \times 9\,\text{m}^2$ office room. Three different phone models: Google Nexus 5, Samsung S3 and Samsung S4 are used to collect the WiFi radio map and also the testing data. Each indoor environment is divided into $1 \times 1\,\text{m}^2$ grids and each grid is sampled 1 min to construct the radio map. The radio map is used as the training data $\mathcal{D}$ to train Gaussian process models for all access points.

4.2 Performance

4.2.1 Error Estimation

Figure 3 shows the CDF distribution of the point-level localization error in both indoor environments. The GP estimations are obtained by Algorithm 2 using the Gaussian processes trained from the radio map. The ground truth error is measured using the testing data. In both cases, the localization algorithm $\mathcal{L}$ is the same nearest neighbor matching (NN1). The CDF graphs in Fig. 3 show the error characteristics of the indoor environment predicted by the GP-based error estimation algorithm and the ground truth respectively. In both environments, the predicted CDF fits the ground truth error distribution very well, which means the predicted floor-level errors for both indoor environment are also very close to the ground truth. The GP-based estimation algorithm provides a smoother result while the error distribution of the ground truth is more scatter, due to the noise in the fingerprints collected in the real phone readings. Figure 3 shows that the GP-based fingerprint sampling algorithm and the error estimation can successfully fit the error characteristics of the indoor environment and provide a close estimation of the localization error.

5 Related Work

In the past two decades wireless indoor localization has attracted a spectrum of research works. Generally they can be divided into either *device-free* or *device-based* localization. In the device free localization [2,21,23,27], the entity being localized does not carry any special purpose devices. Entities are tracked based on their physical properties such as the body interference to the signal propagation. Device-based localization, on the other hand, locates a device or a human subject carrying a the device. Device-based localization gains the most popularity due to the rise of smartphones and the capability of higher accuracy and cheaper deployment. Device-based localization are generally divided into four categories: *infrastructure-based* [15,22], *wireless fingerprint-based* [3,4,26], *propagation model-based* [4,5,7,12] and *SLAM-based* [16,18]. Among all these categories, wireless fingerprint-based localization has the capability to leverage the widely available infrastructures, and has the potential for large scale deployment.

While [7,24] also utilize Gaussian process in the context of localization, they are focusing either on the improving the performance of localization or the GP itself. Different from all these existing works, the accuracy-awareness proposed in this paper requires only the knowledge of the radio map and the localization algorithm used, and provides direct assessment to the accuracies of the fingerprint-based localization system. To further optimize the localization accuracy, more sophisticated optimization techniques [11] can be used and we leave it as a future work to explore.

6 Conclusion

In this paper we propose E^3 for efficient error estimation in fingerprint-based indoor localization systems. Gaussian processes learned from the radio map are used to characterize the fingerprints in the entire indoor environment. Based on the GP-models built, fingerprint sampling and error estimation algorithm are used to estimate the localization errors. Evaluation shows that the E^3 provides a close estimate to the error behaviors of the localization systems. As the efficient error estimation enables direct assessment to fingerprint-based localization systems and has many useful applications, it has the potential to be applied as a standard component in developing future fingerprint-based localization systems.

References

1. Configuring multiple BSSIDs. http://www.cisco.com/web/techdoc/wireless/access_points/online_help/eag/123-04.JA/1100/h_ap_howto_8.html
2. Adib, F., Katabi, D.: See through walls with WiFi!. In: SIGCOMM. ACM (2013)
3. Azizyan, M., Constandache, I., Roy Choudhury, R.: Surroundsense: mobile phone localization via ambience fingerprinting. In: MobiCom. ACM (2009)
4. Bahl, P., Padmanabhan, V.N.: Radar: an in-building rf-based user location and tracking system. In: INFOCOM. IEEE (2000)
5. Chintalapudi, K., Padmanabha Iyer, A., Padmanabhan, V.N.: Indoor localization without the pain. In: MobiCom. ACM (2010)
6. DeGroot, M.H., Schervish, M.J., Fang, X., Lu, L., Li, D.: Probability and Statistics. Addison-Wesley, Reading
7. Ferris, B., Fox, D., Lawrence, N.D.: WiFi-SLAM using Gaussian process latent variable models. In: IJCAI (2007)
8. Ferris, B., Haehnel, D., Fox, D.: Gaussian processes for signal strength-based location estimation. In: Proceedings of Robotics Science and Systems. Citeseer (2006)
9. Kaemarungsi, K., Krishnamurthy, P.: Modeling of indoor positioning systems based on location fingerprinting. In: INFOCOM. IEEE (2004)
10. Kalos, M.H., Whitlock, P.A.: Monte Carlo Methods. Wiley, New York (2008)
11. Li, J., Qiu, M., Ming, Z., Quan, G., Qin, X., Gu, Z.: Online optimization for scheduling preemptable tasks on iaas cloud systems. J. Parallel Distrib. Comput. **72**(5), 666–677 (2012)
12. Lim, H., Kung, L.-C., Hou, J., Luo, H.: Zero-configuration, robust indoor localization: theory and experimentation. In: INFOCOM. IEEE (2006)
13. Liu, H., Gan, Y., Yang, J., Sidhom, S., Wang, Y., Chen, Y., Ye, F.: Push the limit of WiFi based localization for smartphones. In: MobiCom. ACM (2012)

14. Luo, C., Hong, H., Chan, M.C.: Piloc: a self-calibrating participatory indoor localization system. In: IPSN. IEEE (2014)
15. Priyantha, N.B.: The cricket indoor location system. Ph.D. thesis, MIT (2005)
16. Rai, A., Chintalapudi, K.K., Padmanabhan, V.N., Sen, R.: Zee: zero-effort crowd-sourcing for indoor localization. In: MobiCom. ACM (2012)
17. Rasmussen, C.E.: Gaussian Processes for Machine Learning. MIT Press, Cambridge (2006)
18. Shen, G., Chen, Z., Zhang, P., Moscibroda, T., Zhang, Y.: Walkie-markie: indoor pathway mapping made easy. In: NSDI. USENIX (2013)
19. Sun, W., Liu, J., Wu, C., Yang, Z., Zhang, X., Liu, Y.: Moloc: on distinguishing fingerprint twins. In: ICDCS. IEEE (2013)
20. Wang, H., Sen, S., Elgohary, A., Farid, M., Youssef, M., Choudhury, R.R.: No need to war-drive: unsupervised indoor localization. In: MobiSys. ACM (2012)
21. Wilson, J., Patwari, N.: Radio tomographic imaging with wireless networks. TMC (2010)
22. Xiong, J., Jamieson, K.: Arraytrack: a fine-grained indoor location system. In: HotMobile (2012)
23. Xu, C., Firner, B., Moore, R.S., Zhang, Y., Trappe, W., Howard, R., Zhang, F., An, N.: Scpl: indoor device-free multi-subject counting and localization using radio signal strength. In: IPSN. ACM (2013)
24. Xu, N., Low, K.H., Chen, J., Lim, K.K., Ozgul, E.B.: Gp-localize: persistent mobile robot localization using online sparse Gaussian process observation model. In: AAAI (2014)
25. Yang, Z., Wu, C., Liu, Y.: Locating in fingerprint space: wireless indoor localization with little human intervention. In: MobiCom. ACM (2012)
26. Youssef, M., Agrawala, A.: The horus wlan location determination system. In: MobiSys. ACM (2005)
27. Youssef, M., Mah, M., Agrawala, A.: Challenges: device-free passive localization for wireless environments. In: MobiCom. ACM (2007)

Attribute-Based and Keywords Vector Searchable Public Key Encryption

Huiwen Wang, Jianqiang Li, Yanli Yang$^{(\boxtimes)}$, and Zhong Ming

College of Computer Science and Software Engineering,
Shenzhen University, Shenzhen, China
1981525809@qq.com, {lijq,yangyl,mingz}@szu.edu.cn

Abstract. Nowadays, for third-party cloud platforms are not fully trustable, in order to ensure data security and user privacy when storing data on cloud servers, many enterprises and users choose to store their data in the ciphertext form. So it is essential to search the needed cryptographic information in the cloud servers. In this paper, we proposed public key encryption module based on the attributes-multi keyword vector. Firstly, we adopt the prime order bilinear pairings method to improve time efficiency. Secondly, we introduced the attribute and keywords vector to optimize the vector encryption algorithm. Finally, considering conditions of the limited resources, we proposed the storage domain differentiation policy to save the computing resources and improve efficiency. In order to verify the result of this module, we built up a system based on jPBC, a Java class library. The result shows that the system can achieve the strict access control and multi-user shared fast search mechanism. In the future, it will be applied to the new medical cloud scenes.

Keywords: Cloud storage · Search encryption · Attribute-keywords vector · Access control · Differentiation

1 Introduction

With the advent of the era of Cloud Computing and Big Data, more and more enterprises and users have chosen to compute and store their data in the third-party cloud servers, which have enough storing space and computing ability, in order to save local computing and storing overhead costs. The era of "Internet+" has given birth to many new sectors combined with Internet and traditional sectors, such as the new medical cloud which has been a focused spot recently. Patients can upload their health files to the medical cloud platform through applications and software, and specify which doctors can access to those files. Afterwards, through the medical cloud, doctors can read the patient information for diagnosis. In this way, it will appear a problem that the data of patients health files has been completely out of the management of its owner, which may result in a disclosure of the enterprise and users private information [3] and seriously threaten the security of data stored in the cloud.

© Springer International Publishing AG 2017
M. Qiu (Ed.): SmartCom 2016, LNCS 10135, pp. 317–326, 2017.
DOI: 10.1007/978-3-319-52015-5_32

Nowadays, there are more and more data stored in the cloud severs, which caused a problem of how to quickly search the required resources in the massive ciphertext data in the cloud. Thus, the searchable encryption [1,9,14] came into being, and it has at least three advantages:

(1) Resources that has nothing to do with the given keyword wont be downloaded to the local machines, which helps avoiding unnecessary waste of network traffic overhead and storage space.
(2) The Powerful computing ability of cloud servers can be taken advantage of to perform a file searching operation.
(3) Users do not need to decrypt files which do not meet given conditions, which help saving local computing resources.

2   Related Research

Sahai and Water introduced attribute-based encryption (ABE) [10] as a new means for encrypted acess control, which is subject to identity encryption mechanism [8,10] enlightened and developed. In the identity-based encryption, only with the simple access control policy will be based on the size of the threshold value, the owner can not define flexible access control, just counting the number of ciphertext set of attributes. However, the flexible attribute-based encryption can do any more, which basic principle is that a user is able to decrypt a ciphertext if there is a "match" between the private key and the ciphertext. According to the structure in which the access control policy can be divided into two types: ABE on private key policy [6] and ABE on the ciphertext policy [12].

In 2000, Sang et al. [11] put forward a new concept that searchable encryption under symmetric encryption system. In 2004, Boneh et al. have proposed the asymmetric public key searchable encryption schemes [2], which based on the identity-based encryption. There are still some deficiencies in security, for example it only applies to a single scenarios that a keyword ciphertext data can be unique to the user access, which lives in the most complex scenarios are not suitable for. After that, Chang et al. [3] proposed the concept of searchable encryption [4,5,13] for multi-user scenarios and specific detailed implementation program.

3   Our Construction

In this section we provide the construction of our system. We begin by describing the model of searchable encryption and decryption, and attribute-multi keyword vector ciphertext structure for respectively describing our system. Next, we give the description of our scheme. Finally, we follows with a research in storage domain and policy layer.

3.1 Our Model

In this paper, we propose the scheme of based on attribute-multi keyword vector searchable public key encryption is to use prime order bilinear pairings as the algorithm construction, which is asymmetric encryption. In this model, consists of user attributes and keywords set as a vector is encrypted objects rather than user attributes and multi keyword are composed of two vectors, so that the advantage is that you can save cloud storage resource consumption. Figure 1 is a model of this paper.

3.2 System Process

Search encryption and decryption processes of the model:

(1) Using PKG automatic builder generates public key and private key set, the public key (PK) is published, and the private key (SK) are stored and managed to Trust Authority (TA).
(2) Data owners get their own attributes information from the database.
(3) Data owners according to its keywords set and attribute data composed vector, and combined with the public key (PK) to encrypt for attributes-multi keyword vector encryption and generated initial ciphertext (CT).
(4) Into the system according to the policy parameters I and F match computing, and select the corresponding secondary re-encryption scheme to the secondary ciphertext (encodedData), finally the secondary ciphertext and the encrypted file uploaded and stored to the cloud server.

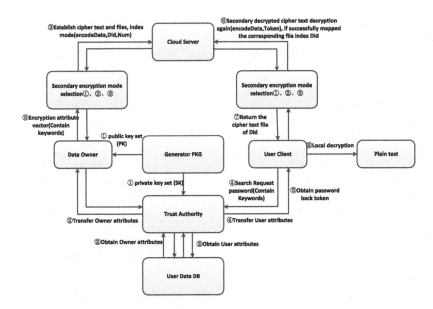

Fig. 1. System model

(5) Users get their own attributes information from the database, and combine with multiple keywords to search together to form their own search password (Query), and combined with the private key (SK) on TA to search request certification.

(6) If the search password (Query) authentication by TA certification, then returned to the requesting user authentication token.

(7) Into the system according to the ciphertext index decrypted vector ciphertext to form decodedData.

(8) User makes a search request to the cloud server based on TA to return a token, the cloud server decrypted token and the secondary decrypted ciphertext decodedData. If the matching decryption is successful, find the vector ciphertext encodedData corresponding Did and ciphertext file.

(9) The cloud server will return the ciphertext files to the client, users locally decrypt the ciphertext forming plaintext. This is what you want.

Encryption and decryption process flow chart shown in Fig. 2:

3.3 Our Construction

Our model is based on the prime order bilinear groups, attributes-multi keyword vector encryption is the prime order algebraic tool to encrypt bilinear pairings. Among them, the encryption vector can allow attribute information with any number of wildcards, so that you can achieve the search classification policy; The last one-dimensional vector is set by multiple keywords component, which is easier to achieve keyword serach, organizational structure of the model shown in Fig. 3:

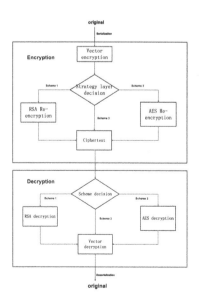

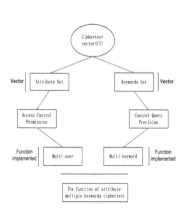

Fig. 2. Encryption and decryption   **Fig. 3.** Ciphertext structure

The implementation main divided into four steps: Setup(), Encryption(), Gentoken(), Test() follows.

3.3.1   Setup (1^n)

PKG automatic builder according to the length of L the vector as input and automatic run an initialization key algorithm, then choose a random entity I=$[p,g_1,g_2,G_1,G_2,GT,e]$, for all $1 \leq i \leq L$, $r \in \sum$, select the security random parameters $t_{i,r}$, $v_{i,r} \in \mathbf{Z_p}$, and set $\overline{T_{i,r}} = g_1^{\frac{1}{t_{i,r}}}$ $\overline{V_{i,r}} = g_1^{\frac{1}{v_{i,r}}}$.

$$PK = [I, (\overline{T_{i,r}} = g_1^{\frac{1}{t_{i,r}}}, \overline{V_{i,r}} = g_1^{\frac{1}{v_{i,r}}})], \quad i \in 1,\ldots,\mathbf{L}, r \in \sum \tag{1}$$

$$SK = [I, (t_{i,r}, v_{i,r})], i \in 1,\ldots,\mathbf{L}, r \in \sum \tag{2}$$

3.3.2   Encryptiopn (PK,x)

The client use public key and vector $\overrightarrow{x} = (x_1, \ldots, x_{L-1}, x_L)$, $x_1, \ldots, x_{L-1} \in \sum_*$, and $x_L \in \sum$, (Respectively the user attribute information and a set of keywords domain information) as input to encrypt and calculate. When $x_i \neq *$ select a random security parameter $a_i \in \mathbf{Z_p}$ so that the sum of the lower limit 0, which is $\sum a_i = 0$. After calculating the vector encryption $CT_{\overline{x}} = (X_i, W_i)_{i=1}^L$, the client sents tuple $(CT_{\overline{x}}, \text{Did})$ to the cloud server, and then indexing. Set as follows:

$$\begin{cases} X_i = \overline{T_{i,x_i}}^{-a_i}, W_i = \overline{V_{i,x_i}}^{-a_i}, & if \quad x_i \neq *; \\ X_i = \varnothing, \quad\quad W_i = \varnothing, & if \quad x_i = * \quad and \quad i \neq L; \end{cases} \tag{3}$$

3.3.3   GenToken (SK,y)

Generating token algorithm is run by the Trust Authority (TA), the private key (SK) and query request vector $\overrightarrow{y}$ as input and the generated token as output. After receiving search vector, Trust Authority (TA) will first authenticate users, if the validation is successful will send the user an access token to retrieve qualifying documents. In order to generate the token, Trust Authority (TA) first select the $L + 1$ random security parameters $s, s_1, \ldots, s_L \in \mathbf{Z_p}$. Set as follows:

$$\begin{cases} T_{i,y_i} = g_2^{t_{i,y_i}}, & V_{i,y_i} = g_2^{v_{i,y_i}}; \\ Y_i = T_{i,y_i}^{S-S_i}, & M_i = V_{i,y_i}^{S_i}; \end{cases} \tag{4}$$

Finally, the token generated by the algorithm: $T_{\overline{y}} = (Y_i, M_i)_{i=1}^L$.

3.3.4   Test ($PK, T_{\overline{y}}, CT_{\overline{x}}$)

The cloud server receives the token to run the Test after decryption matching algorithm, which send the token and server index that all existing vector ciphertext $CT_{\overline{x}}$ as input, if there will be: $Test(PK, T_{\overline{y}}, CT_{\overline{x}}) = 1(true)$, it means the success of the query to find an index $(CT_{\overline{x}}, \text{Did})$ of the corresponding Did, and then find the ciphertext file returned to the user based on Did. The test decryption algorithm Test calculated as follows: If $\overrightarrow{x} = \overrightarrow{y}$, then

$$Test(PK, T_{\overline{y}}, CT_{\overline{x}}) = \prod_{i=1}^{L} e(X_i, Y_i)e(W_i, M_i)$$

$$= \prod_{i=1}^{L} e(g_1^{\overline{t_{i,x_i}}}, g_2^{t_{i,y_i}(s-s_i)})e(g_1^{\overline{v_{i,x_i}}}, g_2^{v_{i,y_i}S_i}) \qquad (5)$$

$$= 1$$

If the Test result for 1, it indicates that the user meets the search criteria and successful search for the resource file, which will return the ciphertext to client, and then decrypted locally.

3.3.5   Differentiation Security Model

In system model, users uploading or sharing any resource are based on attributes-multi keyword vector encryption, but this approach is the lack of personalized protection, especially in the case of the remaining resource are limited. Encryption and decryption services are implemented on the storage domain, so the combination on the basis of differentiation security model to different user identities and the important degree of different files, using AES (128bit) and RSA (1024bit) to achieve a secondary re-encryption with different security levels. It is known RSA encryption is high security, but its slow, time-consuming high, high consumption of resources; AES encryption security degree relatively RSA can only be considered moderate, but its speed, low resource consumption. Figure 4 shows the differential model of the security policy in storage domain.

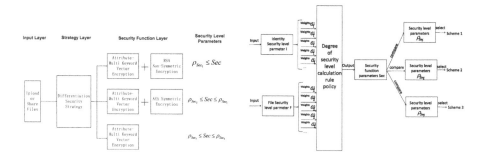

Fig. 4. Differentiation security policy model

Fig. 5. Calculation rule

3.3.6   Calculation Rules

Calculation rules in the policy layer is made up of identity safe level parameter I, file security level parameter F, minimum security level probability parameter ρ_{Sec} and configurable parameters of ω_1, ω_2. The policy layer is set three encryption security service, the main steps of the selection method is as follows:

(1) The system must set the security level probability parameter values ρ_{Sec} in each program execution, primarily for selecting a reasonable level of security encryption service.
(2) When the user uploading or sharing files in the input layer, they need to configure identity security level parameter I and file security level parameter F of the weight value ω_1, ω_2, and $\omega_1 + \omega_2 = 1$.
(3) Into the policy layer, calculating the probability degree of security rules while the output corresponding to the selected security parameter Sec encryption service.

In all three schemes, the security level of probability parameter ρ_{Sec_1}, ρ_{Sec_2}, ρ_{Sec_3} has been determined, policy layer security rule for calculating the degree of probability can be represented by the following formula:

$$Sec = (\frac{\omega_1}{I} + \frac{\omega_2}{F})(1 - \rho_{Sec_3}) + \rho_{Sec_3} \tag{6}$$

The ρ_{Sec_3} represents the actual policy scheme 3 calculates the minimum degree of security ρ. Figure 5 is a complete difference degree calculating the probability of the security rules policy process:

4   Implementation

In this section we mainly comparative experiment algorithm and scheme, including time-consuming between prime bilinear and composite order, and time-consuming in each stage between ours scheme and CPdHVE scheme.

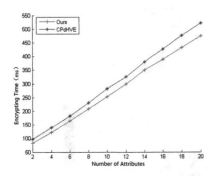

Fig. 6. Time cost of encryption

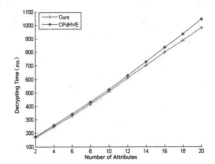

Fig. 7. Time cost of search decryption

4.1 Algorithm Experimental Comparison

On the computational time-consuming, prime order bilinear are 50 times faster than composite order bilinear pairings in computing speed. Assuming that encryption vector length to 10, search decryption calculation Test time-consuming is more seven times than prime order, the most significant is the amount of calculation of the composite order bilinear element ratio of lager prime order. Figures 6 and 7 are composite order bilinear on encryption scheme CPdHVE [7], and ours scheme calculate the cost of encryption and decryption of time. Table 1 and Fig. 7 are the total numbers of files increases with each model search decryption time cost comparison.

Due to public key encryption algorithm inherent defects, does not apply to bulk encryption of data, so the experiment is just in a relatively small scake on the number of files. As shown in Table 1 and Fig. 8 with the increase of data files stored in the cloud server, you can search to the number of files that will

Table 1. Search decryption time comparison of different scheme

Total files	The number of files to search for keywords	Ours search decryption time	CPdHVE search decryption time
5	2	1.423	1.513
10	4	2.758	2.985
15	6	4.27	4.867
20	8	5.983	6.684
25	10	7.01	7.842
30	12	8.394	9.241
35	14	9.968	10.765
40	16	11.13	11.968
45	18	12.545	13.526
50	20	14.128	15.318

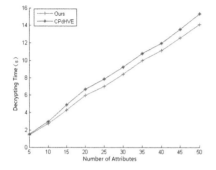

Fig. 8. Time cost of decryption

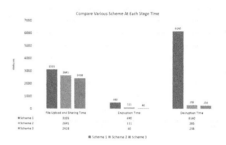

Fig. 9. Compare varsions scheme

increase accordingly, the number of files and search decrypted time is proportional relational. According to the data analysis ours scheme in terms of time efficiency than CPdHVE program to be better.

4.2 Experimental Comparison

In the system model, the introduction of the asymmetric RSA and symmetric AES secondary re-encryption, and the cost of computing time is different. The higher security level of security encryption services are more time consuming for the price. Table 2 and Fig. 9 is in the case of vector attribute number a key initialization time recording the programs calculate the number of vectors attributes in the experiments is 3, get the key the initial time, total encryption time, application token time, total decryption time, and the file upload or share total time.

Table 2. Time-consuming comparison of different schemes at different stages

Scheme	Initialization time	Encryption token time	Application decryption time	Decryption time	File upload and sharing time
Scheme 1	2559	490	236	6140	3103
Scheme 2	2480	111	226	295	2641
Scheme 3	2329	40	220	256	2418

5   Conclusion

Under the background of cloud computing and big data, in order to make the most of cloud storage services and computing resources, while ensuring the security of data stored in the cloud, as well as meet the multi-user scenarios and multiple keywords search. The model is through adding user attributes encryption policy to support multiple data owners for multiple user scenarios, supports multiple keywords high precision search, and make it more flexible and efficiency. Finally, the system model also joined the differentiated security policy, according to different users with different levels of identity and important of files for different levels of security encryption service mechanism.

Acknowledgements. The authors gratefully acknowledge the contribution of the National Science Foundation of China [61572330][61472258], the Natural Science foundation of Guangdong Province [2014A030313554], and the Technology Planning Project (Grant No. 2014B010 118005) from Guangdong Province.

References

1. Baek, J., Safavi-Naini, R., Susilo, W.: Public key encryption with keyword search revisited. In: Gervasi, O., Murgante, B., Laganà, A., Taniar, D., Mun, Y., Gavrilova, M.L. (eds.) ICCSA 2008. LNCS, vol. 5072, pp. 1249–1259. Springer, Heidelberg (2008). doi:10.1007/978-3-540-69839-5_96

2. Boneh, D., Crescenzo, G., Ostrovsky, R., Persiano, G.: Public Key Encryption with Keyword Search. Springer, Berlin, Heidelberg (2003)

3. Chang, Y.C., Michael, M.: Privacy Preserving Keyword Searches on Remote Encrypted Data. Springer, Berlin, Heidelberg (2015)

4. Dong, C., Russello, G., Dulay, N.: Shared and searchable encrypted data for untrusted servers. J. Comput. Secur. **19**(3), 367–397 (2008)

5. Bao, F., Deng, R.H., Ding, X., Yang, Y.: Private query on encrypted data in multi-user settings. In: Chen, L., Mu, Y., Susilo, W. (eds.) ISPEC 2008. LNCS, vol. 4991, pp. 71–85. Springer, Heidelberg (2008). doi:10.1007/978-3-540-79104-1_6

6. Goyal, V., Pandey, O., Sahai, A., Waters, B.: Attribute-based encryption for fine-grained access control of encrypted data. In: Proceedings of ACM Conference on Computer and Communications Security, CCS 2006, Alexandria, VA, USA, 30 October–November, pp. 89–98 (2006)

7. Hattori, M., Hirano, T., Takashi, I., Matsuda, N., Takumi, M., Sakai, Y., Kazuo, O.: Ciphertext-Policy Delegatable Hidden Vector Encryption and Its Application to Searchable Encryption in Multi-user Setting. Springer, Berlin, Heidelberg (2011)

8. Lai, J., Deng, R.H., Liu, S., Weng, J., Zhao, Y.: Identity-Based Encryption Secure against Selective Opening Chosen-Ciphertext Attack. Springer, Berlin, Heidelberg (2014)

9. Michel, A., Mihir, B., Dario, C., Eike, K., Tadayoshi, K., Tanja, L., John, M.L., Gregory, N., Pascal, P., Shi, H.X.: Searchable encryption revisited: consistency properties, relation to anonymous ibe, and extensions. J. Cryptol. **21**(3), 350–391 (2008)

10. Sahai, A., Waters, B.: Fuzzy identity-based encryption. In: Cramer, R. (ed.) EURO-CRYPT 2005. LNCS, vol. 3494, pp. 457–473. Springer, Heidelberg (2005). doi:10. 1007/11426639_27

11. Song, D., Wagner, D., Perrig, A.: Practical techniques for searches on encrypted data. In: IEEE Symposium on Security & Privacy, pp. 44–55 (2000)

12. Waters, B.: Ciphertext-policy attribute-based encryption: an expressive, efficient, and provably secure realization. In: Catalano, D., Fazio, N., Gennaro, R., Nicolosi, A. (eds.) PKC 2011. LNCS, vol. 6571, pp. 53–70. Springer, Heidelberg (2011). doi:10.1007/978-3-642-19379-8_4

13. Yang, Y., Lu, H., Weng, J.: Multi-user private keyword search for cloud computing. In: Proceedings of IEEE International Conference on Cloud Computing Technology and Science, Cloudcom 2011, Athens, Greece, 29 November–December, pp. 264–271 (2011)

14. Zhu, B., Zhu, B., Ren, K.: Peksrand: providing predicate privacy in public-key encryption with keyword search. In: ICC, vol. 2010(4), pp. 1–6 (2011)

PADS: A Reliable Pothole Detection System Using Machine Learning

Jinting Ren and Duo Liu$^{(\boxtimes)}$

College of Computer Science, Chongqing University, Chongqing, China
`jintingren@gmail.com`

Abstract. In recent years, with the popularity of private cars, a increasing number of people prefer autos as their way to travel. However, poor road conditions may cause damages to vehicles and have drawn great concern of governments all over the world. The extreme weather conditions, heavy traffics and low road quality are worsen the situation, making it is a challenging task to keep roads in good conditions. Therefore, frequent repairs are required to avoid damages to vehicles. In this paper, we propose a reliable pothole detection system using machine learning (PADS) to facilitate the road pothole detection and road conditions maintenance. The proposed system provides low latency in potholes detecting, thereby shortening the time for road maintainers to identify poor conditions roads. To make our system easy to deploy, we reduce monetary cots and simplify system architecture. To improve accuracy in potholes detection, we use K_MEANS algorithm based on basic threshold algorithm. Our results display a plot of z-axis accelerations on one road and a pothole-marked map. At last, we show the pothole detection accuracy comparison between our algorithm and basic threshold algorithm.

Keywords: Pothole detection · Road monitoring · Machine learning · IoT · Tri-axial accelerometer

1 Introduction

Poor road surface conditions may cause damages to vehicles, high maintenance cost and even traffic accidents. According to the survey of U.S. Federal Highway Administration [1], thousands of people are hurt or killed each year on roads and highways due to poor road quality and conditions. Therefore, keeping roads in good conditions is important to reduce the traffic accidents. However, maintaining road surface is a challenging task due to extreme weather condition, heavy traffic and the low road quality. To maintain good road condition, frequent repairs are required and make that a reliable and low-latency road conditions monitoring system is much required. This kind of system will be especially useful in reminding road maintainers to repair poor road surfaces.

However, traditional monitoring systems are not suitable in such scenario for the following two reasons: (1) Traditional monitoring systems aim at a overall monitor on roads conditions including pavement images collection, pavement shape detection, road stereo image drawing and etc. Often, this kind of overall

© Springer International Publishing AG 2017
M. Qiu (Ed.): SmartCom 2016, LNCS 10135, pp. 327–338, 2017.
DOI: 10.1007/978-3-319-52015-5_33

monitoring will cost lots of time and can not be low-latency apparently [2,3]. (2) Traditional monitoring systems collect data from two main sources–statistical sensors in the pavement and cars equipped with specific road monitoring sensors like Ground Penetrating Radar (GPR) [4]. Limited by cost, these systems can not scale up to monitor most roads in the city.

In this paper, we proposed PADS a simplified, low-cost and reliable road monitoring system with low latency in detecting potholes. Considering the drawbacks of traditional systems and needs of keeping roads in good conditions, there are three main concerns in our system – achieving low latency in detecting potholes, reducing monetary cost and improving detection accuracy. First, we make only pothole detection function left in our system to achieve low-latency goal. The number of Potholes on road is a convincing evidence to identify whether a road is in good conditions and we can use this feature to get a general picture of road conditions. We can also use the potholes location report to help road maintainers fix poor road surfaces quickly. Second, PADS is implemented using 'dynamic' sensors — sensors placed on autos to reduce cost [5]. Due to the mobility, autos can be randomly distributed in city roads and we can monitor almost every road theoretically if we have enough autos equipped with sensors. Using 'dynamic' sensors instead of statistical senors on roads can reduce monetary cost because placing sensors on several vehicles is cheaper than placing sensors on every road. Finally, wrong potholes detection will make system unauthentic and increase the maintenance cost. To improve detection accuracy, our approach decides to use machine learning methods [6,7]. The K_MEANS algorithm [8,9] is used to cluster road data into two classes and then computes a more accurate threshold to improve detection accuracy. For evaluation, we show a pothole-marked map produced by our system and compare the accuracy of our potholes detecting algorithm with basic threshold algorithm.

In summary, we make the following major contributions:

- We propose a reliable pothole detection system with low latency in detecting potholes.
- Our design uses machine learning methods to improve the accuracy int detection potholes compared of threshold-based pothole detection algorithm.
- We deploy our system on private cars and detect potholes on roads in low latency.

The rest of this paper is organized as follows: Sect. 2 describes the background and the motivation. Section 3 discusses on our system design including hardware platform and architecture. Main algorithm description and its complexity analysis are also included in this section. Section 4 talks about the evaluation of our approach consisting of a potholes marked map, a plot of z-axis accelerations on one road and accuracy analysis. In Sect. 5, the related work will be discussed on. At last, we draw a conclusion in Sect. 6.

2   Background and Motivation

In this section, we first introduce the dangers caused by poor road conditions. We then give the drawbacks of traditional monitoring systems and explain why it can not be applied to keep roads in good conditions. Finally, we discuss our motivation.

2.1 Dangers Caused by Poor Road Conditions

The U.S. Congress passed in a rare bipartisan effort (late 2015) the Surface Transportation Reauthorization and Reform Act of 2015, which provides $233 billion for federal highway maintenance over five years. That is $46 billion per year [10]. The fact shows that every year governments need to spend much money on maintaining road conditions. However, the road conditions of U.S is still poor even though it has cots government so much to maintain road conditions. According to the research, in the cities with worst road conditions in U.S, the ration of poor roads is over 50%. It is a really challenging task to keep roads in good conditions but governments need to do this because poor road conditions can cause many dangers.

There are several dangers that may caused by poor road conditions. First of all, hitting a pothole may cause damages to vehicles. The damages may not only occur in automotive chassis but also in tire puncture and wire rim. In most cases, these damages may just make you pay for repair charge. However, the worse case is traffic accidents. According to statics, one-third accidents involve poor road conditions of approximately 33,000 traffic fatalities each year [10]. Moreover, the number of accidents is predicted to become larger with increases in vehicle traffic in next years.

2.2 Drawbacks of Traditional Monitoring Systems

Traditional monitoring systems is not useful in helping road maintainers fix road surfaces rapidly. These systems are designed to do an overall check on roads. Therefore, it will include many unnecessary check for just fixing the road surface and result in long latency. Moreover, traditional monitoring system often use sensors placed on roads or specific monitoring cars. Limited by the cost, tradition monitoring systems are hard to scale up and then can not monitor most roads in cities. For example, one current equipment used in measuring road condition, which is composed by accelerometers, distance measuring instruments and graphic displays is quite expensive [11]. This road condition systems may cost 8,000 to 220,000 dollars. These two drawbacks determine that traditional road monitoring system can not be applied in fixing road surfaces rapidly to keep roads in good conditions.

2.3 Motivation

Private vehicles have been more and more popular in recent years. To guarantee driving safety, governments take frequent repairs to keep roads in good conditions. Therefore, how to identify roads in poor conditions quickly is important since road maintainers cannot repair any poor conditions roads without identifying them. Fortunately, a road monitoring system may help road maintainers achieve this goal. However, traditional road monitoring systems are long-latency and high-cost. Therefore, they can not be applied to identify poor condition roads quickly. This motivates us to propose a reliable pothole detecting with low latency in detecting potholes. To reduce the cost, we choose only 3-axis accelerometers sensors and GPS module as our data-collecting equipment. To improve detection accuracy, we use machine learning methods to compute a more accurate threshold in identifying potholes.

3   PDSML Design

In this section, we first discuss the hardware platform used by our system. We then introduce the architecture of our system. Finally, we talk about the main algorithm and its complexity analysis.

3.1   Hardware Platform

We reduce monetary costs especially those spent on hardware to make our system easier to scale up. To achieve this goal, only cheap sensors and inexpensive boards are used in our system. One acceleration sensor and one GPS module are used in collecting raw data. For preprocessing raw data and sending filtered data to center servers, we use raspberry pi as our router. We choose raspberry rather than other boards due to its relatively higher computing performance as we decide to run filtering algorithm on router. Another benefit of preferring raspberry is that raspberry board has a built-in wifi module for sending data and has no need for any external data transmitting module. The last component of our system hardware platform is a common center data server. Ignoring the cost of the server, the total hardware platform cost is less than $60, a acceptable price comparing with the cost of traditional road condition monitoring systems. Figure 1 shows the top view of our hardware platform.

3.2   Software Architecture

PADS uses a basic IoT architecture consisted of four layers [12]. Figure 2 shows architecture of our system. The first layer is the sensing layer contains acceleration sensors which are used to collect accelerations in 3-axis and locations respectively. Routers (i.e., raspberry pi 3) in second layer — the network layer — receive raw data and filter out noisy data. After that, routers will send filtered

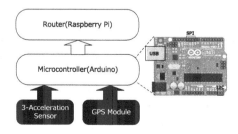

Fig. 1. Top view of our hardware platform.

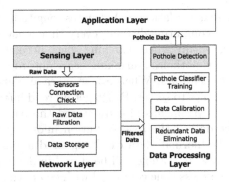

Fig. 2. Architecture of PADS.

data to center data server using built-in wifi module. The most critical part of our system is the third layer — the data processing layer. In this layer, PADS will remove replicated data since we have deployed multiply device on roads. With deduplicated dataset, we still have to 'calibrate' it. Because our equipment is not placed flat on front tyres, our location of potholes will be inaccurate and we need to 'calibrate' the location data. Data-collecting latency and the speed of car equipped our sensors will also affect the accuracy in pothole location. Taking all these factors into account, we 'calibrate' our data to eliminate errors. Once our system get enough simplified and accurate data, a potholes classifier can be trained and then we can detect potholes using this classifier. At last, in application layer we mark all the detected potholes on Google map it the APIS provided by Google.

3.3   Data Processing

There are four main algorithms in our system. However, the filtering algorithm, a Z-DIFF algorithm (using the difference value in Z-axis acceleration to judge if this data should be ignored), is so simple that we have no need to discuss on it in detail. Thus, algorithm part will be mainly divided into three parts–redundancy solving algorithm part, calibrating algorithm part and pothole detection algorithm part.

Redundant road data eliminating algorithm: The first part is about redundant road data eliminating algorithm. This algorithm is used in reducing redundant data in same position collected by different vehicles or one vehicle in different periods. We use distance calculated by longitude and latitude data to judge if a new data is redundant. The computational formula is:

$$a = \sin^2(\Delta\varphi/2) + \cos(\varphi_1)\cos(\varphi_2)\sin^2(\Delta\lambda)$$

$$c = 2 * atan2(\sqrt{a}, \sqrt{1-a})$$

$$Distance = R * c. \tag{1}$$

In above equation, φ denotes latitude, λ denotes longitude and R denotes the earths radius. However, computing distance every time a new data comes can be a time-wasting task especially when searching space becomes large. To achieve low latency in detecting potholes, shrinking searching room is necessary. We give the road label to each data. When a new 'normal' data (no visual z-axis acceleration changes) comes to server, PADS get its road information using map APIS and check if the system has already stored the data of this road. Our system will ignore the data if the road condition information has been stored in our system. Only the 'normal' data can be ignored in our system and every data that possibly represents pothole will be stored in database. Unfortunately, though system will spend less time compared with computing distance for every data, our system still needs to search road information when a new data comes.

Since searching road information every time is still a rime-consuming task, our system has to find a new way to label our data. In fact, vehicles may often stop in the crossroad and the speed of the vehicle will become zero. As a result, a zero-speed data collected by GPS may represent that the vehicle has arrived at the crossroad. The idea is simple, PADS will store the 'normal' data in buffer at first without searching its road information (we will search the road information for the first data) until a new data that has zero value in speed comes. Then PADS compares the road information between the first data and the 'zero-speed' data. After that, our system will ignore the 'normal' data between them if they are on the same road. The drawback of this solution is that we still have to search the road information of each data when the zero-speed data is actually collected in another road. However, the worst case will cost nearly the same time compared with unoptimized algorithm.

Calibrating algorithm: The second part is about calibrating algorithm. Calibrating algorithm is used in making location more accurate. For calibrating data, we should take data-collecting latency and speed into consideration to fix the measurement error caused by the distance between seats and front tyres. We can computer the direction of vehicles using three-axis accelerations. Then we can use the direction and the distance to get the accurate location. However, this direct method can also be time-consuming. We need a simpler method to achieve the low-latency goal. Based on the fact that the distance is so short and we can assume the speed and the direction of the vehicle keep steady during this period. With that assumption, we can replace the location information of current data with one of later collected data to get a more accurate location. All our system do is to compute the latency for our vehicles to drive the distance between seats and front tyres. We define ν for the speed of the vehicle and ψ for the frequency of GPS in collecting data(GPS module have the same frequency as acceleration sensors in collecting data). The τ represents the distance between seats and front tyres. The latency can be expressed as:

$$Latency = \left\lfloor \frac{\psi\tau}{\nu} \right\rfloor. \tag{2}$$

With latency computed, we can calibrate our data.

Pothole detection algorithm: The last part is about pothole detection algorithm. The simplest way to detect pothole on road is using a threshold algorithm [13]. However, assuming that roads in different areas have their own features, using one certain threshold to detect potholes in all areas is not realistic. We solve this problem by building a more accurate model for potholes detecting using machine learning [14]. Indeed, using machine learning to detect pothole will affect performance and make latency longer. We still need to prove accuracy because inaccurate pothole detection will make system unauthentic. In that context, a simple but suitable machine learning algorithm should be chosen to balance accuracy and latency. Finally, our system chooses the K-MEANS algorithm [15]. Using this algorithm, we can simply cluster the data into two clusters using three features x-axis, z-axis and speed. In related work, the threshold only uses z-axis as its feature. However, a pothole will cause changes both in x-axis acceleration (decrease in x-axis acceleration) and z-axis (increase in z-axis acceleration) Using two features will make detection more accurate. We do not use speed directly to cluster data. In contrast, we define two scenes for clustering—high speed scene and low-speed scene. Since acceleration data changes will be quite different between high-speed and low-speed scenes, we need to treat them separately (in our experiment, we define over 25 kmph is high-speed and below it is low-speed). Supposed potholes data as abnormal data, it should be smaller in number of two clusters. However, the results show the two clusters often have similar size. To remedy this issue, we use original threshold (only use z-axis data as threshold) [13] to firstly label the data (pothole label and normal label) and then cluster them. After clustering, we choose the data set having more pothole labels to be the cluster represents pothole data. Then we use the x-axis and z-axis accelerations data of pothole data set to calculate our detection threshold. The computational formula is:

$$Threshold = \frac{\sum_{i=1}^{n} (\chi - z)^2}{n}. \tag{3}$$

Let χ denotes x-axis acceleration and z denotes z-axis acceleration. We use z to minus χ in equation because large z-axis acceleration and small x-axis acceleration mean a pothole.

3.4   Complexity Analysis

The most time-consuming part of our algorithm is pothole detection part. To reduce latency, our system decides to update pothole threshold periodically. In most cases, PADS just uses computed threshold to judge if the vehicle has faced a pothole. Therefore, in that situation, the complexity of detecting algorithm is $\mathcal{O}(n)$. When the system needs to update threshold, the complexity of detecting potholes is mostly dependent on the complexity of $K\_MEANS\_CLUSTER$ algorithm [16,17]. The general Euclidean space d and clusters k will decide $K\_MEANS\_CLUSTER$'s complexity. For our system, both d and k equal to 2 (i.e. 2 attributes x_axis and z_axis). Therefor, the complexity is $\mathcal{O}(n^5 \log n)$. Though the performance of $K\_MEANS\_CLUSTER$ is

---

Algorithm 3.1 Pothole detection algorithm

Input: *train_data_vector*:road condition data vector;*SIZE*:size of road condition vector;*OT*:original threshold provided by other work;*ST*:speed threshold

Output: *Threshold_vector*:pothole threshold in high-speed and low-speed

1: $cluster \leftarrow 2, iterations \leftarrow 100$. //one cluster for pothole data and another for normal data
2: $cluster1\_vector \leftarrow \emptyset, cluster2\_vector \leftarrow \emptyset, TH \leftarrow 0$.
3: **for** $i \leftarrow 1$ **to** $SIZE$ **do**
4:     **if** $train\_data\_vector[i].z\_axis > OT$ **then**
5:         $train\_data\_vector[i].label \leftarrow$ pothole.
6:     **else**
7:         $train\_data\_vector[i].label \leftarrow$ normal.
8:     **end if**
9: **end for**
10: $result \leftarrow$ **K_MEANS_CLUSTER**$(train\_data\_vector, cluster, iterations)$.
11: **if** $result[0].pothole\_label\_count > result[1].pothole\_label\_count$ **then**
12:     $cluster1\_vector \leftarrow result[0]$.
13: **else**
14:     $cluster1\_vector \leftarrow result[1]$.
15: **end if**
16: $high\_speed\_sum \leftarrow 0$.
17: $low\_speed\_sum \leftarrow 0$.
18: $n \leftarrow$ **length**$(cluster1\_vector)$.
19: **for** $i \leftarrow 1$ **to** n **do**
20:     **if** $cluster1\_vector[i].speed > ST$ **then**
21:         $high\_speed\_sum \leftarrow high\_speed\_sum + (cluster1\_vector[i].z\_axis - cluster1\_vector[i].x\_axis)^2$.
22:     **else**
23:         $low\_speed\_sum \leftarrow low\_speed\_sum + (cluster1\_vector[i].z\_axis - cluster1\_vector[i].x\_axis)^2$.
24:     **end if**
25: **end for**
26: $Threshold\_vector$.**add**$(high\_speed\_sum/n)$.
27: $Threshold\_vector$.**add**(low_speed_sum/n).
28: **return** $Threshold\_vector$

---

not so good, the complexity of detecting algorithm is acceptable since updating thresholds occurs in low frequency.

4   Evaluation

In this section, we firstly describe the experimental environments to evaluate our system. Secondly, we display a pothole-marked map and a plot of z-axis accelerations on one road as our experimental results. At last, we compare the accuracy between our detection algorithm and basic threshold algorithm.

Table 1. Item model and detailed features of components

Name	Model	Interface	Resolution	Sample rate	Accuracy	Power supply
3-axis accelerometer	ADXL345	SPI I2C	up to 16 g	up to 3200 HZ	4 mg/LSB	2.0 V to 3.6 V
GPS module	UBLOX NEO-6M	UART USB SPI DDC(I2C compliant)	N/A	up to 5 Hz V	2.5 m CEP	2.7 V to 3.6 V
Logic level converter	5 V to 3.3 V Logic Level Converter	N/A	N/A	N/A	N/A	5 V or 3.3 V
Data collecting board	Genuino UNO	UART USB SPI I2C TWI	N/A	N/A	N/A	5 v
Router	Raspberry Pi 3	UART USB SPI I2C WIFI	N/A	N/A	N/A	2.5 A@5 V

4.1 Experimental Setup

As shown in Table 1, we use ADXL345 and UBLOX NEO-6M to comprise our sensing layer. We use Arduino as a middle data collecting board instead of connecting sensors directly to raspberry due to the limit of raspberry pins amount. Another significance of this design is making system loosely-coupled. Separating router and sensing layer can make it easier to replace sensors without modifying codes on router. We use SPI protocol to connect board with ADXL345 and I2C protocol to connect board with GPS module. The benefit of using two different protocols is to avoid conflicts on pins using. Without sharing pins, we can easily implement our system using only one board. The ADXL345, however, can not be directly connected to Arduino due to its 5 V-system. Therefore, a logical level converter is necessary to convert 5 V to 3.3 V. At last, we connect Arduino to raspberry with a USB cable. Fixing all our equipment on the seat of vehicle, we will finish the setup for our experimental platform.

4.2 Results

The user interface of our system is a pothole-marked map on web page. Moreover, a plot of vehicle's accelerations data on one road which uses sensor data directly has been drawn as a by-product. Figure 3 shows the z-axis accelerations during the vehicle drive on one road. In the figure, time represents data collecting time. From the figure we can see there visual changes on vehicle's z-axis accelerations.

To evaluate our system, we first use z-axis accelerations data and its road information to find which area may have more potholes. We optimize the workload to improve experimental efficiency [18]. After that, we drive the car equipped with experimental equipment on this area for several times. Using these

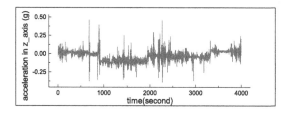

Fig. 3. A plot of z_axis accelerations on one road.

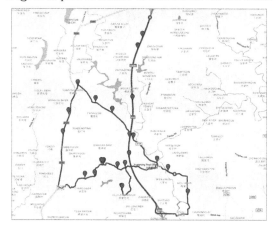

Fig. 4. A pothole-marked map which represents the road surface conditions near G93 highway.

data, our system detects potholes in this area. To make our map more clear, we eliminates all potholes that do not belong to main stem by hand. Finally, we get a pothole-marked map. Figure 4 shows the final pothole-marked map produced by our system. In that map, we have marked potholes detected by our system around G93 highway with read markers.

After analysing results, we find that our system actually can not distinguish pothole from occasional road anomalies on roads like emergency brake. In fact, our system can only distinguish potholes from normal road data. However, these occasional road anomalies have little chance to occur in the same place for several times. Therefore, our system only take the potholes that has been detected for several times to distinguish potholes from occasional road anomalies.

At last, we will evaluate detecting accuracy of our system. The simplest but most time-consuming way to get test data set is using a field trip. However, we are prohibited to stop and check potholes on many roads. To solve it, we drive our test car on certain area roads for several times. Then, we take these potholes data that have been detected every time as actual pothole on road to consist our test data set. Let $N_{non\_pothole\_detection}$ be the number of the pothole that has not been detected and $N_{pothole\_wrong\_detection}$ be the number of normal road data that has been judged pothole wrongly. $N_{detection}$ represents the total number of test data set. We define accuracy of detecting potholes as:

$$Accuracy = 1 - \frac{N_{non\_pothole\_detection} + N_{detection}}{N_{detection}} \qquad (4)$$

The results show that our pothole detection algorithm has truly improved detecting accuracy compared with simple threshold algorithm in z-axis.

5   Related Work

Road Quality Clustering System. Automatic road quality clustering system has ranked road quality utilizing tri-axial accelerometer [19]. This system has used accelerometer data to cluster roads into three classes dependent on its quality. However, it can not locate the potholes on road and has little use in fixing road surface quickly.

Pothole Detection Systems Based on Smart Phones. Current pothole detection systems using vehicles equipped with sensors are often based on smart phones. These systems have used tri-axial accelerometer and GPS of smart phones as our equipment on board to detect potholes [13,20–23]. However, this kind of systems need more human interaction and can not be automatic. Moreover, differences in mobile phones hardware make systems based on it hard to develop.

Pothole Detection using Threshold Algorithm. The simplest way to detect potholes on roads is using threshold algorithm in z-axis [13]. Detecting potholes dependent on certain threshold is simple and has lowest latency. However, the accuracy of this kind of detection is doubted. Moreover, the certain threshold can not be applied to different road conditions.

6   Conclusion

This paper has described the design, implementation and evaluation of a pothole detection system. We have used 3-axis accelerometer sensors and GPS module to collect raw data. Using inexpensive sensors and boards, our equipment costs only about $60. To filter data, we have used a simple Z-DIFF algorithm. We have already solved the redundancy of collected data and calibrated measurement errors. To improve detection accuracy, we have used machine learning methods. We have analysed the complexity our detection algorithm. In evaluation part, we have displayed a pothole-marked map produced by our system and shown z-axis accelerations of one road. Finally, we have evaluated the accuracy of detecting potholes by comparing it with simple threshold algorithm. Overall, we have developed a reliable pothole detection with low latency in detecting potholes.

Acknowledgements. This work is partially supported by grants from the National Natural Science Foundation of China (61672116, 61601067), Research Fund for the Doctoral Program of Higher Education of China (20130191120030), Chongqing High-Tech Research Program cstc2016jcyjA0332, Fundamental Research Funds for the Central Universities (CDJZR14185501, 0214005207005), Chongqing University (2012T0006).

References

1. Federal Highway Administration (1966). https://www.fhwa.dot.gov/
2. Cucchiara, R., Piccardi, M., Mello, P.: Image analysis and rule-based reasoning for a traffic monitoring system. IEEE Trans. Intell. Transp. Syst. **1**(2), 119–130 (2000)
3. Takagi, M., Masaki, I.: Road surface condition monitoring system using sensors disposed under the road, 27 August 2002. US Patent 6,441,748
4. Roadscanners. Roadscanners oy (1998). http://www.roadscanners.fi/
5. Gai, K., Qiu, M., Zhao, H., Tao, L., Zong, Z.: Dynamic energy-aware cloudlet-based mobile cloud computing model for green computing. J. Netw. Comput. Appl. **59**, 46–54 (2016)
6. Rudin, C., Wagstaff, K.L.: Machine learning for science and society. Mach. Learn. **95**(1), 1–9 (2014)
7. Kodratoff, Y., Michalski, R.S.: Machine Learning: An Artificial Intelligence Approach, vol. 3. Morgan Kaufmann, San Francisco (2014)
8. Wagstaff, K., Cardie, C., Rogers, S., Schrödl, S., et al.: Constrained k-means clustering with background knowledge. In: ICML, vol. 1, pp. 577–584 (2001)
9. Hartigan, J.A., Wong, M.A.: Algorithm as 136: A k-means clustering algorithm. J. Roy. Stat. Soc. Ser. C (Appl. Stat.) **28**(1), 100–108 (1979)
10. POTHOLE (2002). http://www.pothole.info/
11. Budras, J.: A synopsis on the current equipment used for measuring pavement smoothness (2001). http://www.fhwa.dot.gov/pavement/smoothness/rough.cfm
12. Gubbi, J., Buyya, R., Marusic, S., Palaniswami, M.: Internet of things (IoT): a vision, architectural elements, and future directions. Future Gener. Comput. Syst. **29**(7), 1645–1660 (2013)
13. Prashanth Mohan, R.R., Padmanabhan, V.N.: Nericell: rich monitoring of road and traffic conditions using Mobile Smartphones (2008)
14. Witten, I.H., Frank, E.: Data Mining: Practical Machine Learning Tools and Techniques. Morgan Kaufmann, San Francisco (2005)
15. Jain, A.K.: Data clustering: 50 years beyond k-means. Pattern Recogn. Lett. **31**(8), 651–666 (2010)
16. Velmurugan, T., Santhanam, T.: Computational complexity between k-means and k-medoids clustering algorithms for normal and uniform distributions of data points. J. Comput. Sci. **6**(3), 363 (2010)
17. Ghosh, S., Dubey, S.K.: Comparative analysis of k-means and fuzzy c-means algorithms. Int. J. Adv. Comput. Sci. Appl. **4**(4), 34–39 (2013)
18. Gai, K., Du, Z., Qiu, M., Zhao, H.: Efficiency-aware workload optimizations of heterogeneous cloud computing for capacity planning in financial industry (2015)
19. Hsu, J. Y.-J., Tai, Y.-C., Chan, C.-W.: Automatic road anomaly detection using smart mobile device (2010)
20. Zviedris, R., Kanonirs, G., Selavo, L., Mednis, A., Strazdins, G.: Real time pothole detection using android smart-phones with accelerometers (2011)
21. Eriksson, J., Girod, L., Hull, B., Newton, R., Madden, S., Balakrishnan, H.: The pothole patrol: using a mobile sensor network for road surface monitoring. In: Proceedings of the 6th international conference on Mobile systems, applications, and services, pp. 29–39. ACM (2008)
22. Ghose, A., Biswas, P., Bhaumik, C., Sharma, M., Pal, A., Jha, A.: Road condition monitoring and alert application: using in-vehicle smartphone as internet-connected sensor. In: 2012 IEEE International Conference on Pervasive Computing and Communications Workshops (PERCOM Workshops), pp. 489–491. IEEE (2012)
23. Friedlander, R.R., Kraemer, J.R.: Evaluating road conditions using a mobile vehicle, 17 September 2013. US Patent 8,538,667

A Quantitative Approach for Memory Fragmentation in Mobile Systems

Yang Li[1], Duo Liu[1(✉)], Jingyu Zhang[1], and Linbo Long[2]

[1] College of Computer Science, Chongqing University, Chongqing, China
liuduo@cqu.edu.cn
[2] College of Computer Science and Technology,
Chongqing University of Posts and Telecommunications, Chongqing, China
longlb@cqupt.edu.cn

Abstract. Mobile devices are equipped with many hardware accelerators to improve the performance and there are a bunch of third-party applications with rich-features in the application market. However, these applications always request large and contiguous physical memory as IO buffers and we observe that physical memory is severely fragmented after the mobile system runs for several hours. As a result, the memory allocation for such large and contiguous IO buffers will result in high latency and power consumption. Thus, this paper proposes a global memory fragmentation quantification approach that summarizes memory blocks access pattern and measures the allocation time of different order's memory block dynamically. Our evaluation on Android Kitkat shows that the global memory fragmentation is very precise to reflect the fluency of whole system.

Keywords: Mobile systems · Memory fragmentation · Quantification · Defragmentation · Anti-fragmentation

1 Introduction

There is an exploding demand for mobile devices, such as smartphones, tablets and wearable devices. According to a recent report related to mobile user behavior [1], mobile users spend 3.3 h a day on average on their smartphones and they usually do switches frequently between screen-on and screen-off. Immediate interactivity is critical when the device is invoked. This clearly indicates the importance of improving the user experience. Memory fragmentation is a serious obstacle preventing efficient memory allocation usage in Android memory management subsystem. Slow memory allocation causes high delay because of fragmented physical memory layout. Previous research is mainly classified into two categories: defragmentation and anti-fragmentation. However, neither of them can deal with fragmentation well. So we propose a global memory fragmentation quantification method to reflect system fluency ahead of time.

Memory fragmentation refers to external fragmentation. In the Linux Kernel [2], the buddy system memory allocator frequently allocates and deallocates memory blocks of different order, which results in fragmented memory

© Springer International Publishing AG 2017
M. Qiu (Ed.): SmartCom 2016, LNCS 10135, pp. 339–349, 2017.
DOI: 10.1007/978-3-319-52015-5_34

to the end. And conventional Linux Kernel avoids requesting contiguous physical memory by page table mapping scheme on the Memory Management Unit (MMU). The hardware-supported feature eliminates the need for allocating contiguous physical memory by providing a virtually contiguous address space over physically scattered memory.

However, more and more hardware accelerators will be adopted into modern smartphones, such as graphics, video encoder, video decoder, imaging, Camera [3]. These accelerators implement specific functionalities very efficiently, and hence offload the task from CPU. But they always require tens of MB contiguous physical memory for IO buffers. The IOMMU (Input Output MMU) maps physical memory block to IO address dynamically, and hence it can eliminate the need for allocation of contiguous memory. Each memory block used by IOMMU is 4 KB, 64 KB and 1 MB [4] and the allocation for IO buffers using large pages (64 KB, 1 MB) instead of small pages (4 KB) is more efficient [5]. However, only small pages are available in the highly fragmented memory, and therefore IOMMU allocation works slowly.

In order to quantify the degree of the whole system memory fragmentation, previous approaches are unusable index [6,7]. It specifies the fraction of free memory that is unusable for the memory allocation for a specific size. However, the unusable index can only identify the fragmentation of single order because the unusable index is computed based on a set of value obtained from the file /proc/buddyinfo. Then this approach to measuring fragmentation neglects the memory access pattern of realistic applications. Request weight of different order's memory block should be used as a hint of improving memory allocation efficiency.

Thus, in this paper we present a global memory fragmentation quantification approach guided by realistic applications' memory access characteristic in mobile systems. We collect the request weight of different order's memory block from applications and then compute the weighted memory allocation time, called WMAT. We implement this scheme with Android Kitkat on Google Nexus 5. Experimental results show that the request weight of different order's memory block and our fragmentation measurement formula can accurately identify the highly fragmented memory state. In summary, we intend to solve this problem by making the following contributions:

- We propose WMAT, a comprehensive approach to quantifying the degree of memory fragmentation;
- We present an efficient implementation of the request frequency of different order's memory block and dynamically measure the WMAT.

The rest of the paper is organized as follows. Section 2 presents the problem statement and motivation. Section 3 presents our proposed approach. Experiments and analysis are presented in Sect. 4. Section 5 discusses the related work. Section 6 concludes this work and discusses future work.

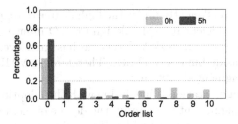

Fig. 1. Percentage of different order's memory to total free memory when the device is just on and runs for 5 h relatively.

2   Background and Motivation

In this section, we first analyse the reasons for memory fragmentation. We then distinguish the relationship between Android Low Memory Killer scheme and defragmentation. Finally we discuss our motivation.

2.1   Memory Fragmentation

The buddy system allocation algorithm is a classical and efficient memory allocation technique. The buddy allocator in each Zone (Kernel divides the memory into Zone) manages a linked-list of memory blocks and each of them has an order, where the order is an integer ranging from 0 to 10 [8]. To be more specific on Android platform, there are five migrate-types' pages in each *free_area* array [8] and they are designed to do anti-fragmentation to some extent.

Usually there are more higher order's memory blocks only when the device boots up and they get decreased while the device runs. Figure 1 shows the change in the percentage of different order's memory block to total available memory in the term of sizes when the device just boots up and runs for 5 h respectively. We evaluate this change by running a series of realistic applications and the details are described in Sect. 4. After 5 h, there are nearly no high order's memory blocks whose order is from 5 to 10, indicating the request for high order's memory blocks is time-consuming, even failed. And this is the cause of memory fragmentation. Chen et al. [9] unveil that smartphone users takes 100 times switches on average between screen-on and screen-off every day. The frequent spawn and exit of applications deplete high order's memory blocks and break their continuity. The lack of high order's memory blocks is the main constraint of the system fluency.

Though virtually contiguous addresses can be translated to physically scattered addresses via MMU hardware, which seems that it is not a major concern that the memory is fragmented with single page's granularity (or, order = 0), there are more and more demands for large physically contiguous memory. For example, IOMMU supports a similar dynamic mapping to MMU between contiguous IO address and physically scattered page blocks. IOMMU depends three kinds of memory block size and their orders are 0, 4, 8 respectively. However, fragmented memory still lacks memory blocks whose orders are 4 and 8, and hence IOMMU works inefficiently under the high degree of fragmentation. Malka et al. [10] propose an efficient IOMMU design for I/O intensive workloads. Pfeffer et al. [11] verify that the IOMMU scheme has high overhead in terms of IO translation look-aside buffers (IOTLB). Therefore, IOMMU does not behave well in the highly fragmented system.

Since memory fragmentation is inevitable with the adoption of buddy allocation algorithm, it is significant to quantify the degree of memory fragmentation. Equation (1) shows the approach to identifying the degree. The value can be obtained from the file /sys/kernel/debug/extftag/unusable_index. It specifies the proportion of unavailable free memory to current total free memory for a specified size. When the $UI(x)$ is 0, all memory blocks in free memory can be used to allocate for the specific size and when the $UI(x)$ is 1, none of the memory block in free memory can satisfy the allocation. However, this formula does not take into account the applications' memory access characteristic and we cannot obtain the sense of the degree of fragmentation of the whole system. The characteristic is that applications request for different order's memory block with different frequency.

$$UI(x) = 1 - \frac{\sum_{i=x}^{n} 2^i \times f_i}{\sum_{i=0}^{n} 2^i \times f_i} \tag{1}$$

2.2   Android *Lowlemorykiller*

To improve system's responsiveness, Android platform caches as many as possible launched applications in memory until the free memory is in pressure [12]. In order to cache more applications, Kwon et al. [13] propose to manage GPU buffers to increase memory utilization. Sunwook Bae et al. [14] propose to identify the topmost caching process in the background to improve the user interactivity.

However, caching makes it easy to reach the threshold of process reclamation. And *lowmemorykiller* in Android platform takes charge of killing processes to deallocate memory resource. Figure 2 shows the change of sizes of memory block whose order is from 0 to 10 when the system runs out of memory and *lowmemorykiller* is invoked to reclaim processes. After a few processes get terminated, the size of high order's memory block from 5 to 10 does not get increased and only the low order gets increased. One *lowmemorykiller* call can only deallocate a certain amount of memory blocks, limited by the thresholds: /sys/module/lowmemorykiller/parameters/adj, /sys/module/lowmemorykiller/parameters/minfree. These two thresholds are specified by userspace to select victim processes with a range of priority values. The higher the *minfree* is set, the more the page blocks are reclaimed. So adjusting these two thresholds can control the reclaimed page amount and probability of securing high order's memory blocks.

However, the default *minfree* array value is set as 12288,15360,18432, 21504,24576,30720 in Android Kitkat, which means the *lowmemorykiller* is called when the available memory runs under the 120 MB (30720 * 4 KB).

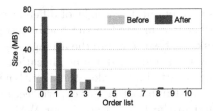

Fig. 2. Sizes of different order's memory block in MB when Android LMK is invoked to reclaim processes. The *minfree* threshold is default.

Such little reclaimed blocks can hardly be merged into higher order blocks. Hence, we can see that *lowmemorykiller* with small *minfree* threshold setting cannot alleviate the degree of memory fragmentation. Modifying the value *minfree* maybe destroy the responsiveness. Moreover, the available free memory, compared with *minfree* to determine the trigger time of *lowmemorykiller* is always scattered and so cannot be used to count the fragmentation degree of the whole system.

2.3 Motivation

To quantify the degree of memory fragmentation of the whole system more precisely, we find that userspace applications request for different order's memory block with different frequency. Guided by the memory access characteristic, we can measure the average memory allocation time dynamically and the allocation time can give us explicit indication of the memory fragmentation. Therefore, we focus on measuring the request weight of different order and allocation time for every request.

3 Design of WMAT

In this section, we first give a description of WMAT. Then we discuss the key techniques of WMAT: (1) the weight of eleven orders and (2) dynamic memory allocation time cost.

3.1 Design Methodology

Since memory fragmentation is inevitable, the key approach to controlling it is to quantify the degree of global fragmentation accurately and then take some defragmented operations ahead of time. To this end, we modify the buddy system allocation algorithm in the Linux Kernel memory management and subsequent defragmented operations are described in Sect. 6.

WMAT consists of two parts: (1) profiling applications' memory access behavior at run time, (2) measuring the weighted memory allocation time dynamically. Applications' run-time behavior is collected during each interval and needs to be analysed to obtain the weight of different order. Then the allocation time for each order is computed in the buddy system "heart" function __alloc_pages_nodemask. Last we compute the average memory allocation time based on the obtained weights and the dynamic allocation time of each order.

We do not change the "heart" buddy allocator algorithm to preserve portability. And our scheme does not impact the user experience. Because our proposed approach is implemented in the Linux Kernel, our WMAT can be applied to systems using Linux Kernel besides Android, such as wearable computing and IOT.

3.2 Weight Identification and Allocation Time Cost

We identify applications' run-time memory access behavior based on the frequency of the order of memory block request from userspace applications. And we collect a bunch of request order from various processes at run time and distinguish userspace processes and kernel threads. Because kernel threads always request memory for unmoveable objects, such as slab pages, TLB pages and we only need to focus on userspace processes. We summarize the frequencies and convert them to the request weight.

P. Kumar [15] reveals that many applications frequently request order-4 memory blocks besides order-0 when they are launched. High request weight and low available memory blocks obviously result in slow allocation. However, combining the request weight with conventional *unusable_index* of each order cannot reflect the allocation time cost accurately. So we design to dynamically measure the allocation time of each order in a predefined interval.

Combining with the request weight, average memory allocation time cost is computed dynamically and periodly. Equation (2) denotes the average memory access time cost for one request. $P(i)$ represents the request weight of order-i memory block from userspace applications. We conduct a durable traces several times and each time we run the device for 5 h. The obtained results are stable and therefore they are constant in the Equation. $C(i)$ denotes the memory allocation time in the typical allocation mask and varies from several microseconds (us) to a few thousand us under different degrees of fragmentation. MAX_ORDER is usually 10 in Android platform. The weighted memory allocation time of eleven orders represents the average memory allocation time for one arbitrary request.

$$WAMT = \sum_{i=0}^{MAX\_ORDER} P(i) \times C(i) \tag{2}$$

3.3 Implementation Details

We modify one file of the original Linux Kernel code, page_alloc.c and add one system call called MemroyRequestWeight into the memory management module. The system call collects the weight from memory request of different order's memory block. Since the GFP_KERNEL is the common allocation flag, we adopt

Table 1. Android experimental workloads.

Category	Applications
Browser	Firefox, Chrome, Opera, UC Browser, Next Browser
Social	Facebook, Pinterest, QQ, Sina Weibo, Instagram
Multimedia	Google Play Music, MX Player, TTpod Player, Youtube, KMPlayer
Shopping	Amazon, Ebay, Fancy, Google Play, TaoBao
News	BBC News, Flipboard, NetEase News, TED, Zaker

Table 2. Weight of order.

Order	0	1	2	3	4	5	6	7	8	9	10
Access weight (%)	97.495	0.527	0.144	0.009	1.812	0.001	0	0	0.012	0	0

Table 3. Standard allocation time.

Order	0	1	2	3	4	5	6	7	8	9	10
Allocation time (us)	2.0	3.3	3.8	4.8	5.5	7.3	12.5	25.4	29.6	51.7	94.9

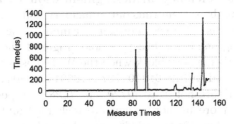

Fig. 3. Weighted memory allocation time when the device runs from booting up till 5 h. We count 160 measurements and each measurement interval is 2 min.

it as variable `gfp_mask` to represent most of the memory allocation types. We distinguish kernel threads and userspace processes by comparing process's flag with $PF\_KTHREAD$. The flag bit of $PF\_KTHREAD$ denotes that a process belongs to one kind of kernel threads. We count the request times for different order's memory block in the kernel function `__alloc_pages_nodemask`.

4   Evaluation

To evaluate our proposed scheme, we measure the degree of global memory fragmentation of a Google Nexus 5 running Android 4.4 Kitkat and Linux Kernel 3.4.0. To make clear, we use five different categories of realistic application workloads in Table 1 and we compare the *lowmemorykiller*'s effect on the reclaim of high order's memory blocks with the hint of WMAT.

Weight of order. Table 2 shows request weight of different order memory block when we run the target device for 5 h to collect several million memory access traces. From the table, we observe that the request for order-0 and order-1 dominates the whole memory access. And the weight of order-4 and order-8 is relatively higher in the high order and the request for them is performance-critical, especially existing in the I/O memory management.

Weighted memory allocation time. Before showing the dynamically measured allocation time, we show a set of standard allocation time when the device just boots up as presented in the Table 3. Using the weight of different orders, we compute a standard average memory allocation time, which is 2.04 us. Figure 3 demonstrates the changes in the weighted memory allocation time when the device runs from booting up till 5 h. From the figure, we observe that since the device boots up the WMAT is kept under the 10 us at most time. There are three peak situations, in which the kernel log shows Android *lowmemorykiller* and *OutOfMemorykiller* are invoked to reclaim the process. After WMAT reaches the peak, it decreases dramatically. Our proposed scheme monitors these extremely high WMAT to indicate that the system is highly fragmented and needs some proactive defragmented approaches.

Lowmemorykiller's efficiency in defragmentation with WMAT. High WMAT indicates that the system runs out of memory and in highly fragmented state. As analysed in Sect. 2, *lowmemorykiller* cannot make a contribution to defragmentation using small *minfree* threshold. Figure 4 shows the change in sizes of different order's memory block when the WMAT is very high and adjust the *minfree* to threshold 49152,61440,73728,86016,98304,122880. Here we mark the two sets of thresholds as *30720* and *122880* respectively. Therefore we can observe that more memory pages are reclaimed and more high order memory blocks are produced in the *122880* case. And subsequent large physically contiguous memory request can be satisfied easily.

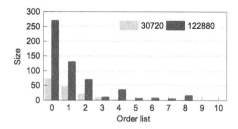

Fig. 4. Size of different order's memory block when the *minfree* is adjusted. *30720* denotes the threshold array set 12288,15360,18432,21504,24576,30720 and *122880* denotes the threshold array set 49152,61440,73728,86016,98304,122880.

5   Related Work

Optimization of buddy system allocation algorithm. Various studies have been conducted to fight the memory fragmentation. A number of studies focus on optimizing buddy system allocation algorithm [16]. In particular, Delvin et al. [17] consider that optimization of garbage collector can reduce the fragmentation caused by the buddy system via comparing several page allocation algorithms. Other researches try to redesign the size relationship between the buddy memory blocks. The previous work is not added into the Linux Kernel mainstream.

Defragmentation. Though Mauerer and Wolfgang [18] shows that the defragmentation approach is complicated by the fact that many physical pages cannot be moved to an arbitrary location, there are some efficient defragmentation ways to mitigate memory fragmentation. P. Kumar et al. [15] analyse the memory allocation path and manage to invoke the kernel function, (e.g. __alloc_pages_direct_ compact(), __alloc_pages_direct_reclaim()) as background thread. They reveal that the thread works to decrease the degree of memory fragmentation when the degree is over 90 %. P. Kumar et al. [19] discover that kernel function shrink_all_ memory() takes charge of reclaiming some pages when the device is in suspend state and they design a background thread to reclaim some pages in case of memory defragmentation bases on this kernel function. Craciunas et al. [20] propose the Compact-Fit scheme in a real-time memory management system and they demonstrate that the scheme can provide predictable memory fragmentation and control the performance versus fragmentation tradeoff via their partial compaction. Kim et al. [21] present a new page allocation scheme to reduce fragmentation of anonymous page and secure more physically contiguous pages.

Anti-fragmentation. The way of anti-fragmentation in the kernel is to try to prevent fragmentation as well as possible from the phase of memory allocation. As the development of kernel 2.6.24, the kernel partitions memory pages into three category: non-movable pages, reclaimable pages and movable pages [18] and memory fragmentation is reduced by grouping pages together depending on their mobility. Kim et al. [22] analyse the memory usage of Android, then group pages with the same lifetime and store them contiguously in fixed-size contiguous region. Their proposed region-based physical memory management can satisfy the large contiguous physical memory block request well and alleviates fragmentation greatly. Jeong et al. [23] reveal that some device vendors statically reserve some proportion of the memory or adopt similar approaches to provide a physically contiguous memory region to its integrated devices, such as a camera and a video decoder. And they propose lazy-migration and adaptive-activation techniques to increase the memory utilization because of large idle time of integrated devices and the experimental results show the return time of rental memory can decrease to 0.77 s. Gorman et al. [24] design an area reclaim algorithm based on the LRU linked-list and reclaim a certain number of page blocks for a specified memory request.

6   Conclusion and Future Work

In this paper, we propose a userspace application memory request-oriented approach to measuring weighted memory access time dynamically. We count the
memory request frequency for different order's memory block and assign the
frequency to the request weight of different order. Then we combine the weight
with the real memory allocation time of each order to compute the average memory allocation time. Our proposed WMAT can indicate the global fragmentation
ahead of time. With the help of WMAT, we dynamically adjust the *minfree*
threshold to get more free memory and more high order's memory blocks.

However, our approach can only quantify the global memory fragmentation
and it still needs some defragmented techniques as presented in Sect. 5. We
expect more advanced defragmented skills to optimize the system fluency ahead
of time based on our computed WMAT.

Acknowledgement. This work is partially supported by grants from the National
Natural Science Foundation of China (61672116, 61601067), Research Fund for the Doc-
toral Program of Higher Education of China (20130191120030), Chongqing High-Tech
Research Program cstc2016jcyjA0332, Fundamental Research Funds for the Central
Universities (CDJZR14185501, 0214005207005), Chongqing University (2012T0006).

References

1. Salesforce, Mobile Behavior Report (2014). https://www.exacttarget.com/sites/
   exacttarget/files/deliverables/etmc-2014mobilebehaviorreport.pdf
2. Love, R.: Linux Kernel Development. Addison-Wesley, Boston (2010)
3. T.I.T.R.: Itrs 2008 edition. Technical report, ITRS (2008). http://www.itrs.net
4. Google.Inc, Android AOSP (2013). https://source.android.com/source/
   building-kernels.html
5. Ben-Yehuda, M., Xenidis, J., Ostrowski, M., Rister, K., Bruemmer, A., Van Doorn,
   L.: The price of safety: evaluating IOMMU performance. In: The Ottawa Linux
   Symposium, pp. 9–20 (2007)
6. Gorman, M., Healy, P.: Measuring the Impact of the Linux Memory Manager,
   Libre Software Meeting (2005)
7. Gorman, M., Whitcroft, A.: The what, the why and the where to of anti-
   fragmentation. In: Ottawa Linux Symposium, vol. 1, pp. 369–384 (2006)
8. Buddy System Allocation Technique. https://en.wikipedia.org/wiki/Buddy_
   memory_allocation
9. Chen, X., Jindal, A., Ding, N., Hu, Y.C., Gupta, M., Vannithamby, R.: Smartphone
   background activities in the wild: origin, energy drain, and optimization. In: Proceedings of the 21st Annual International Conference on Mobile Computing and
   Networking, pp. 40–52 (2015)
10. Malka, M., Amit, N., Ben-Yehuda, M., Tsafrir, D.: rIOMMU: efficient IOMMU
    for I/O devices that employ ring buffers. In: Proceedings of the Twentieth Inter-
    national Conference on Architectural Support for Programming Languages and
    Operating Systems, pp. 355–368 (2015)
11. Pfeffer, Z.: The virtual contiguous memory manager. In: Proceedings of OLS, vol.
    10, pp. 225–230 (2010). Qualcomm Innovation Center

12. Alliance, Open Handset, Android overview, Open Handset Alliance (2011)
13. Kwon, S., Kim, S.-H., Kim, J.-S., Jeong, J.: Managing GPU buffers for caching more apps in mobile systems. In: Proceedings of the 12th International Conference on Embedded Software, pp. 207–216 (2015)
14. Bae, S., Song, H., Min, C., Kim, J., Eom, Y.I.: EIMOS: enhancing interactivity in mobile operating systems. In: Murgante, B., Gervasi, O., Misra, S., Nedjah, N., Rocha, A.M.A.C., Taniar, D., Apduhan, B.O. (eds.) ICCSA 2012. LNCS, vol. 7335, pp. 238–247. Springer, Heidelberg (2012). doi:10.1007/978-3-642-31137-6_18
15. Kumar, P.: Controlling memory fragmentation and higher order allocation failure: analysis, observations and results (2012). http://elinux.org/images/a/a8/ControllingLinuxMemoryFragmentation.pdf
16. Page, I.P., Hagins, J.: Improving the performance of buddy systems. IEEE Trans. Comput. **100**(5), 441–447 (1986)
17. Defoe, D.C., Cholleti, S.R., Cytron, R.K.: Upper bound for defragmenting buddy heaps. ACM SIGPLAN Not. **40**(7), 222–229 (2005)
18. Mauerer, W.: Professional Linux kernel architecture (2010)
19. Kumar, P.: System-wide defragmenter (2015). http://www.elinux.org/File:Tizen-_System-Wide_Memory_Defragmenter_Without_Killing_Any_Application.pdf
20. Craciunas, S.S., Kirsch, C.M., Payer, H., Sokolova, A., Stadler, H., Staudinger, R.: A compacting real-time memory management system. In: USENIX Annual Technical Conference, pp. 349–362 (2008)
21. Kim, J., Min, C., Kim, J., Kang, D.H., Kim, I., Eom, Y.I.: Page allocation scheme for anti-fragmentation on smart devices. In: 2014 IEEE 3rd Global Conference on Consumer Electronics (GCCE), pp. 512–513 (2014)
22. Kim, S.-H., Kwon, S., Kim, J.-S., Jeong, J.: Controlling physical memory fragmentation in mobile systems. In: Proceedings of the 2015 ACM SIGPLAN International Symposium on Memory Management, pp. 1–14 (2015)
23. Jeong, J., Kim, H., Hwang, J., Lee, J., Maeng, S.: Rigorous rental memory management for embedded systems. ACM Trans. Embed. Comput. Syst. **12**(1), 43 (2013)
24. Gorman, M., Whitcroft, A.: Supporting the allocation of large contiguous regions of memory. In: Ottawa Linux Symposium, pp. 141–152 (2007)

A DDoS Detection and Mitigation System Framework Based on Spark and SDN

Qiao Yan$^{(\boxtimes)}$ and Wenyao Huang

College of Computer Science and Software Engineering,
Shenzhen University, Shenzhen, China
yanq@szu.edu.cn, 2150230426@email.szu.edu.cn

Abstract. Distributed Denial of Service (DDoS) attack is a serious threat to commercial service network. DDoS attack has been studied for years. However, detecting and relieving DDoS attacks are still a problem. Especially, nowadays more and more DDoS attacks produce heavy network traffic, it is hard to response rapidly because that needs high processing performance to process massive traffic data. With big data technology, volumes of network traffic data can be processed much faster. Apache Spark can process a great amount of data in a reasonable time so that DDoS attack can be detected in time. Besides, it is difficult to modify the network configuration in traditional network. With Software-Defined Networking (SDN), a new paradigm in networking, networking can be controlled by programs, which makes modifying the network configuration easier. In this paper, a DDoS detection and mitigation system framework in SDN is introduced, a framework that can control network based on analyzing the network traffic data. Comparing to the traditional defense methods of DDoS attack, the framework can response to DDoS attack by rules automatically.

Keywords: Software-defined networking(SDN) · Distributed denial of service (DDoS) · Apache spark

1 Introduction

A Denial of Service (DoS) attack is hacker's attempt in taking up server and network resources in order to harm normal users access to resources. Distributed Denial of Service (DDoS) flooding attacks is the main method to destroy the availability of the server or the network [1]. DDoS is a type of DOS attack where multiple compromised systems, which are often infected with a Trojan, are used to target a single system causing a DoS attack [2].

As the Internet continues to grow and prosper, DDoS attacks continue to increase in severity and frequency. In Q1 2016, the largest anti-DDoS Service Provider (SP), Prolexic released [3] the akamai's state of the internet, stating that comparing to Q1 2015, they observe 280% increase in attacks more than 100 Gbps and 22.47% increase in total DDoS attacks. There is a continuous rising trend in the heavy traffic attack. These are challenges and opportunities.

© Springer International Publishing AG 2017
M. Qiu (Ed.): SmartCom 2016, LNCS 10135, pp. 350–358, 2017.
DOI: 10.1007/978-3-319-52015-5_35

Big data technologies have been paid more and more attention. Using big data [4] represents a new way of defensing DDoS attack. Large volumes of network data are helpful for us to comprehensively understand the context of normal behavior of people and the flow of data over networks so that we can discover DDoS attack in an early stage. To achieve this functionality, the prevention system needs to copy all traffic data of the protected network and implements a comparative, packet-specific, statistical analysis using big data technologies.

Among those big data technologies, Apache Hadoop used to be most important one. Apache Hadoop [5] is an open-source software framework for distributed storage and distributed processing of large data sets on computer clusters, which takes advantage of the method of MapReduce and Hadoop Distributed File System (HDFS) to process large data much faster and more efficiently than conventional supercomputer architecture. An other rising big data technology is Apache Spark [6], which was developed in response to limitations in the MapReduce cluster computing paradigm. Apache Spark provides programmers with an application programming interface centered on a data structure called the resilient distributed dataset (RDD), a read-only multiset of data items distributed over a cluster of machines. Spark's RDDs function as a working set for distributed programs that offers a deliberately restricted form of distributed shared memory. Apache Spark [7] can run programs up to 100x faster than Hadoop MapReduce in memory, or 10x faster on disk. With the help of Apache Spark, a very large amount of network traffic data can be processed in a reasonable time interval so that DDoS defense system can respond to DDoS attack more quickly.

With the effect of big data technology, DDoS defense system can only detect DDoS attack much faster and more efficient. But it is difficult to changing a network setting to defense DDoS attack in traditional network. Every traditional network device combines data layer and control layer. Changing a traditional router setting to defense a DDoS attack is time consuming and not efficient. Besides, traditional devices don't have a global view of the protected network. DDoS attack can be detected early if more information of network have been taken into consideration when DDoS attack happens.

SDN has attracted great interests as a new paradigm in networking [8]. SDN is a network architecture with dynamic, manageable, cost-effective, and adaptable properties. With the help of decoupling control and forwarding functions, SDN architecture is centrally managed and it can be programmatically configured [9]. SDN is helpful to defense DDoS for its global view of the whole protected network and its programmability of the network.

In this paper, DDoS Detection and Mitigation Framework (DDMF) is proposed. DDMF combines big data technologies and SDN to construct a detection and mitigation system. The proposed framework uses Apache Spark to process network traffic data and uses programmability of SDN to control network, which can make detecting and mitigating DDoS attack in a much faster and smarter way.

The rest of this paper is organized as follows. Section 2 describes related works in the field of using big data technologies to detect DDoS attack and using SDN to control network. Section 3 presents our framework. Section 4 simulation of DDMF will be discussed. Finally conclusion of the paper is in Sect. 5.

2 Related Work

In the past decades, the volume and the frequency of DDoS attack have shown a tendency of increase. Different methods have been developed to detect and prevent DDoS attack.

Yeonhee Lee and Youngseok Lee [10] presented us with processing a large number of data packets with MapReduce algorithm in Hadoop. According to the experiment on their test bed, their method can process 500 gigabyte traffic data in 25 min and process 1 terabyte traffic data in 47 min, which is 8 times faster than processing on one worker node and 2.9 times faster than processing on three worker nodes. Two years later, they presented us another more powerful traffic measurement and analysis system [11]. Their proposed system processes 1 TB traffic data within 20 min on 30 worker nodes. Their work shows us the Hadoop Mapreduce can reduce the processing time of big traffic data, which is useful for nowadays as network traffic is getting larger. But their work only analyzed the off-line traffic data in their experiment.

Sufian Hameed and Usman Ali [12] presented us a live DDoS detection system with Hadoop (HADEC). HADEC uses Hadoop MapReduce to process traffic data. The traffic data processed by HADEC is live. HADEC can finish DDoS detection in a affordable time. The research of Sufian Hameed and Usman Ali uses Hadoop MapReduce to process live network data, but their research only detects DDoS attack.

Laizhong Cui, F. Richard Yu, and Qiao Yan [8] state that SDN and Big Data can benefit each other a lot. One of those benefits is that Big data can help SDN to deal with security problems and help SDN with traffic engineering. Qiao Yan et al. state that SDN is a good tool to defeat DDoS attacks. There are some good features of SDN make it easier to detect and react DDoS attacks, such as separation of the control plane from the data plane, a logical centralized controller and programmability of the network by external applications [13].

SDN [14] is an emerging networking paradigm that gives hope to change the limitations of current network infrastructures. SDN breaks the vertical integration by separating the network's control logic (the control plane) from the underlying routers and switches that forward the traffic (the data plane). With the separation of the control and data planes, SDN provides us the programmability of the network, which is helpful to control networking in a intelligent way. A DDoS detection and mitigation system can defense DDoS attack in an intelligent way if it is based on SDN.

Openflow [15] is first standard communication protocol between the control and forwarding layers of an SDN architecture. OpenFlow allows direct access to and manipulation of the forwarding plane of network devices in a SDN controller program. Control layer of SDN architecture communicates forwarding layer's devices with OpenFlow protocol message, such as request for devices status of forwarding layer, configuring device. OpenFlow message mechanism is the method of programming SDN network. DDoS defense system can control the topology of network and limit hosts' accessing by processing the OpenFlow message.

3 Framework

In this section, the components of DDMF is described. As shown in Fig. 1, DDMF consists of three major components: SDN Router Application, Capture Server and Detection Server(Cluster).

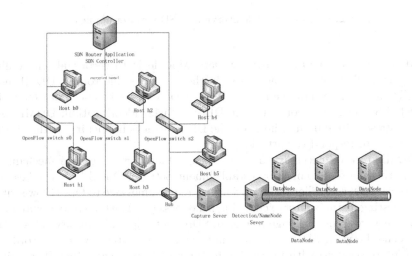

Fig. 1. DDMF: DDoS detection and mitigation framework

3.1 SDN Router Application

SDN router application is application layer of the SDN architecture, which is in charge of setting up the logic of forwarding packet. DDMF uses router application to control SDN network and to block DDoS attack packet flows. The DDMF router application's overview of function module can be designed like Fig. 2.

Main function of Routing Logic Module is to make different network segments can communicate with each other. The router's logic is build on IP layer. OpenFlow supported switches can match IP layer, it is the basis of controlling a available IP networking. Routing Logic Module will use Flow Table Controlling Module to configure forwarding layer's switches to make network traffic complied with our routing logic. Besides, this is the basis of SDN global view. All the switches can share information like different processes in one computer.

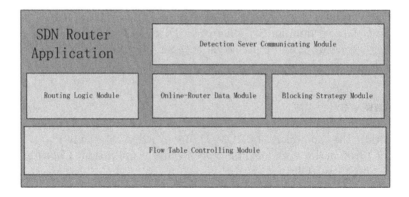

Fig. 2. Function module overview of SDN router application

Main function of Online-Router Data Module is to record all the online routers and all the network addresses in the corresponding router. DDMF uses data set to represent a online router. When DDoS attack has been detected, the Detected Server returns one or more DDoS attacks' identification, DDMF uses these data sets to find out where are the DDoS attacks happening so that it can take action to respond the attack.

DDMF records necessary data by using OpenFlow's message mechanism. During the phase of connection establishment between OpenFlow switches and controller, SDN controller will send features request message to OpenFlow switch for requesting features after exchanging hello message. The OpenFlow switch will reply a features reply message to inform SDN controller of its ability. This is the Handshake between SDN controller and OpenFlow switch. We can record the online router(switch) with processing features reply message. OpenFlow switch may crach for some reasons, DDMF can discover invalid switches by using echo request message. When invalid router is found, DDMF updates the online router data set. Router's data set have its according network addresses. DDMF uses router's network addresses to find out where are the DDoS attacks.

The main function of Detection Sever Communicating module is receiving the commands from DDoS Detection Server and using Blocking Strategy Module to block the detected DDoS attack packet flows. Different communication methods can be used in this module, such as TCP message, UDP message or HTTP message. The communicating message has to specify the identification of the DDoS attack packet flow. In this case, DDMF uses IP address to identify attack packet flow. More information can be provided through the communication if you want to make some complicated strategies. The communicating protocol is based on defense need. A severity or something else can be offered to determine attack level and then Blocking Strategy Module can take corresponding stop action.

The main function of Blocking Strategy Module is to apply some strategies to block DDoS attack packet flow. Defending DDoS is challenging because DDoS attack has a great diversity of methods. Based on the detection result, different strategies are used to prevent different DDoS attacks. DDMF has a block strategy module, which makes defending DDoS flexible. Different blocking actions are taked based on detection server's message. DDMF provides this module because no one secret method can solve all the DDoS attack perfectly.

The main function of Flow Table Controlling Module is to communicate with OpenFlow switches. This module consists of all the necessary functions to configure OpenFlow switches, which is the basic of other modules. Routing Logic Module uses it to configure switches to activate the IP network. Blocking strategy module uses it to modify flow tables to block DDoS attack packet flow. Online-Router Data Module uses it to record online routers.

3.2 Capture Server

When a SDN network is ready, Capture Server starts capturing network traffic and generating network traffic log.

DDMF needs to analyze network traffic data in order to detect DDoS attacks. DDMF can use all kind of technologies to capture and analyze network traffic data. One solution is shown as below.

DDMF can use Tshark to capture the network traffic and use dpkt [16] to read all tshark output packets into a log file. DDMF only outputs the relevant information required during detection phase. The output information can be tuned because redundant information can slow down detection speed. For example, ICMP packet can have timestamps, source IP, destination IP, packet length, ICMP id, sequent number and so on. But if the detection method only uses source IP and destination IP, it is encouraged to just output the source IP and destination IP.

Once the log file is generated, DDMF uses scp tool to transport log file to Detection Server. The scp transfer tool guarantees integrity of the transfer files.

3.3 Detection Server

Detection Server also is the name node of Spark Cluster in DDMF. After transfering log files to Detection Server and uploading log files to HDFS, Detection Server submits the detection job. Detection result will be saved to HDFS.

The Detection Sever has three major modules according to their functionality. The first one is the preparation module that receives log files and submits the detection job. Another module is the detection module that analyzes log files and then determines which packet flow is DDoS attack. The last one is the notification module, which will notify SDN router application to block the DDoS packet flows.

Major preparation work is to receive log files and upload them to HDFS. It is more efficient to store files in HDFS than in local file system and then DDMF starts a detection job. DDoS dection module is the main part of DDMF. After years of development, lots of detecting DDoS methods have been developed. There are detection methods based on neural network, based on entropy, based on counting and so on. DDMF can adapt these detection methods only if they can offer an identification to identify DDoS attack flow and the other information to complete the blocking strategy. Notification Module will inform the SDN router application to take action to stop DDoS attack flows. Different communication technologies can be used, such as TCP message, UDP message, HTTP message and so on.

4   Simulation

In this section, DDMF simulation in the test bed will be discussed. DDMF is a automatic DDoS defense system framework that can stop DDoS attacks based on analysis result of network traffic. The simulation is going to prove DDMF can defense DDoS attacks based on analysis result and regulated policies.

The DDMF simulation environment is shown in Fig. 3, there is 4 hosts named h1, h2, h3 and h4 in the simulation. H2 and h3 are deployed to simulate DDoS attack flows by flooding ICMP to h4. H1 is using ping tool to test h4. After the detection of DDoS, SDN router application is able to stop the DDoS attack flows.

As shown in Fig. 4, DDoS attack flows (host 2 and host 3) are not linear growth because SDN router application can stop the the packet flow based on

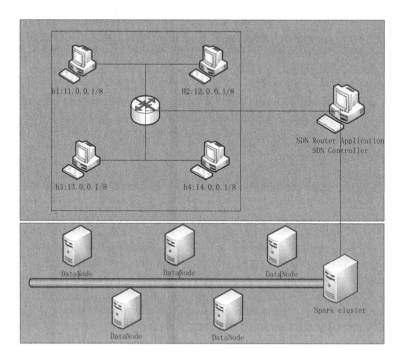

Fig. 3. The DDMF simulation environment

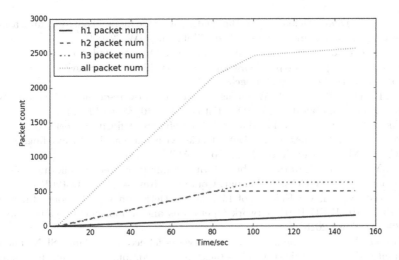

Fig. 4. ICMP packet count over time

detection result automatically. At this point, DDMF stops simulative DDoS attack packet flows based on analysis result of network traffic automatically.

5 Conclusion

In this paper, DDMF is proposed, which is a DDoS defense system framework that is capable of stopping DDoS attack based on analyzing network traffic. DDMF captures live SDN network traffic, processes it to log relevant information in brief form and uses Spark to run detection job. The simulation of DDMF is a proof of feasibility of automatic DDoS defense. Big data is very useful in detecting DDoS attacks because analyzing volumes of traffic data can comprehensively understand context of normal behavior and the flow of data over networks. A DDoS defense system using big data technologies can process massive data in a affordable time so that it can spot and stop DDoS attack in early stage. DDMF has done a beneficial attempt to combine big data technologies and SDN to build a automatic DDoS defense system. In future work, we plan to do advanced experiments on DDMF in order to perfect its design and performance.

References

1. Yan, Q., Gong, Q., Deng, F.-A.: Detection of DDoS attacks against wireless SDN controllers based on the fuzzy synthetic evaluation decision-making model. Ad Hoc Sens. Wireless Netw. **33**, 275–299 (2016)
2. DDoS Attack. http://www.webopedia.com/TERM/D/DDoS_attack.html
3. Middleton, D., Spaulding, A.D.: Q1-2016-state-of-the-internet-security-report. Internet Security and DDoS Attack Report 11-34, Akamai, Q1 2016

4. Hua Wei: DDoS Defense and Big Data. http://e.huawei.com/de/publications/global/ict_insights/hw_331605/industry%20focus/HW_327000
5. Apache Hadoop. https://en.wikipedia.org/wiki/Apache_Hadoop
6. Wiki Apache Spark. https://en.wikipedia.org/wiki/Apache_Spark
7. Apache Spark. http://spark.apache.org/
8. Cui, L., Yu, F.R., Yan, Q.: When big data meets software-defined networking: SDN for big data and big data for SDN. IEEE Netw. **30**, 58–65 (2016)
9. Wiki SDN. https://en.wikipedia.org/wiki/Software-defined_networking/
10. Lee, Y., Lee, Y.: Detecting DDoS attacks with hadoop. In: Proceedings of The ACM CoNEXT Student Workshop, p. 7. ACM (2011)
11. Lee, Y., Lee, Y.: Toward scalable internet traffic measurement and analysis with hadoop. ACM SIGCOMM Comput. Communi. Rev. **43**(1), 5–13 (2013)
12. Hameed, S., Ali, U.: Efficacy of Live DDoS Detection with Hadoop. In: NOMS 2016 - 2016 IEEE/IFIP Network Operations and Management Symposium, pp. 488–494, April 2016
13. Yan, Q., Yu, F.R., Gong, Q., Li, J.: Software-defined networking (SDN) and distributed denial of service (DDoS) attacks in cloud computing vironments: A survey, some research issues, and challenges. IEEE Commun. Surv. Tutorials **18**(1), 602–622 (2016)
14. Kreutz, D., Ramos, F.M.V., Verssimo, P.E., Rothenberg, C.E., Azodolmolky, S., Uhlig, S.: Software-defined networking: a comprehensive survey. Proc. IEEE **103**, 14–76 (2015)
15. OpenFlow. https://www.opennetworking.org/sdn-resources/openflow
16. DPKT. https://pypi.python.org/pypi/dpkt

Based on Cloud Computing and GIS in the Smart Yellow River Emergency System Design and Its Key Technology Research

Kuan He[1], Xu Chen[1], Yuntong Liu[1], Bo Hu[1], and Zhimin Zhou[2(✉)]

[1] Department of Surveying and Mapping Engineering, Yellow River Conservancy
Technical Institute, Kaifeng 475004, Henan, China
[2] College of Environment and Planning, Henan University, Kaifeng 475004, Henan, China
zzm@henu.edu.cn

Abstract. This paper briefly introduces the technical design and development based on Cloud-based GIS in the Smart Yellow River emergency system basic situation, focusing on system design and development process in the data interface, GPS mobile positioning, spatial database and other technical issues in-depth study and exploration, that the Yellow River GIS must be designed to closely follow the development of GIS, System goal is to solve the technical problem in GIS design.

Keywords: Smart yellow river · Emergency system · Data interface · Mobile positioning · Spatial database

1 Introduction

With the rapid development of geographic information system (GIS), global positioning system (GPS), remote sensing (RS), computer storage technology, network technology and mobile communication technology, GIS-based disaster information system, disaster prevention and reduction of data sources has become increasingly diverse. Satellite remote sensing data, aerial image data, GPS ground tracking data, based communication systems and terminal equipment to obtain the location information data, real-time monitoring of the disaster is not the same state data and attribute information data disaster. The multi-source data resources with the full range of disaster prevention and mitigation system display, disaster analysis of the data base. Therefore, the effective use of these data, to achieve the goal of building disaster prevention and mitigation system, how to achieve efficient data transfer and integration is crucial in this.

2 Smart Yellow River Emergency System

"Smart Yellow River Emergency Response System" is "Smart Yellow River" the most important construction projects of geographic information systems technology (GIS technology) is to establish "Smart Yellow River emergency system" and the most

© Springer International Publishing AG 2017
M. Qiu (Ed.): SmartCom 2016, LNCS 10135, pp. 359–366, 2017.
DOI: 10.1007/978-3-319-52015-5_36

important technology are in the computer hardware and software support, the Yellow River space-related data collection, management, operation, analysis, simulation and display, and analysis using geographic model, also will be timely provision of a variety of space, dynamic geographic information for geographic research and geographic decision-making services and set up computer systems. Many topics of the Yellow River watershed of a regional GIS system integration and resource sharing, it will actually build the Yellow River basin-wide information processing, "Smart Yellow River." "Smart Yellow River" projects in flood mitigation, water regulation, water conservation, soil conservation, project management, e-government applications, go through the GIS professional decision support and information services, and through virtual simulation of GIS technology decisions consultation. The overall framework of project design "Smart Yellow River Emergency Response System" shown in Fig. 1.

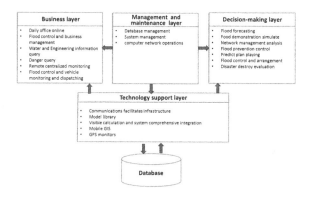

Fig. 1. The overall framework of "Smart Yellow River Emergency Response System" project design

Cloud computing is the development of Distributed Computing, Parallel Computing and Grid Computing. "Cloud" is a computer cluster; each group includes hundreds of thousands, even millions of computers. The ultimate goal of cloud computing is the computer as a public utility to a wide audience, so that people can use, like water, electricity, gas and telephone as the use of computing resources. In the past few years, people's computing paradigm, including the grid computing service basis, P2P computing.

Cloud computing network to provide easily control and powerful messaging capabilities for user, a combination with both, making the prospect of geographic information system applications become more widespread. Cloud computing will be used in the construction of distributed geographic information systems, create the user program can maximize the functionality and efficiency, but also for the massive user base can provide a more stable, fast and secure service.

With LAN technology, the wireless network technology (Wi-Fi), satellite positioning technology (GPS) and the continuous development of computer processing speeds continue to increase, based on data mining techniques, decision optimization, computer-aided decision support technology on emergency response system stood at the forefront of technology. It need to design systems according to the disaster emergency functions, and geographic information systems technology and cloud technology in emergency response system design Smart Yellow River, Yellow River to build a digital emergency system function structure, shown in Fig. 2.

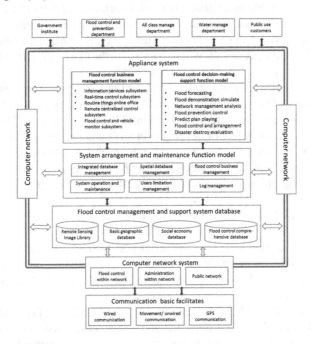

Fig. 2. Smart Yellow River system function structure of the emergency system

Smart Yellow River Emergency System Based on GIS and Cloud Computing is series of emergency systems to achieve this feature, it must address some key GIS technology, this paper will focuses on the data interface technology, GPS mobile positioning technology and spatial database technology three issues.

3   Data Interface Technology

GIS data interface to the external environment of the Yellow River and other systems to provide access to its internal data manipulation interface. The interface can be request/response mode to accept or provide data, the degree of interoperability capabilities across the interface to reflect the size, but has nothing to do with the internal structure of the data. Data with the data provider will usually provide the API, data users can access the system via the API to the internal data. API can be complex data structures or hiding the complexity of the operation, and can be programmed by the API and data servers together to form a more powerful data server to respond to external data service requests. The API in order to reduce dependence on the specific application environment, users, data providers and system developers need to establish a common industry-wide interface. The following tests with a GPS device interface with hydrologic process to analyze the technology.

Advanced hydrological equipment through a computer or PDA (the PDA, for example), collecting GPS data through the serial port, and the GPS electronic map on the location of the GIS data, the updated map data or analysis of spatial properties of PDA and mobile networks or via WEB connected the data request to the server, the server receives a request, the results will be returned to the PDA user. GIS + GPS + Wireless Internet integration, constitutes a "Mobile GIS".

PDA interface software is generally developed by using Java language. Formed in the three versions of Java technology: Micro Micro Edition (J2ME), Standard Standard Edition (J2SE), Enterprise, Enterprise Edition (J2EE). One major J2ME PDA and embedded devices for a variety of development tools and runtime environment. J2ME Java technology has many features, it can all run on Java-enabled devices, short code, safety, implemented with J2ME applications can be easily upgraded to J2SE, J2EE.

The GPS communication protocol NMEA0183 protocol, serial communication parameters are:

Porter law = 4800, data bits = 8 bits, stop bits = 1,

parity = none

GPS and PDA communicate via the serial port to send 10 data per second. GPS navigation to read the actual application of the spatial positioning data, can be updated every few seconds, latitude, longitude and time data. NMEA 0183 data format of position data are as follows:

$GPRMC, 204700, A, 3403.868, N, 11709.432, W, 001.9, 336.9, 170698, 013.6, E*6E.

J2ME for PDA's GPS serial data read and write can be used in two ways, one is using the serial port for a single byte of data read and write raw, another byte array buffer can used to read and write. Using the first method inefficient slow to read and write, to read every 3–7 s to the desired GPS positioning data, which read and write speed, you can read the required GPS data per second, there is no GPS data be lost. J2ME provides a serial read and write the class Protocol, by constructing a Protocol instance serialPort, use serialPort.openInputStream() to obtain the input stream InputStream, InputStream using the GPS serial data read into a buffer byte array into a byte array string, to determine the GPS coordinates of signs "$ GPRMC", the interception of coordinate data. The PDA and GPS receivers will achieve the interface communication.

4   GPS Mobile Positioning Technology

GPS mobile positioning technology applications in the Yellow River is mainly vehicles and engineering vehicles, flood control and scheduling and statistics, these vehicles collectively referred to as specialty vehicles. Car mobile information system is the heart of its communication unit, the dispatch center with wireless connectivity the Yellow River, Yellow River through the cellular GSM network or microwave network as the vehicle to provide mobile information services. One advantage of using the GSM network is that you can use the triangulation feature to solve the problem of GPS to work blind, for example, in the forest of buildings or metal buildings, you can integrate GPS positioning with the network, and with an A-GPS devices achieve high accuracy in any place (5–20 m) of the positioning.

To achieve the Yellow River in GIS GPS mobile positioning technology, a key question is to determine the best path search method with the study.

The best path to achieve the GIS electronic map search is to achieve the efficient dispatching of vehicles, to provide mathematical algorithms to reach the destination quickly. At present the idea to use graph theory to design its storage topology.

Firstly, process the network topology relationship. In the way of network in the mathematical model, only need to use nodes and nodes and road connectivity, that is, nodes and node and topological relations between line segments. And follow the best path algorithm, the results of this data point number. As a general application on the PDA is a Windows CE, not only to consider the time complexity, but also consider the space complexity, we need the classical Dijkstra's algorithm from the perspective of time and space optimization. Mainly rely on the number of intersections, undirected weighted graph and adjacency matrix storage structure. Because it is undirected weighted graph, the adjacency matrix is symmetric matrix, so you can use a one-dimensional array to store its lower triangular matrix to store the graph structure. This avoids the calculation of independent nodes, improving the classical Dijkstra algorithm has the blindness and reduce the number of intermediate points, the search efficiency is improved.

Improvement of the Dijkstra algorithm. It is marked as temporary node t into T grade point and the final selection of the set of p labels points, and the starting point as the first p label (and released from T set), then follow the shortest path selected one by one principle point of p t label points, and set it into the P from the set T (P is the shortest path has been seeking a collection of nodes, the initial value is null), until the target point (or all of the points) are labeled p label point. From the beginning to the end point of the p label sequence represents the best path to the request. Optimize the algorithm, the Dijkstra algorithm, based on the shortest straight line distance between two points using the principle of selecting source and destination in the mid-point between; only choose from the finish line to join the nearest point to P concentration. Specific methods:

(1) Model building. In the weighted graph G, the right side between two points on behalf of the length of the two specified vertices u0 and v0 the shortest distance between the paths to find a problem in the network model has the following characteristics:

① In the graph G in the coordinates of the vertex u can be expressed as (xu, yu).
② the shortest line between two points, any two points in G u, v denote the straight line distance:

$$J(u, v) = ((xu - xv)2 + (yu - yv)2) - 2\,\Pi \tag{1}$$

And J (u, v) <= d (u, v), in which d (u, v) that point from u to v point weight.

Let u0 to ui have obtained the shortest path distance, identified as L (ui), then u0 through ui distance to reach v0 the lower limit should be L (ui) + J (ui, v0).

(2) Description of the algorithm

① u0 as a starting point, each vertex v, compute L(v) values (v ∈ T), and calculate v and the target point v0 the linear distance between J(v, v0), that the current from v to v0 u0 through the distance the lower limit of L(v) + J(v, v0).

② T in the vertex set of rules join P: find one o'clock in the set T ui, ui distance from u0 through the lower limit for the minimum, which satisfy (2), the vertex ui (ui ∈ T) added to the collection P.

$$L(ui) + J(ui, \ v0) = \min\{L(v) + J(v, \ v0)\}(v{\in}T) \tag{2}$$

Through this algorithm can be determine the optimal driving path in the flood control time.

5 Yellow Spatial Database

Spatial database is stored digital terrain maps, digital terrain model DEM, watershed 3-dimensional model of the database. Spatial data with other data, the difference is that it contains the attribute data, but also contains the geometric data, such as point, line, surface, geometry and so on. The traditional approach is based on spatial data file stored in the multi-user concurrent access, data update, data access efficiency, there are some shortcomings, in order to be able to use mature relational database management systems (such as ORACLE, SQL SERVER), Spatial Database Engine (SDE) technology can achieve the spatial data stored in relational databases, and can be the same as ordinary data access data using SQL statements query, delete, and modify. The application layer between the logical servers are used XML for data exchange and communication.

Yellow layered spatial database storage and management in the form of the Yellow River basin, water conservancy, water environment, flood control and drought, water resources, land use, soil and water conservation, water-saving irrigation and socio-economic information, building different applications to meet the digital needs of the Yellow River accuracy requirements of a variety of different spatial databases Yellow River.

Taking into account flood management is a complicated systematic project, in addition to the use of modern advanced network technology, computer technology, GIS technology, in addition, if we can use GPS, wireless communications, personal digital assistant (PDA) technology, will expand communication channels to improve emergency response capabilities. Site information such as flood control, flood control vehicle location, these requests or return information is usually encoded in accordance with certain communications server decodes it; it will forward it to the application server for processing. Communication server is also responsible for the command center of the various instructions or information sent wirelessly to the mobile terminal. Communication between server and mobile terminals will be through the GSM short message or GPRS data service to communicate.

Yellow spatial database contains geometry and attribute information integration framework that provides and supports spatial data types, query languages and interfaces, and space efficient spatial index joint and so on. The current implementation of spatial database there are two main ways: object-oriented database object-relational database approach and methods. The former will be objects of spatial data and non-spatial data and operating packaged together, unified by the object database management and support of nested objects, inheritance and aggregation of information, which is a great way for spatial data management. But the technology is not yet mature, and especially queries optimization more difficult. Object-relational database is the main spatial database technology, which combines a relational database and the advantages of object-oriented database that can directly support complex object storage and management. GIS software directly in the object-relational database definition of spatial data types, spatial operations, spatial indexing, spatial data can be easily managed to complete multi-user concurrency, security, consistency/integrity, transaction management, database recovery, seamless spatial data management operations. Therefore, the use of object-relational database implementation of the GIS data management is to achieve an ideal spatial database approach. At present, some manufacturers have introduced a spatial database, data management, application-specific modules, such as the IBM Informix Spatial DataBlade Module, IBM DB2 Spatial Extender and Oracle's Oracle Spatial, etc., although its function is to be further improved, it has brought to the GIS software development a great convenience.

Important function of spatial database is mainly reflected in the spatial indexing and spatial query. Commonly used index in the current four-tree and R-tree. In general spatial database query using a two-step mechanism, first check out the candidates with the index set, and then using the exact geometry, the candidates focused on the exact solution obtained. Provides spatial query language is an important feature of the spatial database, spatial database in the current general use of relational data in the "select-from-where" pattern to build a query, through the expansion of the SQL language to support spatial object types, spatial relationships and space operations. Especially SQL3 Multimedia specification (SQL3/MM) in part and OpenGIS for SQL Spatial Implementation Specification defines a set of spatial data types, spatial relationships and spatial operations, spatial query language for the design and development provides a framework.

Yellow spatial data warehouse has four main characteristics: the spatial database is based on data from a particular spatial location; themes and subject-oriented complex; the high water data integration expertise; strong time-series data.

6   Conclusion

The "Smart Yellow River" into a leading digital engineering, the Yellow River emergency system must be planning standards and design ideas to a higher height and cutting-edge technology. The current GIS is toward standardization of data (Interoperable GIS), data multi-dimensional (3D & 4D GIS), system integration (Component GIS), system intelligence (Cyber GIS), platform networking (Web GIS) and application of social (digital Earth) direction, that GIS is entering a post-PC era, the era of the Internet age, and integration. The GIS-based line of the GPS positioning, wireless communications, wireless video transmission, remote automatic control and other related technologies combine to enhance the system's real-time, safety and practicality. Therefore, the systematic study of the Yellow River and address critical emergency systems must GIS technology is important. This paper aims to initiate research and development of the Smart Yellow River Emergency Response System provides suggestions and reflection.

References

1. Guo, R.: Spatial Analysis. Wuhan University of Surveying and Mapping Press, Wuhan (1997)
2. Gong, J.: GIS space-time object-oriented data model. Surveying Mapp. **26**, 289–298 (1997)
3. Xu, S.: GPS measurement principle and application. Wuhan University of Surveying and Mapping Press, Wuhan (1998)
4. Liu, Y., Hong, N., et al.: China's urban disaster risk management study of the comprehensive evaluation index system. J. Nat. Disasters **05**, 62–66 (1999)
5. Chen, S., Lu, X., et al.: Introduction to Geographic Information Systems. Science Press, Beijing (2000)
6. Peng, G.: Practical GIS - Geographic Information System Success and Management. Science Press, Beijing (2001)
7. The OpenGIS Abstract Specification. Topic 12: OpenGIS Service Architecture (ISO19119), Version 4.3., January 2002. http://www.opengis.org/techno/abstract/02-112.pdf
8. Tian, Z., Wu, F., Zhou, Z.: Research on emergency support system of sudden water environmental pollution event based on GIS. J. Irrig. Drainage **04**, 131–134 (2009)
9. Yang, J., Wu, S.: Studies on application of cloud computing techniques in GIS. In: 2010 Second IITA International Conference on Geoscience and Remote Sensing, pp. 156–158 (2010)
10. Chen, R.: Cloud computing, intelligent emergency linkage and pragmatic development of smart city strategy. Mobile Commun. **03**, 5–10 (2012)
11. Wang, M., Liu, J.: The emergency monitoring system based on cloud computing research. Environ. Eng. **18**, 139–142 (2014)
12. Li, X., Song, H., He, Z., Yue, Q., Zhao, F.: Research on comprehensive emergency management platform based on cloud computing. Inf. Technol. **38**, 18–20 (2014)
13. Zhao, H.: Design and implementation of the emergency information service system for the sudden water pollution incident in the Yellow River. Proc. Third China Water Conservancy Inf. Digital Water Conservancy Technol. Forum **04**, 287–290 (2015)
14. Zhang, L., Yang, G., Liu, Y.: Application research progress of 3S technology in water pollution monitoring and emergency treatment. Jilin Normal Univ. J. (Nat. Sci. Ed.) **02**, 152–156 (2016)

Based on the MapX Automatic Contour Drawing Method and Accuracy Analysis

Kuan He[1], Dan Zhang[1], Yuntong Liu[1], Sa Huang[1], and Zhimin Zhou[2(✉)]

[1] Department of Surveying and Mapping Engineering, Yellow River Conservancy Technical Institute, Kaifeng 475004, Henan, China
[2] College of Environment and Planning, Henan University, Kaifeng 475004, Henan, China
zzm@henu.edu.cn

Abstract. With the extensive application of GIS, MapX has become the focus of research on its advantages of convenient and powerful, and the contour line synthesis algorithm is one of the research. This article is focuses on the contour based on MapX synthesis algorithm in Microsoft VC++ implementation in accordance with construction TIN, contour tracking, smooth contours by using the MapX control, manual removal of the TIN network structure edge of the triangle to draw smoothness, accurate, aesthetically pleasing contour. In this paper, the experimental results of the analysis and comparison, not only to ensure the accuracy of drawing analysis, but also greatly shorten the drawing and analysis time.In addition the technology can also be used for the environment, resources, weather and other related fields.

Keywords: Triangulated irregular network (TIN) · Contour · Mapx

1 Introduction

In recent years, with the digital globe, smart city and digital Olympics the operation and concept of geographic information system (GIS) is increasingly wide range of applications. The MapInfo Professional map are based on MapX control of the emergence of technology, will enable government and enterprises to facilitate the embedding map, and also to enhanced spatial analysis and implementation of information management. Therefore, a series of questions based on MapX will become a keynote, structure contours is the part of aspect.

Contour line (Isoline) is often used to represent the distribution of topography and spatial distribution of objects a smooth curve, in terrain analysis, a variety of meteorological, hydrological analysis, and many other aspects have a wide range of applications, to provide a powerful tool in the space display and analysis of geographic objects.

In this article, the first method by using the shortest side is constructed from a triangulated irregular network, and then generated triangulated irregular contour line drawing, and using five-point smooth method, the weighted average is smooth axis parabolic law and MapX provide their own method of smooth contour line was smooth, and finally combined with MapX, was shown on the contour, and also by artificial means will be treated as concave polygon TIN, make the contours line drawn more accurately.

M. Qiu (Ed.): SmartCom 2016, LNCS 10135, pp. 367–373, 2017.
DOI: 10.1007/978-3-319-52015-5_37

2 Contour Drawing

Contour as an important GIS data in the curve, it is important to draw, and many scholars have carried out considerable research in this area. In the computer, Contours are usually stored as an ordered coordinate pair of points, which can be thought of as a simple polygon or a polygonal arc of an elevation value attribute. This article is constructed in accordance with this TIN, contour tracking, smooth contour interpolation process on the three coordinates of the operation ordered to draw contour lines.

2.1   Triangulated Irregular Network Structure

Triangulated irregular network (TIN) through the data points from the irregular distribution of the resulting triangles to approximate a continuous terrain surface. In the expression of topographic information, TIN model can describe different levels of resolution of the terrain surface. Now, it can be automatically create a TIN from the discrete points are two main ways: Delaunay triangulation method and the minimum distance method. Given the low computational complexity, fast computation time, this method uses the minimum distance network structure.

The Principle of Recent Distance Constructio: Mainly based on the triangular cosine theorem $\cos C = (a^2 + b^2 - c^2)/2ab$. Two vertices of the triangle by the known A, B (corresponding diagonal edges c its also known), the other vertices of the triangle using the law of cosines to determine the C, C by calculating the maximum angle (the point c, the shortest distance from the edge), guaranteed by the three form a triangle adjacent to the nearest point. As shown in Fig. 1.

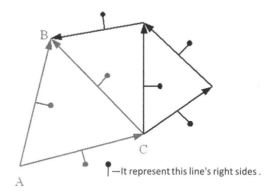

Fig. 1.   Shorter works from the network configuration method

In this paper, to achieve the first define for two structures _Edge and _Triangle network structure were stored there and the emergence of the triangular segments. Secondly, to make sure that all points of the left to determine a vector AB, AB as vector-based, by the rule of cosines to determine the vertex C, then determine a triangle ABC, the other two sides of the triangle as based vectors AC and CB (at right: Note direction), to find its outward expansion of the right triangle to meet the conditions, in this triangle into _Triangle and trilateral stored simultaneously recorded into the _Edge the number of occurrences, in one side can only because of the public side of two triangles, if the number of times over two sides, that edge is invalid, repeat until the end of traversal, and the final structure is to be stored _Triangle triangle network.

2.2 Xontour Tracking

Currently, the contour of the track are based on a rectangular grid of contour tracking, the triangular grid-based terrain contour tracking and contour tracking feature line. In this paper, This triangular grid is based contour tracking method (improved) for the following reasons: 1 point of the discrete sampling; 2, the resolution of this method with a variable that changes when the surface roughness or transformation severe, TIN can contain a large number of data points; 3, this method does not have to go through inter-polation calculation so as to maintain the original point of accuracy.

Base on triangular grid contour tracking algorithm, including striking contour of the plane through the point location and tracking by adjacent contour points in two steps. In this paper, the previous method was improved by adding a _Edge array used to store the order of the contour of the triangle through the side, and the emergence of the five cases has been simplified processing complex, to be followed in this array of side H calculated on the elevation of the point.

First of all, define the required data structures. Here defined structure _Triangle and _Edge the three coordinates are stored, and the triangle logo and contour of the triangle on the side and through the identification.

Secondly, for a given height H, the triangle will be made to meet the conditions. This involves five conditions: 1. A vertex of the triangle H, 2. Two vertices of the triangle H, 3. Triangle of three vertices of H, 4. On both sides of the triangle point H, 5. Triangular points on one side and a vertex of H. In order to simplify this situation the same time guarantee accuracy required in the circumstances, can the vertices of H plus a small variable (in addition to three vertices of H of circumstances), then one of the last four points on both sides of the situation will become the case of H. For the three vertices of the triangle of H is directly deposited into the contour of the data structure.

Then, traverse the proposed triangle, the triangle contains the H-point of the two sides a, b for storage, then as two sides of the side of a triangle to find the corresponding, respectively, based on the other side of c, then the edge c as a side of a, to find c side, iteration it, if it finds the edge c and the beginning of the last two edges of another b-side is equal, then the loop ends, the generate a closed contour. Contour must be closed curves, but due to the size of the display area will be the case is not closed, then the problem is to find the actual operation and the beginning of the last two c-side edge of the other side of b are not equal, then b-side to find places based on the corresponding sides of a triangle d, until the cycle until the end of the last to get a open contour. (As shown below, an array of bold part _Edge structure to be stored in side, left side of closed contours, contours of the right to open). As shown in Fig. 2.

Finally, to elevation contours in store for the H side of the array through which, according to the following formula to find the corresponding elevation of H-point (as shown

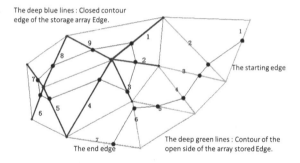

Fig. 2. Base on triangular grid contour tracking algorithm

above the black dot is asked for), followed by the formation of this point will be stored, then to value line data by sequence.

$$
\begin{cases}
x = x_a + \dfrac{(x_b - x_a) \times (H - z_a)}{z_b - z_a} \\
y = y_a + \dfrac{(y_b - y_a) \times (H - z_a)}{z_b - z_a}
\end{cases}
$$

2.3 Contour Smoothing

For more contour to topographic features, it is need to be smoothed. Currently, the smooth contour segmentation algorithms are cubic polynomial interpolation method, the weighted average method is axis parabola, and inclined axis parabolic weighted average method, the tension spline interpolation and semi-parabolic weighted average method. In this paper, we will use smooth with a five-point method, the weighted average is smooth axis parabolic law and provide a smooth MapX own way, and compared. Using the two methods is mainly to save time when in operation, when the data is heavy, will be displayed in the MapX spent a lot of time without having to waste time in the smooth processing.

The smooth five-point method: For a given point (preferably five or more) through the establishment of a cubic polynomial curve equation, requiring the entire curve with continuous first, then derivative data to ensure the smoothness of the curve. The implementation is mainly through the middle of five points to calculate the value of t3, to obtain other points on the curve.

Axis of the parabola is the weighted average method: for a given sequence of n curves points from the first one, in order to take four points, had four points in the first three points can be used as a positive axis parabola, had four points in the last three point to a positive-axis parabola can be used for the two points over the middle axis of the parabola is taking the weighted average, as the middle of two points over the final curve, and ultimately the formation of five polynomials. As shown in Fig. 3.

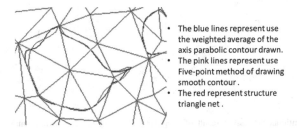

- The blue lines represent use the weighted average of the axis parabolic contour drawn.
- The pink lines represent use Five-point method of drawing smooth contour.
- The red represent structure triangle net.

Fig. 3. The same method in two triangular contour drawing

By contrast, found that positive axis parabola with the five-point weighted average method form a smooth curve is almost smoothed, but smooth for the five-point method if given a five-point position relationship between good and sharp edges may appear situation.

3   Combined with Optimized MapX

3.1   MapX Introduction

MapX is Maplnfo Company's cost performance product, for the development of a GIS system powerful ActiveX control products. It retains the MapInfo Professional workspace concept in its workspace file is Geoset (.Gst) file. For map display, the programmer can use this file, by creating an object based on MapX, and call the Create function and SetGeoSet read.Gst file, which will map the development of embedded program, and out of MapInfo's software platform to run.

3.2   Draw Contours with MapX

In this paper, Microsoft VC++ 2012 as a development platform by using MapX control of the powerful features of the obtained data to show the contour processing.

Because of the TIN network building are convex polygons, concave polygons in some cases more suitable for terrain analysis, we can use MapX control of the selection function, manual do not delete the TIN on the edge of the triangular network, and then re-draw the contours. As shown in Fig. 4(a), the triangular network of convex polygons, as shown in red triangular network structure deformation is too large or out of bounds from the surface does not meet the requirements, through treatment can be seen in Fig. 4(b), the red triangle is removed, while still in this concave polygon TIN network for the contour drawing. This feature enables contour drawing is more beautiful, then this object by calling the function SetSmooth for smooth processing. Results by proving its smooth axis parabola by weighted average method is smooth and the effect is similar.

(a) Conventional processing result of irregular (b) Using MapX drawing contour processing
triangulation method   results

Fig. 4. Base on triangular grid contour tracking algorithm

4   Results of Experimental Treatment

Through the above description, this process in their own editing software available to
run the following diagram:

Colored part of the figure as part of the map area in Henan Province according to
each site, the amount of rainfall collected to draw contour lines. Vertices of the triangle
is that these sites collect rainfall, the red line is at some point rainfall of 120 mm of
rainfall curve, green in this time rainfall of 60 mm of rainfall curve. As shown in Fig. 5.

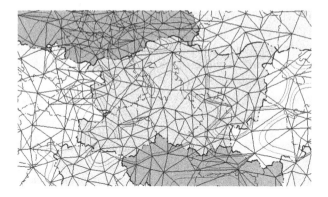

Fig. 5. Results of experimental treatment

5   Conclusions and Prospect

Contour based on MapX synthesis algorithms are use a powerful display of MapX it
will automatically draw the contour on the map. The results of show that the smoothness
of a good way to ensure a high precision, but also greatly reduce the labor intensity
artificially drawn, saving manpower and material resources. The other hand, to solve
the triangulation can be drawn concave polygon problem. As the contours wide range
of applications, I believe that the method has broad application prospects.

References

1. Lu, M., Koike, T., Hayakawa, N.: Distributed hydrology simulation system coupled hydrology model and GIS//LI Jiren. Application of Geographic Information Systems in Hydrology and Water Resources Management. Hohai University Press, Nanjing (1999)
2. Zhao, D.: Contours drawing and analyses based on MapX. J. Tsinghua Univ. (Sci. Technol.) **42**, 1023–1026 (2002)
3. Li, Z., Zhu, Q.: The digital elevation model. Wuhan University Press, Wuhan (2003)
4. Qi, R., Shaolin, Q., Yang, L.: MapX developed using geographic information system. Tsinghua University Press, Beijing (2003)
5. Lun, W., Liu, Y., Zhang, J., et al.: Principles of geographic information systems, methods and applications. Science Press, Beijing (2003)
6. Liu, J., Li, S., Chen, J.: Research on the algorithm of generating equal rainfall line based on MapX component. Geospatial Information **05**, 14–16 (2005)
7. Liu, J., Zhao, Y., Zhang, K.: MapX-based isoline mapping for hydrologic data. Adv. Sci. Technol. Water Resour. **26**, 70–72 (2006)
8. Mapinfo Corporation. MapX v5.0 Developer's Guide, November 18 2009. http://www.pbinsight.com/support/product-documentation/details/mapinfo2mapx
9. Li, M., Zhang, X.: Study on the generation algorithm and its application of irregular triangular mesh. Geomatics Spat. Inf. Technol. **02**, 44–45 (2010)
10. Tang, A., Chen, X., Caicong, W.: Best path analysis based on MapX. Geogr. Geo-Inf. Sci. **01**, 29–33 (2010)
11. Cheng, L., Meng, Z., Liang, M., Yang, X.: Concave polygon fast decomposition algorithm based on Mapx components. J. Agric. Mechanization Res. **07**, 26–29 (2010)
12. Sui, D., Song, A.: The study of data quality based on MAPX. Comput. Program. Skills Maintenance **10**, 39–40 (2011)
13. Zhao, N., Xue, G.: A technique and application of rapid rainfall contour drawing based on MapX. **24**, 20–21 (2013)
14. Zhang, Y., Wang, C., Zhan, J.: Research and application of MapX based on Visual C++ environment. Comput. Knowl. Technol. **08**, 86–88 (2016)

The Function of GIS in the Smart City Construction

Xu Chen[✉] and Kuan He

Department of Surveying and Mapping Engineering, Yellow River Conservancy
Technical Institute, Kaifeng 475004, Henan, China
xuchen006@163.com

Abstract. China's 13th Five - Year Plan of national development clearly put
forward that government will greatly support the construction of smart city during the
period of China's 13th Five - Year Plan. And many places have also started the staged
construction of smart city. GIS (Geographic Information System) plays a very important and fundamental role in supporting the smart city construction. This paper
expounds the connotation of the smart city construction, the important function of
geographic information service in the smart city construction and the application
advantages of GIS in smart city, and analyses the smart city construction which based
on GIS, and also puts forward opinions on the smart city construction, hoping to have
a positive meaning in promoting the construction of smart city.

Keywords: GIS · Smart city · Geographic information service · Spacial data

1 Introduction

Since the reform and opening up, China's urbanization speed is very fast. The urbanization rate has increased from 42.99% in 2005 to 56.1% in 2015 and it will reach around
65% in 2030. It means that there will be more than 10 million people to enter the city
every year. The rapid development of urbanization has brought many problems such as
population management, traffic congestion, environmental protection, safety and so on.
How to solve the outstanding contradictions in the development of the city and how to
achieve the reasonable allocation of limited resources are the problems that each city
manager must face and give overall consideration [1]. The city needs "smart city" and
other new methods to get rid of the dilemma of the development [2]. Smart city is, on
the basis of the existing city information, to achieve a more secure, more efficient, more
responsive and more intelligent urban management.

2 The Connotation of Smart City Construction

2.1 The Definition of Smart City

Since the IBM Company put forward the new idea of smart city in 2008, many experts
have had different understandings of the definition. Academician Li Deren believes that
the smart city is based on the digital earth. It integrates the real world and virtual digital
world effectively through the Internet of Things and it establishes a visible, measurable,

M. Qiu (Ed.): SmartCom 2016, LNCS 10135, pp. 374–380, 2017.
DOI: 10.1007/978-3-319-52015-5_38

controllable and intelligent city management and operation mechanism, which can be used to perceive the state and change of people and things in the real world. It uses the cloud computing center to complete its massive and complex calculation and control, providing a variety of intelligent services for urban management and the public [3]. It can be expressed in a simple formula: Smart City = Digital city + Internet of Things + cloud computing.

2.2 The Connotation of Smart City Construction

Digital city exists in the Cyber space, the mutual mapping between the virtual digital city and the realistic physical city is a digital representation of the physical city of real life in Cyber World [4]. Digital city is the basis of the smart city construction. Smart city relies on digital city technology to organize the people and things in accordance with the geospatial location. It obtains and transmits data and information through the Internet of Things. And it uses cloud computing to deal with massive real-time operation, gives feedbacks to the control system and then through the Internet of Things carry on the intelligent and automatic control. Finally, the city achieves "smart" state. It applies intelligent services to the process of urban construction and management, and this method can be better to meet the requirements of people's life and work [5].

3   Geographic Information in Smart Cities

More than 75% of human activities are related to the geographical location. Therefore, the construction of cities including digital cities and smart cities all need GIS [6]. Digital city organizes the economic, cultural, transportation, energy and educational resources in different fields and different geographical locations according to the specification of geographic coordinates, which provides a basic framework for the smart city [7]. The geographic information of smart city is to provide the various walks of life with all the needs of the geographical location related service. Therefore, the application of geographic information in smart city can be summarized as the following aspects: The first is to display the geographical space; the second is to provide location reference; the third is to assist spatial analysis; the fourth is to support the business reconstruction [8].

3.1   Display the Geographical Space

The identity sign of GIS is the geographical space; the spatial expression and the spatial cognition are the ideological basis of GIS [9]. The geospatial data is the operational objectives of GIS. It is the substantive content of real world which is expressed by GIS through abstracted model. Based on the earth's surface spatial position, geographic spatial data is the data of nature, society and human economic landscape [10, 11]. Therefore, the display of the geographical space is the groundwork of GIS [12]. In the smart city, the display of geographical space is comprehensive, dynamic, multidimensional and interactive.

3.2 Provide Location Reference

In the smart city, the information that needs to be expressed, more like the socio-economic data of the non-spatial data, is varied and not limited to the spatial data. Under the support of GIS, the non-spatial data can be expressed by the geographic information framework, which is called the thematic map in GIS.

3.3 Assist Spatial Analysis

Spatial analysis is considered to be the main function of geographic information system, which is different from general information system, CAD or electronic map system. And it is also the main evaluation index of a geographic information system; it is the core and soul of GIS [13]. The ultimate goal of GIS is to support geographic decision making. Similarly, the ultimate goal of GIS spatial analysis is to provide reference for decision making. For example, we can use 3D GIS spatial analysis to evaluate the effectiveness of urban planning and designing to achieve the best in the smart city construction, which can provide decision-making basis for urban construction [14].

3.4 Support the Business Reconstruction

Supporting the business reconstruction is a special application. It, through the GIS, improves the business practices related to the industry and helps achieve business restructuring and reconstruction [15]. In the smart city, there will be more industries supported by GIS, and the business of these industries will be reconstructed and eventually replaced by a new form of business. With GIS supporting, the smart city will have unlimited possibilities, and the city will be more "Smart".

4   The Application Advantages of GIS in the Smart City Construction

GIS laid the foundation for the transformation of digital city to smart city, and it will play an application advantage in the construction of smart city.

4.1 Abundant Data Resources

Smart city is a new type of city, which is established on the basis of the massive, multi-source, precise and dynamic geographic information data. After many years of efforts and accumulation, China has completed the construction of basic geographic information data, and it is actively constructing the high precision and real-time geographic information acquisition capability, which depends on the trinity of ocean, space, sky and earth [16]. It makes the geographic information resources coverage in time and space wider, data size larger, accuracy higher and currency stronger. The formation of a full range of geographic information data sources and the establishment of a huge dynamic geographic information database are to provide a solid foundation for the smart city construction.

4.2 Leading Technology Advantage

The technical advantages of GIS in smart city are mainly embodied in the following aspects, such as the real time rapid acquisition of urban geographic information, the intelligent processing of massive multi-source geographic information data, the network geographic information intelligence service, and so on [17]. In addition, we developed large-scale distributed geographic information system platform software with independent intellectual property rights. And the software has the function of rapid acquisition and updating of massive data, the geographic information data management function, the true three-dimensional dynamic modeling and visualization, and the function of geographic information network service, etc. which can meet the needs of national infrastructure construction and major information construction.

4.3 Open Sharing Cloud Platform

Smart City needs open sharing cloud platform and GIS in this regard has inherent advantages. GIS establishes the urban geographical space foundation frame for the digital city construction. This framework includes modern surveying and mapping datum construction, basic geographic information database construction, geographic information data acquisition system, geographic information public service platform construction, policies and regulations standard system construction, technical support system construction, organization and operation system construction etc. And it provides the basis for the National Digital City Geospatial Framework to upgrade and transform, as well as constructing the subsequent large-scale smart city cloud platform. It has laid a solid foundation for the smart city, the smart region and the smart China.

4.4 Huge Industrial Development Potential

GIS industry is a strategic emerging industry in China; it is developing rapidly and has great potential. It is an important business support force to promote the transformation of digital city to smart city. And GIS enterprises also become the main body of surveying and mapping geographic information science and technology innovation. Many GIS companies have begun to develop or have developed GIS advanced equipment, high-end equipment, software platforms, etc. which have independent intellectual property rights [18]. These enterprises will become the fusion point of China's high-tech industries, strategic emerging industries and modern services, and will become an important force to promote the construction of smart city.

5 The Smart City Construction and Development Measures Based on GIS

Smart city construction based on GIS is not a simple expansion of digital city, but a huge system composed of digital city, Internet of Things, big data, cloud computing, etc. We should focus on the goal of sustainable development of the city and support the new trend of smart earth and the development of science and technology with coordinated action. And we should take forward-looking vision and strategic thinking, pay high attention to and promote the smart city construction [19].

5.1 Accelerate the GIS Industry Intensive Development

In China, GIS industry development history is short, industrialization is not mature, and there is no perfect market mechanism. The government should strengthen the support and guidance of the GIS industry and encourage enterprises in accordance with the principles of voluntary combination, self-financing, self-management, self-restraint, self-development to establish their own survival mechanism. At the same time, we should standardize the market under the guidance of the government, establish the GIS product standards, gradually implement GIS software product testing, certification and recommendation system and advocate the industry-standard domestic GIS software to create an environment conducive to the construction of national industry and to the fair competition of enterprises at home and abroad. The ultimate goal is to form a more comprehensive GIS Enterprise Architecture combining the large, medium and small ones, and to achieve large-scale and intensive production and operation.

5.2 Attach Importance to Technological Innovation, Speed up the Application of New Technologies

Innovation is the first power of development; technological innovation has a leading role in the overall innovation. We should implement the strategy of innovation driven development and enhance the capability of technological innovation; establish and perfect market-oriented technological innovation system which takes the enterprises as the mainstay and features production-study-research combination; eliminate the "isolated island phenomenon" in technological innovation and accelerate the conversion of the achievement of technological innovation to actual productivity.

5.3 Strengthen the Promotion of Smart City Achievement

As of September 2015, more than 500 cities(accounted for more than half of the total number of smart cities in the world) in China in total had put forward the construction of smart city or were building smart city, including more than 95% of sub-provincial or above cities and 76% of prefecture-level or above cities [20]. Although the constructions of these smart cities are not the same, but some of them have a preliminary intelligent level. These cities can be built as pilot smart cities, to promote the intelligent management level of other cities with their construction and development model.

5.4 Emphasize the Training of GIS Related Professional Talents

Talent is the driving force of innovation and development. Smart city construction requires a large number of professional talents, especially GIS related professionals. The training of GIS related professional talents needs two things. The first is to adhere to the production-study-research cooperation, use the means of "bring in, go out", promote good cooperation between enterprises and institutions of higher learning with GIS major, promote the relevant government departments to do a good job in the operation of the industry norms and establish a good market order. The second is to create the conditions for the establishment of enterprise alliance, to establish the larger institute or talent training base through the enterprise alliance, and to combine with the introduction of talent from the institutions of higher learning to form a talent training mechanism suitable for the enterprise (Wang Jia-yao 2010), hoping to establish a multi-level, diverse GIS education system.

6   Conclusion

Smart city construction is different from the traditional mode of urban construction. It is established on modern technology that is more advanced, more comprehensive, more efficient, more convenient, more intelligent, and more secure. Smart city construction based on GIS has abundant data resources advantages, leading technology advantages, open sharing platform advantages and huge industrial development potential advantages. We should change the concept and carry out the smart city construction through accelerating the GIS industry intensive development, attaching importance to technological innovation, speeding up the application of new technologies, strengthening the promotion of smart city achievement and emphasizing the training of GIS related professional talents etc.

References

1. Boyle, D., Yates, D., Yeatman, E.: Urban sensor data streams: London 2013. IEEE Internet Comput. **17**(6), 1 (2013)
2. Domingue, J., et al. (eds.): FIA 2011. LNCS, vol. 6656, pp. 431–446. Springer, Heidelberg (2011). doi:10.1007/978-3-642-20898-0
3. Deren, L., Jie, S., Zhenfeng, S., et al.: Geomatics for smart cities – concept, key techniques, and applications. Geo-spatial Inf. Sci. **16**(3), 13–24 (2013)
4. Deren, L., Yuan, Y., Zhenfeng, S., et al.: From digital earth to smart earth. Chin. Sci. Bull. **59**(8), 722–733 (2014)
5. Mohanty, S.: Everything you wanted to know about smart cities. IEEE Consum. Electron. Mag. **6**(3), 60–70 (2016)
6. Townsend, A.: Smart Cities: Big Data, Civic Hackers, and the Quest for a New Utopia. W.W. Norton & Company, New York (2013). ISBN 0393082873
7. Batty, M., et al.: Smart cities of the future. Eur. J. Phys. Spec. Top. **214**, 481–518 (2012)
8. Renzhong, G.: The demand and application of geographic information in the smart city. Land Resour. Herald **9**, 47–48 (2014)
9. Yun-feng, K., Xiao-jian, L., Jia-jun, Q., et al.: Analysis on the fundamental issues in the discipline of geographic information systems. Geogr. Geo-Inf. Sci. **22**(5), 1–9 (2006)
10. Deakin, M.: From city of bits to e-topia: taking the thesis on digitally-inclusive regeneration full circle. J. Urban Technol. **14**(3), 131–143 (2007)

11. Deakin, M., Allwinkle, S.: Urban regeneration and sustainable communities: the role of networks, innovation and creativity in building successful partnerships. J. Urban Technol. **14**(1), 77–91 (2007)
12. Wu Xin-cai, X., Shi-wu, W.B., et al.: Principles and Methods of Geographical Information Systems, 3rd edn. Publishing House of Electronics Industry, Beijing (2014)
13. Ren-zhong, G.: Spacial Analysis, 2nd edn. Higher Education Press, Beijing (2001)
14. Solanas, A., Patsakis, C., Conti, M., Vlachos, I., Ramos, V., Falcone, F., Postolache, O., Perez-Martinez, P., Pietro, R., Perrea, D., Martinez-Balleste, A.: Smart health: a context-aware health paradigm within smart cities. IEEE Commun. Mag. **52**(8), 74 (2014)
15. Coe, A., Paquet, G., Roy, J.: E-governance and smart communities: a social learning challenge. Soc. Sci. Comput. Rev. **19**(1), 80–93 (2001)
16. Smart Cities Mission. Ministry of Urban Development, Government of India (2015). Accessed 3 August 2016
17. Wei, X.: A brief analysis of key geo-information technology supporting smarter city construction. Geomatics Spat. Inf. Technol. **38**(7), 97–99 (2015)
18. Graham, S., Marvin, S.: Telecommunications and the City: Electronic Spaces, Urban Place. Routledge, London (1996). ISBN 9780203430453
19. Zai-gao, Y.: The development strategy of smart city. Sci. Technol. Manage. Res. **7**, 20–24 (2012)
20. Zhen-qiang, X.: Key methods of new intelligent city to serve urban management and social governance. Shanghai Urban Manage. **2**, 24–29 (2016)

A Novel PSO-DE Co-evolutionary Algorithm Based on Decomposition Framework

Shaoqiang Yang, Wenjun Wang[✉], Qiuzhen Lin, and Jianyong Chen

Research Institute of Network and Information Security, Shenzhen University, Shenzhen, China
reveriewwj@163.com

Abstract. Comparing with single-population optimization, co-evolution has more benefits in tackling multi-objective optimization problems, as different evolutionary algorithms can work collaboratively. This paper presents a new co-evolution algorithm which employs three populations and integrates particle swarm optimization (PSO) and differential evolution (DE) into the framework of decomposition, named MODEPSO. The main contribution is that the elite solutions got by PSO and archive evolution are considered as evolutionary candidates which will be further evolved by DE operation, so PSO and DE operators can work collaboratively. Experimental results indicate that MODEPSO has better performance than the compared algorithms.

Keywords: Co-evolution · MOEAs · MOEAD · PSO · DE

1 Introduction

Multi-objective optimization problem (MOP) can be defined as simultaneously optimizing more than one objective. A typical MOP can be formulated as follows:

$$\text{minimize:} F(x) = (f_1(x), f_2(x), \ldots, f_m(x))^T \tag{1}$$

where $x = (x_1, x_2, \ldots, x_n)$ is an n-dimensional vector bounded in decision space and m denotes the number of objectives. No single solution can optimize all the objectives simultaneously, so the optimization tries to find out a set of non-dominated solutions, which are well known as Pareto-optimal solutions (PS). Accordingly, the set of their corresponding mapping in objective space is called Pareto-optimal front (PF).

Recent years, many multi-objective evolutionary algorithms (MOEAs) have been proposed to tackle MOPs by using the Pareto domination based or the decomposition based approaches as their selection mechanisms. The representative algorithm of former is NSGA-II [1] which adopts a fast nondominated sorting approach and a crowded-comparison operator to select the potential solution. Now the Pareto domination based method has been improved by using more efficient approaches, such as SPEA2 [2], SDE [3] and NSGA-III [4]. The most famous decomposition based method is MOEA/D [5] which employs a weighted aggregation of all the objectives to achieve decomposition. There are still many enhanced MOEA/D variants presented, such as MOEA/D-DE [6], ENS-MOEA/D [7] and MOEA/D-FRRMAB [8]. In MOEA/D-DE, the Differential

© Springer International Publishing AG 2017
M. Qiu (Ed.): SmartCom 2016, LNCS 10135, pp. 381–389, 2017.
DOI: 10.1007/978-3-319-52015-5_39

Evolution (DE) [9] operator is employed to replace the simulated binary crossover (SBX) operator and performs well. Actually, most of MOEA/D algorithms adopt DE and polynomial mutation (PM) as their evolutionary operators. As particle swarm optimization (PSO) [10] has a fast convergence speed, some representative multi-objective PSO algorithms (MOPSOs) are designed based on Pareto ranking method (e.g., OMOPSO [11] and CMPSO [12]) and based on the decomposition approach (e.g., MOPSO/D [13] and DMOPSO [14]). Recently, some MOPSOs are designed by combining those two approaches, such as D^2MOPSO [15] and MMOPSO [16].

The main contributions of MODEPSO can be summarized as follows: first, PSO and SBX+PM operations are used to find the particles which have higher potential to be further improved. Then, the DE operation is run on these potential candidates for further evolution. Therefore, the PSO and SBX+PM operations can get and transmit some helpful information among different populations, which helps DE to centralize the computing resource on some potential individuals.

2   Related Works

2.1   MOEA/D Algorithm

A general framework of MOEA/D has been proposed in [5], in which an MOP can be decomposed into many single-objective sub-problems. The objective of each sub-problem is a weighted aggregation of all the objectives. The popular decomposition approaches include the weighted sum, Tchebycheff and boundary intersection approaches. Based on the distances among the weight vectors, a neighborhood relations can be defined. Then, each sub-problem is optimized using information mainly from its neighboring sub-problems.

2.2   Differential Evolution

Differential Evolution [9] is a stochastic and population-based search strategy, which is specially designed for continuous optimization problems. It evolves the current individual through mutation and crossover operators, and then selects the better one into the next generation. At each generation, the mutation and crossover operations are executed as follows:

$$v_i = x_{r_1} + F \times (x_{r_2} - x_{r_3}) \tag{2}$$

$$u_{ij} = \begin{cases} v_{ij} & \text{if } r < CR \, or \, j = j_{rand} \\ x_{ij} & \text{otherwise} \end{cases} \tag{3}$$

where F and CR are two control parameters; r is a uniformly distributed random number in $[0,1]$; r_1, r_2 and r_3 are three uniformly distributed random integers for selecting three individuals in the specified population; j stands for the j - th component of decision variable and j_{rand} is a random dimension to ensure that at least one component of u_{ij}

would be replaced by v_{ij}. At last, the individual which has better fitness is chosen to evolve in the next generation.

2.3 Particle Swarm Optimization

In PSO, a swarm consists of a certain number of particles, which represents a potential solution. Each particle i will memorize its historically best position as noted by $pbest_i = (p_{i1}, p_{i2}, \ldots, p_{in})$ and mark the best position as $gbest$ among all $pbest_i$ positions. In each generation, each particle i has its position x_i and velocity $v_i = (v_{i1}, v_{i2}, \ldots, v_{in})$, which can be updated by the following equations:

$$v_i(t+1) = wv_i(t) + c_1 r_1 (x_{pbest_i} - x_i(t)) + c_2 r_2 (x_{gbest} - x_i(t)) \tag{4}$$

$$x_i(t+1) = x_i(t) + v_i(t+1) \tag{5}$$

where t is the iteration number, w is the inertial weight, c_1 and c_2 are two learning factors from the personal and global best particles respectively, r_1 and r_2 are two random numbers generated uniformly in the range [0,1].

3   The Proposed Algorithm: MODEPSO

The framework of MODEPSO performs similarly with genetic algorithm [17, 18], and its flow chart can be presented by Fig. 1.

Fig. 1. Flow chart of MODEPSO

The details of MODEPSO are introduced as follows:
Input:

- Problem;
- Termination criterion;
- N: the size of the PSO population, the DE population and the archive; the number of the sub-problems.
- $\lambda_1, \ldots, \lambda_N$: a set of N weight vectors;
- T: the number of the weight vectors in the neighborhood of each weight vector;

- δ: the probability to select parent solutions in **DE-based Search** and select velocity update equation in **PSO-based Search**.
- nr, ad: the maximal number of solutions replaced by each child solution in **DE-based Search** and **Archive evolution**.

Output:

- Archive A.

Step 1 Initialization

Step 1.1 Calculate the Euclidean distances between any two weight vectors to get the T closest weight vectors to each weight vector. For each $i = 1,\dots,N$, set $B_i = \{i_1, \dots, i_T\}$ where $\lambda_{i_1}, \dots, \lambda_{i_T}$ are the T closest weight vectors to λ_i.

Step 1.2 Generate an initial PSO population $P = (x_1, x_2, \dots, x_N)$ by uniformly and randomly sampling in the search space Ω. Add those individuals to archive A by the archive update method to form the initial archive.

Step 1.3 Initialize $z = (z_1, \dots, z_m)$ by setting $z_j = \min_{1 \le i \le N} f_j(x_i)$.

Step 1.4 Rank the PSO population to ensure $g(x_i | \lambda_i, z)$ is minimal and copy it to the DE population D.

Step 1.5 Initialize the velocity of each particle in P with 0.

Step 2 PSO-based Search

Step 2.1 Randomly choose 20% particles of P as the evolutionary population PE. For $i = 1,\dots,|PE|$, do

Step 2.2 Velocity update: Uniformly and randomly generate a number r from $[0, 1]$. Then calculate the velocity by Eq. (6),

$$v = \begin{cases} wv_i(t) + c_1 r_1 (x_{pbest_i} - x_i(t)) & \text{if } r < \delta \\ wv_i(t) + c_2 r_2 (x_{gbest} - x_i(t)) & \text{otherwise} \end{cases} \tag{6}$$

where x_{gbest} is a random particle from A and $x_{pbest_i} = \arg\ \min g(x_i | \lambda_i, z)$.

Step 2.3 Position update: Update the position of individual using Eq. (5).

Step 2.4 Evaluate: Evaluate the objective functions of new particle y;

Step 2.5 Update of z: For each $j = 1, \dots, m$, if $z_j > f_j(y)$, then set $z_j = f_j(y)$.

Step 2.6 Archive Update: For each solution A_j in A, do

(1) Find the Pareto dominance relationship between A_j and y.

(2) If A_j is dominated by y, record its index in R. If y is equal to or dominated by A_j, the solution y will be discarded, and it means y cannot update A.
   Delete the solutions which are recorded in R. Add y to A.
   If $|A| > N$, delete the most crowded solution by calculating the crowding distances.

Step 2.7 Update DE population: If $g(y | \lambda_i, z) < g(d_i | \lambda_i, z)$, then set $d_i = y$ and record the index i in DI.

Step 3 Archive evolution

Step 3.1 Choose top 50% bigger crowding-distance solutions in A to construct the elitist subset E.
For $i = 1,\dots,|A|$, do
Step 3.2 Generate a random integer j in $[1, |E|]$.
Step 3.3 Generate two child solutions by SBX operator on A_i and E_j.
Step 3.4 Randomly execute the PM operator on one of the child solution to generate a new solution y and add to the temporary subset S.
Step 3.5 Update DE population by y

(1) Find the weight vector λ_j by minimizing $g(y|\lambda_j, z)$.

(2) For $j_t \in B(j)$, replace solution d_{j_t} and record the index j_t as **Step 2.7**, if the solutions number is smaller than ad.

Step 3.6 For each solution in S, update A as **Step 2.6**.

Step 4 DE-based Search
For $i = 1,\dots, |DI|$, do

Step 4.1 Selection of Mating/Update Range:
Uniformly and randomly generate a number $rand$ from $[0, 1]$.
Then set DP by the following equation:

$$DP = \begin{cases} B(i) & \text{if } rand < \delta \\ \{1, \dots, N\} & \text{otherwise} \end{cases} \qquad (7)$$

Step 4.2 Reproduction: Generate a new solution y using Eqs. (2) and (3).
Step 4.3 Update of z as Step 2.5.

Step 4.4 Update of Solutions: For $j \in B(i)$, replace no more than nr solutions d_j with $g(y|\lambda_j, z) < g(d_j|\lambda_j, z)$.
Step 4.5 Update A by y as Step 2.6.

Step 5 Stopping Criterion
If the stopping criterion is satisfied, then stop and output A. Otherwise go to **Step 2**.

As mentioned above, three main procedures are run in MODEPSO, i.e., PSO-based search, DE-based search and archive update, and the PSO-based search and DE-based search are executed alternately to update the external archive during the whole process of evolution, that is why we call MODEPSO a co-evolutionary algorithm.

4 Experimental Results and Analysis

In order to assess the performance of MODEPSO, we compare it with three competitive multi-objective algorithms, i.e., CMPSO, MMOPSO and MOEA/D-DE. We adopted 12 test problems (i.e., ZDT1–ZDT4, ZDT6 [19] and DTLZ1–DTLZ7 [20]), whose true PFs have different characteristics including convexity, concavity, disconnections and multi-modality.

In MODEPSO, the archive update and PSO-based search are performed using the same manner as MMOPSO except that only 20% particles of P are chosen randomly to execute the PSO operation in each generation. For other compared algorithms, their parameter settings are recommended by their original references, as summarized in Table 1. The setting of N in Table 1 is only applicable for bi-objective ZDT problems and the maximum number of function evaluations is set to 25000.

Table 1. The parameter settings for all the algorithms

Algorithms	Parameter settings
MOEA/D	$N = 100, CR = 1.0, F = 0.5, p_m = 1/n, \eta_m = 20, T = 20, \delta = 0.9, n_r = 2$
CMPSO	$N = 20, \omega = 0.9 \to 0.4, c_1 = c_2 = c_3 = 4.0/3$
MMOPSO	$N = 100, \omega \in [0.1, 0.5], c_1, c_2 \in [1.5, 2.0],$ $P_c = 0.9, p_m = 1/n, \eta_c = 20, \eta_m = 20, \delta = 0.9$
MODEPSO	$N = 100, S = 0.2, \omega \in [0.1, 0.5], c_1, c_2 \in [1.5, 2.0], ad = 2$ $CR = 1.0, F = 0.5, p_m = 1/n, \eta_m = 20, T = 20, nr = 2, \delta = 0.9$

For the triple-objective DTLZ problems, their population sizes and maximum number of function evaluations are set to 300 and 75000, respectively. All the algorithms are run 30 times independently for each test instance. As the inverted generational distance (IGD) [6] can examine both of convergence and diversity, it is adopted in this paper to assess the optimization performance.

Table 2 lists the median values and the corresponding inter quartile range (iqr) on IGD result of all the algorithms. The results with underline are statistically similar with that obtained by MODEPSO according to the Wilcoxon's rank sum test with significant level $a = 0.05$, and the best results are identified with bold font. The last row in Table 2 gives the final comparison results of all the compared algorithms, where $-/+/\approx$ indicates the number of test problems that the performances of MODEPSO are respectively better than, worse than and similar to the compared algorithm.

According to Table 2, for the ZDT test problems, MODEPSO obtained the best results on all the ZDT test problems except ZDT6. Although MOEA/D achieved the best result on ZDT6, it performed very poor on ZDT1–ZDT4. As ZDT3 has sectionalized local PFs (see Fig. 2), both MOEA/D and CMPSO could not approach the true PFs, but MODEPSO approximated the true PFs quite well, which indicated that the hybrid operators works better than the single one.

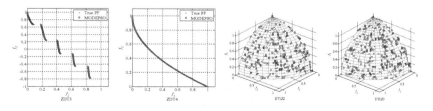

Fig. 2. Partial nondominated solution sets found by MODEPSO

Table 2. The IGD comparison results on the test problems

Algorithms Problems Problems		MOEA/D	CMPSO	MMOPSO	MODEPSO
ZDT1	$\bar{x}$	1.528E-02	4.713E-03	3.888E-03	**3.717E-03**
	iqr	6.180E-03	5.969E-04	5.657E-05	**4.497E-05**
ZDT2	$\bar{x}$	1.267E-02	4.626E-03	3.938E-03	**3.829E-03**
	iqr	3.803E-03	4.905E-04	1.650E-04	**5.078E-05**
ZDT3	$\bar{x}$	5.070E-02	2.524E-02	4.396E-03	**4.375E-03**
	iqr	2.304E-02	8.815E-03	5.280E-05	**4.348E-05**
ZDT4	$\bar{x}$	2.105E-01	5.872E-01	4.150E-03	**3.858E-03**
	iqr	3.454E-01	2.434E-01	8.266E-04	**1.630E-04**
ZDT6	$\bar{x}$	**2.378E-03**	3.239E-03	3.625E-03	2.966E-03
	iqr	**9.685E-06**	3.259E-04	1.946E-04	1.681E-04
DTLZ1	$\bar{x}$	1.653E-02	2.126E-01	2.485E-02	**1.394E-02**
	iqr	1.127E-03	1.201E-01	2.829E-01	**5.251E-04**
DTLZ2	$\bar{x}$	3.752E-02	**3.606E-02**	3.835E-02	3.683E-02
	iqr	2.454E-04	**9.582E-04**	8.341E-04	1.177E-03
DTLZ3	$\bar{x}$	3.980E-02	4.039E-01	3.993E-02	**3.791E-02**
	iqr	5.577E-03	2.245E-01	2.961E-03	**9.725E-04**
DTLZ4	$\bar{x}$	**2.819E-02**	4.967E-02	3.873E-02	3.894E-02
	iqr	**1.383E-03**	1.378E-01	3.434E-03	3.138E-03
DTLZ5	$\bar{x}$	4.785E-03	1.956E-03	**1.316E-03**	1.332E-03
	iqr	3.544E-05	1.134E-04	**3.610E-05**	3.327E-05
DTLZ6	$\bar{x}$	4.483E-03	1.484E-03	1.316E-03	**1.314E-03**
	iqr	1.237E-05	1.010E-04	6.001E-05	**6.087E-05**
DTLZ7	$\bar{x}$	1.117E-01	**3.929E-02**	4.233E-02	4.179E-02
	iqr	1.498E-03	**1.427E-03**	1.884E-03	1.995E-03
$-/+/\sim$		10/2/0	9/2/1	8/1/3	

For the DTLZ test problems, MODEPSO achieved the best results on DTLZ1, DTLZ3 and DTLZ6. It is noted that MODEPSO also obtained the second best performances on other problems. As revealed by the Wilcoxon's rank sum test, MODEPSO got similar results to CMPSO on DTLZ4 and to MMOPSO on DTLZ4 and DTLZ7, while MOEA/D performed worst on DTLZ7. These results also validated again that the co evolution of PSO and DE has presented some advantages over each single one.

The time complexity analysis of MODEPSO is provided and compared with the other algorithms. Based on the pseudo-code of MODEPSO in Sect. 3, the time complexity of MODEPSO is determined by the evolutionary loop in Step 2–4. As the decision variables and objectives are much smaller than the sizes of population and external archive N, so that their impact are generally ignored. In Step 2, the time complexity is $O(N^2)$ when calculating the velocity and updating the archive; other operations' time complexity are all $O(N)$. The time complexity of archive evolution and DE operation are also $O(N^2)$. Thus, the maximum time complexity of MODEPSO is $O(N^2)$. The computational complexities of MOEA/D,

CMPSO and MMOPSO are all $O(N^2)$ as discussed in [16]. Therefore, MODEPSO has the similar time complexity with the compared algorithms.

5   Conclusion

In this paper, we present a novel co-evolutionary algorithm based on decomposition framework. The PSO and DE operators are performed cooperatively in MODEPSO to achieve a good approximation of the true PFs. The experimental results confirmed that MODEPSO performed better on most of test problems adopted when compared to MOEA/D, CMPSO and MMOPSO.

Acknowledgements. This work was supported by the National Natural Science Foundation of China under Grant 61402291, Seed Funding from Scientific and Technical Innovation Council of Shenzhen Government under Grant 0000012528, Foundation for Distinguished Young Talents in Higher Education of Guangdong under Grant 2014KQNCX129, and Natural Science Foundation of SZU under Grant 201531.

References

1. Deb, K., Agrawal, S., Pratap, A., Meyarivan, T.: A fast and elitist multiobjective genetic algorithm: NSGA-II. IEEE Trans. Evol. Comput. **6**, 182–197 (2002)
2. Zitzler, E., Laumanns, M., Thiele, L.: SPEA2: improving the strength Pareto evolutionary algorithm. Computer Engineering and Networks Laboratory, Swiss Federal Institute of Technology (ETH), Zurich, Switzerland, Technical report 103 (2001)
3. Li, M., Yang, S., Liu, X.: Shift-based density estimation for pareto-based algorithms in many-objective optimization. IEEE Trans. Evol. Comput. **18**, 348–365 (2014)
4. Deb, K., Jain, H.: An evolutionary many-objective optimization algorithm using reference-point-based nondominated sorting approach, part I: solving problems with box constraints. IEEE Trans. Evol. Comput. **18**, 577–601 (2014)
5. Zhang, Q., Li, H.: MOEA/D: a multi-objective evolutionary algorithm based on decomposition. IEEE Trans. Evol. Comput. **11**, 712–731 (2007)
6. Li, H., Zhang, Q.: Multiobjective optimization problems with complicated Pareto sets, MOEA/D and NSGA-II. IEEE Trans. Evol. Comput. **13**, 284–302 (2009)
7. Zhao, S., Suganthan, P.N., Zhang, Q.: Decomposition-based multiobjective evolutionary algorithm with an ensemble of neighborhood sizes. IEEE Trans. Evol. Comput. **16**, 442–446 (2012)
8. Li, K., Fialho, A., Kwong, S., Zhang, Q.: Adaptive operator selection with bandits for a multiobjective evolutionary algorithm based on decomposition. IEEE Trans. Evol. Comput. **18**, 114–130 (2014)
9. Storn, R., Price, K.: Differential evolution - a simple and efficient heuristic for global optimization over continuous spaces. J. Global Optim. **11**, 341–359 (1997)
10. Kennedy, J., Eberhart, R.: Particle swarm optimization. In: Proceedings of IEEE International Conference on Neural Networks, pp. 1942–1948 (1995)
11. Sierra, M.R., Coello Coello, C.A.: Improving PSO-based multi-objective optimization using crowding, mutation and epsilon-dominance. Evol. Multi-Criterion Optim. **3410**, 505–519 (2005)

12. Zhan, Z., Li, J., Cao, J., Zhang, J., Chuang, H., Shi, Y.: Multiple populations for multiple objectives: a coevolutionary technique for solving multiobjective optimization problems. IEEE Trans. Cybern. **43**, 445–463 (2013)
13. Peng, W., Zhang, Q.: A decomposition-based multi-objective particle swarm optimization algorithm for continuous optimization problems. In: IEEE International Conference on Granular Computing, pp. 534–537 (2008)
14. Martinez, S.Z., Coello Coello, C.A.: A multiobjective particle swarm optimizer based on decomposition. In: Genetic and Evolutionary Computation Conference, pp. 69–76 (2011)
15. Moubayed, N.A., Petrovski, A., McCall, J.: D^2MOPSO: MOPSO based on decomposition and dominance with archiving using crowding distance in objective and solution spaces. Evol. Comput. **22**, 47–77 (2014)
16. Lin, Q., Li, J., Du, Z., Chen, J., Ming, Z.: A novel multi-objective particle swarm optimization with multiple search strategies. Eur. J. Oper. Res. **247**, 732–744 (2015)
17. Gai, K., Qiu, M., Zhao, H.: Cost-aware multimedia data allocation for heterogeneous memory using genetic algorithm in cloud computing. IEEE Trans. Cloud Comput. (2016)
18. Qiu, M., Zhong, M., Li, J., et al.: Phase-change memory optimization for green cloud with genetic algorithm. IEEE Trans. Comput. **64**, 3528–3540 (2015)
19. Zitzler, E., Deb, K., Thiele, L.: Comparison of multiobjective evolutionary algorithms: empirical results. Evol. Comput. **8**, 173–195 (2000)
20. Deb, K., Thiele, L., Laumanns, M., Zitzler, E.: Scalable test problems for evolutionary multi-objective optimization. In: Abraham, A., Jain, L., Goldberg, R. (eds.) Evolutionary Multiobjective Optimization, Advanced Information and Knowledge Processing, pp. 105–145. Springer, Heidelberg (2005)

Rhythm Authentication Using Multi-touch Technology: A New Method of Biometric Authentication

Nakinthorn Wongnarukane[✉] and Pramote Kuacharoen

Department of Computer Science, Graduate School of Applied Statistics,
National Institute of Development Administration,
118 Serithai Road, Bangkapi Bangkok 10240, Thailand
nakinthorn.n@gmail.com, pramote@as.nida.ac.th

Abstract. A keystroke authentication method has a lower cost and is more powerful and easier to use than other biometric authentication methods. However, traditional keystroke authentication has many weaknesses and is easy to attack by criminals. Attacks can include shoulder surfing attacks, eavesdropping attacks and key-logger attacks. When users try to access their computer or portable device by using keystroke authentication method, the users must push the correct buttons with the correct rhythm in order to be authenticated. If the users make several failed authentication attempts, the system will lock their account. As a result, the users usually use a simple password and rhythm for accessing their account which will make the risk even higher. This research proposes a new method of a keystroke authentication by using multi-touch technique on touchpad which is embedded on a laptop computer. The users can register their rhythm using their fingers on the touchpad to the system as a biometric authentication. An attacker will have difficulties conducting a shoulder surfing attack. This is because the users have no need to type their password and can use one hand to cover the other hand which is used to make their rhythm for the touch. Furthermore, the users can quickly make the rhythm. An eavesdropping attack is rendered useless since the touchpad can get event data when the users touch it without making any sound. Even though some users may not be vigilant and make tapping sounds, an eavesdropper cannot know how many fingers the users use to tap on the touchpad to make one beat. The research results show that the purposed multi-touch rhythm authentication performs better than the traditional keystroke method and provides better security, usability, and faster authentication.

Keywords: Rhythm authentication · Multi-touch authentication · Biometric authentication · Touch screen authentication · Touch pad authentication

1 Introduction

Authentication is one of the most important issues in computer security management and information systems. Many countries have legislation regarding computer crime to control and regulate their citizens. Many criminals are usually arrested by tracing back to their IP address which is used when committing the crime. The criminals often use the victims' computers or accounts (username and password) to access social media

© Springer International Publishing AG 2017
M. Qiu (Ed.): SmartCom 2016, LNCS 10135, pp. 390–399, 2017.
DOI: 10.1007/978-3-319-52015-5_40

such as Facebook and Twitter to conduct illegal activities. Therefore, the computer owners and the account owners must be aware of this issue and defend themselves against criminals who try to hack, attack, and use their computers to commit crimes. For this reason, many computer authentication methods are developed to detect and defend against the criminals from unauthorized access. The identity verification consists of three main methods [1].

- What you know: the secret, only owner knows such as password and PIN
- What you have: the equipment that is used to access something such as ID-card
- What you are: the singularity, only owner is such as fingerprint and signature

From above three methods, the mostly used method is password-based authentication (what you know) because it easiest and most convenient to use. Nowadays, a majority of web application systems such as Google deploy single sign-on (SSO) where the users sign in once and are able to use all products associated with their accounts [2]. It is easier for the users to remember only one password. But, this system is susceptible to a higher risk from criminal attacks since if the password is compromised, all of systems that are using this password will be breached altogether. One Time Password (OTP) [3] may be used for a stronger security method as a two-factor authentication. However, some implementation of OTP has already been hacked. Alternatively, a possession method can be used (what you have). It is based on items that users possess such as ID cards, RFID cards, and USB tokens. However, this method also has problems. Since the users have to carry the authentication items around, they may be lost or stolen. Criminals may be able to obtain an authentication item which may lead to an unexpected security incident. The last method is based on what the users are in the form of biometric data such as fingerprint, handwriting signature, or characteristics of the user behaviors.

The biometric authentication that is widely used is fingerprint authentication [4]. Fingerprint can be used to access a personal computer, open the gate, access a smart phone, and the like. However, fingerprints can easily be forged or copied [5] which make it unsafe to deploy by itself. Other biometric authentications are retina authentication [6], hand-gesture authentication, heart rate authentication, human body print authentication as well as the characteristic behavior such as movement and keystroke authentication. These types of authentication are more secure than fingerprint authentication. However, they are more expensive. If we need better security, we may have to pay more for the device. Some devices may not be easy to use in real life. For example, retina authentication device is very expensive and may be subjected to infection when the eyes are close to the device.

Keystroke authentication [7] may be an alternative since it is not costly and easy to implement. Using they keystroke authentication, the system detects the rhythm of the users pressing and releasing their fingers on keyboard. Keystroke authentication does not require any additional peripheral devices. Although the keystroke authentication is less expensive and easier to use than other methods of biometric authentication, it is susceptible to shoulder surfing attacks. If an attacker has enough time to detect the victim's rhythm and password, the attacker can impersonate the victim. Every authentication method has advantages and disadvantages depending on the situation and how it is deployed.

This research presents another method of authentication and verification of the users who need to access their laptop computer, mobile device, and computer system by using touchpad to produce rhythm print as the biometric authentication. We try to solve the weaknesses of the traditional keystroke authentication without increasing cost and make it easy to use. This paper describes how to using rhythm authentication without using a password. Since many devices have an embedded touchpad or a touch-screen and support a multi-touch system, they can be used as an input device for authentication. The multi-touch system is used to detect the rhythm and the number of fingers that users press to create the rhythm when they tap their fingers on touchpad or screen of smartphone. Attackers would have difficulties trying to conduct shoulder surfing attacks because users do not expose their fingers during tapping by covering the tapping hand with the other hand. Tapping action can be made so that it does not produce any sound. Lightly touch on the touchpad or the touch-screen is enough. This makes it difficult for attackers to perform eavesdrop attacks. Even though a tapping action may produce detectable sound, the attackers cannot know how many fingers are used to produce the sound.

This paper consists of five sections. The next section provides background information and related work that is relevant to the paper. Section 3 describes the design and implementation. Section 4 presents the experimental results. The last section, Sect. 5 concludes the paper.

2   Background and Related Work

Biometric authentication relies on the unique biological characteristics of individuals to determine identity, which consists of two methods, namely, physical biometric and behavioral biometric. The physical biometric uses human body parts such as fingerprint, retina, face, DNA, and hand geometry while the behavioral biometric uses measurable patterns in human activities such as keystroke dynamic, voice ID, and gait analysis. As previously mentioned, fingerprint can be easily forged and other physical biometrics is expensive. One of the most used behavioral biometrics is keystroke dynamic.

Keystroke authentication measures the manner and rhythm in which each individual types. The user must enter their password with the correct rhythm to verify themselves [7, 8]. The user must type his/her password on keyboard with the previously registered rhythm; the system extracts features of the user password and rhythm, and then compares them with user's template in database. The performance of biometric authentication can measure three factors including False Acceptance Rate (FAR), False Rejection Rate (FRR) and Equal Error Rate (EER). FAR is the measure of the possibility that the biometric system incorrectly permits an access attempt by an unauthorized user. FAR is typically calculated by dividing the number of false acceptances by the number of identification attempts. In contrast, FRR is the measure of the likelihood that the biometric system incorrectly denies an access attempt by an authorized user. FRR can be computed by using the ratio of the number false rejections and the number of identification attempts. When FAR and FRR are equal, it is called EER. For creating the keystroke template, various measurements can be taken. However, there are popular measurements, namely, holding time, latency time, and pressure. The time which starts when users push their fingers on the button of the keyboard and ends when they release their finger off button, is called holding time. The time when users switch their finger from the current button to a new button is called latency time as shown in Fig. 1.

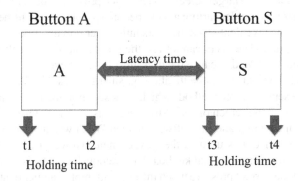

Fig. 1. The measurements of holding time and latency time

From Fig. 1, we can calculate holding time:

$$Holding\ time\ of\ A = t2 - t1$$
$$Holding\ time\ of\ S = t4 - t3$$

We can also calculate latency time:

$$Latency\ time\ of\ A\ switch\ to\ S = t3 - t2$$

Pressure can only be measured on the system which is equipped with pressure feedback. A pressure template can be created by detecting the finger weight when the user pushes or tabs their finger on a touchscreen device or a touchpad on laptop. All three measurements are usually combined to provide accurate verification.

In [9], a keystroke authentication method on smartphones with touchscreen device using virtual keyboard is presented. Holding time, latency time and password are used to produce the keystroke pattern template by applying the following equations.

$$Pattern\ Holding\ time\ /\alpha < Attempt\ Holding\ time < Pattern\ Holding\ time \times \alpha$$

and

$$Pattern\ Latency\ time\ /\alpha < Attempt\ Latency\ time < Latency\ time \times \alpha$$

Attempt Holding time and Attempt Latency time come from when users try to access the system by typing their password with the same rhythm that was previously registered. Attempt Holding time and Attempt Latency time must satisfy the above conditions in order to be authenticated. The paper states that $\alpha = 4$ is the best choice.

Using a virtual keyboard for keystroke authentication on small screen devices is difficult since buttons on the virtual keyboard are very small. A near-by button is usually pressed. An auto-correction feature is usually favorable in typing on a virtual keyboard.

However, for authentication, the users have to enter the correct password within a time limit based on the registered rhythm which is challenging to accomplish. Thus, the users opt to use a simple password and a simple rhythm which is vulnerable to attacks. Although using a tablet or a larger screen device which provides a larger virtual keyboard reduces typing errors, shoulder surfing attacks become problematic. Furthermore, a long password may be inconvenient since the users must enter it many times a day due to the screen locking feature. The users cannot use the other hand to cover the screen when typing on a virtual keyboard because the user must look at keyboard.

In [10], another keystroke authentication on smartphones with Android operating system is also presented. Three methods which consist of holding time, latency time and pressure of fingers on the screen of a smartphone are used. A new virtual keyboard was developed in their research and the software named Weka was used as a classification tool. Euclidian distance was used in their experiment. However, this research has a similar weakness for using a virtual keyboard as previously mentioned. Moreover, it is difficult to detect the correct pressure when the users attempt to authenticate while doing various activities such as walking, standing, and sitting.

Saevanee and Bhatarakosol [11] proposed a keystroke authentication using the touchpad on a laptop with features including holding time, latency time, and pressure. The proposed method divides the touchpad into grids representing a numeric keypad. The experiments were conducted by gathering data while participants entered their 10-digit phone numbers for 30 times continuously. Features were extracted using k-nearest neighbors (k-NN) algorithm and Euclidian distance. The authors claimed that using only the finger pressure with the k-NN analytical method can identify the users with an accuracy rate of 99%. Even though the results seem impressive, measuring pressure is impractical as discussed earlier.

3   Design and Implementation

By using our proposed method, the authentication of multi-touch rhythm can be verified. From related work, we found that a keystroke dynamic authentication can be applied to touchpad and touch-screen devices in a similar fashion as keystroke dynamic authentication on a traditional hardware keyboard. Although previous work used a pressure feature, measuring pressure for authentication is not suitable. For this reason, we propose a new method of authentication using keystroke dynamic method which provides an accurate verification, flexibility, and usability.

Our proposed method is called a multi-touch rhythm authentication which utilizes multi-touch features including three attributes, namely, holding time, latency time, and number of fingers. The proposed method of a keystroke authentication uses the rhythm of the user tapping a touchpad or a touch screen. This method can replace passwords. Users can verify themselves by simply touching on a touchpad on a laptop computer or a touch-screen on a smartphone with their fingers with the enrolled rhythm. Each beat can be made by one or more fingers. Shoulder surfing attacks become impractical since fingers can be covered. The users do not need to look at the keyboard while tapping.

3.1 Registration and Creation of Template Process

Users must register before using the authentication system. The user makes a rhythm of the tapping the touchpad or a touch-screen. The registration UI records holding time and number of figures. For the subsequent taps, latency times are also recorded. After that, the extraction system creates a template of the user and stores it in the biometric database. Figure 2 shows the steps of the user registration and sequence diagram of the template creation process.

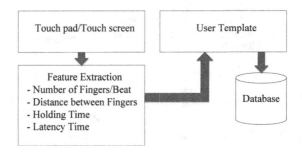

Fig. 2. The flowchart shows the template creation process

3.2 Authentication Process

When the user needs to login to use a computer or device, the user can touch on touchpad of laptop with same rhythm and the same number of fingers per beat. The system will check the holding time, the latency time, and the number of fingers per beat against the template in biometric database. If they match, the user is authenticated. Figure 3 shows the steps of the authentication process.

Table 1. The dataset of user that enter to the system for authentication

Number of Fingers/Beat	Beat	Holding Time (s)	Latency Time (s)
1233	4	0.09400000000005093, 0.08800000000064756, 0.07999999999992724, 0.07999999999992724	0.0000000000000000, 0.4319999999997890, 0.2359999999998763, 0.1679999999996653

4   Experimental Results

To verify our design, we chose Java programming language for the development and we using touchpad's of Mac Book Pro for our experiment and in this experiment we allow up to five fingers for creating one beat. The user must enter the rhythm to register the template in the creation process. This enrolled template will be used to authenticate the user. The authentication process is quite simple. The user taps or touches his/her fingers on touchpad with the correct rhythm to login. After system receives the rhythm

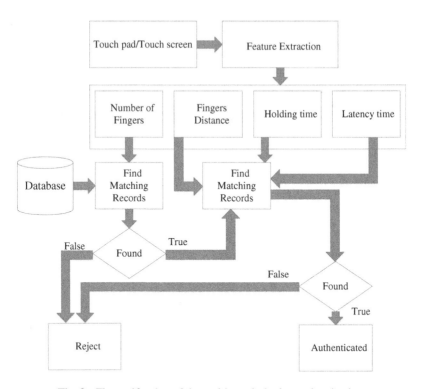

Fig. 3. The verification of the multi-touch rhythm authentication.

from user, the authentication process will start and split to three phases. First, we calculate the number of beats and the number of fingers of each beat.

Table 2. Example set of store data in database

ID	Number of Fingers/Beat	Beat	Holding Time (s)	Latency Time (s)
1	1233	4	0.09400000000005093, 0.10400000000026921, 0.07999999999992724, 0.08800000000064756	0.00000000000000000 0.40799999999944700, 0.22400000000016007, 0.15999999999985448
2	1233	4	0.17799999999988358, 0.16799999999966530, 0.15200000000004366, 0.15200000000004366	0.00000000000000000 0.51200000000062570, 0.50799999999981080, 0.53600000000005820
3	1234	4	0.09400000000005093, 0.09600000000045839, 0.11200000000008004, 0.09600000000045839	0.00000000000000000 0.36799999999948340, 0.15999999999985448, 0.17599999999947613

Table 1 shows the number of fingers per beat which the user entered to the system by touching on his/her touchpad to authenticate. Table 2 shows the stored data pattern in database. For the number of fingers per beat, we can directly compare the value with the stored data in database to reduce number of records in the first phase. After classifying by using number of fingers per beat, we will obtain the minimum number of records which are shown in Table 3.

Table 3. The set of the stored data in database after classifying with number of fingers/beat

ID	Number of Fingers/Beat	Beat	Holding Time	Latency Time
1	1233	4	0.09400000000005093, 0.10400000000026921, 0.07999999999992724, 0.0880000000064756	0.00000000000000000 0.40799999999944700, 0.22400000000016007, 0.15999999999985448
2	1233	4	0.17799999999988358, 0.16799999999966530, 0.15200000000004366, 0.15200000000004366	0.00000000000000000 0.51200000000062570, 0.50799999999981080, 0.53600000000005820

After classifying the number of fingers, the system analyzes the sequence of holding times and latency times of the rhythm. This system compares holding times and latency times directly with the stored data in database within an error range. In this experiment, we define error range is not more than 0.1 s per beat (Fig. 4).

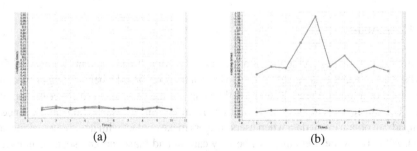

(a)   (b)

Fig. 4. (a) The sequence of fingers in the rhythm matches the enrolled template. (b) The sequence of fingers in the rhythm does not match the enrolled template

Figure 5(a) shows a line graph when an authorized user enters the correct rhythm. On the other hand, Fig. 5(b) shows an example when an attacker passed the first step by tapping with the correct sequence of the number of fingers, but the attacker does not know holding time of each beat.

If both sequence of the number of fingers and the holding times match the enrolled templates, the system checks the latency times between beats in the rhythm. Figure 6(a) shows a line-graph when an authorized user enters the correct rhythm. Figure 6(b) shows an example when an attacker knows both sequence of the number of fingers and the holding times, but not the sequence of latency times in the rhythm.

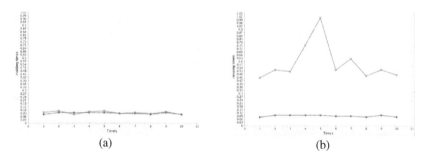

(a)   (b)

Fig. 5. (a) The holding times match the enrolled template (b) Holding times do not match the enrolled template.

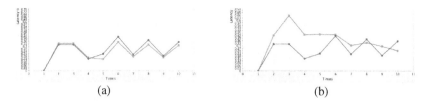

(a)   (b)

Fig. 6. (a) The latency times match the enrolled template. (b) The latency times do not match the enrolled template.

5   Conclusion

The multi-touch rhythm authentication increases security. It can prevent shoulder surfing attacks. This is because users do not need to look at their keyboard when tapping or touching a touchpad or a touch screen. Therefore, they can use the other hand to cover the tapping hand. The attackers will have difficulties performing an eavesdropping attack since a touchpad can obtain event data when users' fingers make contact with it without making any sound. Although some users may not be very careful and make tapping sounds, an eavesdropper cannot know how many fingers the users use to tap on the touchpad to make one beat.

The results show that the proposed multi-touch keystroke authentication is simple and easy to use. Furthermore, it is better than the traditional keystroke methods because it provides improved security and can verify the user more quickly.

References

1. Kim, J.J., Hong, S.P.: A method of risk assessment for multi-factor authentication. J. Inf. Process. Syst. 7(1), 187–198 (2011)
2. Mainka, C., Mladenov, V., Guenther, T., Schwenk, J.: Automatic Recognition, Processing and Attacking of Single Sign-On Protocols with Burp Suite (2015)
3. Subpratatsavee, P., Kuacharoen, P.: Transaction authentication using HMAC-based one-time password and QR code. In: Park, J.J., Stojmenovic, I., Jeong, H.Y., Yi, G. (eds.) Computer Science and its Applications. Lecture Notes in Electrical Engineering (LNEE), vol. 330, pp. 93–98. Springer, Heidelberg (2015)

4. Kant, C., Nath, R.: Reducing process-time for fingerprint identification system. Int. J. Biometric Bioinform. **3**(1), 1–9 (2010)
5. Bhattacharyya, D., Ranjan, R., Alisherov, F., Choi, M.: Biometric authentication: a review. Int. J. of u- e-Serv. Sci. Technol. **2**(3), 13–28 (2009)
6. Galbally, J., Cappelli, R., Lumini, A., Gonzalez-de-Rivera, G., Maltoni, D., Fierrez, J., Maio, D.: An evaluation of direct attacks using fake fingers generated from ISO templates. Pattern Recogn. Lett. **31**(8), 725–732 (2010)
7. Maiorana, E., Campisi, P., González-Carballo, N., Neri, A.: Keystroke dynamics authentication for mobile phones. In: Proceedings of the 2011 ACM Symposium on Applied Computting, pp. 21–26. ACM (2011)
8. Shanmugapriya, D., Padmavathi, G.: A survey of biometric keystroke dynamics: approaches, security and challenges. arXiv preprint arXiv:0910.0817 (2009)
9. Huang, X., Lund, G., Sapeluk, A.: Development of a typing behaviour recognition mechanism on Android. In: 2012 IEEE 11th International Conference on Trust, Security and Privacy in Computing and Communications, pp. 1342–1347. IEEE (2012)
10. Antal, M., Szabó, L.Z., László, I.: Keystroke dynamics on android platform. Procedia Technol. **19**, 820–826 (2015)
11. Saevanee, H., Bhatarakosol, P.: User authentication using combination of behavioral biometrics over the touchpad acting like touch screen of mobile device. In: Computer and Electrical Engineering, pp. 82–86. IEEE (2008)

An Improved Ciphertext-Policy Attribute-Based Encryption Scheme

Hua Zheng[✉], Xi Zhang, and Qi Yang

College of Computer Science and Software Engineering,
Shenzhen University, Shenzhen, China
zhenghuamail@126.com

Abstract. According to characteristics of mass encryption service in cloud storage, this paper proposed an improved SCP-OOABE scheme which uses the access control tree as a control structure and decomposes key generation and encryption into online and offline phase respectively that could be finished in a short time in online phase. In addition, it also avoids the problem that CP-OOABE scheme can't generate keys when constructing access control structures. What's more, it meets the demand of the complex access control structure in cloud storage and was proved that the scheme is against chosen-plaintext attack secure. In a word, our scheme was security, efficient and universal, which suits for the demands of cloud storage.

Keywords: Cloud storage · Access control · Attribute-based encryption · Online/offline · Security

1 Introduction

Cloud storage as a basic function in cloud computing has resolved the problem of mass data of video storage successfully, but it brings with security issues more often. Cloud storage providers are often not fully trusted who exposed the data in untrusted environment and made it vulnerable to various attacks, including malicious attackers, internal staff's snooping, etc. [1]. Therefore, it brought a big challenge of protecting the data security and privacy of users.

Encryption is the most widely used security technology in cloud storage at present. In the cloud storage environment [2], Attribute-Based Encryption is a promising encryption technology. The policy uses cryptographic or key associated access control strategy, so that the encryption policy becomes flexible and the users' permissions become easy to explain and describe. It is an efficient, dynamic, flexible and private encryption.

ABE (Attribute-Based Encryption) is a fuzzy Identity-Based Encryption which proposed by Sahai and Waters [3] on EUROCRYPT in 2005. Fuzzy Identity-Based Encryption was expanded into Attribute-Based Encryption by Goyal et al. [4] in 2006. They clarified the concept and significance of Attribute-Based Encryption. However, the scheme is based on key strategy which is fit for query applications, but not access control occasions. In 2007, Benthencour et al. [5] first proposed a Ciphertext-policy Attribute-Based Encryption Scheme that supported more complex access control. But calculating amount is quite large in encryption and key generation stage, therefore, it was

© Springer International Publishing AG 2017
M. Qiu (Ed.): SmartCom 2016, LNCS 10135, pp. 400–411, 2017.
DOI: 10.1007/978-3-319-52015-5_41

not suitable for mass encryption service in cloud storage. Rouselakis et al. [6] proposed Ciphertext-policy Attribute-Based Encryption in a large space in 2013 which was based on bilinear pairing of prime order. Any string could be directly used as an attribute without digitizing beforehand. However, it didn't pretreat encryption action. With the increasing demand of the users, plenty of calculations will be repeated execution.

Based on this, Hohenberger et al. [7] proposed Ciphertext-Policy Online/Offline [8, 9] Attribute Based Encryption (CP-OOABE) in 2014. Key generation and encryption were decomposes into two stages respectively, which reduce the response time and avoid the repetitive work in the online phase. It is possible to make the users truly enjoy the benefits of cloud storage [10]. However, the scheme uses LSSS matrix [11] as access control structure. When constructing matrixes it is important to find a suitable matrix which will cost a lot of time and was not suit for complex access control requirements in cloud storage. If the rank of attribute matrix A is less than its corresponding augmented matrix B it will turn out to be no solution. The corresponding key would not be solved in the decryption stage. Based on the above problems, we proposed Super Ciphertext Policy Online/Offline Attribute-Based Encryption which proofed CPA security formally. [12] The scheme is based on online/offline encryption which uses access control tree as access control structure and support complex AND/OR operations. It satisfies fine-grained access control and suits for diverse demands of users in cloud storage.

2 Preparation Knowledge

2.1 Access Control Tree

Access control tree τ on behalf of logic expression which is comprised of given attribute. The internal node of the tree was associated with (t, n) and the leaf node was associated with an attribute [13, 14]. Logic AND can be express as (n, n) and logic OR can be express as (1, n). The threshold of intermediate nodes in the tree was expressed by its child nodes and a threshold, such as (k_x, num_x). Among them, k_x represent node threshold and num_x represent the number of k_x's child nodes. The attribute of leaf node in the tree was denote by an attribute and a threshold $k_x = 1$. Such as Fig. 1, only for users aged 20, Grade 1 or 2 meet the access control tree τ.

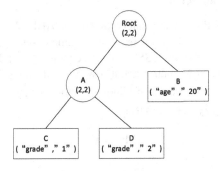

Fig. 1. Access control tree τ

2.2   Bilinear Map

Bilinear map connects two multiplicative cyclic group [15–17] in the process of attribute encryption. Suppose G_0 and G_1 are two multiplicative cyclic group, their order are the prime number p, g are the generator of G_0, $e : G_0 \times G_0 \rightarrow G_1$ is Bilinear mapping. Bilinear mapping e satisfy the following three conditions:

(1) bilinearity: for any $\mu, v \in G_0$ and $a, b \in Z_p$, we have $e(\mu^a, v^b) = e(\mu, v)^{ab}$.
(2) non-degeneration: $e(g, g) \neq 1$.
(3) computability: for any $\mu, v \in G_0$, $e(\mu, v)$ be able to calculate quickly and efficiently.

Lemma 1. For prime number p graded groups G_0, generator is g, Decisional Bilinear Diffie-Hellman (DBDH) issue [18] as follows: for a given input $e(g, g)^z \in G_1$, $(a, b, c \in Z_p)$, determine $e(g, g)^z = e(g, g)^{abc}$.

3   Analysis of CP-OOABE Mechanism

3.1   CP-OOABE Mechanism

CP-OOABE which was proposed by Hohenberger et al. [7] improved Attribute-Based Encryption and increased efficiency. For Attribute-Based Encryption, they proposed a "connect and correct" mechanism that divided key generation and encryption into two steps respectively. That is the first step offline (preprocessing) stage. At this stage, system did not know the users' attribute set and access structure, but accomplished a lot of calculation. The second step online stage, when we know the users' attribute set and access structure, key generation and encryption could be accomplished quickly. The scheme has a great advantage in the mass of services in cloud storage: reducing response time, the time of online key generation and encryption is only 1% of the entire key generation and encryption algorithms. So that the algorithm based on the attribute pooling avoid duplication of work in online phase. The users can better enjoy the convenient of cloud storage.

3.2   Hohenberger et al. Scheme

Hohenberger et al. [7] Proved the safety of CP-OOABE. This paper analyzes the structure of CP-OOABE which is not feasible in some cases.

Hohenberger et al. based on LSSS access scheme (M, ρ) M is a $l \times n$ matrix. Function ρ is a mapping of M matrix about rows and attributes. The scheme multiply matrix M, key α and a Vector bunch of blinding factor y: $(\alpha, y_2, y_3 \ldots y_n)^T$ together and get the vector λ: $(\lambda_1, \lambda_2, \lambda_3 \ldots \lambda_l)^T$, that is:

$$My = \lambda \tag{1}$$

When decrypting, first, for access structures (M, ρ), work out a set of $\omega : (\omega_1, \omega_2, \omega_3 \ldots \omega_l)^T$, and satisfy the relations: $\sum_{i \in I} \omega_i \cdot M_i = (1, 0, 0, \ldots 0)$.

Among them, M_i is i row of matrix M. that is:

$$a_{1,1}\omega_1 + a_{2,1}\omega_2 + \ldots a_{l,1}\omega_l = 1$$
$$a_{1,2}\omega_1 + a_{2,2}\omega_2 + \ldots a_{l,2}\omega_l = 0$$
$$\vdots$$
$$a_{1,n}\omega_1 + a_{2,n}\omega_2 + \ldots a_{l,n}\omega_l = 0$$

Simplification and merger, that is to solve a set of solutions ω : $(\omega_1, \omega_2, \omega_3 \cdots, \omega_l)^T$ satisfy: $M^T \cdot \omega = (1, 0, 0, \cdots 0)^T$.

To do transpose operation on both sides of (1): $(\alpha, y_2, y_3, \ldots y_n) \cdot M^T = (\lambda_1, \lambda_2, \ldots \lambda_l)^T$.

Multiply both sides by $\omega : (\omega_1, \omega_2, \omega_3 \cdots, \omega_l)^T$:

$$(\alpha, y_2, y_3, \ldots y_n) \cdot M^T \cdot \omega = (\lambda_1, \lambda_2, \ldots \lambda_l)^T \cdot \omega$$

$(\alpha, y_2, y_3, \ldots y_n) \cdot (1, 0, 0, \ldots 0)^T = (\lambda_1, \lambda_2, \ldots \lambda_l)^T \cdot \omega$. Thus, $\alpha = \sum \omega_i \lambda_i$

Consider the situation: when the rank of G_0 is not equal to the rank of corresponding augmented matrix, solving matrix $M^T \cdot \omega = (1, 0, 0, \ldots 0)^T$ may be no solution and at this time we was unable to get the $\omega : (\omega_1, \omega_2, \omega_3 \cdots, \omega_l)^T$ vector, so we cannot use $(\lambda_1, \lambda_2, \lambda_3 \ldots \lambda_l)^T$ to decrypt the value of α. The access structure of the scheme have some certain infeasibility: When the range of attribute n is larger, in the access structure the rows of attribute l may be less than the value of n.

4 Improved SCP-OOABE Scheme

4.1 Systems Model

Suppose cloud servers are semi-credible, which complete their work on the one hand, on the other hand be attacked from internal attackers. Attribute authority is regarded as completely credible, which never leak the users' information and accomplish key generation and distribution honestly. Suppose that the communication between Date Provider, Data User, and Cloud Server was security under the protection of some security protocols. Each party of the agreements adopt a pair of public and private key to encrypt communications.

The system model described in this article contains the following four participants: Data Provider, Data User, Cloud Server, Authorities, as shown in Fig. 2:

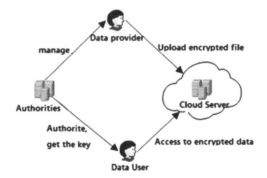

Fig. 2. Cloud storage system model

4.2 Security Model

Suppose SCP-OOABE scheme is: = (Setup, Offline.Keygen, Online.Keygen, Offline. Encryption, Online.Encryption, Descryption). We define the security of SCP-OOABE scheme by the following CPA game. C is the challenger, A is the adversary [6–8].

- *Setup:* C execute Setup algorithm and get the public parameter PK and then put them intact to A.
- *Phase 1*: In this step, A can only query the key SK. At first C initializes an empty table T, an empty set D and an integer count = 0. For each time, A query in the set of attributes S, if the table T exists a tuple (count, S, SK), $D = D \cup S$ count = count + 1, and then passed SK to A. Otherwise, executed the algorithm Offline. Keygen and passed SK to A and then put the record (count, S, SK) into table T.
- *Challenge*: A select an access control structure τ. For all the $S \in D, S \notin \tau$ C execute Offline.Encryption and Online.Encryption algorithm to get the ciphertext CT^* and selected a random value $b \in \{0, 1\}$ if b = 0, put CT^* to A, otherwise put a random value to A.
- *Phase 2*: A repeat *Phase 1*, but cannot be queried the attribute of S when $S \in \tau$.
- *Guess*: A output the value of b that he guessed: b' and wins the game, if $b' = b$.

We defined that the advantage A win the CPA game is: $Adv_{\Pi}^{CPA}(A) = |\Pr[b' = b] - \frac{1}{2}|$.

4.3 SCP-OOABE Scheme

Through the analysis, we know that in the CP-OOABE scheme, since the matrix operation may have no solution, it would lead to α cannot be decrypted. To solve this problem, we improved the CP-OOABE scheme. After that it called SCP-OOABE scheme. In our scheme we introduced access control tree which support AND/OR operation and greatly satisfied the fine-grained access control. The most important is that it meets the diverse need of the user in cloud storage.

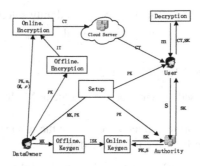

Fig. 3. Interaction among different algorithms

SCP-OOABE contains six algorithms: Setup, Offline.Keygen, Online.Keygen, Offline.Encryption, Online.Encryptiion and Decryption. Each of the algorithms interacted as shown in Fig. 3. Since the short online time in this scheme, once it got users' attribute set and access structure, data owners could quickly accomplish encryption work and upload encrypted files. At the same time the Authorities could complete key generation in a short time and deliver keys to the users. Compared with traditional attribute-based encryption it could satisfy the massive encryption services preferably in cloud storage. Specific steps:

1. Setup $(1^\lambda) \rightarrow$ PK, MK

The algorithm input a safety factor λ and randomly generate two multiplicative cyclic groups G_0 and G_1. Their order is prime number p and bilinear map is e: $G_0 \times G_0 \rightarrow G_1$. Randomly chose the generator of G_0 as g, h, μ, v, ω and an index $\alpha \in Z_p$ and a Hash function: $H : \{0,1\}^* \rightarrow Z_q^*$

The main pubic parameter of the system PK and the master key MK are: $PK = (G_0, H, p, g, h, \mu, v, \omega, e(g,g)^\alpha)$ $MK = (PK, \alpha)$.

2. Offline.Keygen (MK) $\rightarrow$ ISK

The algorithm input master key and generate an intermediate key module *ISK*. Randomly select $r \in Z_p$ then $K_0 := g^\alpha \omega^r$, $K_1 := g^r$, $K_v := v^{-r}$. The intermediate key module is: $ISK = (K_0, K_1, K_v)$.

3. Online.Keygen(PK, ISK, S) $\rightarrow$ SK

The algorithm input master key, intermediate key module and the users' attribute set $S = (A_1, A_2, A_3, \cdots A_k)$. Randomly select $r_1, r_2, \cdots r_k \in Z_p$, and calculate: $K_{i,2} = g^{r_i}, K_{i,3} = (\mu^{H(A_i)}h)^{r_i}K_v$. Then private key of the users becomes $SK = (S, K_0, K_1, \{K_{i,2}, K_{i,3}\}_{i \in [1,k]})$.

4. Offline.Encryption (PK) $\rightarrow$ IT

Input: pubic parameter PK. Randomly select $s \in Z_p$, $C_0 := g^s$ then $key_{main} := (e(g,g)^{\alpha s}, C_0)$.

For each attribute i in the attribute pooling randomly select $\lambda_i, x_i, t_i \in Z_p$ and calculate: $C_{i,1} := \omega^{\lambda_i} v^{t_i}$, $C_{i,2} := (\mu^{-x_i} h)^{t_i}$, $C_{i,3} := g^{t_i}$ then $IT_{att} := (\lambda_i, x_i, t_i, C_{i,1}, C_{i,2}, C_{i,3})$ $IT = (key_{main}, s, IT_{att})$.

5. Online.Encryption (PK, IT, τ, M) $\rightarrow$ CT

This mechanism is used to encrypt the plaintext M, Specifically described as follows:

Construct access control tree: given a tree τ, use the following method to choose a polynomial for each node from the top to the down:

(a) For the root node: randomly select a polynomial of which the degree is k_r-1 and k_r is the threshold value of the root which satisfies $q_{root}(0) = s$ and then distribute a unique index value for each child node r_c, x = index(r_c).

(b) With regard to the intermediate nodes N except the root node, randomly select a polynomial of which the degree is k_N-1, and k_N is the threshold value of N and satisfies the formula $q_N(0) = q_{parent(N)}(index(N))$, then distribute a unique index value for each child node N_c, x = index(N_c).

(c) With regard to the leaf node L, randomly select a polynomial of which the degree is 0 and satisfies $q_L(0) = q_{parent(L)}(index(L))$, then $C_{i,4} = q_L(0) - \lambda_i$, $C_{i,5} = t_i(x_i - H(val_i))$ and i is the index of the corresponding attribute in attribute pooling and val_i is the attribute value of i. Generate ciphertext CT:

$$(\tau, \bar{M} = Me(g,g)^{\alpha s}, C_r = H(\bar{M}\|e(g,g)^{\alpha s}), C_0, \{C_{i,1}, C_{i,2}, C_{i,3}, C_{i,4}, C_{i,5}, \}_{\forall i \in [1,k]})$$

6. Decryption (CT) $\rightarrow$ M

Constructing a recursive algorithm: DecryptionNode (CT,SK,N) which regarded the ciphertext CT, User's private key SK and the node of access control tree N as input.

(a) If N is the leaf node, let i = att(N), if i is contained in attribute set S, then:

$$res_N = DecryptionNode(CT, SK, N)$$
$$= e(C_{i,1}, K_1)e(C_{i,2}\mu^{C_{i,5}}, K_{i,2})e(C_{i,3}, K_{i,3})e(\omega^{C_{i,4}}, K_1)$$
$$= e(\omega^{\lambda_i} v^{t_i}, g^r)e((\mu^{x_i} h)^{-t_i} \mu^{t_i(x_i - H(val_i))}, g^{r_i})$$
$$e(g^{t_i}, (\mu^{H(A_i)} h)^{r_i} v^{-r})e(\omega^{q_N(0)-\lambda_i}, g^r) = e(\omega, g)^{rq_N(0)}$$

otherwise, DecryptionNode (CT,SK,N) = $\perp$.

(b) If N is an intermediate node then each child node of N call DecryptionNode

(CT, SK, N_c). All of the nodes of which the value of index satisfied $res_{N_c} \neq \perp$ construct a set I_c, if $|I_c| \geq K_N$, K_N is the threshold value of node N, then:

$$res_N = \text{DecryptionNode}\,(CT, SK, N)$$

$$= \prod_{\forall index(N_c)=z\in I_c} res_{N_c}^{\Delta z,I_c(0)} = e(\omega, g)^{rq_N(0)}$$

Otherwise, DecryptionNode(CT, SK, N) = ⊥.

(c) After calculating all the nodes recur to the root. $res_{root} = e(\omega, g)^{rs}$

(d) Calculate M: $M = \dfrac{Me(g,g)^{\alpha s}}{\frac{e(C_0,K_0)}{res_{root}}} = \dfrac{Me(g,g)^{\alpha s}}{\frac{e(g^s,g^{\alpha}g^{\omega r})}{e(\omega,g)^{rs}}}$

(e) Verify the signature. Calculating the expression $C'_r = H(M||e(g,g)^{\alpha s})$ if it is equal to the signature value of CT, supposing unequal, the ciphertext has been tampered. What should be done is to abandon it and request service towards authorities again.

5   Security Analysis

Theorem 1. if the scheme of Hohenberger et al. [7] is against chosen-plaintext attack secure, so do the scheme of SCP-OOABE.

Prove: if the adversary A which is polynomial-bounded could win the SCP-OOABE game with great advantages, then the simulator B which is polynomial-bounded could break through the CPA security scheme of Hohenberger et al. [7]

- *Initialization:* B gain the attribute set of A $S := (A_1, A_2, A_3, \cdots, A_k)$ And deliver it to HW (the scheme of Hohenberger et al. [7]), which is the challenger.
- *Setup:* B get the pubic parameters from HW:
  $PK := (G_0, H, p, g, h, \mu, v, \omega, e(g,g)^{\alpha})$ And deliver them to A intactly.
- *Phase 1*: Query the key which is corresponding to the attribute of A, but it has been invariable in the game.
- *Challenge*: B randomly select two isometric messages m_0, m_1 and deliver them to HW. HW randomly select $b \in \{0, 1\}$ and then deliver ciphertext CT^* to B:

$$((Mx, \rho), \bar{M} = Me(g,g)^{\alpha s}, C_r = H(\bar{M}||e(g,g)^{\alpha s}), C_0, \{C_{i,1}, C_{i,2}, C_{i,3}, C_{i,4}, C_{i,5},\}_{\forall i\in[1,k]})$$

And (Mx, ρ) is the LSSS matrix of HW scheme.

B construct an access control tree τ which contain a root node Root, l leaf nodes NL and l is the number of rows in matrix Mx. Randomly selected $s', \alpha', x_1, x_2, \ldots x_l \in Z_p$ which satisfied: $q_{Root}(0) = s', q_{N_i}(0) = x_i\alpha'$ New ciphertext CT_s^* are:

$$(\tau, \bar{M} = m_b e(g,g)^{\alpha s}e(g,g)^{\alpha's'}, C_r = H(\bar{M}||e(g,g)^{\alpha s}), C_0, \{C_{i,1}, C_{i,2}, C_{i,3}, C_{i,4}, C_{i,5}\}_{\forall i\in[1,k]})$$

If trying to testify whether the information has been encrypted correctly, one only need to use Lagrangian interpolation operation to all the leaf nodes, and decrypt $e(g,g)^{\alpha's'}$. Which is the message B guessed be encrypted $\eta_B \in \{0,1\}$ and Calculate $key_{guess} = \bar{M}/\eta_B$ and deliver tuple (key_{guess}, CT_s^*) to A.

- *Phase 2:* B repeatedly operate the inquiry of the first stage.
- *Guess:* A output $\eta_A \in \{0,1\}$. If $\eta_A = 0$ then A consider that key_{guess} is the key of CT^* and let B output η_B. If $\eta_A = 1$ then A consider that key_{guess} is a random key. The challenge perfectly end.

If the PPT adversary A can win the game with great advantages, B can break through the system of HW.

6 Performance Analysis

As shown in Fig. 1, comparing the calculated performance, E_0 represents an exponential operation of G_0, M_0 represents a multiplicative operation of G_0, M_1 represents a multiplicative operation of G_1, E_1 represents an exponential operation of G_1, P represents the complexity of attribute or access control structure, l represents the number of attributes (rows of the matrix/the leaf nodes of the tree), n represents columns of the matrix and x represents intermediate nodes of the tree. It is observed that the computational complexity of key generation and encryption of this paper is basically the same with the HW. However, in our scheme, the access structure is flexible with high practicability and never impose restrictions on generating the access control matrix (Table 1).

Table 1. Performance comparation

	Hohenberger's	Our's
Offline.keygen	$(3P + 4)E_0 + (P + 1)M_0$	$(3P + 4)E_0 + (P + 1)M_0$
Online.keygen	$P\,M_1$	$P\,M_1$
Offline.encrypt	$E_1 + (5P + 1)E_0 + 2P\,M_0$	$E_1 + (5P + 1)E_0 + 2P\,M_0$
Offline.encrypt	$l \times n$	$l + x$
Decrypt	$(E_1 + E_0 + M_0)P + E_0$	$(2\,E_0 + M_0)P$
Access control	Low	High
Practicable		

Simulation experiments about the scheme of Hohenberger et al. [7] and ours had been done. JPBC [18] (Java Pairing-Based Cryptography Library) is developed by Ben Lynn which is a Java encryption library based on paired. It could realize bilinear map and was a frequently-used library in the developing test of attribute encryption. In the experiment, the structure of the linear secret sharing matrix of Hohenberger et al. [7] is based on the scheme of Ruj et al. [19].

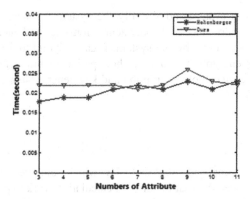

Fig. 4. Comparison of online encryption time

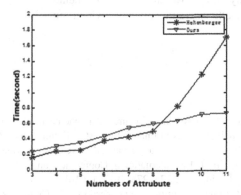

Fig. 5. Comparison of decryption time

As shown in Fig. 4, the encryption times are very close. Since our plan used recursive, the encryption time is slightly longer than the plan of Hohenberger et al. [7]. As shown in Fig. 5, when the number of attributes is less than 8, the decryption time is slightly longer than the plan of Hohenberger et al. [7]. But when the number of attributes is greater than 8, the decryption time is obviously less than the plan of Hohenberger et al. [7]. The reason is that looking for the submatrix which satisfied the access structure and solving the matrix equation will consume much time. It can be seen that in the process of decryption, when the level of recursion is low, access structure tree is superior to LSSS matrix with the increasing number of attributes.

Through the analysis, the online/offline attribute encryption scheme based on access structure tree support any attribute value: the scheme supports the attribute of variable length strings and AND/OR operation. In comparison to LSSS matrix, the access structure becomes more accurate and exact, and the decryption time is less than the plan of Hohenberger et al. [7] What's more, there is no problem when constructing the LSSS matrix the key cannot be formed. Besides, it is possible to reduce the number of public parameters in system and simplify the system management. The response time of our scheme is short, since in the key generation stage and encryption stage, a lot of pretreatment had been done by the system. It can respond to a variety of attribute structures and user's encryption in a very short period of time. It is suitable for the thought of the mass service about encryption and key generation and high concurrency in cloud storage [20].

7   Conclusion

In the research we proposed SCP-OOABE which inherited the advantage of CP-OOABE. It transferred the vast majority of calculating work to preprocessing stage, guaranteed the performance of cloud storage and relieved the bottlenecks of key generation and encryption. In addition, the problem that the access structure matrix may have no solution was avoid. What's more, it proved that the security of SCP-OOABE formally and be validated by experimental comparison. In a word, our scheme was security, efficient and universal, which suits for the demands of cloud storage.

References

1. Qiu, M., et al.: Proactive user-centric secure data scheme using attribute-based semantic access controls for mobile clouds in financial industry. Future Gener. Comput. Syst. (2016)
2. Gai, K., Qiu, M., Zhao, H., et al.: Privacy-aware adaptive data encryption strategy of big data in cloud computing. In: IEEE International Conference on Cyber Security and Cloud Computing, Beijing (China), pp. 273–278. IEEE (2016)
3. Sahai, A., Waters, B.: Fuzzy identity-based encryption. In: Cramer, R. (ed.) EUROCRYPT 2005. LNCS, vol. 3494, pp. 457–473. Springer, Heidelberg (2005). doi:10.1007/11426639_27
4. Goyal, V., Pandey, O., Sahai, A., et al.: Attribute-based encryption for fine-grained access control of encrypted data. In: Proceedings of the 13th ACM Conference on Computer and Communications Security, New York (USA), pp. 89–98. ACM (2006)
5. Bethencourt, J., Sahai, A., Waters, B.: Ciphertext-policy attribute-based encryption. In: Security and Privacy, Los Alamitos, pp. 321-334. IEEE (2007)
6. Rouselakis, Y., Waters, B.: Practical constructions and new proof methods for large universe attribute-based encryption. In: Proceedings of the 2013 ACM SIGSAC Conference on Computer & Communications Security, New York (USA), pp. 463–474. ACM (2013)
7. Hohenberger, S., Waters, B.: Online/offline attribute-based encryption. In: Krawczyk, H. (ed.) PKC 2014. LNCS, vol. 8383, pp. 293–310. Springer, Heidelberg (2014). doi:10.1007/978-3-642-54631-0_17
8. Chow, S.S.M., Liu, J.K., Zhou, J.: Identity-based online/offline key encapsulation and encryption. In: Proceedings of the 6th ACM Symposium on Information, Computer and Communications Security, New York (USA), pp. 52–60. ACM (2011)
9. Guo, F., Mu, Y., Chen, Z.: Identity-based online/offline encryption. In: Tsudik, G. (ed.) FC 2008. LNCS, vol. 5143, pp. 247–261. Springer, Heidelberg (2008). doi:10.1007/978-3-540-85230-8_22

10. Li, Y., Dai, W., Qiu, M., Ming, Z.: Privacy protection for preventing data over-collection in smart city. IEEE Trans. Comput. (2015)

11. Beimel, A.: Secure schemes for secret sharing and key distribution, Technion-Israel Institute of technology, Faculty of computer science, Haifa, Isral (1996)

12. Wang, Z., Chen, F., Xia, A.: Attribute-based online/offline encryption in smart grid. In: Computer Communication and Networks (ICCCN), pp. 1–5. IEEE, Berlin (2015)

13. Fan, C.I., Huang, V.S.M., Ruan, H.M.: Arbitrary-state attribute-based encryption with dynamic membership. IEEE Trans. Comput. **63**(8), 1951–1961 (2014)

14. Zhang, X., Chen, M., Liu, H., et al.: Practical identity-based threshold decryption scheme without random oracle (in Chinese). J. ShenZhen Univ. Sci. Eng. **37**(3), 340–346 (2010)

15. Lewko, A.: Tools for simulating features of composite order bilinear groups in the prime order setting. In: Pointcheval, D., Johansson, T. (eds.) EUROCRYPT 2012. LNCS, vol. 7237, pp. 318–335. Springer, Heidelberg (2012). doi:10.1007/978-3-642-29011-4_20

16. Lewko, A., Okamoto, T., Sahai, A., Takashima, K., Waters, B.: Fully secure functional encryption: attribute-based encryption and (hierarchical) inner product encryption. In: Gilbert, H. (ed.) EUROCRYPT 2010. LNCS, vol. 6110, pp. 62–91. Springer, Heidelberg (2010). doi:10.1007/978-3-642-13190-5_4

17. Zhang, X., Chen, M., Yang, L., et al.: Cryptanalysis of an identity-based multi-recipient signcryption scheme (in Chinese). J. ShenZhen Univ. Sci. Eng. **37**(4), 408–412 (2010)

18. Galindo, D.: Boneh-Franklin identity based encryption revisited. In: Caires, L., Italiano, Giuseppe, F., Monteiro, L., Palamidessi, C., Yung, M. (eds.) ICALP 2005. LNCS, vol. 3580, pp. 791–802. Springer, Heidelberg (2005). doi:10.1007/11523468_64

19. Ruj, S., Stojmenovic, M., Nayak, A.: Decentralized access control with anonymous authentication of data stored in clouds. IEEE Trans. Parallel Distrib. Syst. **25**(2), 384–394 (2014)

20. Qiu, M., Ming, Z., Li, J., Gai, K., Zong, Z.: Phase-change memory optimization for green cloud with genetic algorithm. IEEE Trans. Comput. **64**(12), 3528–3540 (2015)

An Information-Centric Architecture for Server Clustering Towards 3D Data-Intensive Applications

Longjiang Li$^{(\boxtimes)}$, Jianjun Yang, and Yuming Mao

School of Communication and Information Engineering,
University of Electronic Science and Technology of China, Chengdu, China
{longjiangli,jjyang,ymmao}@uestc.edu.cn

Abstract. Data-intensive applications, such as three-dimensional (3D) virtual reality, typically involve a large amount of data interaction, so that the server becomes a system performance bottleneck, as users become larger. In order to overcome the shortcomings of existing service models, this paper proposes an information-centric architecture (ICA) for server clustering towards data intensive applications, which ensures transparent and effective access, and supports various types of services with different dynamic requirements of storage and bandwidth by quantitatively partitioning service functions and using the modular realization. Finally, a data-intensive 3D virtual reality project was tested and the results show that the proposed scheme has good scalability and efficiency.

Keywords: Information centric architecture · Publishsubscribe mode · Cluster server framework · Data-intensive applications · 3D virtual reality

1 Introduction

With the rapid development of information technology such as the Internet, data-intensive applications [5,14], which require large volumes of data and devote most of their processing time to I/O and manipulation of data, are gaining more and more attention [10]. As typical instances of data-intensive applications, three-dimensional graphical applications, such as 3D online game [3,9], 3D web [7], stereoscopic display [11], have been widely used in military, aerospace, distance education, medical and entertainment areas, etc.

At present, most of the 3D graphical network applications are still using the client-server (C/S) (or the browser-server (B/S) mode) [1,2], which is easy for deployment and maintenance but is not scalable for scenarios which relies on rendering graphical applications on remote systems with specialized resources

This work was supported by the National Natural Science Foundation of China (61273235) and the National Key Technology Research and Development Program of China under Grant No. 2015BAH08F01.

M. Qiu (Ed.): SmartCom 2016, LNCS 10135, pp. 412–418, 2017.
DOI: 10.1007/978-3-319-52015-5_42

and fast access to huge data models [4]. Cluster server architecture, which tries to organize a cluster of workstations and PCs as an Internet server [8], can overcome the scalability issues of C/S model. However, most researches related to cluster server focuses on workload balancing over the similar tasks and have no good support for heterogeneous multimedia downloading and interaction, which are just essential for 3D graphical network applications [6,12,13] that may involve interactions with textures, animations, messages, movies, and audios.

In order to solve this problem, this paper proposes an information-centric architecture (ICA) for server clustering towards data intensive applications, by using an information-centric controller for managing all server's functions, so that it is transparent for end users to deploy or maintain these server's functions. By borrowing the publishsubscribe idea, the informationcentric controller can ensure transparent and effective access of users, and supports various types of services with different dynamic requirements of storage and bandwidth by quantitatively partitioning service functions and using the modular realization.

2   The Proposed Scheme

By observing the process of users accessing the server, we found that the accessing process of users to three-dimensional services usually experience three steps, i.e., to login, to localize 3D contents, and to download contents. Here, 3D contents may be textures, animations, messages, movies, or audios, which have various latency and bandwidth requirements. When users are large in scale, the process of downloading usually occupies almost all bandwidth, so that other users are not able to access the server, which damages the user's experience. In order to improve the performance of user access, we propose an information-centric architecture (ICA) for three-dimensional graphical services that separates user connections and other functions, so that the system can be scaled easily and transparently for end users.

As shown in Fig. 1, the architecture introduces an information centric controller (ICC) to manage all other servers, such as Authentication and Accounting Server (CAS), User Connection Servers (UCS), Content Servers (CS), Message Servers (MS), Application Servers (AS) and Relay Servers (RS). In the whole system, ICC is responsible for coordinating all other servers, so there should be only one ICC in the whole system, though a backup of ICC is possible. The amount of all other servers can be increased as needed.

2.1   Information Centric Architecture

The whole system allows only one ICC, whose main function is responsible for coordination between all other servers. Although from the structural point of view, the use of a single ICC for centralized management, does not seem reliable, but in fact, the main function of ICC is to do role assignment for other servers and redirection, so the actual amount of communication data is not too large. By borrowing the publish-subscribe idea, all other servers are required to register

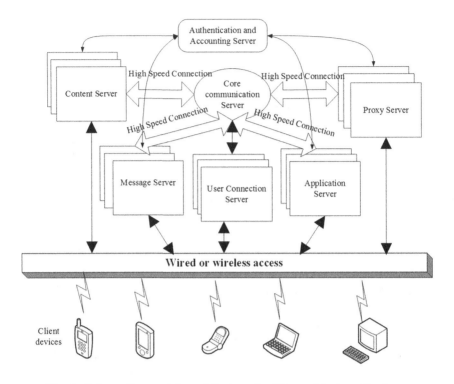

Fig. 1. Information centric architecture for 3D graphical applications

with ICC before providing services, to deregister itself from ICC when shutdown or maintenance is needed. ICC maintains a directory of all other servers, and commits a predetermined policy to redirect client requests to a particular server. For security purposes, a backup for ICC is necessary, but a further discuss is out of scope of this paper.

The primary function of UCS is to accept the client first connects and provides the client address the query other servers. In order to make the client always knows how to access the server, UCS using public addresses. When there are multiple types of user terminal access for example, a class of users use 3G connected server, and the other user connected using WIFI server, you can typically deploy two servers, each with a corresponding physical communication interface. The main function of CAS is to administrate user rights and support for VIP authentication. If a user visits a fee service, user access should be recorded for traffic billing, and CAS also supports for other server user authentication queries. The main function of CS is responsible for content management, all relevant data and information services, are stored on the CS, and allows the client to download or update the information content. Any client access to the CS, must obtain CAS certification, it has to decide whether to provide appropriate services. The main function of MS is to deal with various system messages, including: Scene synchronization message, content update message, a notification message authentication, the server automatically switches, fault notification. When users log in, such as the need to elevate privileges, such as VIP authentication message also needs to be managed by MS. The main function of AS is to provide application-related services, such as system configuration needs, announcement, deployment of new features. To support load balancing between servers, manage users to switch between different server, RS is responsible to redirect connection requests and resource location.

2.2 Gradual Deployment of Services

Functionally, all server functions can be installed on the same server equipment, but for best performance, each server can be installed on an independent device. After installation, each server will automatically start registration with ICC, and ICC maintains a ready list of servers ($rList$), and no need of manual configuration. As it can be seen, the proposed architecture can greatly simplify the configuration of the clustering servers, and can easily expands the system capacity by adding more servers according to the performance requirements.

The process of a client's login is as follows:

1. The user uses a URI to access a web service via one UCS by a GET command.
2. The UCS queries the ICC with the URI to get an AS and a CS address. At the same time, the UCS provides a Logon Page to the user.
3. The user inputs the logon information to the AAS via the UCS by a POST command, and AAS grants an user ID pass and returns it to the UCS.
4. The UCS uses the user ID pass to keep the session for the user and App contents that AS returns and 3D contents that CS returns will be pushed to the user via the user ID pass.

After the successful login, the user can keep interaction with CS and AS by using the user ID pass. If some servers fail, UCS cannot get App contents or 3D contents and an error message which contains the URI will be notified to ICC. ICC will redirect the service to new AS or CS. If some server needs upgrading, it can send an EXIT command to ICC. ICC will redirect the existing services to other AS or CS. Once the new server is ready, an JION request will be sent to the ICC, which will add these servers to the $rList$.

3   Experiments

In order to verify the feasibility and performance of the proposed framework, we test a three dimensional virtual reality prototype system based on the proposed architecture. Clients need to access CSs to display a 3D virtual reality scenario consisting of text, figures and models with textures. Particularly, model data and texture data have maximal storage and bandwidth occupancy (including figures and scenes, about 4 MB). In a small-scale testing phase, we installed several content servers for storing 3D content and all other services run on the same server.

To test the system scalability, we changed the number of CSs and kept other conditions remain unchanged. System log files records the process and the metrics of a variety of operations (a connection request begins, the download is complete, etc.), and corresponding timestamps.

We use three performance metrics: logon latency, positioning model delay (Shapebox latency) and downloading delay (Download latency). Logon latency means the time from the first time to initiate a connection request until a successful login. Because in the three-dimensional virtual reality applications, a Shapebox model is usually used to determine the position of the model, and thus before downloading a model, the application needs to know the Shapebox of the model, so that the client does not need to download all models but only those models that are located in the field of view. The length of time required to download the Shapebox information is called Shapebox latency. Download latency refers to the duration between the time to issue a download request and the time when the download is complete.

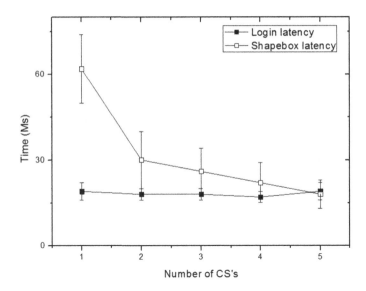

Fig. 2. Logon latency and shapebox latency

Figures 2 and 3 show the relationship between the various parameters and the CS index number between, showing an average of 10 client access, a confidence level of 90%. The results showed that with the increase of the number of CS, logon delays almost stay no change. This is because the logon process is independent of CS. Shapebox latency and Download latency are gradually reduced, because newly added CS shares the pressure on the existing CS. Therefore, both latency delay decreases as the number of CS increases. In addition, the test also showed that it is not necessary to restart the whole system while increasing or decreasing CSs, because ICC can redirect existing connections to other servers without

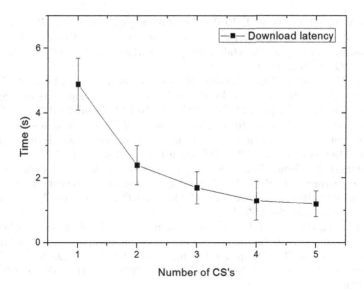

Fig. 3. Download latency of 3D models

interrupting the client download, which validates the flexibility and efficiency of the proposed architecture.

4   Conclusion

This paper presents an information-centric architecture (ICA) for server clustering towards data intensive applications, which can meet special performance requirements of three dimensional graphical online applications. By using publish-subscribe mode, an information-centric controller can manage multiple servers with difference bandwidth and storage requirements in a scalable way. The deployment of more other servers is transparent to the end users, which is convenient for service provider to upgrade the system without restarting the whole system.

The further work is to research on transparent segmentation of content resources and automatic load balancing mechanism, so that a wider range of data-intensive applications can be applied.

References

1. Calonego, N., da Silva, P.L., de Freitas, W., Matthiesen, R.C.: Distributed system for collaborative authorship: integrating "client/server" and mobile agents. In: Proceedings of the XVI Symposium Virtual and Augmented Reality (SVR), pp. 333–336 (2014). doi:10.1109/SVR.2014.56

2. Caro, A.O., Sarmiento, W.J.: Evaluation of 3D applications on mobile gaming consoles using client-server architecture. In: Proceedings of the Brazilian Symposium on Games and Digital Entertainment (SBGAMES), pp. 138–145 (2010). doi:10. 1109/SBGAMES.2010.20

3. Carter, C., Rhalibi, A.E., Merabti, M.: Development and deployment of cross-platform 3D web-based games. In: Proceedings of the Developments in E-Systems Engineering (DESE), pp. 149–154 (2010). doi:10.1109/DeSE.2010.32

4. Gorgan, D., Capatana, O.: Remote visualization of 3D graphical models. In: Proceedings of the 36th International Information Communication Technology Electronics Microelectronics (MIPRO) Convention, pp. 263–268 (2013)

5. Kijsipongse, E., U-ruekolan, S.: Scaling HPC clusters with volunteer computing for data intensive applications. In: Proceedings of the 10th International Computer Science and Software Engineering (JCSSE) Joint Conference, pp. 138–142 (2013). doi:10.1109/JCSSE.2013.6567334

6. Kopinski, T., Handmann, U.: Touchless interaction for future mobile applications. In: Proceedings, 2016 International Conference on Computing, Networking and Communications (ICNC), pp. 1–6 (2016). doi:10.1109/ICCNC.2016.7440589

7. Leavitt, N.: Browsing the 3D web. Computer **39**(9), 18–21 (2006). doi:10.1109/ MC.2006.296

8. Liu, J., Xu, L., Gu, B., Zhang, J.: A scalable, high performance internet cluster server. In: Proceedings of the Fourth International High Performance Computing in the Asia-Pacific Region Conference/Exhibition, vol. 2, pp. 941–944 (2000). doi:10. 1109/HPC.2000.843575

9. Liu, S.F., Lin, C.I., Su, C.Y., Tseng, S.P.: A 3D on-line golf game with motion sensing technology. In: Proceeding of the International Conference on Orange Technologies (ICOT), pp. 149–152 (2015). doi:10.1109/ICOT.2015.7498498

10. Moazzami, M.M., Phillips, D.E., Tan, R., Xing, G.: ORbit: a platform for smartphone-based data-intensive sensing applications. IEEE Trans. Mob. Comput. **PP**(99), 1 (2016). doi:10.1109/TMC.2016.2573825

11. Plewinski, P., Makowski, D., Mielczarek, A., Napieralski, A., Sztoch, P.: Remote control of 3D camera rig with embedded system. In: Proceedings of 22nd International Mixed Design of Integrated Circuits Systems (MIXDES) Conference, pp. 144–147 (2015). doi:10.1109/MIXDES.2015.7208499

12. Sato, Y., Saito, M., Koike, H.: Real-time input of 3D pose and gestures of a user's hand and its applications for hci. In: Proceedings of the IEEE Virtual Reality, pp. 79–86 (2001). doi:10.1109/VR.2001.913773

13. Vormoor, O.: Quick and easy interactive molecular dynamics using java3d. Comput. Sci. Eng. **3**(5), 98–104 (2001). doi:10.1109/5992.947113

14. Yu, B., Pan, J.: Location-aware associated data placement for geo-distributed data-intensive applications. In: Proceedings of the IEEE Conference on Computer Communications (INFOCOM), pp. 603–611 (2015). doi:10.1109/INFOCOM.2015. 7218428

Efficient Computation for the Longest Common Subsequence with Substring Inclusion and Subsequence Exclusion Constraints

Xiaodong Wang[2] and Daxin Zhu[1]([envelope])

[1] Quanzhou Normal University, Quanzhou 362000, China
dex@qztc.edu.cn
[2] Fujian University of Technology, Fuzhou 350108, China

Abstract. A generalized longest common subsequence problem is studied in this paper. In the generalized longest common subsequence problem, a constraining sequence of length s must be included as a substring and the other constraining sequence of length t must be excluded as a subsequence of two main sequences and the length of the result must be maximal. For the two input sequences X and Y of lengths n and m, and the given two constraining sequences of length s and t, we present an $O(nmst)$ time dynamic programming algorithm for solving the new generalized longest common subsequence problem. The time complexity can be reduced further to cubic time in a more detailed analysis. The correctness of the new algorithm is proved.

1 Introduction

The longest common subsequence (LCS) problem is a well-known measurement for computing the similarity of two strings. It can be broadly applied in diverse areas, such as file comparison, pattern matching and computational biology [3,4,8,9]. Given two sequences X and Y, the longest common subsequence (LCS) problem is to find a subsequence of X and Y whose length is the longest among all common subsequences of the two given sequences. For some biological applications some constraints must be applied to the LCS problem. These kinds of variants of the LCS problem are called the constrained LCS (CLCS) problem [2,5,6,10]. Recently, Chen and Chao [1] proposed the more generalized forms of the CLCS problem, the generalized constrained longest common subsequence (GC-LCS) problem. For the two input sequences X and Y of lengths n and m, respectively, and a constraint string P of length r, the GC-LCS problem is a set of four problems which are to find the LCS of X and Y including/excluding P as a subsequence/substring, respectively.

In this paper, we consider a more general constrained longest common subsequence problem called STR-IC-SEQ-EC-LCS, in which a constraining sequence of length s must be included as a substring and the other constraining sequence of length t must be excluded as a subsequence of two main sequences and the length of the result must be maximal. We will present the first efficient dynamic programming algorithm for solving this problem.

© Springer International Publishing AG 2017
M. Qiu (Ed.): SmartCom 2016, LNCS 10135, pp. 419–428, 2017.
DOI: 10.1007/978-3-319-52015-5_43

2   Characterization of the STR-IC-SEQ-EC-LCS Problem

A sequence is a string of characters over an alphabet $\sum$. A subsequence of a sequence X is obtained by deleting zero or more characters from X (not necessarily contiguous). A substring of a sequence X is a subsequence of successive characters within X.

For a given sequence $X = x_1 x_2 \cdots x_n$ of length n, the ith character of X is denoted as $x_i \in \sum$ for any $i = 1, \cdots, n$. A substring of X from position i to j can be denoted as $X[i : j] = x_i x_{i+1} \cdots x_j$. If $i \neq 1$ or $j \neq n$, then the substring $X[i : j] = x_i x_{i+1} \cdots x_j$ is called a proper substring of X. A substring $X[i : j] = x_i x_{i+1} \cdots x_j$ is called a prefix or a suffix of X if $i = 1$ or $j = n$, respectively.

An appearance of sequence $X = x_1 x_2 \cdots x_n$ in sequence $Y = y_1 y_2 \cdots y_m$, for any X and Y, starting at position j is a sequence of strictly increasing indexes $i_1, i_2, \cdots, i_n$ such that $i_1 = j$, and $X = y_{i_1}, y_{i_2}, \cdots, y_{i_n}$. A compact appearance of X in Y starting at position j is the appearance of the smallest last index i_n. A match for sequences X and Y is a pair (i, j) such that $x_i = y_j$. The total number of matches for X and Y is denoted by δ. It is obvious that $\delta \leq nm$.

For the two input sequences $X = x_1 x_2 \cdots x_n$ and $Y = y_1 y_2 \cdots y_m$ of lengths n and m, respectively, and two constrained sequences $P = p_1 p_2 \cdots p_s$ and $Q = q_1 q_2 \cdots q_t$ of lengths s and t, the SEQ-IC-STR-IC-LCS problem is to find a constrained LCS of X and Y including P as a substring and excluding Q as a subsequence.

Definition 1. Let $Z(i, j, k, r)$ denote the set of all LCSs of $X[1 : i]$ and $Y[1 : j]$ such that for each $z \in Z(i, j, k, r)$, z includes $P[1 : k]$ as a substring, and excludes $Q[1 : r]$ as a subsequence, where $1 \leq i \leq n, 1 \leq j \leq m, 0 \leq k \leq s$, and $0 \leq r \leq t$. The length of an LCS in $Z(i, j, k, r)$ is denoted as $g(i, j, k, r)$.

Definition 2. Let $W(i, j, k, r)$ denote the set of all LCSs of $X[1 : i]$ and $Y[1 : j]$ such that for each $w \in W(i, j, k, r)$, w excludes $Q[1 : r]$ as a subsequence, and includes $P[1 : k]$ as a suffix, where $1 \leq i \leq n, 1 \leq j \leq m, 0 \leq k \leq s$, and $0 \leq r \leq t$. The length of an LCS in $W(i, j, k, r)$ is denoted as $f(i, j, k, r)$.

Definition 3. Let $U(i, j, k)$ denote the set of all LCSs of $X[i : n]$ and $Y[j : m]$ such that for each $u \in U(i, j, k)$, u excludes $Q[k : t]$ as a subsequence, where $1 \leq i \leq n, 1 \leq j \leq m, 0 \leq k \leq t$. The length of an LCS in $U(i, j, k)$ is denoted as $h(i, j, k)$.

Definition 4. Let $V(i, j, k)$ denote the set of all LCSs of $X[1 : i]$ and $Y[1 : j]$ such that for each $v \in V(i, j, k)$, v excludes $Q[1 : k]$ as a subsequence, where $1 \leq i \leq n, 1 \leq j \leq m, 0 \leq k \leq t$. The length of an LCS in $V(i, j, k)$ is denoted as $v(i, j, k)$.

The following theorem characterizes the structure of an optimal solution based on optimal solutions to subproblems, for computing the LCSs in $W(i, j, k, r)$, for any $1 \leq i \leq n, 1 \leq j \leq m, 0 \leq k \leq s$, and $0 \leq r \leq t$.

Theorem 1. *If $Z[1 : l] = z_1, z_2, \cdots, z_l \in W(i, j, k, r)$, then the following conditions hold:*

1. *If $i, j, k > 0, r = 1, x_i = y_j = p_k = q_r$, then $z_l \neq x_i$ and $Z[1 : l] \in W(i - 1, j - 1, k, r)$.*
2. *If $i, j, k > 0, r > 1, x_i = y_j = p_k = q_r$, then $z_l \neq x_i$ implies $Z[1 : l] \in W(i - 1, j - 1, k, r)$; $z_l = x_i$ implies $Z[1 : l - 1] \in W(i - 1, j - 1, k - 1, r - 1)$.*
3. *If $i, j, k > 0, x_i = y_j = p_k$ and $r > 0, x_i \neq q_r$ or $r = 0$, then $z_l = x_i = y_j = p_k$ and $Z[1 : l - 1] \in W(i - 1, j - 1, k - 1, r)$.*
4. *If $i, j, k > 0, x_i = y_j$ and $x_i \neq p_k$, then $z_l \neq x_i$ and $Z[1 : l] \in W(i - 1, j - 1, k, r)$.*
5. *If $i, j > 0, k = 0, r = 1, x_i = y_j = q_r$, then $z_l \neq x_i$ and $Z[1 : l] \in W(i - 1, j - 1, k, r)$.*
6. *If $i, j > 0, k = 0, r > 1, x_i = y_j = q_r$, then $z_l \neq x_i$ implies $Z[1 : l] \in W(i - 1, j - 1, k, r)$; $z_l = x_i$ implies $Z[1 : l - 1] \in W(i - 1, j - 1, k, r - 1)$.*
7. *If $i, j > 0, k = 0, x_i = y_j$ and $r > 0, x_i \neq q_r$ or $r = 0$, then $z_l = x_i$ and $Z[1 : l - 1] \in W(i - 1, j - 1, k, r)$.*
8. *If $i, j > 0, x_i \neq y_j$, then $z_l \neq x_i$ implies $Z[1 : l] \in W(i - 1, j, k, r)$.*
9. *If $i, j > 0, x_i \neq y_j$, then $z_l \neq y_j$ implies $Z[1 : l] \in W(i, j - 1, k, r)$.*

3 A Simple Dynamic Programming Algorithm

Our new algorithm for solving the STR-IC-SEQ-EC-LCS problem consists of three main stages. The core idea of the new algorithm can be described by the following Theorem 2.

Theorem 2. *Let $Z[1 : l] = z_1, z_2, \cdots, z_l$ be a solution of the STR-IC-SEQ-EC-LCS problem, i.e. $Z[1 : l] \in Z(n, m, s, t)$, then its length $l = g(n, m, s, t)$ can be computed by the following formula:*

$$g(n, m, s, t) = \max_{1 \leq i \leq n, 1 \leq j \leq m, 1 \leq r \leq t} \{f(i, j, s, r) + h(i + 1, j + 1, r)\} \quad (1)$$

where $f(i, j, s, r)$ is the length of an LCS in $W(i, j, s, r)$ defined by Definition 2, and $h(i, j, r)$ is the length of an LCS in $U(i, j, r)$ defined by Definition 3.

The boundary conditions of this recursive formula are $f(i, 0, 0, 0) = f(0, j, 0, 0) = 0$ and $f(i, 0, k, r) = f(0, j, k, r) = -\infty$ for any $0 \leq i \leq n, 0 \leq j \leq m, 1 \leq k \leq s$, and $1 \leq r \leq t$.

Based on this formula, our algorithm for computing $f(i, j, k, r)$ is a standard dynamic programming algorithm. By the recursive formula (1), the dynamic programming algorithm for computing $f(i, j, k, r)$ can be implemented as the following Algorithm 1.

It is obvious that the algorithm requires $O(nmst)$ time and space.

The second stage of our algorithm is to find LCSs in $U(i, j, k)$. The length of an LCS in $U(i, j, k)$ is denoted as $h(i, j, k)$. Chen et al. [1, 7] presented a dynamic programming algorithm with $O(nmt)$ time and space.

Algorithm 1. Suffix

Input: Strings $X = x_1 \cdots x_n$, $Y = y_1 \cdots y_m$ of lengths n and m, respectively, and two constrained sequences $P = p_1 p_2 \cdots p_s$ and $Q = q_1 q_2 \cdots q_t$ of lengths s and t
Output: $f(i, j, k, r)$, the length of an LCS of $X[1:i]$ and $Y[1:j]$ including $P[1:k]$ as a suffix, and excluding $Q[1:r]$ as a subsequence, for all $1 \leq i \leq n, 1 \leq j \leq m, 0 \leq k \leq s$, and $0 \leq r \leq t$.

 1: **for all** i, j, k, r , $0 \leq i \leq n, 0 \leq j \leq m, 0 \leq k \leq s$ and $0 \leq r \leq t$ **do**
 2:     $f(i, 0, k, r), f(0, j, k, r) \leftarrow -\infty, f(i, 0, 0, 0), f(0, j, 0, 0) \leftarrow 0$ {boundary condition}
 3: **end for**
 4: **for all** i, j, k, r , $1 \leq i \leq n, 1 \leq j \leq m, 0 \leq k \leq s$ and $0 \leq r \leq t$ **do**
 5:     **if** $x_i \neq y_j$ **then**
 6:         $f(i, j, k, r) \leftarrow \max\{f(i - 1, j, k, r), f(i, j - 1, k, r)\}$
 7:     **else if** $k > 0$ **and** $x_i = p_k$ **then**
 8:         **if** $r = 0$ **and** $x_i \neq q_r$ **then**
 9:             $f(i, j, k, r) \leftarrow 1 + f(i - 1, j - 1, k - 1, r)$
10:         **else if** $r = 1$ **and** $x_i = q_r$ **then**
11:             $f(i, j, k, r) \leftarrow f(i - 1, j - 1, k, r)$
12:         **else**
13:             $f(i, j, k, r) \leftarrow \max\{1 + f(i - 1, j - 1, k - 1, r - 1), f(i - 1, j - 1, k, r)\}$
14:         **end if**
15:     **else if** $k = 0$ **then**
16:         **if** $r = 0$ **or** $x_i \neq q_r$ **then**
17:             $f(i, j, k, r) \leftarrow 1 + f(i - 1, j - 1, k, r)$
18:         **else if** $r = 1$ **and** $x_i = q_r$ **then**
19:             $f(i, j, k, r) \leftarrow f(i - 1, j - 1, k, r)$
20:         **else**
21:             $f(i, j, k, r) \leftarrow \max\{1 + f(i - 1, j - 1, k, r - 1), f(i - 1, j - 1, k, r)\}$
22:         **end if**
23:     **else**
24:         $f(i, j, k, r) \leftarrow f(i - 1, j - 1, k, r)$
25:     **end if**
26: **end for**

By Theorem 2, the dynamic programming matrices $f(i, j, k, r)$ and $h(i, j, k)$ computed by the algorithms *Suffix* and *SEQ-EC-R* can now be combined to obtain the solutions of the STR-IC-SEQ-EC-LCS problem as follows. This is the final stage of our algorithm.

From the 'for' loops of the algorithm, it is readily seen that the algorithm requires $O(nmt)$ time. Therefore, the overall time of our algorithm for solving the STR-IC-SEQ-EC-LCS problem is $O(nmst)$.

4   Improvements of the Algorithm

S. Deorowicz [3] proposed the first quadratic-time algorithm for the STR-IC-LCS problem. A similar idea can be used to improve the time complexity of our

---

Algorithm 2. STR-IC-SEQ-EC-LCS

Input: Strings $X = x_1 \cdots x_n$, $Y = y_1 \cdots y_m$ of lengths n and m, respectively, and two constrained sequences $P = p_1 p_2 \cdots p_s$ and $Q = q_1 q_2 \cdots q_t$ of lengths s and t

Output: The constrained LCS of X and Y including P as a substring, and including Q as a subsequence.

1: Suffix {compute $f(i, j, k, r)$}
2: SEQ-EC-R {compute $h(i, j, k)$}
3: $i^*, j^*, k^* \leftarrow 0, tmp \leftarrow -\infty$
4: **for** $i = 1$ to n **do**
5:   **for** $j = 1$ to m **do**
6:     **for** $k = 1$ to t **do**
7:       $x \leftarrow f(i, j, s, k) + h(i + 1, j + 1, k)$
8:       **if** $tmp < x$ **then**
9:         $tmp \leftarrow x, i^* \leftarrow i, j^* \leftarrow j, k^* \leftarrow k$
10:       **end if**
11:     **end for**
12:   **end for**
13: **end for**
14: **if** $tmp > 0$ **then**
15:   $back(i^*, j^*, s, k^*)$
16:   $backr(i^* + 1, j^* + 1, k^*)$
17: **end if**
18: **return** $\max\{0, tmp\}, i^*, j^*, k^*$

---

dynamic programming algorithm for solving the STR-IC-SEQ-EC-LCS problem. The improved algorithm is built on dynamic programming with some preprocessing. To show its correctness it is necessary to prove some more structural properties of the problem.

Let $Z[1 : l] = z_1, z_2, \cdots, z_l \in Z(n, m, s, t)$, be a constrained LCS of X and Y including P as a substring and excluding Q as a subsequence. Let also $I = (i_1, j_1), (i_2, j_2), \cdots, (i_l, j_l)$ be a sequence of indices of X and Y such that $Z[1 : l] = x_{i_1}, x_{i_2}, \cdots, x_{i_l}$ and $Z[1 : l] = y_{j_1}, y_{j_2}, \cdots, y_{j_l}$. From the problem statement, there must exist an index $d \in [1, l - t + 1]$ such that $P = x_{i_d}, x_{i_{d+1}}, \cdots, x_{i_{d+s-1}}$ and $P = y_{j_d}, y_{j_{d+1}}, \cdots, y_{j_{d+s-1}}$.

Theorem 3. *Let $i'_d = i_d$ and for all $e \in [1, s - 1]$, i'_{d+e} be the smallest possible, but larger than i'_{d+e-1}, index of X such that $x_{i_{d+e}} = x_{i'_{d+e}}$. The sequence of indices defines the same constrained LCS as $Z[1 : l]$.*

Proof.

From the definition of indices i'_{d+e}, it is obvious that they form an increasing sequence, since $i'_d = i_d$, and $i'_{d+s-1} \leq i_{d+s-1}$. The sequence $i'_d, \cdots, i'_{d+s-1}$ is of course a compact appearance of P in X starting at i_d. Therefore, both components of I' pairs form increasing sequences and for any (i'_u, j_u), $x_{i'_u} = y_{j_u}$. Therefore, I' defines the same constrained LCS as $Z[1 : l]$.

The proof is completed. $\square$

The same property is also true for the jth components of the sequence I. Therefore, we can conclude that when finding a constrained LCS in $Z(i, j, k, r)$, instead of checking any common subsequences of X and Y it suffices to check only such common subsequences that contain compact appearances of P both in X and Y. The number of different compact appearances of Q in X and Y will be denoted by δ_x and δ_y, respectively. It is obvious that $\delta_x \delta_y \leq \delta$, since a pair (i, j) defines a compact appearance of Q in X starting at ith position and compact appearance of Q in Y starting at jth position only for some matches.

Base on Theorem 2, we can reduce the time complexity of our dynamic programming from $O(nmst)$ to $O(nmt)$. The improved algorithm consists of also three main stages.

Definition 5. For each occurrence i of the first character p_1 of $P[1 : s]$ in $X[1 : n]$, lx_i is defined as the index of the last character p_s of a compact appearance of P in X. If $x_i \neq p_1$ or there is no compact appearance of P after i, then $lx_i = 0$. Similarly, for each occurrence j of the first character p_1 of $P[1 : s]$ in $Y[1 : m]$, ly_j is defined as the index of the last character p_s of a compact appearance of P in Y.

In the first stage both sequences X and Y are preprocessed to determine two corresponding arrays lx and ly.

In the second stage two DP matrices of SEQ-EC-LCS problem are computed: $h(i, j, k)$, the reverse one defined by Definition 3, and $v(i, j, k)$, the forward one defined by Definition 4. Both of the DP matrices can be computed by the SEQ-EC-LCS algorithm of Chen et al. [1].

In the last stage, two preprocessed arrays lx and ly are used to determine the final results. To this end for each match (i, j) for X and Y the ends (lx_i, ly_i) of compact appearances of P in X starting at position i and in Y starting at position j are read. The length of an STR-IC-SEQ-EC-LCS, $g(n, m, s, t)$ defined by Definition 1, containing these appearances of P is determined as a sum of three parts. For some indices i, j, k, r, $v(i - 1, j - 1, k)$, the constrained LCS length of prefixes of X and Y ending at positions $i - 1$ and $j - 1$, excluding $Q[1 : k]$ as a subsequence, $h(lx_i + 1, ly_j + 1, r)$ the constrained LCS length of suffixes of X and Y starting at positions $lx_i + 1$ and $ly_j + 1$, excluding $Q[r : t]$ as a subsequence, and the constraint length s. The integers k and r have some relations.

Definition 6. For each integer $k, 1 \leq k \leq t$, the index $\alpha(k)$ is defined as:

$$\alpha(k) = \max_{0 \leq r \leq s-k+1} \{r | P \ includes \ Q[k : k+r-1] \ as \ a \ subsequence\} \quad (2)$$

Since the constrained LCS A of prefixes of X and Y ending at positions $i - 1$ and $j - 1$, excludes $Q[1 : k]$ as a subsequence, the concatenation of A and P will exclude $Q[1 : r]$ as a subsequence, where $r = k + \alpha(k)$. The constrained LCS B of suffixes of X and Y starting at positions $lx_i + 1$ and $ly_j + 1$, excludes $Q[r : t]$ as a subsequence. Therefore, the concatenation of A, P and B excludes Q as a subsequence.

Algorithm 3. STR-IC-SEQ-EC-LCS

Input: Strings $X = x_1 \cdots x_n$, $Y = y_1 \cdots y_m$ of lengths n and m, respectively, and two constrained sequences $P = p_1 p_2 \cdots p_s$ and $Q = q_1 q_2 \cdots q_t$ of lengths s and t

Output: The length of an LCS of X and Y including P as a substring, and excluding Q as a subsequence.

```
 1: SEQ-EC {compute v(i, j, k)}
 2: SEQ-EC-R {compute h(i, j, k)}
 3: Prep {compute lx, ly}
 4: i*, j*, k*, r* ← 0, tmp ← 0
 5: for i = 1 to n do
 6:     for j = 1 to m do
 7:         if lxᵢ > 0 and lyⱼ > 0 then
 8:             for k = 1 to t do
 9:                 r ← k + α(k)
10:                 c ← v(i − 1, j − 1, k) + h(lxᵢ + 1, lyⱼ + 1, r) + s
11:                 if r > t then
12:                     tmp ← ∞
13:                 end if
14:                 if tmp < c then
15:                     tmp ← c, i* ← i, j* ← j, k* ← k, r* ← r
16:                 end if
17:             end for
18:         end if
19:     end for
20: end for
21: if tmp > 0 then
22:     backf(i* − 1, j* − 1, k*)
23:     print P
24:     backr(lx_{i*} + 1, ly_{j*} + 1, r*)
25: end if
26: return max{0, tmp}, i*, j*, k*, r*
```

Theorem 4. *The algorithm STR-IC-SEQ-EC-LCS correctly computes a constrained LCS in $Z(n, m, s, t)$. The algorithm requires $O(nmt)$ time and to $O(nmt)$ space in the worst case.*

Proof.

Let $Z[1 : l] = z_1, z_2, \cdots, z_l$ be a solution of the STR-IC-SEQ-EC-LCS problem, i.e. $Z[1 : l] \in Z(n, m, s, t)$, and its length be denoted as $l = g(n, m, s, t)$. To prove the theorem, we have to prove in fact that

$$g(n, m, s, t) = s + \max_{1 \leq i \leq n, 1 \leq j \leq m, 0 \leq k \leq t} \{v(i − 1, j − 1, k) + h(lx_i + 1, ly_j + 1, k + \alpha(k))\} \quad (3)$$

where $h(i, j, k)$ is the length of an LCS in $U(i, j, k)$ defined by Definition 3, and $v(i, j, k)$ is the length of an LCS in $V(i, j, k)$ defined by Definition 4.

Since $Z[1 : l] \in Z(n, m, s, t)$, $Z[1 : l]$ must be an LCS of X and Y including P as a substring, and excludes Q as a subsequence. Let the first appearance of the string P in $Z[1:l]$ starts from position $l' - s + 1$ to l' for some positive integer $s \leq l' \leq l$, i.e. $Z[l' - s + 1 : l'] = P$.

Let

$$r^* = \max_{1 \leq r \leq t} \{r | Q[1:r] \text{ is a subsequence of } Z[1:l'-s]\}$$

Since $Z[1:l'-s]$ excludes Q as a subsequence, we have $r^* < t$, and thus $Z[1:l'-s]$ excludes $Q[1:r^*+1]$ as a subsequence.

Let

$$(i^*, j^*) = \min_{1 \leq i \leq n, 1 \leq j \leq m} \{(i,j) | Z[1:l'-s+1] \ X[1:i] \ \text{and } Y[1:j]\}$$

Then, $x_{i^*} = y_{j^*} = p_1 = z_{l'-s+1}$, and $x_{lx_{i^*}} = y_{ly_{j^*}} = p_s = z_{l'}$.

Therefore, $Z[1:l'-s]$ is a common subsequence of $X[1:i^*-1]$ and $Y[1:j^*-1]$ excluding $Q[1:r^*+1]$ as a subsequence; $Z[l'+1:l]$ is a common subsequence of $X[lx_{i^*}+1:n]$ and $Y[ly_{j^*}+1:m]$.

It follows from Definition 4 that

$$l' - s \leq v(i^* - 1, j^* - 1, r^* + 1) \tag{4}$$

Since $Q[1:r^*]$ is the longest prefix of Q in $Z[1:l'-s]$, and

$$\alpha(r^* + 1) = \max_{0 \leq r \leq s-r^*+2} \{r | P \text{ includes } Q[r^*+1:r^*+r] \text{ as a subsequence}\}$$

we have, $Z[1:l']$ includes $Q[1:r^*+\alpha(r^*+1)]$ as a subsequence. It follows from $Z[1:l]$ excludes Q as a subsequence that $Z[l'+1:l]$ excludes $Q[r^*+1+\alpha(r^*+1):t]$ as a subsequence. Therefore, we have $Z[l'+1:l]$ is a common subsequence of $X[lx_{i^*}+1:n]$ and $Y[ly_{j^*}+1:m]$ excluding $Q[r^*+1+\alpha(r^*+1):t]$ as a subsequence. It follows from Definition 3 that

$$l - l' \leq h(lx_{i^*} + 1, ly_{j^*} + 1, r^* + 1 + \alpha(r^* + 1)) \tag{5}$$

Combining formulas (4) and (5) we have,

$$l - s \leq v(i^* - 1, j^* - 1, r^* + 1) + h(lx_{i^*} + 1, ly_{j^*} + 1, r^* + 1 + \alpha(r^* + 1))$$

Therefore,

$$l \leq s + \max_{1 \leq i \leq n, 1 \leq j \leq m, 0 \leq k \leq t} \{v(i-1, j-1, k) + h(lx_i + 1, ly_j + 1, k + \alpha(k))\} \tag{6}$$

On the other hand, for any $a \in V(i,j,k)$ and $b \in U(lx_i+1, ly_j+1, k+\alpha(k))$, $1 \leq i \leq n, 1 \leq j \leq m, 1 \leq k \leq t$, let $c = a \oplus P \oplus b$. If $lx_i > 0$ and $ly_j > 0$, then c must be a common subsequence of $X[1:n]$ and $Y[1:m]$ including P as a substring. Furthermore, we can prove c excludes Q as a subsequence.

In fact, since a excludes $Q[1:k]$ as a subsequence, the length of the longest prefix of Q in a is at most $k - 1$, and thus the length of the longest prefix of Q in $a \oplus P$ is at most $k - 1 + \alpha(k)$. Since b is a common subsequence of

$X[lx_i + 1 : n]$ and $Y[ly_j + 1 : m]$ excluding $Q[k + \alpha(k) : t]$ as a subsequence, we have, $c = a \bigoplus P \bigoplus b$ is a common subsequence of $X[1 : n]$ and $Y[1 : m]$ including P as a substring and excluding Q as a subsequence, and thus $|a \bigoplus P \bigoplus b| \leq l$. Therefore,

$$s + \max_{1 \leq i \leq n, 1 \leq j \leq m, 0 \leq k \leq t} \{v(i - 1, j - 1, k) + h(lx_i + 1, ly_j + 1, k + \alpha(k))\} \leq l \quad (7)$$

Combining formulas (6) and (7) we have,

$$l = \max_{1 \leq i \leq n, 1 \leq j \leq m, 0 \leq k \leq t} \{v(i - 1, j - 1, k) + h(lx_i + 1, ly_j + 1, k + \alpha(k))\}$$

The time and space complexities of the algorithm are dominated by the computation of the two dynamic programming matrices $v(i, j, k)$ and $h(i, j, k)$. It is obvious that they are all $O(nmt)$ in the worst case.

The proof is completed. □

5 Conclusions

We have suggested a new dynamic programming solution for the new generalized constrained longest common subsequence with substring inclusion and subsequence exclusion constraints. The first dynamic programming algorithm requires $O(nmst)$ in the worst case, where n, m, s, t are the lengths of the four input sequences respectively. The time complexity can be reduced further to cubic time in a more thorough analysis. Many other generalized constrained longest common subsequence problems have analogous structures. It is not clear that whether the similar technique of this paper can be applied to these problems to achieve efficient algorithms. We will explore these problems further.

Acknowledgement. This work was supported by Intelligent Computing and Information Processing of Fujian University Laboratory and Data-Intensive Computing of Fujian Provincial Key Laboratory.

References

1. Chen, Y.C., Chao, K.M.: On the generalized constrained longest common subsequence problems. J. Comb. Optim. **21**(3), 383–392 (2011)
2. Crochemore, M., Hancart, C., Lecroq, T.: Algorithms on Strings. Cambridge University Press, Cambridge (2007)
3. Deorowicz, S.: Quadratic-time algorithm for a string constrained LCS problem. Inform. Process. Lett. **112**(11), 423–426 (2012)
4. Deorowicz, S., Obstoj, J.: Constrained longest common subsequence computing algorithms in practice. Comput. Inform. **29**(3), 427–445 (2010)
5. Gotthilf, Z., Hermelin, D., Lewenstein, M.: Constrained LCS: hardness and approximation. In: Ferragina, P., Landau, G.M. (eds.) CPM 2008. LNCS, vol. 5029, pp. 255–262. Springer, Heidelberg (2008). doi:10.1007/978-3-540-69068-9_24

6. Gotthilf, Z., Hermelin, D., Landau, G.M., Lewenstein, M.: Restricted LCS. In: Chavez, E., Lonardi, S. (eds.) SPIRE 2010. LNCS, vol. 6393, pp. 250–257. Springer, Heidelberg (2010). doi:10.1007/978-3-642-16321-0_26

7. Gusfield, D.: Algorithms on Strings, Trees, and Sequences: Computer Science and Computational Biology. Cambridge University Press, Cambridge (1997)

8. Peng, Y.H., Yang, C.B., Huang, K.S., Tseng, K.T.: An algorithm and applications to sequence alignment with weighted constraints. Int. J. Found. Comput. Sci. **21**(1), 51–59 (2010)

9. Tang, C.Y., Lu, C.L.: Constrained multiple sequence alignment tool development and its application to RNase family alignment. J. Bioinform. Comput. Biol. **1**, 267–287 (2003)

10. Tseng, C.T., Yang, C.B., Ann, H.Y.: Efficient algorithms for the longest common subsequence problem with sequential substring constraints. J. Complex. **29**, 44–52 (2013)

An Optimal Algorithm for a Computer Game in Linear Time

Daxin Zhu[1] and Xiaodong Wang[2(✉)]

[1] Quanzhou Normal University, Quanzhou 362000, China
[2] Fujian University of Technology, Fuzhou 350108, China
wangxd135@139.com

Abstract. A single player game consisting of n black checkers and m white checkers, called shifting the checkers, is studied. The minimum number of steps needed to play the game for general n and m is proved to be $nm + n + m$. An optimal algorithm to generate an optimal move sequence of the game consisting of n black checkers and m white checkers is presented, and finally, an explicit solution for the general game is given.

1 Introduction

Combinatorial games often lead to interesting, clean problems in algorithms and complexity theory. Many classic games are known to be computationally intractable. Solving a puzzle is often a challenge task like solving a research problem. You must have a right cleverness to see the problem from a right angle, and then apply that idea carefully until a solution is found.

In this paper we study a single player game called shifting the checkers. The game is similar to the Moving Coins puzzle [2,3,7,9], which is played by rearranging one configuration of unit disks in the plane into another configuration by a sequence of moves, each repositioning a coin in an empty position that touches at least two other coins. In our shifting checkers game, there are n black checkers and m white checkers put on a table from left to right in a row. The $n + m + 1$ positions of the row are numbered $1, \cdots, n + m + 1$. We are interested in algorithms which, given integers n and m, generate the corresponding move sequences to reach the final state of the game with the smallest number of steps. In this paper we present an optimal algorithm to generate all of the optimal move sequences of the game consisting of n black checkers and m white checkers.

2 A Linear Time Construction Algorithm

The Decrease-and-Conquer strategy [1,4–6,8] for algorithm design is exploited to design a linear time algorithm.

Without loss of generality, we assume $n \geq m$ in the following discussion. Since there are only 4 possible moves $slide(w, l)$, $slide(b, r)$, $jump(b, w, l)$, and $jump(b, w, r)$, we can simplify our notation for these 4 moves to $slide(l)$, $slide(r)$, $jump(l)$, and $jump(r)$ in the following discussion.

© Springer International Publishing AG 2017
M. Qiu (Ed.): SmartCom 2016, LNCS 10135, pp. 429–438, 2017.
DOI: 10.1007/978-3-319-52015-5_44

2.1   A Special Case of the Problem

We first focus on the special case of $n = m$. If we denote a black checker by b, a white checker by w, and the vacant position by O, then any status of the checker board can be specified by a sequence consisting of characters b, w and O. The special case of our problem is then equivalent to transforming the initial sequence $\overbrace{b\cdots b}^{m}O\overbrace{w\cdots w}^{m}$ to the sequence $\overbrace{w\cdots w}^{m}O\overbrace{b\cdots b}^{m}$ in the minimum number of steps.

We have noticed that a key status of the checker board can be reached from the initial status with minimum number of steps.

Lemma 1. *The initial status of the checker board* $\overbrace{b\cdots b}^{m}O\overbrace{w\cdots w}^{m}$ *can be transformed to one of the status of the checker board* $O\overbrace{bw\cdots bw}^{2m}$ *or* $\overbrace{bw\cdots bw}^{2m}O$ *in* $\frac{m(m+1)}{2}$ *steps.*

Lemma 2. *The key status of the checkerboard* $O\overbrace{wb\cdots wb}^{2m}$ *or* $\overbrace{wb\cdots wb}^{2m}O$ *can be transformed to the final status* $\overbrace{w\cdots w}^{m}O\overbrace{b\cdots b}^{m}$ *in* $\frac{m(m+1)}{2}$ *steps.*

The 3 stages of the algorithms can now be combined into a new algorithm to solve our problem for the special case of $n = m$. The algorithm requires

$$m(m + 1)/2 + m + m(m + 1)/2 = m^2 + 2m$$

steps. It has been known that $m^2 + 2m$ is a lower bound to solve the game consisting of m black checkers and m white checkers. Therefore, our algorithm is optimal to solve the game for the special case of $n = m$. From this point, we can also claim that the algorithms move1 and move4 presented in the proofs of Lemmas 2 and 3 are also optimal. Otherwise, there must be an algorithm to solve the problem in less than $m^2 + 2m$ steps and this is impossible.

2.2   The Algorithm for the General Case of the Problem

We have discussed the special case of $n = m$. In this subsection, we will discuss the general cases $n > m$ of the problem. In these general cases, $n - m > 0$.

We can first use the algorithm move1 to transform the checkerboard to the status $\overbrace{b\cdots b}^{n-m}O\overbrace{bw\cdots bw}^{2m}$ or $\overbrace{b\cdots b}^{n-m}\overbrace{bw\cdots bw}^{2m}O$ in $\frac{m(m+1)}{2}$ steps. Then m jumps are applied to transform the checkerboard to the status $\overbrace{b\cdots b}^{n-m}O\overbrace{wb\cdots wb}^{2m}$ or $\overbrace{b\cdots b}^{n-m}\overbrace{wb\cdots wb}^{2m}O$.

At this point, we have to try to move the leftmost $n - m$ black checkers to the rightmost $n - m$ positions. It is not difficult to do this by a simple algorithm similar to the algorithm move1.

Lemma 3. *The key status of the checkerboard* $\overbrace{b\cdots b}^{n-m}O\overbrace{wb\cdots wb}^{2m}$ *or* $\overbrace{b\cdots b}^{n-m}\overbrace{wb\cdots wb}^{2m}O$ *can be transformed to the status* $\overbrace{wb\cdots wb}^{2m}O\overbrace{b\cdots b}^{n-m}$ *or* $O\overbrace{wb\cdots wb}^{2m}\overbrace{b\cdots b}^{n-m}$ *in* $(n-m)(m+1)$ *steps.*

The 4 stages of the algorithms can now be combined into a new algorithm to solve our problem for the general cases of $n \geq m$. By Lemmas 2, 3 and 4 we know that the algorithm requires $m(m+1)/2 + m + (n-m)(m+1) + m(m+1)/2 = nm + n + m$ steps. It has been known from Theorem 1 that $nm + n + m$ is a lower bound to solve the game consisting of n black checkers and m white checkers. Therefore, our algorithm is optimal to solve the game for the general cases of $n \geq m$. We can also claim that the algorithm move3 is also optimal. Otherwise, there must be an algorithm to solve the problem in less than $nm + n + m$ steps and this is impossible.

Theorem 1. *The algorithm move n,m,d requires $nm + n + m$ steps to solve the general moving checkers game consisting of n black checkers and m white checkers, and the algorithm is optimal.*

3 The Explicit Solutions to the Problem

The optimal solution found by the algorithm move or iterative_move can be presented by a vector x. For $i = 1, 2, \cdots, nm + n + m$, the step i of the optimal move sequence is given by x_i. This means that the checker located at position x_i will be moved in step i to the current vacant positions and leaving the positions x_i the new vacant positions. This can also be viewed that x is a function of i, which is called a move function. In this section we will discuss the explicit expression of function x.

If we denote $x_0 = n + 1$ and

$$d_i = x_{i-1} - x_i, 1 \leq i \leq nm + n + m \tag{1}$$

then the vector d will be a move direction function of the corresponding move sequence.

A related function t can then be defined as $t_i = \sum_{j=1}^{i} d_j, 1 \leq i \leq nm+n+m$. Since

$$t_i = \sum_{j=1}^{i} d_j = \sum_{j=1}^{i}(x_{j-1} - x_j) = x_0 - x_i = n + 1 - x_i$$

we have

$$x_i = n + 1 - t_i, 1 \leq i \leq nm + n + m \tag{2}$$

Therefore, our task is equivalent to compute the function t efficiently.

In this section, the functions x and t will be divided into three parts. The first part is corresponding to the first two stages of the algorithm iterative_move presented in the last section. The second part is corresponding to the stage 3 of the algorithm iterative_move and the third part is corresponding to the stage 4.

3.1 The First Part of the Solution

Similar to the initial value of dir which can be set to l or r, the first move direction d_1 can be set to 1 or -1. If we set $d_1 = 1$, then from the algorithm iterative_move presented in the last section, the move direction sequence for the stage 1 and 2 must be

$$1, -2, -1, 2, 2, 1, \cdots, (-1)^{m-1}, \overbrace{2(-1)^m, \cdots, 2(-1)^m}^{m}$$

This move direction sequence can be divided into m sections as

$$\overbrace{1, -2}^{2}, \overbrace{-1, 2, 2}^{3}, \cdots, \overbrace{(-1)^{m-1}, 2(-1)^m, \cdots, 2(-1)^m}^{m+1}$$

The section j consists of 1 slide and j jumps and thus has a size of $j + 1$.

The total length of the sequence is therefore $s_1 = \sum_{j=1}^{m}(j+1) = m(m+3)/2$. Our task is now to find $t_i = \sum_{j=1}^{i} d_j$ quickly for each $1 \le i \le s_1$.

If we denote the $j + 1$ elements of the section j as $a_{tj}, 1 \le t \le j + 1$, and the sum of section j as $a_j = \sum_{t=1}^{j+1} a_{tj}, j = 1, \cdots, m$, then it is not difficult to see that for each $j = 1, \cdots, m$,

$$a_{tj} = \begin{cases} (-1)^{j-1} & t = 1 \\ 2(-1)^j & t > 1 \end{cases} \tag{3}$$

and for $1 \le k \le j + 1$,

$$\sum_{t=1}^{k} a_{tj} = (-1)^j (2k - 3) \tag{4}$$

Therefore, $a_j = (-1)^j (2j - 1), j = 1, \cdots, m$. These m sums form an alternating sequence

$$-1, 3, -5, \cdots, (-1)^m (2m - 1)$$

For each $1 \le k \le m$, we have,

$$\sum_{j=1}^{k} a_j = \sum_{j=1}^{k} (-1)^j (2j - 1) = (-1)^k k \tag{5}$$

The steps in each section must be

$$\overbrace{1, 2}^{1}, \overbrace{3, 4, 5}^{2}, \cdots, \overbrace{(m-1)(m+2)/2 + 1, \cdots, m(m+3)/2}^{m}$$

If we denote the $j + 1$ steps of the section j as $b_{tj}, 1 \le t \le j + 1$, and the boundary of section j as $b_j = b_{(j+1)j}, j = 1, \cdots, m$, then it is not difficult to see that for each $j = 1, \cdots, m$,

$$\begin{cases} b_j = j(j+3)/2 & 1 \le j \le m \\ b_{tj} = b_{j-1} + t & 1 \le t \le j + 1, 1 \le j \le m \end{cases} \tag{6}$$

For any integer $1 \le i \le b_m$, the corresponding integer k such that the integer i falls into the section k can be computed by a function $f_1(x)$ as follows.

Lemma 4. *Let* $f_1(x) = \frac{\sqrt{8x+1}-1}{2}$. *For any integer* $1 \le i \le b_m$, *it must be a step number in the section* $k = \lfloor f_1(i) \rfloor$.

Proof. It can be seen readily that function $f_1(x)$ is a strictly increasing function on $(0, +\infty)$. For each section $k, 1 \le k \le m$, its first step number is $b_{k-1} + 1 = (k-1)(k+2)/2 + 1$ and it satisfies

$$f_1\left(\frac{(k-1)(k+2)}{2} + 1\right) = \frac{\sqrt{4(k-1)(k+2)+9}-1}{2} = \frac{\sqrt{(2k+1)^2}-1}{2} = k$$

Therefore, for each integer i in the section k, we have, $k \le f_1(i) < k+1$. This means $\lfloor f_1(i) \rfloor = k$.

The proof is completed.   □

From Lemma 4 and formulas (4) and (5), we can now compute $t_i = \sum_{j=1}^{i} d_j, 1 \le i \le s_1$ as follows.

Let $\alpha = \lfloor f_1(i) \rfloor$, then,

$t_i = \sum_{j=1}^{i} d_j = \sum_{j=1}^{\alpha-1} a_j + \sum_{j=b_{\alpha-1}+1}^{i} d_j = (-1)^{\alpha-1}(\alpha - 1) + (-1)^{\alpha}(2(i - b_{\alpha-1}) - 3)$.

It follows that for each $1 \le i \le s_1$,

$$t_i = (-1)^{\alpha}(2i - \alpha(\alpha + 2)) \tag{7}$$

where, $\alpha = \lfloor \frac{\sqrt{8i+1}-1}{2} \rfloor$.

It follows from formula (2) that for each $1 \le i \le s_1$,

$$x_i = n + 1 - (-1)^{\alpha}(2i - \alpha(\alpha + 2)) \tag{8}$$

If we set $d_1 = -1$, a similar result will be obtained. In this case, we have,

$$x_i = n + 1 + (-1)^{\alpha}(2i - \alpha(\alpha + 2)) \tag{9}$$

Combine these two cases, we conclude that,

$$x_i = n + 1 - d_1(-1)^{\alpha}(2i - \alpha(\alpha + 2)) \tag{10}$$

$$\overbrace{1, 2(-1)^{m+1}, \cdots, 2(-1)^{m+1}}^{m}, \overbrace{1, 2(-1)^{m+2}, \cdots, 2(-1)^{m+2}}^{m}, \cdots, \overbrace{1, 2(-1)^{n}, \cdots, 2(-1)^{n}}^{m} \tag{11}$$

3.2   The Second Part of the Solution

If we set $d_1 = 1$, then according to the algorithm iterative_move presented in the last section, the move direction sequence for the stage 3 must be in the form of (11).

This move direction sequence can be divided naturally into $n - m$ sections. The section j consists of 1 slide and m jumps and thus has a size of $m + 1$. The total length of the sequence is therefore $s_2 = (n - m)(m + 1)$. Our task for this part is now to find $t_i = \sum_{j=1}^{i} d_j$ quickly for each $s_1 + 1 \leq i \leq s_1 + s_2$.

If we denote the $m + 1$ elements of the section j as $a_{tj}, 1 \leq t \leq m + 1$, and the sum of section j as $a_j = \sum_{t=1}^{m+1} a_{tj}, j = 1, \cdots, n - m$, then it is not difficult to see that for each $j = 1, \cdots, n - m$,

$$a_{tj} = \begin{cases} 1 & t = 1 \\ 2(-1)^{m+j} & t > 1 \end{cases} \tag{12}$$

and for $1 \leq k \leq m + 1$,

$$\sum_{t=1}^{k} a_{tj} = 1 + (-1)^{m+j}(2k - 2) \tag{13}$$

Therefore, $a_j = 1 + 2m(-1)^{m+j}, j = 1, \cdots, n - m$. These $n - m$ sums form an alternating sequence

$$1 + 2m(-1)^{m+1}, 1 + 2m(-1)^{m+2}, \cdots, 1 + 2m(-1)^n$$

For each $1 \leq k \leq m$, we have,

$$\sum_{j=1}^{k} a_j = k + \sum_{j=1}^{k} 2m(-1)^{m+j} = k + m(-1)^{m+k} - m(-1)^m \tag{14}$$

If we denote the $m + 1$ steps of the section j as $b_{tj}, 1 \leq t \leq m + 1$, and the boundary of section j as $b_j = b_{(m+1)j}, j = 1, \cdots, n - m$, then it is not difficult to see that for each $j = 1, \cdots, n - m$,

$$\begin{cases} b_j = j(m+1) & 1 \leq j \leq n - m \\ b_{tj} = b_{j-1} + t & 1 \leq t \leq m + 1, 1 \leq j \leq n - m \end{cases} \tag{15}$$

For any integer $1 \leq j \leq b_m$, the corresponding integer k such that the integer j falls into the section k can be computed by a function $f_2(x)$ as follows.

Lemma 5. Let $f_2(x) = \frac{x+m}{m+1}$. For any integer $1 \leq j \leq b_m$, it must be a step number in the section $k = \lfloor f_2(j) \rfloor$.

Proof. It can be seen readily that function $f_2(x)$ is a strictly increasing function on $(0, +\infty)$. For each section $k, 1 \leq k \leq n - m$, its first step number is $b_{k-1} + 1 = (k-1)(m+1) + 1$ and it satisfies

$$f_2((k-1)(m+1) + 1) = \frac{(k-1)(m+1)+1+m}{m+1} = \frac{k(m+1)}{m+1} = k$$

Therefore, for each integer j in the section k, we have, $k \leq f_2(j) < k + 1$. This means $\lfloor f_2(j) \rfloor = k$.

The proof is completed.   □

From Lemma 4 and formulas (13) and (14), we can now compute $t_i = \sum_{j=1}^{i} d_j, s_1 + 1 \leq i \leq s_1 + s_2$ as follows.

Let $r = i - s_1, \beta = \lfloor f_2(r) \rfloor$, and $p = r - (\beta - 1)(m+1)$ then, $t_i = \sum_{j=1}^{i} d_j = t_{s_1} + \sum_{j=1}^{\beta-1} a_j + \sum_{j=1}^{p} a_{j\beta}$.

Therefore

$$
\begin{aligned}
t_i - t_{s_1} \\
= \beta - 1 + m(-1)^{m+\beta-1} - m(-1)^m + 1 + (-1)^{m+\beta}(2p - 2) \\
= \beta + (-1)^{m+\beta}(2p - 2) - m((-1)^{m+\beta} + (-1)^{m+2\beta}) \\
= \beta + (-1)^{m+\beta}(2p - 2 - m(1 + (-1)^{\beta}))
\end{aligned}
$$

It follows that for each $s_1 + 1 \leq i \leq s_1 + s_2$,

$$
t_i = t_{s_1} + \beta + (-1)^{m+\beta}(2p - 2 - m(1 + (-1)^{\beta})) \tag{16}
$$

where, $\beta = \lfloor \frac{i - s_1 + m}{m+1} \rfloor$, and $p = i - s_1 - (\beta - 1)(m + 1)$.

It follows from formula (2) that for each $s_1 + 1 \leq i \leq s_1 + s_2$,

$$
x_i = n + 1 - t_{s_1} - \beta - (-1)^{m+\beta}(2p - 2 - m(1 + (-1)^{\beta})) \tag{17}
$$

If we set $d_1 = -1$, a similar result will be obtained. In this case, we have,

$$
x_i = n + 1 - t_{s_1} - \beta + (-1)^{m+\beta}(2p - 2 - m(1 + (-1)^{\beta})) \tag{18}
$$

Combine these two cases, we conclude that,

$$
x_i = x_{s_1} - \beta - d_1(-1)^{m+\beta}(2p - 2 - m(1 + (-1)^{\beta})) \tag{19}
$$

3.3 The Third Part of the Solution

According to the algorithm iterative_move presented in the last section, if $d_1 = 1$, then the move direction sequence for the stage 4 must be

$$
(-1)^{n+1}(1, \overbrace{2, \cdots, 2}^{m-1}, -1, \overbrace{-2, \cdots, -2}^{m-2}, \cdots, (-1)^{m-1}).
$$

This move direction sequence can be divided naturally into m sections. The section j consists of 1 slide and $m - j$ jumps and thus has a size of $m - j + 1$. The total length of the sequence is therefore $s_3 = m(m+1)/2$. Our task for this part is now to find $t_i = \sum_{j=1}^{i} d_j$ quickly for each $s_1 + s_2 + 1 \leq i \leq s_1 + s_2 + s_3 = nm + n + m$.

If we denote the $m - j + 1$ elements of the section j as $a_{tj}, 1 \leq t \leq m - j + 1$, and the sum of section j as $a_j = \sum_{t=1}^{m-j+1} a_{tj}, j = 1, \cdots, m$, then it is not difficulty to see that for each $j = 1, \cdots, m$,

$$
a_{tj} = \begin{cases} (-1)^{n+j} & t = 1 \\ 2(-1)^{n+j} & t > 1 \end{cases} \tag{20}
$$

and for $1 \leq k \leq m - j + 1$,

$$\sum_{t=1}^{k} a_{tj} = (-1)^{n+j}(2k-1) \tag{21}$$

Therefore, $a_j = (-1)^{n+j}(2(m-j)+1), j = 1, \cdots, m$. These m sums form an alternating sequence

$$(-1)^{n+1}((2m-1), -(2m-3), \cdots, (-1)^{m-1})$$

For each $1 \le k \le m$, we have,

$$\sum_{j=1}^{k} a_j = \sum_{j=1}^{k} (-1)^{n+j}(2m - (2j-1))$$
$$= (-1)^n (2m((-1)^k - 1)/2) - (-1)^k k)$$

Therefore,

$$\sum_{j=1}^{k} a_j = (-1)^n ((-1)^k (m-k) - m) \tag{22}$$

If we set $j = i - s_1 - s_2$, then the step numbers in each sections must be

$$\overbrace{1, \cdots, m}^{m}, \overbrace{m+1, \cdots, 2m-1}^{m-1}, \cdots, \overbrace{m(m+1)/2}^{1}$$

If we denote the $m - j + 1$ steps of the section j as $b_{tj}, 1 \le t \le m - j + 1$, and the boundary of section j as $b_j = b_{(m-j+1)j}, j = 1, \cdots, m$, then it is not difficulty to see that for each $j = 1, \cdots, m$,

$$\begin{cases} b_j = j(m+1) - j(j+1)/2 & 1 \le j \le m \\ b_{tj} = b_{j-1} + t & 1 \le t \le m - j + 1, 1 \le j \le m \end{cases} \tag{23}$$

For any integer $1 \le j \le b_m$, the corresponding integer k such that the integer j falls into the section k can be computed by a function $f_3(x)$ as follows.

Lemma 6. Let $f_3(x) = m - \sqrt{m(m+1) - 2x + 9/4} + 3/2$. For any integer $1 \le j \le b_m$, it must be a step number in the section $k = \lfloor f_3(j) \rfloor$.

Proof. It can be seen readily that function $f_2(x)$ is a strictly increasing function on $(0, m(m+1)/2]$. For each section $k, 1 \le k \le m$, its first step number is $b_{k-1} + 1 = (k-1)(m+1) - k(k-1)/2 + 1$ and it satisfies

$$f_3((k-1)(m+1) - k(k-1)/2 + 1) = m + 3/2$$
$$-\sqrt{m(m+1) - 2(k-1)(m+1) + k(k-1) - 2 + 9/4}$$
$$= m + 3/2 - \sqrt{(k-m-3/2)^2} = k$$

Therefore, for each integer j in the section k, we have, $k \le f_3(j) < k+1$. This means $\lfloor f_3(j) \rfloor = k$.

The proof is completed. $\qquad \square$

From Lemma 6 and formulas (21) and (22), we can now compute $t_i = \sum_{j=1}^{i} d_j$, $s_1 + s_2 + 1 \leq i \leq nm + n + m$ as follows.

Let $r = i - s_1 - s_2$, $\gamma = \lfloor f_3(r) \rfloor$, and $q = r - (\gamma - 1)(m + 1) + \gamma(\gamma - 1)/2$ then,

$$t_i = \sum_{j=1}^{i} d_j = t_{s_2} + \sum_{j=1}^{\gamma-1} a_j + \sum_{j=1}^{q} a_{j\gamma}.$$

Therefore

$$t_i - t_{s_2}$$
$$= (-1)^n((-1)^{\gamma-1}(m - \gamma + 1) - m) + (-1)^{n+\gamma}(2q - 1)$$
$$= (-1)^{n+\gamma}(\gamma + 2q - m - 2) - m(-1)^n$$

It follows that for each $s_1 + s_2 + 1 \leq i \leq nm + n + m$,

$$t_i = t_{s_2} + (-1)^{n+\gamma}(\gamma + 2q - m - 2) - m(-1)^n \tag{24}$$

where, $\gamma = \lfloor m - \sqrt{m(m + 1) - 2(i - s_1 - s_2) + 9/4} + 3/2 \rfloor$, and $q = i - s_1 - s_2 - (\gamma - 1)(m + 1) + \gamma(\gamma - 1)/2$.

It follows from formula (2) that for each $s_1 + s_2 + 1 \leq i \leq nm + n + m$,

$$x_i = n + 1 - t_{s_2} - (-1)^{n+\gamma}(\gamma + 2q - m - 2) + m(-1)^n \tag{25}$$

If we set $d_1 = -1$, a similar result will be obtained. In this case, we have,

$$x_i = n + 1 - t_{s_2} + (-1)^{n+\gamma}(\gamma + 2q - m - 2) - m(-1)^n \tag{26}$$

Combine these two cases, we conclude that,

$$x_i = x_{s_2} - d_1((-1)^{n+\gamma}(\gamma + 2q - m - 2) - m(-1)^n) \tag{27}$$

Summing up, the explicit optimal solutions for solving the general game of shifting the checkers consisting of n black checkers and m white checkers can be given in three parts as shown in the following Theorem.

Theorem 2. *In the general game of shifting the checkers consisting of n black checkers and m white checkers, its optimal move steps $x_i, 1 \leq i \leq nm + n + m$, can be expressed explicitly in formulas (29) and (30). where, d is the first move direction.*

$$x_i = \begin{cases} n + 1 - (-1)^\alpha d(2i - \alpha(\alpha + 2)) & 1 \leq i \leq s_1 \\ x_{s_1} - \beta - (-1)^{m+\beta} d(2p - 2 - m(1 + (-1)^\beta)) & s_1 + 1 \leq i \leq s_1 + s_2 \\ x_{s_2} - d((-1)^{n+\gamma}(\gamma + 2q - m - 2) - m(-1)^n) & s_1 + s_2 + 1 \leq i \leq nm + n + m \end{cases} \tag{28}$$

$$\begin{cases} s_1 = m(m + 3)/2 \\ s_2 = (n - m)(m + 1) \\ \alpha = \lfloor \frac{\sqrt{8i+1}-1}{2} \rfloor \\ \beta = \lfloor \frac{i-s_1+m}{m+1} \rfloor \\ \gamma = \lfloor m - \sqrt{m(m + 1) - 2(i - s_1 - s_2) + 9/4} + 3/2 \rfloor \\ p = i - s_1 - (\beta - 1)(m + 1) \\ q = i - s_1 - s_2 - (\gamma - 1)(m + 1) + \gamma(\gamma - 1)/2 \end{cases} \tag{29}$$

It requires $O(1)$ time to compute $(-1)^k$ for any positive integer k, since

$$(-1)^k = \begin{cases} -1 \text{ if } k \text{ odd} \\ 1 \text{ if } k \text{ even} \end{cases}$$

Therefore, for each $1 \leq i \leq nm + n + m$, x_i can be computed in $O(1)$ time by using the formula (27), and then the optimal move sequence of the general game consisting of n black checkers and m white checkers can be easily computed in optimal $O(nm + n + m)$ time.

Acknowledgement. This work was supported by Fujian Provincial Key Laboratory of Data-Intensive Computing.

References

1. Bird, R.: Pearls of Functional Algorithm Design. Cambridge University Press, Cambridge (2010). pp. 258–274
2. Demaine, E.D.: Playing games with algorithms: algorithmic combinatorial game theory. In: Sgall, J., Pultr, A., Kolman, P. (eds.) MFCS 2001. LNCS, vol. 2136, pp. 18–33. Springer, Heidelberg (2001). doi:10.1007/3-540-44683-4_3
3. Demaine, E.D., Demaine, M.L.: Puzzles, art, and magic with algorithms. Theor. Comput. Syst. **39**(3), 473–481 (2006)
4. Levitin, A., Levitin, M.: Algorithmic Puzzles. Oxford University Press, New York (2011). pp. 3–31
5. Kleinberg, J., Tardos, E.: Algorithm Design. Addison Wesley, Boston (2005). pp. 223–238
6. Kreher, D.L., Stinson, D.: Combinatorial Algorithms: Generation, Enumeration and Search. CRC Press, New York (1998). pp. 125–133
7. Gray, J.S.: The shuttle puzzle a lesson in problem solving. J. Comput. High. Educ. **10**(1), 56–70 (1998)
8. Goodrich, M.T., Tamassia, R.: Algorithm Design and Applications. Wiley, Hoboken (2014). pp. 239–325
9. Fibonacci numbers. In: Hazewinkel, M. (ed.) Encyclopedia of Mathematics, vol. 2, pp. 55–56. Springer (2001)

AUDITOR: A Stage-Wise Soft-Error Detection Scheme for Flip-flop Based Pipelines

Hong Zhang$^{(\boxtimes)}$, Ying Li, Hongfeng Sun, and Yanchun Yang

School of Information Technology, Shandong Women's University,
Jinan 250300, China
zhh6856@126.com

Abstract. The shrinking feature sizes make transistors increasingly susceptible to soft errors, which can severely degrade the systems' RAS (Reliability, Availability, and Serviceability). The tough challenge results from not only increasing SER (soft error rate) of storage cells, but also the increasing susceptibility of combinational logics to soft errors. How to efficiently detect soft errors becomes the primary problem in the Backward Error Recovery (BER) schemes that are cost-effective in soft error tolerance. This paper presents a soft error detection scheme, AUDITOR, for flip-flop based pipelines. The AUDITOR copes with both types of soft errors—**single event upset** (SEU) and **single event transient** (SET). We propose a "local-audit" fault detection mechanism, by which each pipeline stage is verified independently and the verifying result registers with a dedicated "audit" bit (V-bit). All the V-bits are distributed across the whole pipeline and synergically monitor the pipeline execution. To relax the constraint of SET detection capability imposed by the inherent fully synchronous operation mode in flip-flop based pipelines, we firstly propose using path-compensation technique to address this constraint. Furthermore, a reuse-based design paradigm is employed to reduce the implementation complexity and area overhead. The AUDITOR possesses robust detection capability and short detection latency, at the expense of about 29% and 50% increase in area and power consumption, respectively.

Keywords: Soft-error detection · Flip-flop based pipelines · Reliability

1 Introduction

The maturing nano-electronic technologies enable tens of billions transistors with lower voltage, higher frequency and higher integrated density to be packed into a single chip. These trends, however, pose some challenges. One major challenge is that the IC (Integrated Circuit) chips are increasingly susceptible to soft errors, thereby seriously threatening systems' RAS (Reliability, Availability, and Serviceability). The situation gets much worse beyond 45 nm. Not only do soft errors impact the storage units, such as registers, flip-flops, latches, RAM-cells, but also combinational logics [1]. Therefore, it is imperative to design an efficient

© Springer International Publishing AG 2017
M. Qiu (Ed.): SmartCom 2016, LNCS 10135, pp. 439–448, 2017.
DOI: 10.1007/978-3-319-52015-5_45

reliability-assuring scheme for both sequential and combinational logic circuits. In this paper, we focus on the soft error detection-the primary support for the Backward Error Recovery (BER) schemes, which are cost-efficient to deal with rare-happened soft errors. In addition, we just pay attention to the mainstream **flip-flop based** pipelines.

For a typical pipeline, protecting it against soft errors implies (1) protecting the pipeline flip-flops from unintentional bit-flip induced by single event upset (SEU) [2]; (2) preventing the pipeline flip-flops form capturing the single event transient (SET) [2] in combinational logics. Both SEUs and SETs are induced by soft errors[1].

So, we straightforwardly employ the DFD scheme for pipeline protection. Primarily, a necessary modification has to be made to make it work in the pipelined context—the latches have to be substituted for flip-flops, as shown in Fig. 1. The reason for this modification is the latches fail to "shadow" the main flip-flops during transparent mode, thereby could causing timing violation between neighboring stages and detection "blind zone".

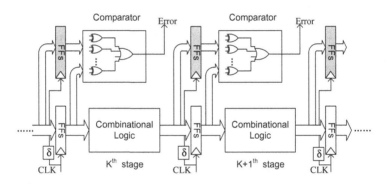

Fig. 1. A "Half-Baked" scheme—employ modified delay-fault detection scheme for pipeline protection

We propose a new scheme, AUDITOR, to remedy the above deficiencies. The main contributions of this paper are as follows: We propose a "local-audit" mechanism to synchronize the distributed asynchronous "Error" signals, and Short-Path Compensation (SPC) technique to relax the SET detection constraint. Furthermore, a reused-based flip-flop design paradigm was employed to eliminate the extra design complexity and reduce area overhead.

The evaluation based on a commercial IEEE 754-compatible pipelined FPU (float Point Unit) shows that the AUDITOR can achieve robust soft error detection capability, at the expense of about 29 % and 50% increase in area and

---

[1] Some researchers view the terminology SEU as the general soft errors: both transient voltage pulses in combinational logics and unintentional bit flip in sequential logics [3]. As many others researchers did, we just take the SEU as unintentional bit flip in this paper, which will not be problematic.

power consumption, respectively. This overhead is far more efficient than most Multiple-Module-Redundancy (e.g. DMR, TMR) based schemes.

2 Protecting Pipeline Structures from Soft Errors

2.1 Synchronize Distributed "Error" Signals

To address the first deficiency (presented in Sect. 1), we propose a "local-audit" mechanism. Each set of pipeline flip-flops (PFFs) is appended with a valid-indicating bit (V-bit), as shown in Fig. 2. The state of the Kth V-bit indicates whether the computation of the Kth stage is valid or not. The detection results of the upstream PFFs of the Kth stage is sent to the downstream PFFs and stored in the Kth V-bit, rather than sent out immediately as an asynchronous signal. This change creates the chance that the pipeline execution can be monitored in every stage and no one detection results will be lost until this operation completed. Those asynchronous error-indicating signals in the "half-backed" pipeline are synchronized after being captured by these synchronous-updated V-bits. We just need to check the last stage V-bit of a pipelined computation to verify the whole execution. The last verifying function can be implemented as a appended extra pipeline stages or, to avoid extending pipeline latency, incorporated into the last pipeline stage. In addition, the global routing for the original "aggressive" OR can be eliminated.

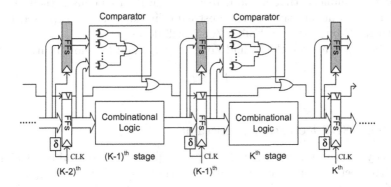

Fig. 2. Local-audit for synchronizing error-indicating signals

Generally, the delay of the Comparator will not result in critical path; however, which may incur some detection "Blind Zone". Several timing-relevant parameters are denoted as:

t_{pd}, the propagation delay of the Comparator logic;

t_{hold}, the flip-flop hold time;

t_{setup}, the flip-flop setup time;

t_{cd}, the contamination delay (also known as short-path delay) of the combinational logic.

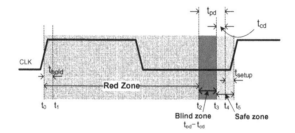

Fig. 3. Red Zone, Blind Zone, and Safe Zone

If $t_{pd} > t_{cd}$, a whole cycle could be divided into several zones by some typical timing parameters. Figure 3 shows the division. If a soft error strikes a upstream flip-flop at the period of (t_0, t_2), then the comparator is able to detection the fault and set up the error-indicating signal to the downstream V-bit. We call this region as "Red Zone". If the soft error happens during (t_3, t_5), the effects of the soft error have no chance to set up at the downstream flip-flops; so does the error-indicating signal. Therefor, (t_3, t_5) is called "Safe Zone". But if the soft error happens during (t_2, t_3), the effects of the SEUs can definitely set up at the downstream flip-flops, but the detection result signal can not because of the propagation delay of the detection logic relatively too long to the coming effective clocking edge. That is why this period is called "Blind Zone".

It can be induced that if the detection logic is fast enough, then any soft error happened at any time in the upstream flip-flops can be captured by the downstream V-bit, then the "Blind Zone" will do not exist.

In brief, to eliminate the "Blind Zone", the t_{pd} and t_{cd} should satisfy

$$t_{cd} > t_{pd}. \tag{1}$$

2.2 Relax SET Detection Constraint

To remedy the second deficiency, firstly, we explain the essential constraints of the SET detection. Suppose that the duration of the SET pulse is t_{set}, and the clock skew between the main FFs clock and the shadow FFs clock is δ. If

$$\delta > t_{set}, \tag{2}$$

then we can assert that if a SET is captured by one of the main flip-flops, then the according shadow flip-flops do not, and vice versa. Therefore, the comparison logic dedicated for SEU detection now can be reused for SET detection. The only required modification is that an intentional positive phase skew, δ, is settled. The proposed scheme, however, imposes some subtle timing constraints when it is embedded in a pipeline context. If the contamination delay of the Kth stage is less than δ, then the Kth set of shadow latches may capture the $(K-1)^{th}$ stage output data which should not have been set before the effective edge of the delayed clock. This phenomenon (a.k.a "short path" effect) will make the AUDITOR generate considerable false positive, even significantly disturb the pipeline normal execution. So we need to make some delay compensation to these "short paths" in the combinational logics.

Based on above analysis, we get another substantial constraint:

$$\delta < t_{cq} + t_{cd} + t_{setup}, \tag{3}$$

which ensures that the timing of "shadow" flip-flops will not be violated by short-path signals.

From (2) and (3), the following equation

$$t_{cd} > t_{set} - t_{cq} - t_{setup} \tag{4}$$

can be obtained. The implication of (4) sheds some light on how to modify the original pipeline logics to make it efficiently cooperate with the fault detection logics: pay some attention to the contamination delay (t_{cd}) of combinational logics.

This goal can be achieved by compensating the delay of some short paths; we name this technique as **Short-Path Compensation**.

Mathematically, the path compensation process can be modeled as an Linear Programming (LP) problem: For a given combinational circuit topology and the expectation of contamination delay, t_{exp}, determining a set of path segments, S_c, which should be inserted with delay units, to achieve the goal of using the fewest number of delay units to remedy all of the paths whose propagation delay, t_{pd}, is less than t_{exp}, but without incurring a performance penalty. The detail model can be find in Appendix. How to solve this LP problem is beyond the scope of this paper. According the practical situation, we at least can conclude that the problem is solvable. Furthermore, abundance of approach have been proposed to address this kind of LP problems in polynomial time complexity. In brief, from the algorithm-level, we conclude that the Path-Compensation process can be efficiently performed.

2.3   Eliminate Intrusive Complexity and Ensure Nominal Performance

To address the third deficiency, the detection logics should be involved into the staged combinational logics as less as possible, if can not be totally avoided. We employ the reuse-based flip-flop design paradigm proposed by S. Mitra [11] to approach the goal. In this design, the scan portion of the scannable flip-flop are reused as redundancy of the flip-flops working in functional mode to detect SEU. Mitra's experimental results show that the reused-design paradigm can construct high-reliable flip-flops at the expanse of about doubled-power consumption, but just impose negligible performance penalty. Comparing against the original Master-Slave flip-flops, the designs impose not only **zero** D-to-Q time overhead, but also **zero** clock-to-Q time overhead.

From our AUDITOR design point of view, besides the **zero** performance overhead, there are two distinct benefits from the adoption of this design: (1) the detection logic complexity can be absorbed into the flip-flops design that is optimization high-efficient due to "one fits to all", thereby significant reducing the over all complexity contributed by fault detection logics; (2) the area overhead can be reduced from reusing on-chip real estate taken by *de facto* DFT (Design for Testability) infrastructures.

Furthermore, the SET detection mechanism can be readily incorporated with SEU detection in the paradigm, and the only modification is that the shadow flip-flops (reused from scan portion) are clocked by a delayed clock (CLKD) which is derived from the main clock (CLK) and locally generated. All the Error signals in a multi-bit flip-flops are OR-ed together.

After these explanation, we can outline the AUDITOR in Fig. 4, where the distributed asynchronous "error" signals are handled in synchronous fashion, the short-path emergencies are resolved by the technique of Short-Path Compensation, and the design complexity and area overhead are minimized by adoption of reuse paradigm.

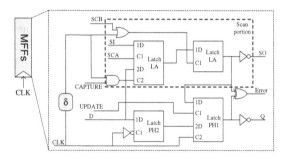

Fig. 4. Auditor scheme

3   Evaluation

We choose a pipelined floating-point unit (FPU) as our target pipeline which implements the SPARC V9 floating-point instructions and supports for all IEEE 754 floating-point data types. The FPU is adopted by OpenSPARC T1—an 8-core processor developed by Sun Microsystem [17]. The FPU comprises three independent pipelines: Multiplier pipeline (MUL), Adder pipeline (ADD) and Divider pipeline (DIV), which shared by 8 cores through a high performance crossbar interconnection. The basic organization of the three pipelines are depicted in Table 1. More design details can be found in [17].

We implemented the AUDITOR in Verilog-HDL modified from the target FPU pipelines and synthesize it using the Synopsys® Design Compiler with a target UMC 0.18um technology. The timing-related computations are conducted by PrimeTime—a popular commercial Timing Analysis tools, provided

Table 1. Pipeline Organization and Performance

Type	# Stage	Execution latency	# State bit
ADD	4	4	1697
MUL	6	7	780
DIV	7	32 or 61*	677

by Synopsys® Inc. A linear delay model is adopted when conducting timing calculation. The short-path compensation is realized by set some timing constrains. After this, we using the post-simulation to verify the AUDITOR pipelines functionality and timing, and dump the according VCD (Value Change Dump) format data for power evaluation. We use PrimePower, a gate-level power simulation and analysis tool provided by Synopsys®, for power evaluation.

Besides the slightly complicated timing calibration which can incur a little extra design effort, the more substantial overhead may lie in 3 aspects: area, power consumption, and performance.

3.1 Area

The area overhead consists of two parts: (1) the sequential (non-combinational) logic overhead and combinational logic overhead.

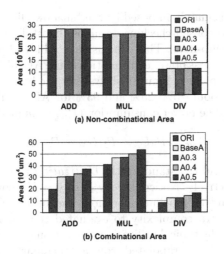

Fig. 5. Area of different pipelines

The sequential logic overhead is mainly incurred by the shadow flip-flops. Fortunately, since the scan-based DFT techniques have become a *de facto* standard in IC design [11], through reusing the scan portion of scannable flip-flops, the overhead imposed by shadow registers will not be substantial. Another source of sequential logic overhead is taken by these V-bits, but this overhead is very negligible due to the limited number of the V-bits (one bit per stage). Figure 5(a) shows the sequential area overhead across the three pipelines: ADD-pipe, MUL-pipe, and DIV-pipe, respectively, where "ORI" stands for the original scannable pipeline, "BaseA" for the pipeline armed by the AUDITOR but without conducting short-path compensation, "A0.x" for the armed pipeline with the short-path compensated to 0.x cycle. The over all sequential logic overhead is only about negligible 1%.

The combinational logic overhead results from two sources: (1) the Auditor infrastructures, including the XOR gates reside in the modified self-checking flip-flops, and OR-trees for auditor bits generation; area overhead referred to this source can be viewed as "fixed area overhead" since this portion is must-have under the AUDITOR framework. (2) the short-path compensation process; area overhead referred to this source can be viewed as "variable area overhead" since this portion depends on the compensation degrees.

Figure 5(b) shows that the *fixed area overhead* is about 30.5% on average; the *variable area overhead* is increasing with the promoting compensation degrees. Compared with the Base-Auditor, the most intense compensation, A0.5, increasing the combinational area by 21%, 14%, and 31% for ADD-pipe, MUL-pipe, and DIV-pipe, respectively.

3.2 Power Consumption

The power overhead is mainly imposed by the "shadow" flip-flops. As [11] shows, the power consumption of self-checking flip-flops almost doubled compared to the original master-slave flip-flops. This conclusion is still held in our modified flip-flops. We compare the gross power consumption of the AUDITOR armed pipelines against the original pipelines. Figure 6 shows that the base AUDITOR increases power consumption about 70%, 25% and 93% for ADD, MUL, and DIV pipe, respectively. The average power consumption (P_{avg}) can be described as:

$$P_{avg} = P_A U_A + P_M U_M + P_D U_D, \tag{5}$$

where, the U_A, U_M, and U_D stand for the *Utilization* ratio of the ADD, MUL, and DIV pipe, respectively. Notice, here $U_A + U_M + U_D$ could be greater than 1 since some independent pipeline operations can be parallelized in the FPU, which incurs some extra complexity to evaluated the whole FPU power consumption. Supposing the typical utilization proportion ($U_A : U_M : U_D$) is around 10 : 5 : 1 (which can be extrapolated from a benchmark), then, based on (5), the overall power overhead can be computed as about 50%.

After short-path compensation, the power consumption should increase since some delay buffers (or gate) are introduced. We study the sensitivity of the power increasing against the compensation degrees.

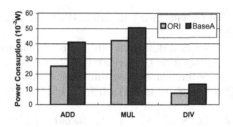

Fig. 6. Power comparison between original pipelines and base AUDITOR

3.3   Performance

A pipeline performance mainly depends on (1) the pipeline flip-flops time overhead and (2) the critical path delay of the combinational logics. A pipeline armed by AUDITOR will not significantly sacrifice performance due to the two reason: (1) the performance of pipeline flip-flops do not be degraded (which have been studied in [11]); (2) The short path compensation to the combinational logics will not exacerbate the critical path if the critical timing be hold tightly. Form our experiments, because we just require the delay compensation to the short-path to been increased to a half of the critical path delay, this ideal compensation process can always be achieved by using PrimeTime. Therefore, we conclude that the AUDITOR results in zero performance overhead.

4   Conclusions

The AUDITOR scheme presented in this paper is high-efficient to soft error detection for flip-flops based pipeline structures. The AUDITOR employs a local-audit detection scheme which can synchronize the asynchronous error-indicating signals, thereby providing a base for accurate controls of error recovery. We proposed the short-path compensation technique to remedy the deficiency of SET detection capability. The experimental results shows that the area overhead incurred by this technique is insensitive to the compensation intensity. Through comprehensive investigation to a pipelined Floating Point Unit (FPU) adopted by UltraSPARC T1, we conclude that the AUDITOR scheme can achieve perfect SEU and SET fault coverage, short detection latency at the expense of modest area overhead, while about 70%, 25%, 93% power overhead for ADDer, MULtiplier, and DIVider pipeline, respectively.

References

1. Shivakumar, P., Kistler, M., Keckler, S.W., Burger, D., Alvisi, L.: Modeling the effect of technology trends on the soft error rate of combinational logic. In: Proceedings of the 2002 International Conference on Dependable Systems and Networks (DSN), pp. 389–398 (2002)

2. Nicolaidis, M.: Design for soft error mitigation. IEEE Trans. Device Mater. Reliab. **5**(3), 405–418 (2005)
3. Heidergott, W.: SEU tolerant device, circuit and processor design. In: Proceedings of the 42nd Design Automation Conference (DAC), pp. 5–10 (2005)
4. Nicolaidis, M.: Time redundancy based soft-error tolerance to rescue nanometer technologies. In: Proceedings of the 17th IEEE VLSI Test Symposium (VTS), pp. 86–94 (1999)
5. Zyuban, V., Brooks, D., Srinivasan, V., Gschwind, M., Bose, P., Strenski, P.N., Emma, P.G.: Integrated analysis of power and performance for pipelined micro-processors. IEEE Trans. Comput. **53**(8), 1004–1016 (2004)
6. Ernst, D., Kim, N.S., Das, S., Pant, S., Rao, R., Pham, T., Ziesler, C., Blaauw, D., Austin, T., Flautner, K., Mudge, T.: Razor: a low-power pipeline based on circuit-level timing speculation. In: Proceedings of the 36th Annual IEEE/ACM International Symposium on Microarchitecture (MICRO), pp. 7–18 (2003)
7. Nicolaidis, M.: GRAAL: a new fault tolerant design paradigm for mitigating the flaws of deep nanometric technologies. In: Proceedings of the IEEE International Test Conference (ITC), pp. 1–10, October 2007
8. Sutherland, I., Sproull, R.F., Harris, D.: Logical Effort: Designing Fast CMOS Circuits. Morgan Kaufmann, San Francisco (1999)
9. Weste, N., Harris, D.: CMOS VLSI Design: A Circuits and Systems Perspective, 3rd edn. Addison Wesley, Boston (2005)
10. Dodd, P.E., Massengill, L.W.: Basic mechanisms and modeling of single-event upset in digital microelectronics. IEEE Trans. Nucl. Sci. **50**(3), 583–602 (2003)
11. Mitra, S., Seifert, N., Zhang, M., Shi, Q., Kim, K.S.: Robust system design with built-in soft-error resilience. IEEE Comput. **38**(2), 43–52 (2005)
12. Das, S., Roberts, D., Lee, S., Pant, S., Blaauw, D., Austin, T., Flautner, K., Mudge, T.: A self-tuning DVS processor using delay-error detection and correction. IEEE J. Solid State Circ. **41**(4), 792–804 (2006)
13. Wang, N.J., Patel, S.J.: ReStore: symptom-based soft error detection in micro-processors. IEEE Trans. Dependable Secure Comput. **3**(3), 188–201 (2006)
14. Eaton, P., Benedetto, J., Mavis, D., Avery, K., Sibley, M., Gadlage, M., Turflinger, T.: Single event transient pulsewidth measurements using a variable temporal latch technique. IEEE Trans. Nucl. Sci. **51**(6), 3365–3368 (2004)
15. Saggese, G.P., Wang, N.J., Kalbarczyk, Z.T., Patel, S.J., Iyer, R.K.: An experi-mental study of soft errors in microprocessors. IEEE Micro **25**(6), 30–39 (2005)
16. Manne, S., Klauser, A., Grunwald, D.: Pipeline gating: speculation control for energy reduction. In: Proceedings of the 25th Annual International Symposium on Computer Architecture (ISCA), pp. 132–141 (1998)
17. Sun Microsystem Inc.: OpenSPARC T1 Microarchitecture Specification (2006)

Aggregating Heterogeneous Services in the Smart City: The Practice in China

Fangping Li[1,2(✉)] and Bo Li[3]

[1] Digital China Information System Co., Ltd., Beijing 100085, China
dclifp@126.com
[2] Key Laboratory of High Confidence Software Technologies,
Ministry of Education, Beijing 100871, China
[3] School of Computer Science and Engineering,
Beihang University, Beijing, China
libo@act.buaa.edu.cn

Abstract. In the past fifteen years, each department of Chinese government and business service provider has established their own information service systems to publish services for citizens. However, these systems are isolated from each other and the services they provided are heterogeneous. It is more beneficial to build a common infrastructure to connect, aggregate and utilize all these services. In this paper, a Smart Citizen Fusion Service Platform Software Product Line, called CrowdService, is introduced as a solution. Based on open data, crowd sourcing, user profile and application framework, the architecture of CrowdService supports the development of personalized applications in smart cities. It consists of resource gathering components, open API component, PAAS components, service recommendation, etc. which supports rapid development of personalized applications in different cities. In terms of system management, CrowdService provides minimized total product development cost, increased productivity, on-time delivery, full control over all lifecycles, decreased defect rate both in assets and products, compliance on mission focus, architectural conformance, process compliance, high quality and customer satisfaction. Moreover, our one practice of smart citizen fusion service platform in China is introduced to demonstrate the effectiveness of the CrowdService.

Keywords: Smart city · Fusion service · Citizen centric · Human-centric · Open data · User profile · Crowd sourcing · Application frame work

1 Introduction

In the past fifteen years, each department of Chinese government and business service provider have established their own information service systems to publish services for citizens. People is getting used to manage their daily life on the internet. We hope that all the problem can be solved on the internet easily. Now indeed is the case, the government launched a number of services on the network such as online payment, household management, online appointment. These services systems indeed help citizens a lot.

© Springer International Publishing AG 2017
M. Qiu (Ed.): SmartCom 2016, LNCS 10135, pp. 449–458, 2017.
DOI: 10.1007/978-3-319-52015-5_46

However, these systems are isolated from each other and the services they provided are not well organized. To apply for affordable housing, for example, applier should fill in a lot of forms on the internet, then, government needs to know the income level and many other information of the applier. After checking, the results should be in the online publicity. The present situation is that the information is heterogeneous and isolated. Neither the applier nor the government cannot get the information in a single platform. We hope all the services can be fused into one platform, where people can obtain as much information. In China, more than 100 cities have proposed to build smart city and provided fusion services to citizens.

The goal of this paper is to develop a platform which can merge as much information and provide application programming interface (API). Based on this platform, many current tasks can be polymerized. People can access the services easily, and do not need to access a variety of information in different platforms. Meanwhile, the service provider can easily provide services due to aggregated information. We call this platform CrowdService, that can help government and enterprises even individual develop smart city fusion service for a city.

The contribution of this paper is summarized as follow:

1. Unified Service Model. Since we aim at building a general platform, we need to develop a standard for this platform. All the services running on the platform need to comply with the standard. The advantages of a unified standard are obvious, which can reduce the develop difficulty and form a good ecology system.
2. Develop-time and Runtime platform. On this platform, all the resources can be polymerized. During the develop time, providers can use API to develop their service. In the running time, The service provider does not need to worry about the running environment. Platform itself is robust enough to schedule the running services.
3. Real-world system in China. We have built a few real-world systems in some cities of China. In this paper, we take two cities, i.e., Weihai and Benxi, as examples. The platform in real world does play a role. Many citizens and services providers really benefit from it.

The rest of this paper is organized as follows. Sections 2 and 3 presents The Overview of CrowdService and The Management of CrowdService. Section 4 gives a practice to demonstrate effectiveness of our platform in the context of the smart city. Finally, we conclude our work and discuss future work in Sect. 5.

2   The Overview of CrowdService

2.1   The Specific Market Demands of CrowdService

According to our survey, there are more than 1500 public services and countless business services in a city [1]. As described above, the concept of Smart City that shift the way from the delivery of specific services to a citizen centric approach, so the scale of challenges is forcing cities to rethink their strategies and innovate to improve service levels, in particular:

1. A smart city requires innovative services that provide information, knowledge, transactions capabilities to citizens about all aspects of their life in the city [2]. In China, the State Department requires government at all levels to open their data to public by 2020. So it's possible to release data to enable new services to develop and citizens to make informed decisions e.g. providing real-time information on traffic to assist citizens in planning journeys;
2. Crowd Sourcing is the process of obtaining needed services, ideas, and/or contents by soliciting contributions from a large group of people, especially from an online community, rather than from traditional employees or suppliers 3. There are so many service requirements in a city, that we need numerous self-identified volunteers or part-time workers to work to meet these needs;
3. The main reason for emphasizing service delivery was that the public sector has been slow and unresponsive to the citizens needs in nowadays. According to the International Bank for Reconstruction and Development/The World Bank (2005), public service delivery has been inconsistent with citizens' preferences and considered feeble in developing countries [4]. So the modern platform needs the ability of online service delivery;
4. E-government advancements have not fully resolved the challenge of providing citizens with a single entry point for services that involve different government entities. So, we need a novel platform for the service management and application development in smart cities. The most important part of the platform is service integration which can integrate both back office and increasingly front line services.

2.2   The Feature Model of CrowdService

CrowdService is a platform independent SPL including core infrastructure based on Open Data, Human-centric Service, User Profile, Crowd Sourcing and Application Framework Model plus development environments, software lifecycle management, operation and maintenance techniques and tools. In order to eliminate unnecessary coding efforts, CrowdService provides components for every tier of web and mobile application development. The Feature Model of CrowdService is depicted in Fig. 1.

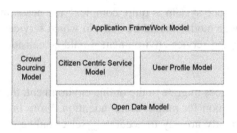

Fig. 1. The feature model of CrowdService

- Open Data Model

Human-centric service is an important domain of the smart city. Including rich applications that help residents with shopping, dining, transportation, entertainment, and other daily activities. While offering services, these systems have generated a large amount of hierarchical data. Usually, data from each system is incomplete, and one system complements another [5]. In order to build fusion service, data should be gathered and opened to innovative application, hence The Open Data Model of CrowdService is set up, as shown in Fig. 2.

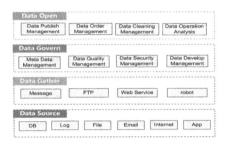

Fig. 2. The open model of CrowdService

The Open Data Model of CrowdService is divided into four tiers : (i) Data Source; the information of smart city comes from a wide variety of data sources, including DB, Log, File, Email, Internet and App. (ii) Data Gather; the key of smart city is the breakthrough of integration of multi-source heterogeneous urban information, the way of integration including message, ftp, web service and robot etc., (iii) Data Govern; through meta data management, data quality management, data security management and data development management, the product of CrowdService could take good advantage of the multisource and multi-temporal data to make high spatial and temporal analysis to assist decisions on urban management. (iv) Data Open; As smart city involves many sectors and industries, we need to break trade barriers so as to achieve information sharing and information exchanging among client, business and government, this tier includes data publish management, data order management, data cleaning management and data operation analysis.

- Citizen Centric Service Model

The human-centric service system is quite complex, and needs modeling technology. A human-centric service model should be built, in which service lifecycle, service life event, service subject and service environment should be considered, shown in Fig. 3.

- User Profile Model

As more and more services migrate to the online environment, there is a corresponding increase in the competition for online users' attention. From the user experience perspective, there is currently no way to accurately select online service, to present to a given online user that would most likely be of interest to that user, such as targeted

Fig. 3. The citizen centric service model

services. One way to alleviate this disconnect is to match user profile information to specific available online content. A user profile model is built in CrowdService, including user identity, user behavior, user related information and user information, shown in Fig. 4.

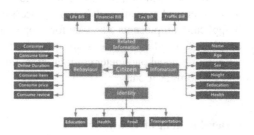

Fig. 4. The user profile model

- Crowd Sourcing Model

Crowdsourcing-processes involve three different stakeholders: the individuals forming 'the crowd', the companies or organizations looking to benefit from the crowd input, and an intermediation platform, the so-called 'crowdsourcing enabler'.

In collaboration with the smart city we design the crowd sourcing model. The model includes four parts and four kinds of tools. Developer Team & Coordination Environment provides team organization and cooperation supports. Toolkit provides the interfaces and management functions of tools. Workspace is the store for mediate products. App Store is the place for products trading. Four kinds of tools include SAAS application develop tools, service assemble and develop tools, component develop and assembled tools and software resource manage tools.

- Application Framework Model

The CrowdService application framework model provides a clear separation of business logic from the presentation and data persistence.

The presentation layer has been derived from HTML5. HTML5 is an open standard for user interface description language. It is rendering device and user interface metaphor independent.

The component layer communicates via interfaces called "services" that are the contracts between the components and the outer world.

The business framework layer mainly addresses the use of SpringFrameWork for seamless integration of SPL and rule-based business process management from architectural viewpoint.

The persistence layer has been shaped to integrate data and service sources to the business framework layer, we use MyBatis as the framework implementation of this layer.

2.3   The Development Model of CrowdService

The Development Model of CrowdService defines how to develop a smart citizen service platform for a city.

The Development Model of CrowdService is divided into three programs, the first program is Definition, which includes two processes: Mainframe definition and Application definition, Mainframe definition process defines the whole frame of smart citizen service platform of a city, Application definition process defines each application. Generally, each application represents a service for people; the second program is configuration and development, which includes six processes: Mainframe import, common parameter configuration, local parameter configuration, domain configuration, custom development and Platform API call.

2.4   The Core Asset Development of CrowdService

The core asset development establishes a production capability for Smart Citizen Fusion Service Platform in a city. It can help developers build a platform that can provide services to citizens rapidly. This chapter introduces the necessary input and output of CrowdService Core Asset Development.

The Input of CrowdService Core Asset Development. Inputs to CrowdService Core Asset Development include:

(1) Product constraints:
     Commonalities among Smart Citizen Fusion Service includes:
     (a) Depending on existing system information to provide service product
     (b) The public service pattern is public service guide, reading of public policy, public service strategy, booking of public service online, agent of public service, review of public service
     (c) Integrating existing internet service as business service
     (d) providing the payment of service
     (e) providing the authentication of service
     Variations among Smart Citizen Fusion Service includes:
     (a) different organization of services
     (b) different service provider
     (c) different service interface

    (d) different business process of service

    (e) regional characters of service

    (f) services should be provide according to different time

(2) Style, patterns and framework

    (a) SOA

    (b) Spring framework, Mybatis framework

(3) Production constraints

    (c) the national standard

(4) Production strategy

    (a) the national standard

(5) Inventory of preexisting assets

    (a) CMS

    (b) pay component

    (c) SMS gateway

    (d) visit analysis system

The Output of CrowdService Core Asset Development. Besides core assets, the outputs of CrowdService software product line core asset development include CrowdService software product line scope, which describes the Smart Citizen Fusion Service Platform that will constitute the CrowdService software product line or that the CrowdService software product line is capable of including, and a Smart Citizen Fusion Service Platform production plan, which describes how a Smart Citizen Fusion Service Platform is produced from the core assets. All three outputs must be present to ensure the production capability of a CrowdService software product line.

The most important core asset of CrowdService is its architecture. CrowdService presentation layer includes Web App and Mobile App, they utilize HTML5 rendering power that can present very dense, responsive and graphically rich user interfaces. CrowdService Runtime Container layer acts like a bridge from the presentation layer to the Open Data Platform layer meeting also instant querying and reporting needs and this layer can easily connect to Outer System. The CrowdService Runtime Container organization is SOA-compliant, the User Profile and the Service Push component can provide user the personalized service.

Among the other important assets are the components of CrowdService, we develop a lot of components in each layer of CrowdService, including Application layer components, Channel Frame components, Application Support components, Runtime Environment components, Information Storage components, Operation platform components, Maintenance platform components and Open platform components, we can easily build a CrowdService product utilizing these components.

3   The Management of CrowdService

Management at the technical (or project) and organizational (or enterprise) levels must be strongly committed to the software product line effort and the product line's success [6].

3.1   The Technical Management of CrowdService

Technical management oversees the core asset development and the product development activities, ensuring that the groups building core assets and those building products engage in the required activities, following the processes defined for the product line, and collecting data sufficient to track progress [6].

CrowdService Software Product Line Team is responsible for CrowdService Software Product Line Development and provide technique support to CrowdService Product Development Team. CrowdService Product Development Team is responsible for CrowdService Product Development.

3.2   The Organizational Management of CrowdService

Organizational management must set in place the proper organizational structure that makes sense for the enterprise and ensure that organizational units receive the right resources (for example, well-trained personnel) in sufficient amounts [6].

CrowdService Leader Team is set for coordination between CrowdService Software Product Line Team and CrowdService Product Team.

4   Two Practical Systems

This section briefly describes the products/projects developed using CrowdService SPL. As will be noticed, the common characteristics of these projects are that they have gathered a lot of information and services from existing system to the open data platform, for public service, these projects provide functions of public service guide, reading of public policy, public service strategy, booking of public service online, agent of public service, review of public service, for business service, these projects integrate internet service. Whether public service or business service, they need authentication service. Systems must be available 7 × 24, loads will certainly increase over the time, they need B2B and B2C integration, etc. Actually these characteristics are fully compliant with the scope of CrowdService.

4.1   The Case of WEIHAI City

WEIHAI Smart Citizen Fusion Service Platform has been developed in approximately 6 months and completed in 100 man-months. It is in operation for more than two years now.

The system is designed to provide full life cycle service to citizens. It has been integrated with more than 20 external systems using XML/Web Service based adapters and gathered more than 1000 categories of data into open data platform, this data constructs Weihai Open Data Model; we abstract personal information from external systems to construct Weihai User Profile Model, these two models are two key features of CrowdService, as shown in Fig. 5. The human-centric service model has been built

Fig. 5. WEIHAI open data and user profile model

on product and process concept. The system contributed to the asset library significantly in such a way that administrative approval service, health services, social services, provident fund services and other business services have been developed.

WEIHAI Smart Citizen Fusion Service Platform open data to small and medium-sized Enterprises, person to encourage them develop innovative applications, this forms crowd sourcing model for the platform and at the same time WEIHAI Smart Citizen Fusion Service Platform provides components to CrowdService Software Product Line.

WEIHAI Smart Citizen Fusion Service Platform provides services through web and mobile, the UI is shew in Fig. 6.

Fig. 6. WEI HAI smart citizen fusion service platform

4.2   Summary of Cases

Developing on CrowdService has dramatically changed the organization of Smart City Projects related to human-centric services. They have been organized based on the software development processes of CrowdService, and specialized teams are responsible for the requirement management, change management, configuration and release management, test management and issue tracking and quality management. On the other hand, the dedicated development teams are responsible for the development of application modules.

5   Conclusions

In this paper, we have introduced the CrowdService Software Product Line. CrowdService provides a platform and core assets designed for human-centric services systems in smart city. The special emphasis has been given to CrowdService feature model.

The paper addresses CrowdService in three different perspectives: within the core asset development, CrowdService scope has been defined, and core assets are named including the system architecture and components. Product development defines the development activity as an integration process and defines management processes in CrowdService. The Product Line management defines both the technical and organizational aspects including project setup, project organization, resource planning, process management.

Acknowledgements. This work is funded by the National High Technology Research and Development Program of China under Grant No. 2016YFB1000801.

References

1. Li, F., Wang, J.: Citizen fusion service platform for smart cities: architecture, technologies and practice. In: International Workshop on Requirements Engineering in the Global Context (2014)
2. Lee, J., Lee, H.: Developing and validating a citizen-centric typology for smart city services. Government Inf. Q. **31**, S93–S105 (2014)
3. Crowdsourcing - definition and more. http://merriam-webster.com
4. Rafia, N.A.Z.: E-governance for improved public services in Fiji. J. Serv. Sci. Manage. **3**, 190–203 (2009)
5. Xia, D., Cui, D., Wang, J., Wang, Y.: A novel data schema integration framework for the human centric services in smart city. ZTE Commun. **13**(4) (2015)
6. Northrop, L.M.: SEI's software product line tenets. IEEE Softw. **2** (2007)
7. Gai, K., Qiu, M., Zhao, H., Tao, L., Zong, Z.: Dynamic energy-aware cloudlet-based mobile cloud computing model for green computing. J. Netw. Comput. Appl. **59**, 46–54 (2016)
8. Gai, K., Du, Z., Qiu, M., Zhao, H.: Efficiency-aware workload optimizations of heterogeneous cloud computing for capacity planning in financial industry. IEEE Computer Society (2015)

A Hybrid Memory Hierarchy to Improve Cache Reliability with Non-volatile STT-RAM

Naikuo Chen[✉], Zhilou Yu, and Ruidong Zhao

Inspur Inc., Jinan, People's Republic of China
{chennk,yuzhl,zhaord}@inspur.com

Abstract. With the development of manufacturing technology, chips have more integration density. However, soft errors have become a sensitive concern in the design of computer systems. So, in this paper, it aims to find the potential vulnerable data and allocates it into a reliable *Non-Volatile Memory* (NVM) with the assistance of complier on a hybrid memory hierarchy with NVM. It has proposed a word level lifetime model of data cache for the purpose of vulnerability estimation and critical data protection. Then, it has abstracted the NVM-assisted cache vulnerability factor to evaluate the reliability of data cache and used to measure the impact on reliability of data cache. Since STT-RAM is used as NVM to build an on-chip SPM then the traditional compile-time data allocation method. The data has been better protected, and the reliability of the whole system can be improved naturally.

Keywords: Non-Volatile Memory · Shared transistor technology random access memory · Cache reliability · Hybrid memory hierarchy

1 Introduction

As manufacturing technology progresses by reducing feature size, providing more integration density and increasing device functionality, soft errors have become a sensitive concern in the design of computer systems [1]. Meanwhile, the novel resistive *Non-Volatile Memory* (NVM), such as *Phase Change Memory* (PCM) [2] and *Magnetic RAM* (MRAM) [3], is more reliable than the traditional storage chips in the severe environments because of the different physical mechanism. So, this paper aims at finding the potential vulnerable data and allocates it into the reliable NVM with the software programming methods and the assistance of complier on a hybrid memory hierarchy with NVM. Since the data has been better protected, the reliability of the whole system can be improved naturally.

For a long time, SRAM-based traditional caches have become one of the most vulnerable micro-architectural components in the processor [4]. As a conventional charge memory, SRAM requires discrete amounts of charge to induce a voltage state with the hard problems as charge leakage, bit flip, and it gets worse in the strong particle environments. For example, each bit in SRAM is stored with a bitable latching circuitry and its state is fully dependent on the supply voltage, then any noise of the voltage may cause a fatal bit flip [5].

© Springer International Publishing AG 2017
M. Qiu (Ed.): SmartCom 2016, LNCS 10135, pp. 459–468, 2017.
DOI: 10.1007/978-3-319-52015-5_47

Then, the statistical models have been proposed to measure the cache vulnerability quantitatively. In [6], *architectural vulnerability factor* (AVF) is introduced to measure the soft error rate, which is defined as the probability that a fault will result in a visible error in the final output of a program. It is estimated by employing an *architecturally correct execution* (ACE) [13] analysis of each bit in a component, and for a time interval, all bits are divided into two groups, ACE or un-ACE, then ACE bits are regarded as the vulnerable data. In [6], it studies a bit's whole lifetime in a component and makes a detailed interval division including more than 14 divisions as fill-to-read, read-to-read, write-to-read, fill-to-write, etc. Then, it classifies these detailed intervals to be ACE, un-ACE, or unknown (such as fill-to-end). For example, read-to-write is un-ACE because its value can be corrected by the latter write operations. Especially for data cache, in [7], it defines *cache vulnerability factor* (CVF) as the probability that a soft error in cache can be propagated to the processor or memory hierarchy, and concludes six different cases: read-read (R-R), write-read (W-R), dirty-replacement (D-Repl), read-write (R-W), write-read (W-W), clean-replacement (C-Repl). Then, R-R, W-R and D-Repl are ACE. Recent techniques choose only the most vulnerable data value in cache to protect, reducing the performance degradation. In [10], it selectively marks some cache lines dirty in L1 cache to make the vulnerable data invalid in advance, while this increases the cache miss rate. Other technologies with the help of main memory, such as selectively writing the specific vulnerable cache lines back to main memory, have also been studied.

So, in this paper, it aims to find the potential vulnerable data and allocates it into a reliable *Non-Volatile Memory* (NVM) with the assistance of complier on a hybrid memory hierarchy with NVM. It has proposed a word level lifetime model of data cache for the purpose of vulnerability estimation and critical data protection. One approach is to reduce the vulnerable data's residency time in cache. In [9], it proposes to refresh the cache line by the selective re-fetching from the low level cache or memory, while this needs additional data transmission bandwidth between the memory hierarchy. Then, it has abstracted the NVM-assisted cache vulnerability factor to evaluate the reliability of data cache and used to measure the impact on reliability of data cache. Since STT-RAM is used as NVM to build an on-chip SPM then the traditional compile-time data allocation method. The data has been better protected, and the reliability of the whole system can be improved naturally.

2   Related Work

Bit flips in SRAM and DRAM memory system caused by alpha particles and high-energy neutrons from cosmic radiation may not lead to visible errors in the final output. Only faults occurring on bits that will be consumed by the processor or stored to lower memory system may produce an error in program outcome. So data in cache can be clarified based on whether or not it has an effect on the program outcome. In [13], those data that may impact the final outcome are

defined as required for architecturally correct execution (ACE) while the others
are assumed to be unnecessary for architecturally correct execution (un-ACE).
Then the possibility that a raw fault in a component will result in a user-visible
error is defined as the AVF.

For the purpose of computing AVF, the ACE and un-ACE intervals for each
single bit needs to be identified. Based on this identification, the AVF for a single
bit is the fraction of time the bit contains ACE intervals, and thus the AVF for
a hardware unit is simply the average AVF for all its bits in that unit. Then,
the AVF of a hardware unit is computed as:

$$AVF = \frac{Average\ number\ of\ ACE\ bits\ resident\ in\ a\ unit\ in\ a\ cycle}{Total\ number\ of\ bits\ in\ the\ hardware}$$

The above equation can be rewritten as:

$$AVF = \frac{\sum Residency(in\ cycles)\ of\ all\ ACE\ bits}{Total\ number\ of\ bit\ \times\ Total\ execution\ cycles}$$

The key issue here is to determine which part is ACE and which part is un-
ACE during a bits lifetime. Previous research have developed detailed cache data
lifetime model to estimate the vulnerability of cache. In [6], it first introduces
lifetime analysis to compute the AVF of a processor's instruction queue and
execution unit. Through lifetime analysis, it divides up a bit's lifetime during a
program execution into ACE and un-ACE components. Then in [14], it proposes
a lifetime model for data cache array with the purpose of computing the un-ACE
fraction of a bit's lifetime. It identifies all the activities during a bit's lifetime,
including fill, read, write and eviction. The summary of non-overlapping lifetime
intervals and its classification into ACE or un-ACE states is listed in the Table 1.

In the lifetime analysis in [6], it doesn't consider the state (like dirty or
clean) of the cache data item. For example, in write-back data cache, whether
the read-to-evict interval of a data item is ACE depends on the cache line's state.
If the cache line is clean, then this interval is un-ACE; otherwise, the cache line
will be written back to lower memory, so it is ACE. For the purpose of cache
vulnerability characterization, in the lifetime model proposed in [6], it takes the
cache line state into account and divides the lifetime of a data item.

Table 1. Classification of lifetime intervals

	ACE	un-ACE	Unknown
Write-through data cache	fill-to-read, read-to-read, write-to-read	idle, fill-to-write, fill-to-evict, read-to-write, read-to-evict, write-to-write, write-to-evict, evict-to-fill	fill-to-end, read-to-end, write-to-end
Write-back data cache	fill-to-read, read-to-read, write-to-read, write-to-evict, write-to-end	idle, fill-to-write, fill-to-evict, read-to-write, read-to-evict, write-to-write, evict-to-fill	fill-to-end, read-to-end

3   Cache Vulnerability Model for NVM

3.1   Word-Level Lifetime Analysis

In this study, a detailed cache lifetime is analyzed at word level. Our word-level lifetime analysis modeled the L1 write-back data cache. We perform lifetime analysis at word level because:

(1) our profile information is used to direct complier to allocate data and the compiled program data is word (4-bytes, generally) address alignment generally;

(2) although the basic unit size of CPU access is in byte, most CPU data accesses are word.

Figure 1 shows four typical access cases and its lifetime intervals. It is regarded as an example to explain our lifetime model. Compared with the lifetime analysis described in Sect. 2, our model takes both cache line status and the phase between fill and the first access into account. In write-back data cache, the cache line may be in two statuses: clean or dirty. The lifetime intervals of words are different in clean cache line from that of in dirty cache line, since this difference may lead to different vulnerability contribution. Then it has a focus on all typical access activities to these words in data cache, including fill, read, write, discard and write back.

For clean word in clean cache line (as shown in Fig. A), all access activities to the word are read operations. The phase is defined from fill to the first read access as fill-to-read (Fill-R) interval, and define the following access intervals during two consecutive read operations as read-to-read (R-R). These Fill-R and R-R intervals are illustrated in Fig. 1A as red color parts. The last interval is Clean-Repl (green part), which is the phase from the last read to discard. The Fill-R and R-R intervals are ACE, because errors that occur in these two intervals may be loaded in to CPU. Obviously, Clean-Repl is un-ACE, because any errors occur in this interval are discarded and ignored.

For these words in dirty cache line, based upon different CPU access activities are divided into three groups:

– The first group involves clean words, which are not modified in its lifetime (as shown in Fig. 1B). In Fig. 1B, after being filled into cache, the word is accessed with four consecutive read operations and is written back in the end. Its lifetime consists of three types of access intervals: Fill-R, R-R and dirty-write-back (Dirty-WB). Clearly, all three types of intervals are ACE, since incorrect values may be both loaded into CPU and written back to lower memory hierarchy. In another word, the whole lifetime (from Fill to Write-back) of clean word in dirty line is ACE;

– The second group contains words that all the access activities to it are write rather than read. As shown in Fig. 1C, the lifetime includes a fill-to-write (Fill-W) interval, two write-to-write (W-W) intervals and Dirty-WB. Fill-W and W-W are un-ACE, because in this case, the modified word is not used

but again updated, any soft error on the updated word between the two write accesses is not recognized (assume that all write access activities are correct);
- In the third group, there are both read and write operations during the words lifetime. In this group, the lifetime contains the following intervals: Fill-W (or Fill-R), R-R, write-to-read (W-R), W-W, read-to-write (R-W) and Dirty-WB (A possible situation is shown in Fig. 1D). Notice that W-R interval is ACE, since the written word is used by the later read access immediately and the correctness of this word is necessary for the correct execution of the program.

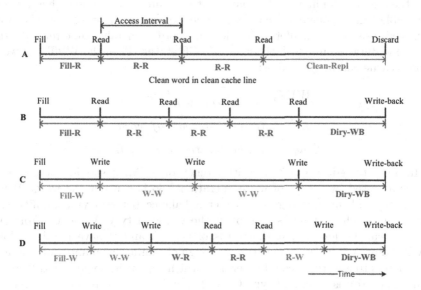

Fig. 1. The lifetime intervals of a word in different cache lines with respect to various access activities. Figure A illustrates a situation where the word is in clean cache line and all the access activities are read. Figure B, C and D illustrate three situations where the words are in dirty cache lines but have different access activities (Colour figure online).

Note that, for word in dirty line, no matter whether the last access to it is read or write, the phase varies from last access to written-back Dirty-WB. Besides, some words may neither be read nor written during its whole lifetime. If the word is in a clean cache line, it would not have any impact on the system, neither CPU nor memory, and its lifetime are regarded as fill-to-eviction (Fill-E) in our model. But if the word is in a dirty cache line, it may affect the lower memory hierarchy since it will be written back when a replacement occurs, called its lifetime fill-to-write-back (Fill-WB). Therefore, words that are never accessed contribute to the vulnerability only if they are written to lower memory. Its clearly that Fill-E is un-ACE while Fill-WB is ACE. Table 2 shows all the lifetime intervals described in our model and its ACE status.

Table 2. Words lifetime intervals classification and its ACE status.

Classification	ACE	un-ACE
Lifetime of word in clean line	Fill-R, R-R	Fill-E, Clean-Repl
Lifetime of word in dirty line	Fill-R, R-R, W-R, Fill-WB, Dirty-WB	Fill-W, W-W, R-W

3.2 Word Vulnerable Time

Word vulnerable time (WVT) is used to evaluate the reliability of each word with the intent of finding out key data that need protection. The higher a words vulnerable time is, the word has more time exposed to errors and needs to be protected. Based on our lifetime model, the word vulnerable time refers to the total time duration in which the word is ACE status. Then the WVT of a word in data cache can be defined as:

$$WVT_{word} = \sum_{LI \in ACE} Length_{LI}$$

$$LI : Lifetime\ Intervals\ of\ word$$

$$Length_{LI} : The\ duration\ time\ of\ LI.$$

In order to reduce the high write latency of NVM, it should not put data that is accessed by plenty of write operations. So it can put the words WVT and write/read amount together to get a balance between vulnerability and performance. Then, it tends to protect the words of type A and B. From the Fig. 1, it can be found that the words of type A and B have relatively long WVT and are not accessed by write. Through obtaining the number of read and write access to a word, it could determine which type it is. Type A and B words number of write is zero, and type C words number of read is zero, and type D is a compromise.

Algorithm 1 shows a methodology to compute the WVT for each word in data cache at runtime. It has modeled a write-back data cache and the lifetime intervals that need to be taken into account include all the ACE intervals listed in Table 2. In the Algorithm 1, WVT is used to stand for a word in cache and a cache line respectively. It also associates with 4 variables (VulnerableTime, WhenAccessd, ReadNum and WriteNum) with each word in the cache. The VulnerableTime represents the WVT of each word and the WhenAccessd represents the time when the word is recently accessed.

WhenAccessed is set or each word on a cache line Fill and update it on a Write access. At this point, the VulnerableTime is not changed since all lifetime intervals ending with Write are un-ACE (see Table 2). While on a Read access, it first changes the VulnerableTime and then update WhenAccessed for each word. On a Replacement or Flush, the VulnerableTime of each word updates in the cache line to add Dirty-WB interval only if the line is dirty. Finally, the cache at the end of the simulation is flushed to update the VulnerableTime of word in dirty lines that are still in cache. During the execution, the read and write amounts of each word also update correspondingly.

Algorithm 1. WVT computation for each word in cache

W_i : a word in data cache
CL_k : cache line
1: **if** W_i *in* CL_k *is accessed by op* **then**
2:      **switch** (*op*)
3:      **case** *write*:
4:        $WhenAccessed[W_i] = now$
5:        $WriteNum[W_i] + +$
6:      **end case**
7:      **case** *read*:
8:        $VulnerableTime[W_i] + = now - WhenAccessed[W_i]$
9:        $WhenAccessed[W_i] = now$
10:        $ReadNum[W_i] + +$
11:      **end case**
12:      **end switch**
13: **end if**
14: **if** CL_k *is filled* **then**
15:      **for** each $W_i \in CL_k$ **do**
16:        $WhenAccessed[W_i] = now$
17:      **end for**
18: **end if**
19: **if** CL_k *is replaced* **then**
20:      **switch** (*dirty bit*)
21:      **case** *dirty*:
22:      **for** each $W_i \in CL_k$ **do**
23:        $VulnerableTime[W_i] + = now WhenAccessed[W_i]$
24:      **end for**
25:      **case** *clean*:
26:      *do nothing*
27:      **end case**
28:      **end switch**
29: **end if**

3.3   Cache Vulnerability Factor of NVM

NVM-assisted cache vulnerability factor (NCVF) can be used to evaluate the reliability of data cache. The concept of NCVF is derived from AVF. The NCVF in this work is defined as the average fraction of all words lifetime during which the words are in ACE intervals over the total execution time. Then, the NCVF of data cache can be computed by:

$$NCVF = \frac{\sum_i^W \left(\sum_j^n ACE_j \right)}{W \times Total\_Exec\_Time}$$

Where W represents the total number of words in data cache, and ACE_j represents the time of j_{th} ACE interval of $word_i$, and Total_Exce_Time is the total simulation time for the benchmark. Based on our Word Vulnerable Time, the above equation can also be calculated as follows:

$$NCVF = \frac{\sum_{i=1}^{W}(WVT_i)}{W \times Total\_Exec\_Time}$$

WVT_i is the Word Vulnerable Time of word i which has been described in Sect. 3.1. Note that our lifetime analysis and WVT is performed at word level, so our NCVF describes the reliability of data cache at word granularity.

4   Experiments and Results

For our experiments, we have implemented the reliability estimation method and NVM component simulator in the SimpleScalar/ARM toolset. To evaluate the vulnerability of L1 data cache and to find the most important words, it has extended the SimpleScalar cache part to integrate our WVT computation algorithm described in Sect. 3. To simulate the static data allocation, the simulator is modified to trace and collect the L1 data cache access information, and then fed the access profile to our key words selection tool. The tool puts the address of key words into a list. After that, the key words address list is fed to the modified SimpleScalar with a NVM simulation unit. Based on the list, the simulator could determine whether an access is a NVM access or a cache access. The simulator models an embedded microprocessor with a 32 KB L1 data cache.

It can collect the output of un-optimized running on simulator without NVM as the baseline. Then the baseline is compared with the results of optimized program running on simulator with NVM. Two types of available NVM space are compared in our work: 32 KB and 64 KB. In this experiment, the STT-RAM is used to realize the on chip NVM SPM. Two types available SPM space are also evaluated: 32 KB and 64 KB. To measure the performance power overhead of the STT-RAM, it has used dynamic energy consumption of read and write operations. And to evaluate the performance, the long write latency of STT-RAM is considered.

Then, it can compute NCVF of the benchmarks for base case and improved case using our formula proposed in Sect. 3.3 below in order to evaluate the reliability of data cache. The NCVF of the benchmarks for base case is shown in Table 3. As shown in the Fig. 2, it can find that NCVF has been reduced in some degree by using our improved NVM based protection architecture. It shows that

Table 3. NCVF of the benchmarks for base case.

basicmath	bitcount	blowfish	crc	dijkstra	fft	patricia
0.0163	0.043	0.0862	0.045	0.7257	0.8187	0.918
quicksort	rijndael	sha	stringsort	susan	jpeg	rsynth
0.7215	0.1583	0.1479	0.25	0.5247	0.7326	0.4785

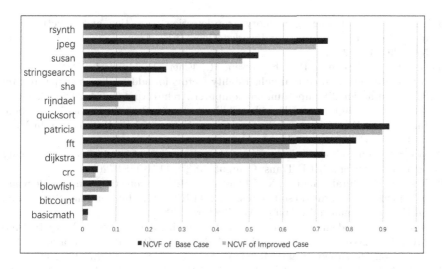

Fig. 2. Comparison of NCVF between traditional case and improved case.

the potential vulnerable data have been protect well with the assistance of compiler on a hybrid memory hierarchy with STT-RAM.

5   Conclusions

In this study, we use a formula to compute the vulnerability of data cache. It needs to point out that our vulnerability model is only focused on computing the reliability of data cache not the full system and it has considered the errors that propagates from data cache to other parts of the system rather than that from other components to data cache. If the data cache has a high NCVF, it is more susceptible to soft errors and more likely to affect the correctness of the program execution. So the objective of our NVM based protection scheme is to reduce the NCVF of data cache. It has used NCVF to measure the impact on reliability of data cache when inducing NVM component in following sections.

References

1. Baumann, R.: Soft errors in advanced computer systems. IEEE Design Test Comput. **22**(3), 258–266 (2005)
2. Raoux, S., Burr, G., Breitwisch, M., Rettner, C., Chen, Y., Shelby, R., Salinga, M., Krebs, D., Chen, S.-H., Lung, H.-L., Lam, C.: Phase-change random access memory: a scalable technology. IBM J. Res. Dev. **52**(4.5), 465–479 (2008)
3. Tehrani, S., Slaughter, J.M., Deherrera, M., Engel, B.N.: Magnetoresistive random access memory using magnetic tunnel junctions. Proc. IEEE **91**(5), 703–714 (2003)
4. Eltawila, A.A., Engelb, M., Geuskensc, B.: A survey of cross-layer power-reliability tradeoffs in multi and many core systems-on-chip. Microprocess. Microsyst. **37**(8, Part A), 760–771 (2013)

5. Slayman, C.W.: Cache and memory error detection, correction, and reduction techniques for terrestrial servers and workstations. IEEE Trans. Device Mater. Reliab. **5**(3), 397–404 (2005)
6. Biswas, A., Racunas, P., Cheveresan, R., Emer, J., Mukherjee, S.S., Rangan, R.: Computing architectural vulnerability factors for address-based structures. In: 32nd International Symposium on Computer Architecture (ISCA 2005) (2005)
7. Zhang, W.: Computing cache vulnerability to transient errors and its implication. In: 20th IEEE International Symposium on Defect and Fault Tolerance in VLSI Systems (DFT 2005) (2005)
8. Wang, S.: On the characterization and optimization of on-chip cache reliability against soft errors. IEEE Trans. Comput. **58**(9), 1171–1184 (2009)
9. Yoon, D.H., Muralimanohar, N., Chang, J.: FREE-p: protecting non-volatile memory against both hard and soft errors. In: 2011 IEEE 17th International Symposium on High Performance Computer Architecture (2011)
10. Kadayif, I., Kandemir, M.: Modeling and improving data cache reliability. In: SIGMETRICS 2007 Proceedings of the 2007 ACM SIGMETRICS International Conference on Measurement and Modeling of Computer Systems, vol. 12 (2007)
11. Nair, A.A., Eyerman, S., Chen, J.: Mechanistic modeling of architectural vulnerability factor. ACM Trans. Comput. Syst. (TOCS) **32**(4), 11 (2015). TOCS Homepage archive
12. Wang, K.L., Alzate, J.G., Amiri, P.K.: Low-power non-volatile spintronic memory: STT-RAM and beyond. J. Phys. D: Appl. Phys. **46**(7), 074003 (2013)
13. Mukherjee, S.S., Weaver, C., Emer, J., Reinhardt, S.K., Austin, T.: A systematic methodology to compute the architectural vulnerability factors for a high-performance microprocessor. In: Proceedings of the 36th Annual IEEE/ACM International Symposium on Microarchitecture, p. 29. IEEE Computer Society, December 2003
14. Biswas, A., Racunas, P., Emer, J., Mukherjee, S.: Computing accurate AVFs using ACE analysis on performance models: a rebuttal. IEEE Comput. Architect. Lett. **7**(1), 21–24 (2008)
15. Guthaus, M.R., Arbor, A., Ringenberg, J.S., Ernst, D., Austin, T.M.: MiBench: a free, commercially representative embedded benchmark suite, WWC-4. In: 2001 IEEE International Workshop on Workload Characterization (2001)

SOA Reference Architecture: Standards and Analysis

Yuan Yuan[1,2], Bo Li[3(✉)], and Heather Kreger[4]

[1] School of Computer Science and Engineering, Beihang University, Beijing, China
[2] Digital China Holdings Ltd., Beijing, China
yuanyuanc@dcholdings.com
[3] State Key Laboratory of Software Development Environment,
Beihang University, Beijing, China
libo@act.buaa.edu.cn
[4] IBM (International Business Machines Corporation), Beijing, China
kreger@us.ibm.com

Abstract. Service-Oriented Architecture (SOA) is an architectural style for designing systems in terms of services and the outcomes of services. In recent years, SOA has become a technology hot spot that is recognized and respected in industry. SOA Reference Architecture (RA) specifies fundamental elements of a SOA solution which be considered and used as a core guide for planning, designing, developing, deploying and managing SOA systems. In this paper, we give an overview of the standards on SOA RA developed by ISO/IEC JTC1/SC38, The Open Group, OASIS and China. The relationship and differences among those SOA RAs were analyzed. Finally a set of guidelines are developed for the application of SOA RAs in common situations.

Keywords: SOA · Reference architecture · RA · Standards

1 Introduction

Service Oriented Architecture (SOA) is an architectural style for facilitating Big Data, Cloud Computing [1], Internet of Things and Mobile Internet etc. SOA is an architectural style for designing systems in terms of services and the outcomes of services. 'Service' is used in SOA as the basic component to constitute or integrate systems that are suitable for a variety of business requirements.

A SOA Reference Architecture (SOA RA) standard enumerates the fundamental elements of a SOA solution or enterprise architecture for solutions and provides the architectural foundation for the solution by specifying the capabilities and architectural building blocks required to implement those capabilities. However, there are several SOA Architecture standards today in 1st Joint Technical Committee of International Standard Organization and the International Electrotechnical Commission (ISO/IEC JTC 1), the Organization for the Advancement of Structured Information Standards (OASIS), the Open Group (TOG) and China.

A joint document was released in 2009 as a result of that collaboration, Navigating the SOA Open Standards Landscape Around Architecture [6]. This paper was intended to help the SOA community at large to understand the myriad of overlapping standards

© Springer International Publishing AG 2017
M. Qiu (Ed.): SmartCom 2016, LNCS 10135, pp. 469–476, 2017.
DOI: 10.1007/978-3-319-52015-5_48

and other technical products produced by these organizations with specific emphasis on the Architecture of SOA. Since 2010, ISO/IEC JTC1/SC38 (Distributed application platforms and services)/WG2 (SOA) started to play a key role in the improvement of the reconciliation among various SOA concepts, definitions, technical frameworks, models, and architectures by establishing of one consolidated SOA RA as an international standard.

This paper describes the background and development processes, and analyses the differences and relationships between the multiple SOA RA related standards in OASIS, Open Group, China and ISO/IEC JTC1/SC38. It then presents guidance for how to adopt these SOA RAs in industries.

2   Standards of SOA RA

2.1   OASIS's SOA RM and RA

The OASIS Reference Model for SOA [2] is intended to capture the "essence" of SOA, as well as provide a core vocabulary and common understanding of SOA concepts. The goals of the reference model include a common conceptual framework that can be used consistently across and between different SOA implementations, common semantics that can be used unambiguously in modeling specific SOA solutions, unifying concepts to explain and underpin a generic design template supporting a specific SOA, and definitions that should apply to all SOA.

The OASIS Reference Architecture for SOA Foundation [3] is a view-based abstract reference architecture foundation that models SOA from an ecosystem/paradigm perspective. It specifies three viewpoints; specifically, the Participation in a Service Ecosystem viewpoint, the Realizing SOAs viewpoint, and the Ownership in a SOA Ecosystem viewpoint. Each of the associated views that are obtained from these three viewpoints is briefly described below.

2.2   Open Group's SOA Ontology and Reference Architecture

The Open Group SOA Ontology Technical Standard [4] is similar to the above OASIS Reference Model for SOA in that it captures a set of related concepts within the SOA space and explains what they are and how they are related to each other. The objectives are to facilitate understanding of these terms and concepts within the context of SOA, and potentially to facilitate model-driven implementation. The ontology is represented in UML [7] and OWL (Web Ontology Language) [8] to enable automation and allow tools to process it; for example, reasoning applications could use the SOA ontology to assist in service consumer and provider matching, service value chain analysis, and impact analysis.

The Open Group SOA Reference Architecture [5] is intended to support the understanding, design, and implementation of common system, industry, enterprise, and solution architectures leveraging the principles of an SOA.

The Open Group SOA Reference Architecture can represent both abstract enterprise scale designs as well as concrete SOA implementations. This SOA reference architecture provides the basis, template, or blueprint, for an enterprise architecture so that the architect can instantiate the standard during each individual project or solution that is being developed within an organization.

2.3 China's SOA RA

In China, SOA has become the primary method and technology extensively exploited in both software development and information system integration. In 2009, China National Information Technology Standardization Technical Committee (NITS) established a SOA Standard Working Group to promote development and application of SOA national standards. In 2012, the national standard of SOA General Technical Requirement was published. This specified a Technical Reference Architecture for SOA applications [5].

Figure 1 illustrates the scope, basic elements and technical capabilities requirements to apply to SOA-based applications.

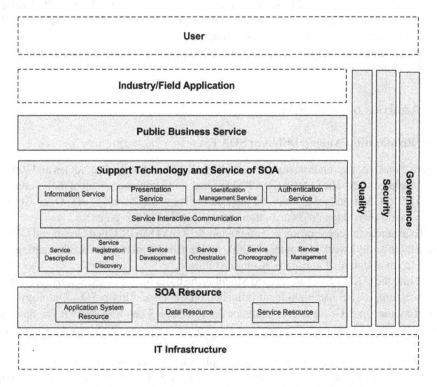

Fig. 1. Technical reference architecture for SOA solutions [5]

2.4 ISO/IEC JTC1's SOA Reference Architecture

In 2015, ISO/IEC 18384, Reference Architecture for Service Oriented Architecture, is published by ISO/IEC JTC 1, based on the contributions from China, U.S. Japan, Ireland, Korea, France, Canada, UK, Germany, The Open Group, OASIS etc.

ISO/IEC 18384 defines the terminology, technical principles, reference architecture and an ontology for SOA. The targeted audience of ISO/IEC18384 includes, but is not limited to, standards organizations; architects, architecture methodologists, system and software designers, business people, SOA service providers, SOA solution and service developers, and SOA service consumers who are interested in adopting and developing SOA. For example, this standard can be used to introduce SOA concepts and to guide to the developing and managing SOA solutions.

The Reference Architecture for SOA contains three parts:

1. Terminology and Concepts – which defines the terminology, basic technical principles and concepts as well as a meta model for SOA
2. Reference Architecture for SOA Solutions – which defines the detailed SOA reference architecture layers, including capabilities, architectural building blocks, as well as a the types of services in SOA solutions.
3. SOA Ontology – which defines the core concepts of SOA and their relationships in an ontology.

3   Analysis of SOA RAs

3.1   Relationship Among Different SOA RAs

Figure 2 outlines the relationship between the Reference Models, Ontologies and Reference Architectures. The OASIS SOA RM has had some influence on all of the standards. The JTC1 SC38 standard is based on the existing standards from OASIS, Open Group, China, and Japan (Japan's contribution is not discussed in this paper).

To promote the cooperation and alignment between OASIS, Open Group and OMG, these three groups worked together in 2009 and finished a joint paper on Navigating the SOA Open Standards Landscape Around Architecture.

While the OASIS SOA Reference Model is the basis for the OASIS SOA Reference Architecture, it also had significant influence on the Open Group SOA Ontology and therefore the Open Group SOA Reference Architecture. Therefore, there is fair amount of alignment on core concepts between the Open Group and OASIS SOA Reference Architecture.

China's RA is based on SOA-based application experience in various industries deployed by main vendors. It also refers to the SOA RM and RAs of OASIS and The Open Group. Most of the layers, key elements and capabilities defined in Chinese RA are consistent with the Open Group and OASIS RA.

The JTC1 SOA RA is intended to accommodate all of the aspects and requirements of all of the input standards, including a contribution from Japan not discussed in this paper.

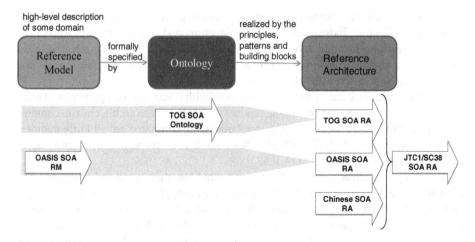

Fig. 2. Influence of standards

3.2 Difference Among Different SOA RAs

There are a number of features we can compare the reference architectures by. In Table 1, we compare the level of abstractness, scope, and focus.

Table 1. Difference analysis among SOA RAs

	OASIS RM	OASIS RA	Open Group RA	Open Group Ontology	Chinese SOA RA	JTC1/SC38 SOA RA
Objective	Concepts	Concepts relationships	Solution basis	Concepts 7 relationships	Solution basis	Concepts solutions
Abstract level	Most abstract	<OASIS RM> TOG RA	<OASIS RA	<OASIS RM	<OASIS RA = TOG RA	<OASIS RA = TOG RA
Type	Reference Model	Foundation	Common systems	Ontology	Common systems	Common systems
Focus	Neutral	Ecosystem	Enterprise	Neutral	Enterprise	Neutral and enterprise
Proposer	OASIS	OASIS	Open group	Open group	China	SC38
Representation	UML	UML	Stack	UML, OWL	Stack	Stack

Given SOA is a style, SOA solutions can be created using any platform or technology. According to TOGAF [9], architectures can be defined at different levels of abstraction ranging from foundation architectures to common systems architectures, and industry and organization-specific architectures. As the architecture becomes more concrete, architectural decisions and assumptions are made. Reference architectures should be evolved and instantiated as an industry architecture or organization-specific architecture for a particular domain of interest or for specific projects. They are useful to inform the architecture, guide the work of the solution team, including constraining choices in developing the solution.

Like architectures, the level of abstractness for reference architectures varies from very abstract defining just concepts, to concrete, where it is a reference architecture for an organization's business solution. Reference Architectures exist along the same continuum: completely abstract conceptual foundation reference architecture, generic reference architecture, industry reference architecture, and concrete reference architecture. Architectures that are more concrete have had more architectural principles, requirements, and decisions made for them. In addition different reference models and ontologies exist at different levels of abstractness.

Another feature we can compare architectures with is scope or completeness. Again, this is a continuum from narrow, single product or design pattern architecture to comprehensive, end to end architectures. The standards being compared are all fairly comprehensive in scope, though the OASIS SOA RA Foundation may be slightly more comprehensive than the Open Group SOA RA.

How the architecture is intended to be used should be understood and compared. The SOA RM, SOA Ontology, and SOA Foundation are to supply definitions of key concepts and the relationships between them. The SOA Ontology formalizes a set of these concepts with UML and OWL representations to enable tools and understanding. The SOA Foundation and SOA Ontology are fairly consistent but not identical in how the concepts are defined and related. The SOA Foundation defines an extensive set of concepts to be used for understanding the SOA ecosystem. The Open Group SOA RA provides a foundation for a technical solution with architectural building blocks that can be selected and implemented. These architectural building blocks and their implementations are used to implement and support services. China's SOA RA defines the foundation layers, basic elements and technical capabilities of SOA-based solutions in Chinese industries. It is applicable to the design, development, operation and governance and also serves as the basis for developing technical, quality, testing and engineering standards in China. The JTC1 SOA RA provides a general SOA foundation framework and common service categories for worldwide SOA solutions which could be adopted and specified in different countries and domains.

4   Guidelines for Adopting Different SOA RAs

First, which standards are appropriate depends on the stakeholders, their needs, particular viewpoints, users' SOA maturity, and the needs of the project. Referring to the development of the 'Navigating the SOA Open Standards Around Architecture [6] ' paper, we believe these guidelines could be taken into consideration and be helpful to architects:

For understanding SOA core concepts: For understanding concepts, all of these standards can be helpful, but provide different depths of concepts.

The OASIS Reference Model for SOA [2] provides a common vocabulary for understanding the "essence" of SOA. It is, by design, a highly abstract model targeting a large, cross-cutting audience that includes non-technical readers as much as it does technical readers.

The Open Group SOA Ontology [4] builds on a set of the concepts in the OASIS Reference Model for SOA and provides additional SOA concepts and relationships taken from the viewpoints of different stakeholders as well as an enterprise-wide perspective. It also provides as a common language for formally describing SOA concepts that can be leveraged by abstract as well as solution-oriented reference architectures. The ISO/IEC JTC1 SC38 Reference Architecture for Service Oriented Architecture provides terminology and the general technical principles underlying SOA, including principles relating to functional design, performance, development, deployment and management.

The Chinese General Technical Requirement of SOA [5] defines SOA conceptual model, reference architecture, and the basic technical capabilities requirements of SOA-based application.

The Open Group SOA Reference Architecture [5] defines the key architectural elements in SOA and the key relationships between these elements relevant to enterprises.

The OASIS Reference Architecture for SOA Foundation [3] does this as well for the SOA ecosystem and ownership viewpoints.

Understanding considerations for cross-ownership boundaries of SOA ecosystems:
While both SOA reference architectures provide guidance that is important for SOA implementations that span ownership boundaries, the OASIS Reference Architecture for SOA Foundation [3] is especially focused on this scenario and provides architectural considerations for interacting with services owned by another company.

Understanding the completeness of SOA architectures and implementations: The OASIS Reference Architecture for SOA Foundation [3] provides models that function as a checklist that can be used to evaluate architectures and implementations of SOA.

Understanding the deployment of SOA in an enterprise: The Open Group SOA Reference Architecture [5] provides a stack organization of SOA architectural building blocks for an enterprise and guidance on the use and deployment of these building blocks.

Understanding the basis for an industry or organizational reference architecture:
The Open Group SOA Reference Architecture [5], the Chinese General Technical Requirement of SOA [5], and the ISO/IEC JTC1 SC38 General Technical Principles [2] provides guidance on refining this SOA reference architecture into an industry or solution SOA reference architecture.

Understanding the implications of architectural decisions: The Open Group SOA Reference Architecture [5] provides guidance to SOA designers and implementers by providing a concrete basis for making architectural and design decisions. It provides a model and framework for evaluating architectural concerns for designing an SOA.

Understanding how to position vendor products in an SOA context: The Open Group SOA Reference Architecture [5] and the Chinese General Technical Requirement of SOA [5] provides a layered stack with architectural building blocks and capabilities that map naturally to vendor products available to support SOA.

Understanding the general baseline for different application scope: The OASIS, Open Group and ISO/IEC JTC1/SC38 RAs could be used in the international level. The Chinese General Technical Requirement of SOA should be the first option for SOA-based application in China.

5 Conclusion

This paper proposed the mapping of SOA RA standards of Open Group, OMG and OASIS to ISO/IEC JTC1 and China. It also presented analysis and suggestions for adopting SOA RA in the real world. The information in this paper will provide useful guidance for SOA based application scenarios, and also provide fundamental input for the future standardization on Big Data service aspect.

Acknowledgement. The authors gratefully acknowledge the anonymous reviewers for their helpful suggestions. The work reported in this paper has been supported by grants from the State Key Laboratory of Software Development Environment (SKLSDE-2015ZX-14), and Research Fund of Guangxi Key Lab of Multi-source Information Mining & Security (grant number MIMS15-02).

References

1. Deutsch, D.R.: Cloud computing: building firm foundations for standards development, ISO Focus+, November–December 2011
2. OASIS Reference Model for Service Oriented Architecture Version 1.0, 12 October 2006. http://docs.oasis-open.org/soa-rm/v1.0/soa-rm.pdf
3. OASIS Reference Architecture Foundation for Service Oriented Architecture Version 1.0, Committee Specification Draft 03, 06 July 2011. http://docs.oasis-open.org/soa-rm/soa-ra/v1.0/csd03/soa-ra-v1.0-csd03.pdf
4. The Open Group SOA Ontology Standard V1, October 2010. http://www2.opengroup.org/ogsys/jsp/publications/PublicationDetails.jsp?catalogno=c104, http://www.opengroup.org/soa/source-book/ontology/index.htm
5. The Open Group SOA Reference Architecture Standard V1, December 2011. http://www2.opengroup.org/ogsys/jsp/publications/PublicationDetails.jsp?catalogno=c119, http://www.opengroup.org/soa/source-book/soa_refarch/index.htm
6. Navigating the SOA Open Standards Landscape Around Architecture, Joint Paper by the Open Group, OASIS, and OMG, November 2009. http://www2.opengroup.org/ogsys/jsp/publications/PublicationDetails.jsp?catalogno=w096, http://www.opengroup.org/soa/source-book/stds/index.htm
7. OMG Unified Modeling Language (OMG UML), Superstructure, Version 2.2, OMG Doc. No.: formal/2009-02-02, Object Management Group (OMG), February 2009. www.omg.org/spec/UML/2.2/Superstructure
8. Web Ontology Language (OWL), World Wide Web Consortium (W3C), April 2009. www.w3.org/2007/OWL/wiki/OWL_Working_Group
9. The Open Group Architecture Framework (TOGAF), Version 9 Enterprise Edition, February 2009. www.opengroup.org/togaf

An Evolutionary Approach for Short-Term Traffic Flow Forecasting Service in Intelligent Transportation System

Fan Fei[1,2], Shu Li[1,2], Wanchun Dou[1,2(✉)], and Shui Yu[3]

[1] State Key Laboratory for Novel Software Technology,
Nanjing University, Nanjing, China
ffwilliam1992@gmail.com, shuli@smail.nju.edu.cn, douwc@nju.edu.cn
[2] The Department of Computer Science and Technology,
Nanjing University, Nanjing, China
[3] School of Information Technology, Deakin University, Melbourne, Australia
shui.yu@deakin.edu.au

Abstract. In recent years, traffic flow prediction has become a crucial technique in ITS (intelligent transportation system), which is helpful for alleviating the congestion in many metropolises and improving the efficiency of public traffic service. On the other hand, with the development of traffic sensors, traffic data are collected with a fantastic scale. It leads ITS into a data-driven application fashion. With this observation, it is a challenge to accurately and promptly forecast the traffic flow by effectively utilizing the big traffic data. In view of this challenge, in this paper, we propose an evolutionary method for short-term traffic flow forecasting service. Concretely, in our method, traffic flow is firstly specified by a model of time series. Then, the model is decomposed into seasonal component and the residual component. The seasonal component reflects the history average condition, while we treat the residual component as the output of a linear filter. The proposed method is evaluated with real bus transaction dataset. The experimental results show the effectiveness of our method.

Keywords: Public traffic service · Traffic flow prediction · Time series · Evolutionary method

1 Introduction

In recent years, the traffic flow prediction has become a crucial area in ITS (intelligent transportation system), which aims to alleviate the more and more deteriorate traffic congestion problem in the metropolises such as New York, Tokyo, Beijing, etc. [1,2] With the rapid development of traditional and new-emerging traffic sensor, such as RFID sensor, ILDs, etc., traffic data are exploding [3,4]. The availability of massive traffic data collected from different sources has transformed the traditional technology-driven ITS to a versatile and powerful data-driven ITS.

© Springer International Publishing AG 2017
M. Qiu (Ed.): SmartCom 2016, LNCS 10135, pp. 477–486, 2017.
DOI: 10.1007/978-3-319-52015-5_49

Much work have been done that focus on the short-term prediction of amount of vehicles passing a freeway or a city arterial street [5,6,13]. The accurate and timely prediction of the passenger flow of some bus line, as a kind of traffic flow, is also crucial in that it can help the bus dispatch center determine how many buses of certain bus line are dispatched and therefore improve the public traffic service. The short-term prediction of traffic flow calls for the real-time accumulation of the recent traffic flow. The amount of vehicles passing certain street aggregating to 15 min, 30 min or 1 h can be counted by the on-road sensor such as inductive loop detectors (ILDs), RFID reader or the HD cameras. These information can be transmitted to the database and make possible the short-term prediction of such kind of traffic flow. In the past, the information about the passenger flow is obtained when the bus arrive at the destination and the transaction records, stored in U disk, are transmitted to the database. Today, the new equipped bus card reader can transmit these transaction records information in real-time, which technically supports the short-term prediction of the passenger flow.

Predicting the traffic flow, in particular the short-term traffic flow, is a complicated problem in that the short-term traffic flow is always subject to chaotic property, especially in the developing country [7]. Temporal effects like festival, traffic accident, road construction, weather can have a drastic impact on the traffic flow. The traffic flow prediction can be usually characterized by a time series of both historical and real-time data collected from multiple sources [5]. Formally, given a series $\{X_1, X_2, \cdots, X_t\}$, where X_t denotes the traffic flow volume from t to $t + 1$, in equal time interval, the problem can be stated as predicting X_{t+1}, X_{t+2}, X_{t+i} where $i \in N^+$.

Previous researches have achieved a great deal in the short-term traffic flow prediction, however, the researches neglect mining the information about the error between prediction and afterwards observation. These information generally reflect some temporal effects and the ignored factors that may affect the traffic flow. As these information is not easy to foresee, utilizing these information timely can help for further prediction. Another aspect is that a model with concrete interpret-ability is more important sometimes when compared with accuracy. For example, the traffic manager may prefer a prediction with error estimation so as to estimate the possibility when the prediction fails and take every measure to curb the worst case.

The remainder of our paper is organized as follows: Sect. 2 provides a concise introduction of the related preliminary knowledge; Sect. 3 presents the algorithm and gives a theoretical analysis about the solution to the problem and the evaluation of the proposed method is presented in Sect. 4; Sect. 5 presents the related work and Sect. 6 concludes this paper and shows our future work.

2   Preliminary Knowledge

2.1   ARIMA Model

ARIMA model, which refers to autoregressive integrated moving average model. ARIMA model is extensively deployed in the area of warehouse management, stock exchange prediction, risk management, etc. The ARIMA model assume that the future value of a variate, in our case the volume of passenger flow, is a linear function of several past observation value and random errors.

A general ARIMA model of order (p, d, q) is written as

$$\phi(B)\nabla^d R_t = \theta(B)e_t, \tag{1}$$

where B is the back-shift operator $B^d R_t = R_{t-d}$; ∇ is the single lag difference operator; $\nabla = 1 - B$ and $\nabla^d = (1 - B)^d$. e_t represents the random error at time t, reflecting the temporal effect of the system. $\phi(B)$ and $\theta(B)$ represent the autoregressive and moving average operator respectively, which are defined as

$$\phi(B) = 1 - \phi_1 B - \phi_2 B^2 - \cdots - \phi B^p \tag{2}$$

$$\theta(B) = 1 - \theta_1 B - \theta_2 B^2 - \cdots - \theta B^q \tag{3}$$

where $\phi_1, \phi_2, \ldots, \phi_p$ are the autoregressive coefficients and $\theta_1, \theta_2, \ldots, \theta_q$ are the moving average coefficients. e_t is generally regarded as the Gaussian white noise with variance σ^2.

Two important property of e_t will be used in the following discussion.

1. Correlation coefficient is zero for any two different time, namely

$$cov(e_t, e_s) = \begin{cases} 0 & if \ s \neq t, \\ \sigma^2 & if \ s = t. \end{cases} \tag{4}$$

2. The conditional expectation of the white noise e_{t+l} at time t is zero, namely

$$E_t[e_{t+l}] = 0, \quad l \in N^+. \tag{5}$$

When the order of d is zero, the ARIMA model degenerates into the ARMA (autoregressive moving average) model, which is a linear time series model. The time series depicted by ARMA model can be regarded as the output of the linear filter $(\psi_1, \psi_2, \ldots, \psi_j \ldots)$ for input $\{e_t\}$, which can be defined as

$$R_t = e_t + \psi_1 e_t + \psi_2 e_t + \cdots = \sum_{j=0}^{\infty} \psi_j e_{t-j}, \tag{6}$$

where $\psi_1, \psi_2, \ldots, \psi_j \ldots$ are the coefficients of the linear filter and $\psi_0 = 1$.

At time t we rewrite the above equation for time $t + l$ as

$$R_{t+l} = \sum_{j=0}^{l-1} \psi_j e_{t+l-j} + C_t(l), \tag{7}$$

where $C_t(l) = \sum_{j=l}^{\infty} \psi_j e_{t+l-j}$. According to Eq. (5), the condition expectation of R_{t+l} is given as

$$E[R_{t+l} | R_t, R_{t-1}, R_{t-2}, \ldots] = C_t(l). \tag{8}$$

2.2   Definition and Notation

Definition 1 A passenger flow time series. $PFTS$ is a sequential series of passenger flow volume, which are collected from, in our case, the card reader and are aggregated in one hour, 24 h/day.

$$PFTS = \{X_t\} = \{\cdots X_{t-1}, X_t, X_{t+1} \cdots\}$$

where X_t is the traffic flow volume from $t - 1$ to t and $t = 0, 1, ..., 23$.

Table 1 introduces the main symbols and notation used in our paper.

Table 1. Table of key notations

Notation	Description
X_t	Passenger flow volume from $t - 1$ to t
S_t	Seasonal component of passenger flow volume from $t - 1$ to t
R_t	Residual component of passenger flow volume from $t - 1$ to t
$\widehat{X_{t+l}}$	Predicted passenger flow volume from $t + l - 1$ to $t + l$
B	Back-shift operator, which means $B^j X_t = X_{t-j}$
∇	Single lag difference operator, $\nabla X_t = X_t - X_{t-1}$
e_t	Gaussian white noise at t, $i = 1, 2, 3 \ldots$
$\widehat{z_t(l)}$	Passenger flow volume from $t + l - 1$ to $t + l$ predicted at time t

3   An Evolutionary Approach for Short-Term Traffic Flow Forecasting

3.1   History Average Method and STL Decomposition

We observe from the dataset described in Sect. 5 that the weekdays passenger flow are similar to each other, while the Saturday and the Sunday show the similarity.

According to the observation above, the $PFTS$ can be sorted to two category, the weekdays $PFTS$ and the weekends $PFTS$. Based on the history record, we can count the passenger flow volume of each hour as the future prediction and treat weekdays and weekends respectively. In the following discussion, we use $PFTS_{weekday}[t]$ and $PFTS_{weekend}[t]$ representing the average passenger flow volume from $t - 1$ to t in weekday and weekend respectively.

We also observe that both the $PFTS$ in weekday and in weekend fluctuate sharply over a day, and does not satisfy the stationary prerequisite. Generally, difference operation is widely used to transform the primary time series stationary. However, the first order autocorrelation coefficient $\{\nabla X_t\}$, in our experiment on the real data depicted in Sect. 6, is only approximately 0.4 and is not remarkable enough to capture any useful linear relationship of the time series.

Therefore, STL decomposition is deployed as a preprocessing step to acquire a stationary time series.

STL decomposition is a widely used method to decompose a time series into seasonal, trend, remainder component.

We assume that the primary $PFTS$ is an addition of the seasonal component $\{S_t\}$ which is deterministic and the residual component $\{R_t\}$, which is a stochastic time series, namely

$$X_t = S_t + R_t. \tag{9}$$

3.2 Evolutionary Method

After STL decomposition, the residual component can be viewed as an output of the linear time-invariant filter Ψ, with coefficient $\psi_1, \psi_2, \psi_3 \ldots$, as is shown in Eq. (6).

The order of the ARIMA model can be determined according to the autoregressive function and AIC criteria. Here, we use the ARIMA model with order $(2, 0, 0)$.

Therefore, according to the Eq. (1) the residual component of $PFTS$ is modeled as:

$$R_t = \phi_1 R_{t-1} + \phi_2 R_{t-2} + e_t. \tag{10}$$

The autoregressive coefficient, namely ϕ_1 and ϕ_2 can be calculated from the training dataset and in this paper, we use the statistical software to determine these parameters. As is the passenger flow pattern in weekend and weekdays are remarkably different, when we use software to determine the autoregressive coefficient, we treat the weekday and weekend respectively.

According to the Eq. (10), to predict R_{t+1}, we need the value of R_t and R_{t-1}. e_{t+1} can be regarded as zero according to the Eq. (5).

The basic l−periods prediction for X_t can be implemented by the following algorithm.

The Algorithm 1 implements the l−periods forecasting. At time $t + 1$, there are more information about passenger flow, therefore, we can improve our prediction of the rest $l - 1$ periods.

Theorem 1. $\widehat{z_{t+1}(l - 1)} = \widehat{z_t(l)} + \psi_{l-1} e_{t+1}.$

Proof. According to the STL decomposition, we get $\widehat{z_t(l)} = S_{t+l} + \widehat{r_t(l)}$ and $\widehat{z_{t+1}(l-1)} = S_{t+l} + \widehat{r_{t+1}(l-1)}$. $\widehat{z_{t+1}(l-1)} - \widehat{z_t(l)} = \widehat{r_{t+1}(l-1)} - \widehat{r_t(l)}$. For $\widehat{r_t(l)}$, namely the conditional expectation of R_{t+l} at time t, we substitute it for $\widehat{C_t(l)}$ according to Eq. (8). $\widehat{r_{t+l}(l-1)}$ therefore can be replaced by $\widehat{C_{t+1}(l-1)}$. $\widehat{r_{t+1}(l-1)} - \widehat{r_t(l)} = \widehat{C_{t+1}(l-1)} - \widehat{C_t(l)} = \psi_{l-1} e_{t+1}$. The theorem is proved.

Following Theorem 1, when we let $l = 1$ and immediately get the following corollary.

Algorithm 1. The basic $l-$periods forecasting

Input: Forecast date D, begin time t, periods l, R_t and R_{t-1}
Output: The $l-$periods forecasted passenger flow volumes $\{\widehat{X_{t+1}}, \widehat{X_{t+2}}, \ldots, \widehat{X_{t+l}}\}$
1: **if** D is weekday **then**
2:      $\phi_1 = \phi_{1,weekday}$; $\phi_2 = \phi_{2,weekday}$;
3: **else**
4:      $\phi_1 = \phi_{1,weekend}$; $\phi_2 = \phi_{2,weekend}$;
5: **end if**
6: $//\widehat{R_t} = R_t$ and $\widehat{R_{t-1}} = R_{t-1}$ at time t
7: **for** $i \leftarrow 1$ to l **do**
8:      $\widehat{R_{t+i}} = \phi_1 \widehat{R_{t+i-1}} + \phi_2 \widehat{R_{t+i-2}}$
9: **end for**
10: **if** D is weekday **then**
11:      **for** $i \leftarrow 1$ to l **do**
12:          $\widehat{X_{t+i}} = \widehat{R_{t+i}} + PFTS_{weekday}[t + i]$;
13:      **end for**
14: **else**
15:      **for** $i \leftarrow 1$ to l **do**
16:          $\widehat{X_{t+i}} = \widehat{R_{t+i}} + PFTS_{weekend}[t + i]$;
17:      **end for**
18: **end if**
19: **return** $\{\widehat{X_{t+1}}, \widehat{X_{t+2}}, \ldots, \widehat{X_{t+l}}\}$

Algorithm 2. update method for the rest $(l - 1)-$periods

Input: begin time t, periods l, X_{t+1}, primary forecasting $\{\widehat{X_{t+1}}, \widehat{X_{t+2}}, \ldots, \widehat{X_{t+l}}\}$
Output: Passenger flow volume $\{\widehat{X_{t+2}}, \widehat{X_{t+2}}, \ldots, \widehat{X_{t+l}}\}$
1: $e_{t+1} \leftarrow X_{t+1} - \widehat{X_{t+1}}$
2: **for** $i \leftarrow 2$ to l **do**
3:      $X_{t+i} = X_{t+i} + \psi_{i-1}e_{t+1}$
4: **end for**
5: **return** $\{\widehat{X_{t+2}}, \widehat{X_{t+2}}, \ldots, \widehat{X_{t+l}}\}$

Corollary 1. $R_{t+1} = \widehat{r_t(1)} + e_{t+1}$.

Equation (6) can be rewritten as $\psi(B)e_t = R_t$ where $\psi(B) = 1+\psi_1 B+\psi_2 B^2+ \cdots + \psi_i B^i + \cdots, i \in N^+$. In our case, the order of the ARIMA model is $(2,0,0)$, that is $\phi(B)R_t = e_t$. We substitute $\psi(B)e_t$ for R_t, deducing $\phi(B)\psi(B) = 1$. That is
$$(1 - \phi_1 B - \phi_2 B^2)(1 + \psi_1 B + \psi_2 B^2 + \cdots + \psi_i B^i + \cdots) = 1$$
Let the coefficient of B with power greater than zero be zero, we can get

$$\begin{aligned} \psi_1 &= \phi_1 \\ \psi_2 &= \phi_1\psi_1 + \phi_2 \\ &\cdots \\ \psi_l &= \phi_{l-2}\psi_{l-1} + \phi_{l-2}\psi_{l-1}. \end{aligned} \qquad (11)$$

Then, following Eq. (11) we can calculate the coefficient of ψ_l, $l \in N$ iteratively. At time $t + 1$, the update for $\widehat{R_{t+i}}, i = 2, 3, .. l$ can be implemented by Algorithm 2.

4   Experimental Evaluation

4.1   Experiment Data Sets and Experiment Context

The dataset used in our experiment is the real transaction data collected from the bus from Foshan to Guangzhou. The bus line is 281, which is an inter-city bus line. There are totally 6020530 transaction records, ranging from 2014/08/01 to 2014/12/31, which consist of the identification of the card, the deal time and the type of the card, etc.

Technically, we conducted our experiment in a HANA cluster environment. The proposed services are distributed in the cluster. As a result of making full potential of the memory database, our method can be implemented in a short response time (Table 2).

Table 2. The experiment settings

	Client	HANA cluster
Hardware	Lenovo ThinkpadT430 machine with Intel i5-3210M 2.50GHz processor, 8GB RAM and 250GB Hard Disk	**Master** (1 node): HP Z800 Workstation Intel(R)Multi-Core X5690 Xeon(R), 3.47GHz/12M Cache, 6cores, 2 CPUs, 128GB (8 × 8GB+4X16GB) DDR3 1066MHz ECC Reg RAM, 2TB 7.2K RPM SATA Hard Drive **Slave** (1 node): HP Z800 Workstation Intel(R) Multi-Core X5690 Xeon(R), 3.47GHz/12M Cache, 6cores, 2 CPUs, 128GB (8 × 8GB+4X16GB) DDR3 1066MHz ECC Reg RAM, 2TB 7.2K RPM SATA Hard Drive
Software	Windows 7 Professional 64bit OS and HANA Studio	SUSE Enterprise Linux Server 11 SP3 and SAP HANA Platform SP07

4.2   Performance Evaluation

The volume in the morning and evening are remarkably high and the nocturnal volume is roughly zero. Therefore, the nocturnal forecasting service is less meaningful and when comparing the predictive ability of the different methods, we only take into account the passenger flow volume from 6:00 to 22:00.

In this research, we explore root-mean-square error (RMSE) and mean absolute percentage error (MAPE) for performance analysis, which are defined as

$$RMSE = \sqrt{\frac{1}{K}\sum_{k=1}^{K}(X_k - \widehat{X_k})^2} \tag{12}$$

$$MAPE = \frac{1}{K}\sum_{k=1}^{K}\frac{\left|X_t - \widehat{X_t}\right|}{X_k} \times 100\%. \tag{13}$$

where X_k denotes the actual passenger volume from 6:00 to 22:00 and $\widehat{X_k}$ denotes the predicted value generated by different prediction model. To measure the predictive ability of the model for each day, the value of K is identically 16 when X_k denotes the passenger volume of one hour.

Without the loss of generality, we take a whole week for performance evaluation, from 2014/12/01 to 2014/12/07. The compared method are:

1. HA (History average method) considers the weekdays and weekend passenger flow volume separately, and use the calculated history average volume as the predicted value.
2. RW (Random Walk) explores the STL decomposition first and assume that $R_t = R_{t-1} + e_t$ where the R_t denotes the residual component and e_t is a random variate that follows the Gaussian distribution. In practice, as the expectation of the random variate e_t is zero, the predicted value $\widehat{R_t}$ is identically R_{t-1}.
3. ANN (Artificial Neural Network) uses single hidden layer feed forward network is deplored to model the passenger flow [8].

Figure 1 shows that there are four days out of seven days that our method outperforms other methods in term of MAPE. Figure 2 shows that there are five days out of seven days that our method outperforms other method in term of RMSE. As is shown in Fig. 3, five predictions from five successive hours illustrate that the trend predicted passenger flow trend approximate the actual trend. Figure 4 compare the prediction of the passenger flow volume from 15:00 to 16:00 and show that the prediction is outperformed by the afterwards prediction and the most recent prediction is remarkably more approximate the actual value.

5   Related Work

Time series analysis has been widely used to model the traffic flow and has achieved great success. The ARIMA model [2] is first utilized to solve the short-term traffic flow problem. A. Abadi et al. [10] proposed an algorithm based on the autoregressive model that can adjust itself to unpredictable events and pay attention to the sparsity of the traffic data. Billy M. Williams et al.

Another group of methods are related with the prosperous area in recent years, machine learning, which have been widely deployed in ITSs field. These methods,

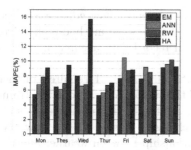

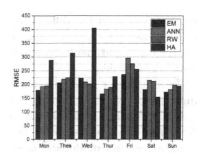

Fig. 1. Comparison of EM, ANN, RW, HA in MAPE (%)

Fig. 2. Comparison of EM, ANN, RW, HA in RMSE

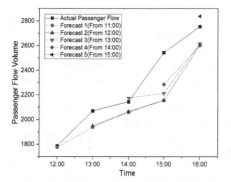

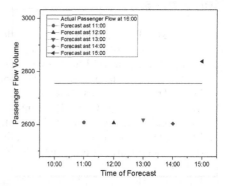

Fig. 3. Monday afternoon prediction showing evolution

Fig. 4. Convergence of 16:00 forecast on a Monday afternoon

unlike ARIMA model, always deploy nonparametric model. The two representative methods are certainly artificial neutral networks (NNs) [10,11] and support vector regression [13]. These methods have strong capacity to handle the complex and uncertain nonlinear traffic time series. Deep learning method [4,6] is also proposed to model the traffic flow in a deep architecture model. Tigran T. Tchrakian *et al.* [12] used spectral analysis method for the implementation of short-term traffic flow prediction with the capacity of real-time updating.

6   Conclusion and Future Work

This paper has proposed an evolutionary method for the passenger flow prediction and has demonstrated that the proposed method is effective. We sort the daily passenger flow pattern into two categories, namely weekdays and weekends and calculate the average passenger flow volume for each time interval, one hour a unit in our paper. A STL decomposition is deployed to make stationary the original passenger flow time series. Then, based on the history record, we give predicted passenger flow prediction of several periods and update the future prediction based on the difference of the actual passenger flow volume and the predicted passenger flow volume. To verify that the proposed method is effective, we carry out experiments on actual transaction record data accumulated from the bus card reader.

We also notice that the proposed evolutionary method requires further improvement. For example the parameter of the model comes from the training data and is fixed for weekday and weekend. Actually, these parameters shift a little as the daily passenger flow pattern changes. Hence, we are now trying optimizing our model and mining the relationship between weather and the passenger flow pattern for a dynamic adjustment of the parameters.

References

1. Bolshinsky, E., Freidman, R.: Traffic flow forecast survey, Technion–Israel Institute of Technology–2012–Technical Report–15p (2012)
2. Kong, Q.-J., Xu, Y., Lin, S., Wen, D., Zhu, F., Liu, Y.: UTN-model-based traffic flow prediction for parallel-transportation management systems. IEEE Trans. Intell. Transp. Syst. **14**(3), 1541–1547 (2013)
3. Chen, C.P., Zhang, C.-Y.: Data-intensive applications, challenges, techniques and technologies: a survey on big data. Inf. Sci. **275**, 314–347 (2014)
4. Lv, Y., Duan, Y., Kang, W., Li, Z., Wang, F.-Y.: Traffic flow prediction with big data: a deep learning approach. IEEE Trans. Intell. Transp. Syst. **16**(2), 865–873 (2015)
5. Lippi, M., Bertini, M., Frasconi, P.: Short-term traffic flow forecasting: an experimental comparison of time-series analysis and supervised learning. IEEE Trans. Intell. Transp. Syst. **14**(2), 871–882 (2013)
6. Huang, W., Song, G., Hong, H., Xie, K.: Deep architecture for traffic flow prediction: deep belief networks with multitask learning. IEEE Trans. Intell. Transp. Syst. **15**(5), 2191–2201 (2014)
7. Mohan, P., Padmanabhan, V.N., Ramjee, R.: Nericell: rich monitoring of road and traffic conditions using mobile smartphones. In: Proceedings of the 6th ACM Conference on Embedded Network Sensor Systems, pp. 323–336. ACM (2008)
8. Hou, Y., Edara, P., Sun, C.: Traffic flow forecasting for urban work zones. IEEE Trans. Intell. Transp. Syst. **16**(4), 1761–1770 (2015)
9. Abadi, A., Rajabioun, T., Ioannou, P., et al.: Traffic flow prediction for road transportation networks with limited traffic data. IEEE Trans. Intell. Transp. Syst. **16**(2), 653–662 (2015)
10. Jun, M., Ying, M.: Research of traffic flow forecasting based on neural network. In: Second International Symposium on Intelligent Information Technology Application, IITA 2008, vol. 2. pp. 104–108. IEEE (2008)
11. Jiang, X., Adeli, H.: Dynamic wavelet neural network model for traffic flow forecasting. J. Transp. Eng. **131**(10), 771–779 (2005)
12. Tchrakian, T.T., Basu, B., O'Mahony, M.: Real-time traffic flow forecasting using spectral analysis. IEEE Trans. Intell. Transp. Syst. **13**(2), 519–526 (2012)
13. Jeong, Y.-S., Byon, Y.-J., Mendonca Castro-Neto, M., Easa, S.M.: Supervised weighting-online learning algorithm for short-term traffic flow prediction. IEEE Trans. Intell. Transp. Syst. **14**(4), 1700–1707 (2013)

The Production Investigation of DOM Based on Full Digital Photogrammetric System VirtuoZo

Dan Zhang[1], Kuan He[1], Xuemin Shi[2(✉)],
Zhengpeng Wu[3], and Hui Zhang[1]

[1] Department of Surveying and Mapping Engineering,
Yellow River Conservancy Technical Institute, Kaifeng 475004, Henan, China
[2] School of Civil Engineering and Architecture, Henan University,
Kaifeng 475000, Henan, China
shixuemin@163.com
[3] Tianjin Institute of Surveying and Mapping, Tianjin 300381, China

Abstract. Based on the airborne LiDAR data from Jiyuan test area, this paper studies the technical scheme and operation flow of the all - digital photogrammetry workstation VirtuoZo, which is used to rapidly produce and update geo-information data DOM for surveying and mapping. It provides the technical support for the production standard formulation of the geographic information.

Keywords: The digital orthogonal map · VirtuoZo · Three-dimensional triangulation

1 Introduction

Since the early 1980s, the production application of the large scale topographic map using aerial survey technology was carried out in some large cities, such as Beijing, Shanghai, Shenyang, Wuhan and so on, in china. But with the development of the national economic construction, the demand of large scale digital orthogonal map (DOM) that used in city planning, city construction and management increase, and meanwhile the currency requirement is increasing day by day.

Commonly, the image data of aerophotograph, MODIS images and SPOT images are used in surveying and mapping production, and these images are obtained by aerial photography or remote sensing satellites. The production process of the digital orthogonal map is described as follows: first of all, some ground control points are measured on the image pair; and secondly the digital elevation model (DEM) within the scope of this image pair are generated using the digital photogrammetric workstation; thirdly the tilt error and projection error of the image are corrected and finally the film map of digital photograph of each single model is generated. Then a number of the digital orthogonal map are jointed together, and after the image smoothing process, the digital orthogonal map is cut out according to a certain range of the image.

In order to meet the demand of the frame construction of basic geographic spatial infrastructure in Jiyuan city, this paper discusses the production method of the digital

M. Qiu (Ed.): SmartCom 2016, LNCS 10135, pp. 487–494, 2017.
DOI: 10.1007/978-3-319-52015-5_50

orthogonal map (DOM) through the 1: 10000 DOM produced examples based on all Digital Photogrammetry Workstation VirtuoZo of Supresoft company.

2 Technical Scheme

The basic process to produce DOM using VirtuoZo digital photogrammetric workstation is shown in Fig. 1.

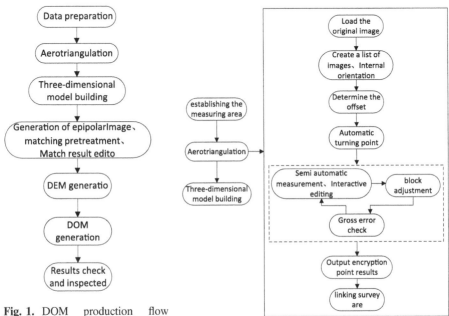

Fig. 1. DOM   production   flow chart

Fig. 2. Aerotriangulation flow chart

2.1 Data Preparation

2.1.1 Data Analysis of Survey Area

Flight quality: flight route curvature to meet the requirements; no aerial absolute vulnerability discovery; course overlap in 65%–75%, side lap general in 35–50%, but because of the overlap is too large, the measurement of photo control point is increased.

Image quality: because of aerial camera focal length, aerial photographic materials and the aerial vertical visibility, there is a phenomenon of color restoration distortion of aerial film which affect external interpretation of the fringe section and the color rendition of the whole image, we should use the central part of the image as much as possible to over come the shortcoming of the image quality.

2.1.2   Data Preparation of Survey Area

Scanning Digital Image: the scanning resolution is 255 μm, and the image is stored in TIFF format without compression. The scanned image should be moderate contrast, full color, full image and frame label clear.

Image Smoothing: color levels adjustment. The computer quantify image samples to 256 levels. After the comparison test, the gray scale output is defined between 30 and 225, which not only meets the requirement of the operator, but also is less lost of image. The colour levels of orthophoto map which is generated by correcting and mosaicking are again adjusted to 0 to 255, and it can improve image clarity and ensure uniform color balance. For single image adjustment, the "histogram" function is used, and for the measurement of different light and shade areas, the "curve" adjustment method is used. The photos should be compared with each other and be consulted with each other (including heading, lateral, area). The visual method is relatively simple, but the accuracy is poor. So local area of the same name can be measured firstly by the "histogram" function, and then the gray level distribution, intermediate value and average value are compared. If the difference is large, the result should be adjusted until it meets the requirements of gray level transition.

Survey area: camera files, control point files, aerial image index, photo area, route, aerial image metadata file.

2.2   Three-Dimensional Triangulation

The survey area are established by using VirtuoZo AATM, and PAT-B is used in regional network adjustment. The three-dimensional triangulation is carried out by zone, and its process is shown in Fig. 2.

The connection strength of each model should be improved to improve the precision of the results. 5 points are chosen near the vertical line which is near the main point in each photograph, and 3 encryption points are chosen near the point; Frame scalar measurement use automatic internal orientation, and inspect artificially piece by piece. The default to the residuals should not be greater than ±20 μm. In order to ensure the stability of the regional network, each photograph should have connection point on the course and lateral in the connection between strips. The connection point that exist on the treetop and connection point of texture difference should be moved to the point which image texture is clear, and can be accurately matched. The connection point and the ground control point are edited until it fully meet the requirements and there is no gross error. Rigorous adjustment of light beam method is carried out by introducing correction of photo deformation system error correction, earth curvature and atmospheric refraction after the gross error detection.

During the three-dimensional triangulation, multi angle image encryption screen test point should be used to interpretate and identify the data when the position is not clear, or ground deformation is too large.

2.3   Image Matching Pretreatment

The nuclear line image range is generated after the three-dimensional model is established. According to the image of the same name, the image is re arranged to form a nuclear line image.

Some necessary feature points, lines or surfaces are picked up according to the terrain feature to decrease the local deformation of digital orthophotoimage and improve the accuracy of digital orthophoto map. So as to control the matching precision of DEM and ensure the accuracy of the digital image. Elevation must be strictly cut to the ground, feature line node uniform and can not cross, node symmetry on the ridge and below the ridge in feature extraction. Nonrelated regional line cut the ground, and the regional difference in line does not differ too much, so as not to affect the accuracy of image matching; The river water characteristic acquisition and general characteristics of the flow lines are the same, but the elevation change collected should be consistent with the river, because the water flow to the low; If ponds, lakes and other water surface elevation is consistent with each other, we can measure the height of the water, and then lock the elevation, water collection characteristics, so as to ensure the water level; If not, you should first determine the surface average height, and then lock the elevation line acquisition characteristics. The following processing, such as image matching, contour editing, will be carried out the nuclear line image. The matching window size and matching interval should be determined by parameter setting, and the point of the same name should be determined along the core line. According to the terrain of the model to edit and ensure that the contour of the surface.

2.4   DEM Generation

Digital terrain elevation model (DEM) is the basis of the production of a positive image. DEM can be generated after the image matching editing; and then according to the image matching effect DEMMaker can be used to edit the DEM. The production process is shown in Fig. 3.

The grid spacing of DEM is more appropriate in the general election between 20–50 m, it can not only reflect the features of the changes, but also ignore certain objects (for example: because there are more city buildings, if the grid spacing is too small, the deformation of buildings may caused and it will increase the workload of editing). The range of the feature image acquisition should not exceed the connection of the encryption point 20 m. Acquisition of characteristic elements are stored by mappable unit as unit, storage, collection range map profile range extension of 10 m. Collection scope is external expansion to m of map profile range.

Figure painting medium grid DEM are generated by using the collected feature images or vector structure TIN. The editing command of DEMMaker is used to check the degree of anastomosis of DEM point and each model. Feature points, feature lines, and feature surfaces should be added for the regions in which DEM point and each model are misfit to make each grid node close to the surface. After checking the quality of each model grid in the graph, the characteristic elements of the map can be derived.

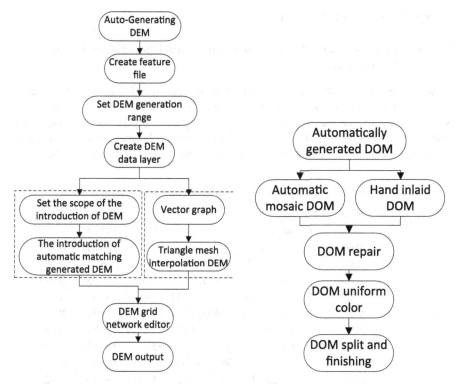

Fig. 3. Make DEM flow chart   **Fig. 4.** Make DOM flow chart

2.5 DOM Generation

Digital orthophoto map (DOM) is obtained by using digital elevation model (DEM) data, applying digital differential correction technique and carrying image photo rectification to eliminate the projection error of image (Fig. 4). The scope of the range of aerial photographs correction should in orientation points, and the maximum range should not exceed the control point of photograph of 20 m.

A standard map of DOM map need more than one image pair ortho image to join and mosaic. Due to the image gray level difference of surface features with the same name that caused by the difference of image photography angle, the complexity of the connection is far greater than the line drawing. The boundary requires not only to meet the accuracy of spatial coordinates of objects, but also to ensure that the image gray level of the same name in the area. Hand stitching must be used when city images are handled, because there are many surface features in the city and there are not DEM correction for tall buildings. Smooth curve of approximate straight line is the best choice and try to avoid small angle line when you select the stitching line. Furthermore, you should try to avoid tall buildings, and you can use the single model image to patch splice defect that cannot be eliminated. Image smoothing may be needed for the boarded area at the edge of the positive image. The method is the same with the above, and there is no longer describe about it.

3 Experiment and Analysis

3.1 Conversion and Post - Conversion Accuracy Statistics of Point Cloud Coordinate System

The method of plane coordinate transformation and elevation fitting is used to transform the point cloud coordinate system. D grade GPS control network results of Jiyuan city is used to calculate the conversion parameters by choosing 10 points in the plane coordinate transformation, and Henan $2' \times 2'$ quasi-geoid refinement results are used in elevation conversion. The elevation differences of checkpoint is shown in Table 1.

Table 1. The elevation difference of the laser data after calibration (unit: m)

Serial number	Point number	X	Y	H_S	H_{LiDAR}	ΔH
001	A001	640605.274	3893035.602	196.992	197.042	+0.050
002	A002	640624.720	3893034.227	196.826	196.817	−0.009
003	A003	640653.458	3893031.985	196.561	196.619	+0.058
004	A004	640681.832	3893029.979	196.350	196.338	−0.012
005	A005	640732.967	3893026.232	195.903	195.899	−0.004
006	A006	640775.201	3893023.083	195.513	195.529	+0.016
007	A007	654190.474	3893318.133	149.874	149.821	−0.053
008	A008	654199.923	3893343.246	150.183	150.132	−0.051
009	A009	654214.488	3893382.056	150.635	150.591	−0.044
010	A010	654230.205	3893423.829	151.190	151.161	−0.029

3.2 DEM Production and Extraction of Landform Elements

According to the measurement accuracy requirements of 1: 10000 DEM, the point cloud data are divided into Ground, Water, Noise, Default and Temp, and the final DEM product is constructed and outputted to the ground point cloud data with fine classification. The field measured elevation is used to check the production of DEM data as a checkpoint. It is calculated that the elevation error is ±0.12 m. The DEM data is shown in Fig. 5.

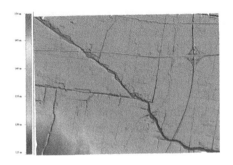

Fig. 5. Flat ground DEM shading effect diagram

Fig. 6. Xiaolangdi Reservoir DOM

3.3 DOM Production

The corrected outer azimuth element is obtained through the regional free network adjustment of the outer orientation element of each photo and the elimination of error between the various photos which are obtained using the camera calibration value calculation. And the DOM of Xiaolangdi Reservoir in Jiyuan city is produced by using accurate DEM data to orthorectification, mosaic, splicing, color and sub-frame output. And then the DOM accuracy are made by actual inspection (Fig. 6). As a result, the error in the plane position is ±0.546 m.

4 Concluding Remarks

This paper discusses the production process and quality control methods of making DOM through the full digital photogrammetric workstation VirtuoZo, connecting with the production practice of the film map of digital photograph in Jiyuan city. Large scale production and application of digital image is the inevitable trend of the development of "digital city", which provides a powerful information support and information security for the building "digital city".

References

1. Lu, X., Pang, X., Wu, Y.-B., Li, Q., Li, G., Yu, H.: Key technology and application of airborne LiDAR basic surveying and mapping. Bull. Surv. Mapp. **0**(9), 26–30 (2014)
2. Fu, X.L., Bao, F., Wang, W.-A.: A method for obtaining the outer outline of buildings from airborne LiDAR point cloud. Eng. Surv. Mapp. **20**(01), 44–47 (2011)
3. Henn, A., Groger, G., Stoh, V., et al.: Model driven reconstruction of roofs from sparse LIDAR point clouds. ISPRS J. Photogrammetry Remote Sens. **76**, 17–29 (2013)
4. Chum, O., Matas, J.: Randomized RANSAC with Td, d test. Image Vis. Comput. (IVCNZ) **2**(10), 837–842 (2004)
5. Shakarji, C.: Least-squares fitting algorithms of the NIST algorithm testing system. J. Res. Nat. Inst. Stand. Technol. **103**(6), 633–641 (1998)
6. Hu, Y.: Automated extraction of digital terrain models, road networks and building using airborne lidar data, Calgary University, Calgary, AB, Canada (2003)
7. Jacobsen, K., Lohmann, P.: Segmented filtering of laser scanner DSMs. Int. Arch. Photogrammetry Remote Sens. **34**(3/W13) (2003)
8. Oda, K., Tadashi, T., Takeshi, D., et al.: Automatic building extraction and 3-D city modeling from LIDAR data based on hough transformation. Int. Arch. Photogrammetry Remote Sens. Spat. Inf. Sci. 277–280 (2004)
9. Kilian, J., Haala, N., English, M.: Capture and evaluation of airborne laser scanner data. Int. Arch. Photogrammetry Remote Sens. **31**(B3), 385–388 (1996)
10. Kim, K., Shan, J.: Building roof modeling from airborne laser scanning data based on level set approach. ISPRS J. Photogrammetry Remote Sens. **66**(4), 484–497 (2011)
11. Kraus, K., Pfeifer, N.: Advanced DTM generation from LiDAR data. Int. Arch. Photogrammetry Remote Sens. Spat. Inf. Sci. **34**(3/W4), 21–30 (2001)

12. Cheng, L., Tong, L., Yanming, C., et al.: Integration of LiDAR data and optical multi-view images for 3D reconstruction of building roofs. Opt. Lasers Eng. **51**(4), 493–502 (2013)
13. Ming, W.G., De Feng, L., Qing, W.S.: Making orthophotoimages based on the VirtuoZo digital photogrammetric system. Eng. Surv. Eying Mapp. **9**(4) (2000)
14. Weili, Z.: Manufacturing of DOM by application of JX4 digital photograph surveying station. Bull. Surv. Mapp. no. 2 (2007)
15. Chunlan, B.: Research on making DEM based on VirtuoZo IGS. Geospatial Inf. **5**(1) (2007)
16. Shuhuang, Z.: Technique of the quality control and evaluation of digital orthophoto maps. Geol. Fujian, no. 1 (2007)

A Congestion Avoidance Algorithm Based on Fuzzy Controlling

Hui Zhang[(⊠)], Dan Zhang, Peng Xu, Yi Wang, and Kuan He

Department of Surveying and Mapping Engineering,
Yellow River Conservancy Technical Institute, Kaifeng 475004, Henan, China
kfzhanghui@163.com

Abstract. Many researches and experiments have shown that delay-based congestion avoidance (DCA) algorithms improve TCP throughputs. However, many factors affect the veracity of DCA algorithms. In order to solve these problems, a fuzzy control based on DCA algorithms (FDCA) is proposed which eliminates these negative influences when gauging network congestion conditions. The performance of the algorithm has been tested and evaluated using NS simulator and the simulation results have demonstrated that the algorithm identifies network congestions more accurately and consequently increases network throughputs.

Keywords: Fuzzy control · Congestion control · Congestion avoidance

1 Introduction

TCP protocol is commonly used by hosts to manage network congestion on the Internet. Many researches have shown that native TCP congestion management mechanism is conservative in nature, affecting the network throughputs in many scenarios. There have been a lot of ongoing researches on TCP congestion management with the aim to increase the throughputs of TCP protocol. A well-known improvement is TCP Reno method. TCP Reno uses ACK messages – the successful arrival or not of them, to perform congestion control. However, this does not take into consideration of the transport latency of ACK messages. Upon the occurring of network congestion, the time latencies in ACK messages will start to increase. But, as long as they are not exceeding TTL and not lost, even if the latencies of ACK messages are getting greater, TCP Reno will continue to increase the transmit window which in turn worsen the network congestion. As a further improvement, TCP Vegas [1] algorism was developed and focused on CWND adjustment algorism improvement – congestion window adjustment in congestion avoidance phase. This is based on observing the changes of Round Trip Time (RTT) in TCP connections to detect potential network congestions, rather than detecting packet losses. Accordingly CWND sizes are adjusted. This is a delay-based congestion avoidance (DCA) algorism. In DCA algorism, changes in RTT are considered as consequences of network congestions. Increases in RTT suggest network congestion and the algorism would reduce sender's CWND size by a defined percentage. Decreases in RTT suggest network load is easing and the algorism would

M. Qiu (Ed.): SmartCom 2016, LNCS 10135, pp. 495–505, 2017.
DOI: 10.1007/978-3-319-52015-5_51

increase sender's CWND size by a ratio. Leading DCA algorisms developed until today include TCP/Vegas and Dual. Researches have shown that TCP Vegas can increase throughput by 40%–70%, comparing with widely used TCP Reno algorism.

Previous researches on DCA focused on low speed data links or networks with high percentage of DCA flows. This research document [2] has shown that in modern high speed data links or where there are less DCA flows, using RTT to manage congestion will result in TCP throughput decreases. After analyzing a large number of tests, it is discovered that only 7–58%. In fact, this document [2] has pointed out that it is neither accurate to only use RTT to indicate 18% of packet losses are relevant to increases in RTT. In these circumstances, using DCA algorism to manage congestion would reduce TCP throughput by 7%–58%. In fact, this document [2] has pointed out it is neither accurate to only use RTT to indicate network congestions nor that it has the ability to detect sudden (bursty) congestions on high speed networks – as such can not to be used to accurately gauge the status of network congestions.

This paper uses ideas in this document [3] to thoroughly analyze the use of RTT as congestion control indicators and has discovered two defects. First, there is no consideration of the impact on RTT due to non congestion delays. Second, on high speed networks sudden bursty congestions can occur within a fraction of a time, traditional RCA algorisms can neither detect accurately nor react in time. This paper discusses a fuzzy control based on DCA algorithm (FDCA). This algorism not only eliminates the impact to RTT due to non congestion delays by using fuzzy theory to describe the fuzzy nature of changes of RTT but also utilizes a combined method based on real time data and historical data of RTT to engage a fuzzy control mechanism to adjust to CWND. This has consequently achieved optimized detection and reaction to network congestions, effectively resolving the issue of reduced throughputs due to bursty network congestions, and greatly increasing TCP throughputs in congestion avoidance phase. Simulation tests have demonstrated that this algorism can significantly increase TCP throughputs and improve network performance.

2   The Fuzzy Control Model of a Congestion Avoidance Algorism

In typical DCA algorism TCP Vegas, congestion window is not continuously increased in the congestion avoidance phase. Rather, the additional network traffic is estimated to reasonably control CWND. First BaseTT is recorded as $BaseRTT = min(RTT_i)$, then calculate $Expected = CWND/BaseRTT$. When a new sampling $NewRTT$ is received, calculate $Actual = CWND/NewRTT$. Compare $Expected$ and $Actual$, we get $Diff = Expected - Actual$. As per definition, $Diff$ *should be* greater than 0. If $Diff < 0$, then $BaseRTT$ is too large and needs to be adjusted to $NewRTT$. Two thresholds are defined: α and β. When $Diff < \alpha$, Vegas increases CWND linearly; When $Diff > \beta$, Vegas decreases CWND linearly; CWND is not adjusted when $\alpha < Diff < \beta$.

In fact, $Diff = Expected - Actual = (NewRTT - BaseRTT)CWND/(BaseRTT*NewRTT)$, so

$$Diff = \Delta RTT * CWND/(BaseRTT * NewRTT) \tag{1}$$

In formula (1), ΔRTT is the difference (D-value) of the minimum RTT and the current RTT, and is a crucial factor influencing the outcomes of DCA algorisms. Usually, RTT is affected by four factors: data transport time Tp (the physical signal transport time of a packet through all data links), processing time T_d (time it taken) for a packet to be processed by all nodes), transmit time T_t (time it taken) by network nodes to place a packet onto data links) and queuing time Tq (time a packet waits in queues at network nodes). Among these, only T_q is good indicator of network congestion. If all other factors stay the same or do not change significantly during the trip of a network packet, ΔRTT would indicate the level of network congestions reasonably well.

However, for TCP traffic, while T_p and T_t can be stable [4] in a given duration of time, it is highly likely that T_d incurs changes. When we examine a router's processing of network packets, the process time variations due to packet sizes are negligible. However, when destination host generates ACKs, some ACKs would incur additional latency of 0–500 ms due to TCP Delayed Acknowledgement TCP ACK combinations. This introduces significant impact to overall RTT. As to T_d, this can be represented as $T_d = T_d' + T_{delay}$. T_d' represents the sum of processing times to a packet by all network nodes and T_{delay} is the ACK combination delay, varies dynamically between 0–500 ms.

We can derive from above analysis, that network congestion detection mechanisms solely relying on RTT are vulnerable to T_d interference. Such interference can be eliminated through further processing. This paper recommends an algorism, which employs a corrector to eliminate errors introduced by T_d, before subjecting RTT to fuzzy control algorism. In this way T_q is obtained which reflects changes in congestion levels more accurately, as per Fig. 1:

Fig. 1. DCA's fuzzy controller model

In high speed networks bursty congestions typically last very short times (去s). Queuing in buffers due to network fluctuation can be cleared up swiftly by high speed data links – disappearing in short period of time. A result of this is there can be sparse occurrence of large latency RTT. In such scenarios, if we use Vegas algorism, it will use linear computation to drastically reduce CWND to avoid generation of congestion. However, the most likely scenario is network has already returned to normal after a large RTT. Vegas algorism in these cases will reduce TCP throughputs. To counter the negative impact that DCA algorisms bring in bursty congestions, this paper discusses the use of not only ΔRTT to identify network congestions, but also the use of the relative changes in RTT as well as the most recent historical changes of RTT for fuzzy control.

In Fig. 1, the inputs to the Fuzzy Controller include not only ΔRTT but also the difference of two consecutive $T_q - \Delta T$. One such input is ΔT_{now} – the difference of current T_q and last T_q. Another such input is ΔT_{old} – the sum of the differences of last n T_q. When large ΔRTT or large ΔT_{now} starts occurring, Fuzzy Controller will not immediately greatly reduce CWND. Rather, it will perform more reasonable processing based on ΔT_{old}, effectively avoids the issue of reduced throughputs due to bursty congestions. The number of the most recent sampled T_q differences should not exceed the number of ACKs within the given RTT time frame, subjecting to real life scenarios.

3   The DCA Algorism Using Fuzzy Control

3.1   The Algorism of the Corrector

From the above analysis we can conclude, that the effective component from TCP RTT that indicating network congestions is T_q. ACK delays cause significant changes in RTT and affect the congestion indication accuracy by direct use of RTT. To increase the accuracy of congestion prediction, we need to overcome the influences by ACK delays.

In a given stable duration of time, RTT of a TCP flow can be calculated using this simplified formula:

$$RTT = T_p + T_t + T_d{}' + T_{delay} + T_q \tag{2}$$

In Formula (2), we need to know T_q to predict the status of network congestions. And finding out T_q is difficult. So to obtain a measurement that reflects the network congestions, we compare the subsequent two RTT values:

$$RTT_1 = T_{p1} + T_{t1} + T_{d1}{}' + T_{delay1} + T_{q1} \tag{3}$$

$$RTT_2 = T_{p2} + T_{t2} + T_{d2}{}' + T_{delay2} + T_{q2} \tag{4}$$

Generally, within the lapse of time of two subsequent RTTs, $T_{p1} \approx T_{p2}, T_{d1}{}' \approx T_{d2}{}', T_{t1} \approx T_{t2}$. Perform Formula (3) – Formula (4), and further transformation, we get:

$$\Delta T_{now} = T_{q2} - T_{q1} = RTT_2 - RTT_1 - (T_{delay2} - T_{delay1}) \tag{5}$$

Using Formula (5), ΔT_{now}, which can indicate queuing delays, can be derived from two consecutive RTTs, and be used as indicator for network congestion levels. This would greatly increase the accuracy of network congestion prediction. However, T_{delay} in Formula (5) is not practically obtainable. We slightly modify the TCP protocol, leveraging a reserved field in TCP Header to transport back the latency (ms) of the ACT combination delays. Every time the receiver sends ACK, TCP will write the value of timer 'ACK delay' to the reserved field. This reserved field in TCP header only has

6 bits so we define each unit as 4 ms, thus, a total of $2^6 \times 4$ ms, i.e., 256 ms can be represented. As to normal TCP algorisms, this reserved field can be totally ignored so (when?) compatible to sender's processing.

Besides, the corrector will store the recent calculated ΔT_{now}. Based on the RTT size, the corrector decides how many ΔT_{now} it will store. In each calculation:

$$\Delta T_{old} = \sum_{i=1}^{n} \Delta T_{nowi} \tag{6}$$

Through this way the corrector can obtain the trending values of the recent changes in queuing delays, together with the current calculated ΔT_{now} are fed to the fuzzy controller for processing. When data link statues incur changes, all stored ΔT are cleared. In the mean time, the corrector also calculates the difference of current RTT and BaseRTT − ΔRTT and feeds to the fuzzy controller, as well as updating BaseRTT. So

$$\Delta RTT = RTT - BaseRTT \tag{7}$$

3.2 The Delay Differences (Jitters) Based Congestion Evaluation

Due to the fuzzy attributes of the very concept of congestion, we can not just use precise queuing time to describe network congestion. Although in previous discussions we have used corrector to process RTT and obtained relatively accurate differences in queuing times, we can only state it reflects congestion status in some duration of time. I.e., with RTT of a particular time, ΔRTT and ΔT_{now} may demonstrate large values, suggesting the possibility of the existence of network congestion, but a very small or negative value of ΔT_{old} would indicate this current congestion status is highly likely to be bursty in nature. As such, we introduce fuzzy theory to describe this fuzzy natured procedure, fully leveraging the advantages that fuzzy theory possesses on handling non linear situations and problems with uncertainty, to achieve the goal of optimizing congestion control mechanism, increasing the throughputs of the whole network.

ΔRTT, ΔT_{now} and ΔT_{old} have values in vast different ranges. For the sake of algorism processing, we first apply normalization to these three parameters to derive them into delay difference (jitter) ratios in closed interval [0, 1]. We use the timed out value in TCP calculation as the maximum RTT to work out:

$$\Delta RTT_{max} = timeout - BaseRTT \tag{8}$$

Then the normalization results to ΔRTT, $\Delta Tnow$ and $\Delta Told$ will be:

$$\Delta RTT' = \Delta RTT / \Delta RTT_{max} \tag{9}$$

$$\Delta T'_{now} = (\Delta RTT_{max} + \Delta T_{now}) / 2\Delta RTT_{max} \tag{10}$$

$$\Delta T'_{old} = (\Delta RTT_{max} + \Delta T_{old}) / 2\Delta RTT_{max} \tag{11}$$

The delay difference ratios are categorized into three language variables {Free, Normal, Congestion} with following detailed definitions: Free (F) – the network is light on congestion, with free capacity; Normal (N) – The network has low level congestion; Congestion (C) – the network is congested. As per Zadeh's definition of fuzzy subset, we define domain of discourse U: delay difference ratios (the ratio of the current jitter to the maximum jitter) and three mappings from U to closed internal $[0, 1]$: $U \rightarrow [0, 1]$, $u \rightarrow \mu F(u)$; $\mu N:U \rightarrow [0, 1]$, $u \rightarrow \mu N(u)$; $\mu C:U \rightarrow [0, 1]$, $u \rightarrow \mu C(u)$. Three fuzzy subsets of U are established: F, N, & C. And $\mu F(u)$, $\mu N(u)$, $\mu C(u)$ are named membership functions to F, N and C, respectively. At same time, we use decision set $V = \{F, N, C\}$ to describe the current congestion status of the whole network.

The selection of membership functions is the crucial step in performing fuzzy comprehensive analysis. Common membership function choices are trigonometric functions, trapezoidal functions and spline functions. We derive the required number of language fuzzy sets to describe every jitter from the quantified magnitude of congestions. Every language fuzzy set maps to a different congestion level and a different membership function. In this discussion we use trigonometric functions to define the three membership functions on U: $\mu F(u)$, $\mu N(u)$, $\mu C(u)$, as shown in Fig. 2:

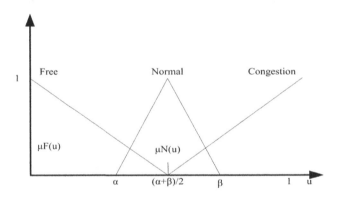

Fig. 2. Membership functions

The Normal Interval in Fig. 2 is defined by α and β, similar to the two thresholds α and β in TCP Vegas, to ensure the network status in system stable durations.

3.3   The Congestion Processing Using the Three Jitter Ratios

For the purpose of accurately obtaining the current congestion status of the network, comprehensive evaluation of the three jitter ratios is required. We use factors set U to represent three fuzzy variables: $U = \{u_1, u_2, u_3\}$, u_1, u_2, u_3 map respectively to three jitter ratios: $\Delta RTT'$, $\Delta T'now$ and $\Delta T'old$. A fuzzy relationship is established from factors set U to decision set V, which we can describe using fuzzy matrix R:

$$R = \begin{bmatrix} \mu F(u_1) & \mu N(u_1) & \mu C(u_1) \\ \mu F(u_2) & \mu N(u_2) & \mu C(u_2) \\ \mu F(u_3) & \mu N(u_3) & \mu C(u_3) \end{bmatrix}$$

In consideration of different significant levels of the jitter ratios in comprehensive evaluation, we distribute weighting values as $w = (w_1, w_2, w_3)$, where weighting coefficient w_k represents the significant level in our view of the jitter ratio k, where $0 \leq w_k \leq 1, k = 1, 2, 3; w_1 + w_2 + w_3 = 1$. Obviously, when $w_1 = 1, w_2 = w_3 = 0$, the result approaches to that of TCP Vegas.

Let the comprehensive evaluation result be $p = (p_1, p_2, p_3)$, then:

$$p = w \circ B \qquad (12)$$

where $\circ$ is the Max-min synthesis operation commonly used in fuzzy matrixes, and p_j represents the different statuses of degree of membership of how the current level of network congestion belongs to V, after comprehensive considerations of the three jitter ratios. When compare $p_i(i = 1, 2, 3)$, there exists a k, so that

$$p_k = \max_j \left\{ p_j \right\} \qquad (13)$$

Based on the principle of maximum degree of membership, we can evaluate the network status defined by p_k, and intelligently control CWND. Table 1 describes the principles of fuzzy congestion processing:

Table 1. The principles of fuzzy congestion processing

Status	Action
Free	Increase congestion window linearly
Normal	Make no change to congestion window
Congest	Decrease congestion window linearly

3.4 Algorism Processing Steps

① Perform correction processing to received RTT, as per formulas (5), (6) and (7) to calculate respectively ΔT_{now}, ΔT_{old} and ΔRTT.

② As per formulas (9), (10) and (11), apply normalization to ΔT_{now}, ΔT_{old} and ΔRTT, to derive $\Delta T'_{now}$, $\Delta T'_{old}$ and $\Delta RTT'$.

③ Fill in Matrix R with the results from normalization processing.

④ As per defined weighting w, calculate p with formula (12).

⑤ Calculate p_k through formula (13) to obtain the current network status.

⑥ Use Table 1 to calculate the newest value of CWND.

⑦ Finish.

4   Performance Simulation Analysis

We used NS to perform simulation analysis to the designed algorism. First, further expansion is made to NS to realize the FDCA algorism which first time defined by this paper. Then the performance of this algorism is compared to (with) the performance of TCP Vegas. The network topology and its parameters used in the simulation are shown in Fig. 3:

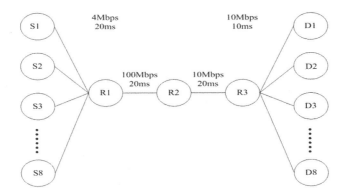

Fig. 3. Network topology

Bottlenecks will not occur between R1 and R2. The data link bandwidth between R2 and R3 is 10 Mbps, when all sending hosts transmitting data to respective receiving hosts, this data link will be the bottleneck. The initial transport latency of each data link is shown in the diagram which can be adjusted during simulation. We also define the sending-receiving success rates of all data links are 100% to avoid any deviation cased by it on the evaluation of the algorism on each node.

When performing simulation, each router was set to FIFO and DropTail queuing management algorism, buffer max queue length 100 KB, and acknowledgement window size 20 message segments. The maximum data packet length was set to 1400 bytes. $S1 \sim S6$ run TCP, use respectively pure TCP Vegas and FDCA (will be referred to as FDCA TCP next) in congestion avoidance phase to compare the congestion avoidance improvements by FDCA algorism. S7 and S8 used to generate background traffic.

During the performance evaluation, default values were set for fuzzy membership function parameter α and β as 0.4 and 0.6; weighting distribution as $w = (0.4, 0.3, 0.3)$. The main focus was to examine the throughputs of FDCA TCP and pure TCP Vegas under various conditions, as explained below.

4.1   Performance Comparison with Different Transport Delays

To demonstrate that FDCA has sound adaptability over transport delays, we setup six Telnet flows, between Si and Di (i = 1, 2, …, 6) respectively. S7 and S8 transmitted

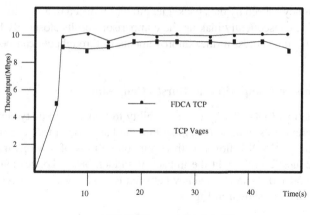

a) R2-R3 Transport delay: 20ms

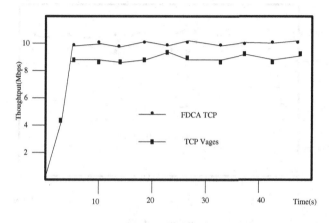

b) R2-R3 Transport delay: 100ms

Fig. 4. (a) R2-R3 Transport delay: 20 ms and (b) R2-R3 Transport delay: 100 ms

CBR background traffic. All TCP flows were two ways, with the host to server traffic being 1/3 of that of from server to host. For comparison purpose, all other settings were the same to FDCA TCP and TCP Vegas. Every simulation run was 50 s. The throughput results of R3 are reported in Fig. 4:

Figure 4a shows the network throughputs with small transport latencies between R2 and R3. We can see that FDCA TCP significantly increases the overall throughputs. We can also find that throughputs of TCP Vegas are inconsistent while FDCA TCP is relatively consistent. This was mainly due to t the changes in transport latencies and ACK delays caused TCP Vegas misjudgment and lowered the congestion prediction accuracy.

Then we slightly adjusted the simulation condition to increase the data link latency between R2 to R3 to 100 ms. The results are shown in Fig. 4b. We found reduced performance differences between TCP Vegas and FDCA TCP. This was mainly because the latency caused by data packet length was reduced to a smaller proportion in the overall RTT, consequently making the comparison results closer. By further adjust FDCA's α and β thresholds, further improvements to throughputs can be achieved.

4.2   Performance Comparison in Bursty Congestions

To demonstrate that FDCA has good adaptability to impacts of bursty congestions, we modified the previous simulation setup so that the background traffic generated by S7 and S8 becoming ON-OFF flows, with continuous 50 ms of 4 Mbps traffic in every second on average. To highlight the impact of sudden bursty congestion, R2's buffer size was set to 50 KB. Data link latency between R2 to R3 was set to 100 ms. The R3 throughput results are shown in Fig. 5:

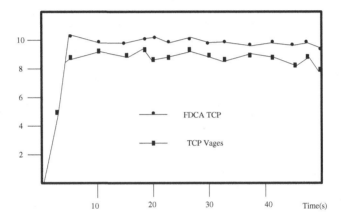

Fig. 5.   Throughputs of Telnet flows

We can see clearly that FDCA TCP significantly increased the throughputs. We can also find that in bursty congestion scenarios neither TCP Vegas' nor FDCA TCP's throughputs were consistent. However, FDCA TCP has less fluctuation amplitude. This is mainly because the use of historical jitters by FDCA TCP has smoothened the large magnitude movements in throughputs caused by bursty congestions.

5   Conclusion

The Fuzzy Control based on DCA Algorism considered the impact of transport latencies and ACK delays, as well as the current and historical changes of RTT, significantly increasing the accuracy of network congestion prediction. Simulation experiments have demonstrated that FDCA can provide significant performance gains in bursty congestion common scenarios. It can also significantly increase network throughputs and can increase the resource utilization of IP networks.

Further research through better understanding of network traffic characteristics could lead to establishment of better membership function models, which will better predict the network congestion status, and further improve the performance of FDCA algorism.

References

1. Brakmo, L.S., Peterson, L.L.: TCP Vegas: end to end congestion avoidance on a global internet. IEEE J. Sel. Areas Commun. **13**(8), 1465–1480 (1995)
2. Martin, J., Nilsson, A., Rhee, I.: Delay-based congestion avoidance for TCP. IEEE/ACM Trans. Netw. **11**(3), 356–369 (2003)
3. Yi, F., Xia, M., Wang, Y., Zeng, J.: An improved TCP congestion avoidance algorithm based on delay. Comput. Sci. (33), 61–64 (2006)
4. Paxson, V: Measurements and analysis of end-to-end Internet dynamics. Ph.D. dissertation, UC Berkeley (1996)
5. Jiang, F., Zhang, Q.: Congestion control in multihop wireless networks. IEEE Trans. Veh. Technol. **56**(2), 863–873 (2007)
6. Mondal, S.A., Luqman, F.B.: Improving TCP performance over wired-wireless networks. Comput. Netw. **51**, 3799–3811 (2007)
7. Tsaoussidis, V., Zhang, C.: TCP-Real: receiver-oriented congestion control. Comput. Netw. **40**, 477–497 (2002)
8. Baucke, S.: Using receiver-based rate matching for congestion detection in rate-based protocols. IEEE (2003)
9. Mehra, P., De Vleeschouwer, C., Zakhor, A.: Receiver-driven bandwidth sharing for TCP and its application to video streaming. IEEE Trans. Multimedia **7**(4), 740–752 (2005)
10. Hasegawa, G., Nakata, M., Nakano, H.: Modeling TCP throughput over wired/wireless heterogeneous networks for receiver-based ACK splitting mechanism. IEICE Trans. Commun. **E90-B**(7), 1682–1691 (2007)
11. Chan, M.C., Ramjee, R.: Improving TCP/IP performance over third-generation wireless networks. IEEE Trans. Mob. Comput. **7**(4), 430–443 (2008)

An Efficient Adaptive-ID Secure Revocable Hierarchical Identity-Based Encryption Scheme

Changji Wang[1]([✉]), Yuan Li[2], Shengyi Jiang[1], and Jiayuan Wu[3]

[1] Collaborative Innovation Center for 21st-Century Maritime Silk Road Studies,
Cisco School of Informatics, Guangdong University of Foreign Studies,
Guangzhou 510006, China
wchangji@gmail.com
[2] School of Computer Science, Fudan University, Shanghai 200433, China
[3] School of Data and Computer Science, Sun Yat-sen University,
Guangzhou 510006, China

Abstract. User revocation is an important functional requirement in hierarchical identity-based encryption (HIBE) scheme from the point of practical applications. In this paper, we propose a revocable HIBE scheme with constant ciphertext size and prove it is semantic secure under the strong adaptive-ID model by employing the recent dual system encryption methodology and predicate encoding technique.

Keywords: Hierarchical identity-based encryption · Revocable identity-based encryption · Revocable hierarchical identity-based encryption · Dual system encryption · Predicate encoding

1 Introduction

Hierarchical identity-based encryption (HIBE) [1] is an important extension of identity-based encryption (IBE) [2], which can offer user more levels of flexibility in sharing sensitive data for cloud computing [3]. In an HIBE scheme, user identities are arranged in an organizational hierarchy, and the trusted private key generator (PKG) can delegate private key generation and identity authentication to its sub-authorities, who in turn can keep delegating private key generation and identity authentication further down the hierarchy to the users. The first fully functional HIBE scheme was proposed by Gentry and Silverberg [4], which is proved to be adaptive-ID secure in the random oracle model. Recently, Lewko and Waters [5] proposed the first HIBE scheme (LW-HIBE) with constant size ciphertext in composite order bilinear groups, which is proved to be adaptive-ID secure in the standard model.

To apply an IBE scheme or an HIBE scheme in real applications, we should revoke the private key of a user if his private key is compromised or his credential is expired. Several efficient revocable IBE (RIBE) schemes have been proposed by using the complete subtree method [6] in recent years ([7–10]). Seo and Emura [11] proposed the first realization of revocable HIBE (RHIBE)

M. Qiu (Ed.): SmartCom 2016, LNCS 10135, pp. 506–515, 2017.
DOI: 10.1007/978-3-319-52015-5_52

scheme in CT-RSA 2013. However, Seo and Emura's RHIBE scheme is only proved to be semantically secure in the relaxed selective-ID model, and the size of ciphertext increases linearly with respect to the depth of the identity hierarchy. Seo and Emura proposed to construct more efficient (in the sense of the ciphertext size) and secure (in the sense of adaptive-ID model) RHIBE scheme as an open question. Tsai et al. [12] proposed an adaptive-ID secure RHIBE scheme. However, it is not scalable. Recently, Seo and Emura [13] presented a new method to construct RHIBE scheme that implements history-free updates, and proposed two RHIBE schemes with shorter secret keys and constant size ciphertexts using the complete subtree revocation method and subset difference revocation method, respectively. Unfortunately, both schemes are only proved to be selectively secure under the decisional ℓ-weak Bilinear Diffie-Hellman Inversion assumption, and both schemes have the inherent limitation that the size of public parameters linearly grows to the maximum hierarchy depth.

To obtain an adaptive-ID secure RHIBE scheme with constant-size ciphertexts, a natural approach is to combine the complete subtree method with LW-HIBE scheme [5]. Unfortunately, this direct combination may result in the exponential growth of the size of subkeys and difficulty in the security proof [11]. In this paper, we propose an efficient adaptive-ID secure RHIBE scheme with constant-size ciphertext, which solves the open problem presented by Seo and Emura [11]. In our RHIBE construction, we apply the *asymmetric trade* strategy [11] to avoid exponentially large secret keys in the corresponding hierarchical level, and prove its security by using Wee's strategy [14], which combines dual system encryption methodology with predicate encoding technique.

The rest of the paper is organized as follows. In Sect. 2, we introduce some preliminary work necessary for our RHIBE construction and security proofs. In Sect. 3, we give formal syntax and security definitions for RHIBE scheme. In Sect. 4, we describe our RHIBE construction from composite bilinear groups. In Sect. 5, we present efficiency analysis and security proofs of our RHIBE construction. Finally, we conclude the paper in Sect. 6.

2   Preliminaries

Let p_1, p_2 and p_3 are distinct primes, and $n = p_1 p_2 p_3$. A composite order bilinear groups generator $\mathcal{G}$ is a probabilistic polynomial-time (PPT) algorithm that on input a security parameter κ outputs a composite order bilinear group $(n, \mathbf{G}, \mathbf{G}_T, \hat{e})$, where $\mathbf{G}$ and $\mathbf{G}_T$ are multiplicative groups of order n, and $\hat{e}$: $\mathbf{G} \times \mathbf{G} \to \mathbf{G}_T$ is a bilinear pairing with the following properties:

- Bilinearity: For all $g, h \in \mathbf{G}$ and $a, b \in \mathbf{Z}_n^*$, we have $\hat{e}(g^a, h^b) = \hat{e}(g, h)^{ab}$.
- Non-degeneracy: If g is a generator of $\mathbf{G}$, then $\hat{e}(g, g)$ is a generator of $\mathbf{G}_T$.
- Computability: There exists an efficient algorithm to compute $\hat{e}(g, h)$ for all $g, h \in \mathbf{G}$.

We denote by $\mathbf{G}_{p_1}$, $\mathbf{G}_{p_2}$ and $\mathbf{G}_{p_3}$ the subgroups of order p_1, p_2 and p_3 in $\mathbf{G}$, respectively. Our RHIBE construction relies on the following two assumptions which are instances of the general subgroup decision assumption in composite order bilinear groups [14].

Denote $(p_1 p_2 p_3, \mathbf{G}, \mathbf{G}_T, \hat{e}) \leftarrow \mathcal{G}(1^\kappa)$, $g_1 \xleftarrow{\$} \mathbf{G}_{p_1}^*$, $g_3 \xleftarrow{\$} \mathbf{G}_{p_3}^*$, $g_{12} \xleftarrow{\$} \mathbf{G}_{p_1}\mathbf{G}_{p_2}$, $D = (g_1, g_3, g_{12})$, $L_1 \xleftarrow{\$} \mathbf{G}_{p_1}$ and $L_2 \xleftarrow{\$} \mathbf{G}_{p_1}\mathbf{G}_{p_2}$. Given two distributions $(\mathcal{G}(1^\kappa), D, L_1)$ and $(\mathcal{G}(1^\kappa), D, L_2)$, the advantage of an adversary $\mathcal{A}$ in breaking the subgroup decision problem 1 is defined as

$$\mathrm{Adv}_{\mathcal{G},\mathcal{A}}^{\mathrm{SGD}_1}(\kappa) = \Big| \Pr[\mathcal{A}(\mathcal{G}(1^\kappa), D, L_1) = 1] - \Pr[\mathcal{A}(\mathcal{G}(1^\kappa), D, L_2) = 1] \Big|$$

Assumption 1. *We say that $\mathcal{G}$ satisfies the subgroup decision assumption 1 if $Adv_{\mathcal{G},\mathcal{A}}^{SGD_1}(\kappa)$ is a negligible function of κ for any PPT adversary $\mathcal{A}$.*

Denote $(p_1 p_2 p_3, \mathbf{G}, \mathbf{G}_T, \hat{e}) \leftarrow \mathcal{G}(1^\kappa)$, $g_1 \xleftarrow{\$} \mathbf{G}_{p_1}^*$, $g_3 \xleftarrow{\$} \mathbf{G}_{p_3}^*$, $g_{12} \xleftarrow{\$} \mathbf{G}_{p_1}\mathbf{G}_{p_2}$, $g_{23} \xleftarrow{\$} \mathbf{G}_{p_2}\mathbf{G}_{p_3}$, $D = (g_1, g_3, g_{12}, g_{23})$, $L_1 \xleftarrow{\$} \mathbf{G}_{p_1}^*\mathbf{G}_{p_3}$ and $L_2 \xleftarrow{\$} \mathbf{G}_{p_1}^*\mathbf{G}_{p_2}^*\mathbf{G}_{p_3}$. Given two distributions $(\mathcal{G}(1^\kappa), D, L_1)$ and $(\mathcal{G}(1^\kappa), D, L_2)$, the advantage of an adversary $\mathcal{A}$ in breaking the subgroup decision problem 2 is defined as

$$\mathrm{Adv}_{\mathcal{G},\mathcal{A}}^{\mathrm{SGD}_2}(\kappa) = \Big| \Pr[\mathcal{A}(\mathcal{G}(1^\kappa), D, L_1) = 1] - \Pr[\mathcal{A}(\mathcal{G}(1^\kappa), D, L_2) = 1] \Big|$$

Assumption 2. *We say that $\mathcal{G}$ satisfies the subgroup decision assumption 2 if $Adv_{\mathcal{G},\mathcal{A}}^{SGD_2}(\kappa)$ is a negligible function of κ for any PPT adversary $\mathcal{A}$.*

The KUNode algorithm was proposed by Boldyreva et al. [7] to achieve efficient revocation for IBE schemes. Denote the root node of the tree $\mathbb{T}$ by root. If η is a leaf node, we denote the set of nodes on the path from η to root by Path(η). If η is a non-leaf node, we denote the left and right child of η by η_L and η_R, respectively. Each user is assigned to a unique leaf node. If a user (assigned to η) is revoked on time T, then (η, T) is added into the revocation list **RL**.

> $\mathbf{X}, \mathbf{Y} \leftarrow \emptyset$;
> For $\forall(\eta_i, \mathsf{T}_i) \in \mathbf{RL}$, if $\mathsf{T}_i \leq \mathsf{T}$, then add Path$(\eta_i)$ to $\mathbf{X}$;
> For $\forall\eta \in \mathbf{X}$, if $\eta_L \notin \mathbf{X}$, then add η_L to $\mathbf{Y}$;
> For $\forall\eta \in \mathbf{X}$, if $\eta_R \notin \mathbf{X}$, then add η_R to $\mathbf{Y}$;
> If $\mathbf{Y} = \emptyset$, then add root to $\mathbf{Y}$;
> Return $\mathbf{Y}$.

To prove adaptive security under simple assumptions for HIBE and attribute-based encryption schemes ([15–17]), Waters [18] first introduced a new proof methodology named dual system encryption. In a dual system encryption scheme, there are two types of keys and ciphertext: normal and semi-functional. A normal private key can decrypt normal or semi-functional ciphertexts, and a normal ciphertext can be decrypted by normal or semi-functional private keys. However, when a semi-functional private key is used to decrypt a semi-functional ciphertext, decryption will fail with high probability.

To simplify the process of designing and analyzing of dual system encryption schemes in composite order bilinear groups, Wee [14] introduced a framework named predicate encoding. A predicate encoding for a boolean predicate $P(\cdot, \cdot)$, is specified by a pair of sender encoding $sE(x, w)$ and receiver encoding $rE(\alpha, y, w; r)$ with a common private input w, which satisfy the following three requirements:

- Reconstruction: If $P(x, y) = 1$, we can recover α from the encodings.
- Privacy: If $P(x, y) = 0$, the encodings hide α perfectly.
- w-hiding: All information about w in the receiver encoding can be hidden by setting the randomness r to some fixed value.

3 Syntax and Security Definitions of RHIBE Scheme

The KEM/DEM hybrid encryption paradigm is a simple way of constructing efficient and practical public key encryption schemes, and it has been successfully adapted in many new standards and recommendations for encryption [19]. In this approach, one first uses a key encapsulation mechanism (KEM) to generate a random symmetric key K together with a ciphertext that encapsulates K, and the symmetric key is then fed into a highly efficient data encapsulation mechanism (DEM), such as AES, to encrypt the actual message. We give the syntax of RHIBE scheme in the framework of key encapsulation as follows.

- **Setup**$(1^\kappa, N, L)$: The probabilistic setup algorithm is run by the PKG. It inputs a security parameter κ, maximal number of users N in each level, and maximum depth L of the hierarchy of identity. It outputs the master public key mpk, the master secret key msk, the initial revocation list $\mathbf{RL} = \emptyset$, and a initial state st_0.
- **KeyGen**$(sk_{ID_{|\ell}}, st_{ID_{|\ell}}, ID_{|\ell+1}, mpk)$: The probabilistic private key generation algorithm is run by a user (or the PKG). It inputs the ℓ-th level user's private key $sk_{ID_{|\ell}}$ (or msk for the PKG), state $st_{ID_{|\ell}}$, a children's identity $ID_{|\ell+1}$ and mpk. It outputs a private key $sk_{ID_{|\ell+1}}$ for $ID_{|\ell+1}$.
- **KeyUp**$(sk_{ID_{|\ell}}, T, \mathbf{RL}_{ID_{|\ell}}, st_{ID_{|\ell}}, ku_{ID_{|\ell-1}, T}, mpk)$: The probabilistic key update algorithm is run by a user. It inputs the ℓ-th level user's private key $sk_{ID_{|\ell}}$, key update time period T, a revocation list $\mathbf{RL}_{ID_{|\ell}}$, state $st_{ID_{|\ell}}$, update key $ku_{ID_{|\ell-1}, T}$ published by $ID_{|\ell-1}$, and mpk. It outputs a update key $ku_{ID_{|\ell}, T}$ or $\perp$ (when $ID_{|\ell}$ is revoked).
- **DKG**$(sk_{ID_{|\ell}}, T, ku_{ID_{|\ell-1}, T}, mpk)$: The probabilistic decryption key generation algorithm is run by a user. It input the ℓ-th level user's private key $sk_{ID_{|\ell}}$, key update time period T, a update key $ku_{ID_{|\ell-1}, T}$, and mpk. It outputs a decryption key $dk_{ID_{|\ell}, T}$ that can be used during time period T or $\perp$ that means the identity $ID_{|\ell}$ is revoked for some period $T' \leq T$.
- **Enc**$(ID_{|\ell}, T, msg, mpk)$: The probabilistic encryption algorithm is run by a sender. It inputs a receiver's identity $ID_{|\ell}$, time period T, a message msg, and mpk. It outputs a ciphtertext ct as well as a session key $K \in \{0, 1\}^\kappa$. Note that the actual message msg is then encrypted using AES algorithm with the session key K.

- **Dec**($\mathsf{dk}_{\mathsf{ID}_{|\ell}},\mathsf{T},\mathsf{ct},\mathsf{mpk}$): The deterministic decryption algorithm is run by a receiver. It inputs decryption key $\mathsf{dk}_{\mathsf{ID}_{|\ell}},\mathsf{T}$, a ciphertext ct, and mpk. It outputs the session key K if $\mathsf{ID}_{|\ell}$ is not revoked at time period T; otherwise, it outputs $\perp$.
- **Revoke**($\mathsf{ID}_{|\ell},\mathsf{T},\mathbf{RL}_{\mathsf{ID}_{|\ell-1}},\mathsf{st}_{\mathsf{ID}_{|\ell-1}}$): The deterministic revocation algorithm is run by a user. It inputs an identity $\mathsf{ID}_{|\ell}$, a time period T, the revocation list $\mathbf{RL}_{\mathsf{ID}_{|\ell-1}}$ managed by $\mathsf{ID}_{|\ell-1}$ and state $\mathsf{st}_{\mathsf{ID}_{|\ell-1}}$. It updates the revocation list $\mathbf{RL}_{\mathsf{ID}_{|\ell-1}}$ by adding $\mathsf{ID}_{|\ell}$ as a revoked user at time period T.

Note that algorithm **KeyGen** and **Revoke** are stateful, and if $\mathsf{ID}_{|\ell}$ is revoked at time period T, then all descendants should be revoked at the time T. Figure 1 explains the hierarchical structure with binary tree structures.

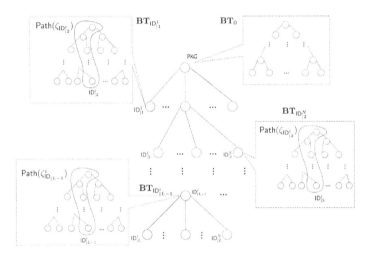

Fig. 1. The hierarchical structures.

We give the security definition for RHIBE scheme by extending Seo et al.'s security definition for RIBE scheme [9], which captures realistic threats including decryption key exposure. Let Π be an RHIBE scheme, we say that Π is adaptively secure if any PPT adversary $\mathcal{A}$ has negligible advantage in this following experiment:

$$\mathsf{Exp}_{\mathcal{A}}^{\Pi}(\kappa):$$
$$(\mathsf{mpk},\mathsf{msk},\mathbf{RL}_0,\mathsf{st}_0) \leftarrow \mathbf{Setup}(1^{\kappa},N,L),$$
$$(\mathsf{ID}_{|\ell}^{*},\mathsf{T}^{*},\mathsf{st}) \leftarrow \mathcal{A}^{\mathcal{O}}(\mathsf{st},\mathsf{mpk})$$
$$b \xleftarrow{\$} \{0,1\}, \mathsf{K}_1 \xleftarrow{\$} \{0,1\}^{\kappa}$$
$$(\mathsf{ct}^{*},\mathsf{K}_0) \leftarrow \mathbf{Enc}(\mathsf{mpk},\mathsf{ID}_{|\ell}^{*},\mathsf{T}^{*}),$$
$$b' \leftarrow \mathcal{A}^{\mathcal{O}}(\mathsf{mpk},\mathsf{st},\mathsf{ct}^{*},\mathsf{K}_b),$$
$$\text{return 1 if } b' = b \text{ and 0 otherwise.}$$

The adversary $\mathcal{A}$'s advantage is defined as

$$\mathrm{Adv}_{\mathcal{A}}^{\Pi}(\kappa) = |\Pr[b' = b] - \frac{1}{2}|.$$

In the above experiment, $\mathcal{O}$ is a set of oracles defined as follows.

- **KeyGen**($\cdot$): It takes as input an identity $\mathsf{ID}_{|\ell}$ of length ℓ. It returns the private key $\mathsf{sk}_{\mathsf{ID}_{|\ell}}$ by running the **KeyGen** algorithm.
- **KeyUp**($\cdot$): It takes an identity $\mathsf{ID}_{|\ell}$ of length ℓ and a time period T as input. It returns the update key $\mathsf{ku}_{\mathsf{ID}_{|\ell},\mathsf{T}}$ by running the **KeyUp** algorithm.
- **Revoke**($\cdot$): It takes as input an identity $\mathsf{ID}_{|\ell}$ of length ℓ, its child identity $\mathsf{ID}_{|\ell+1}$, and a time period T. It revokes $\mathsf{ID}_{|\ell+1}$ and updates $\mathbf{RL}_{\mathsf{ID}_{|\ell}}$ by running the **Revoke** algorithm.
- **DKG**($\cdot$): It takes as input an identity $\mathsf{ID}_{|\ell}$ of length ℓ, and a time period T. It returns $\mathsf{dk}_{\mathsf{ID}_{|\ell},\mathsf{T}}$ by successively running algorithms **KeyGen**, **KeyUp** and **DKG**.

We assume that all oracles share a state, and the adversary $\mathcal{A}$ is allowed to query above oracles with the following restrictions:

- **KeyUp**($\cdot$) and **Revoke**($\cdot$) can be queried on time period which is greater than or equal to the time of all previous queries.
- If **KeyGen**($\cdot$) was queried on $\mathsf{ID}_{|\ell}^*$ ($\ell \leq \ell^*$), then **Revoke**($\cdot$) must be queried to revoke $\mathsf{ID}_{|\ell}^*$ or one of its ancestors at a time period T for some $\mathsf{T} \leq \mathsf{T}^*$.
- **DKG**($\cdot$) cannot be queried on time period T before **KeyUp**($\cdot$) was queried on T. In addition, **DKG**($\cdot$) cannot be queried on $(\mathsf{ID}_{|i}^*, \mathsf{T}^*)$ for $1 \leq i \leq \ell^*$.

4   Our RHIBE Construction

Table 1 summarizes the notations that will be used in our RHIBE construction. Each intermediate level user $\mathsf{ID}_{|\ell} = (\mathsf{I}_1, \cdots, \mathsf{I}_\ell)$ has its own binary tree $\mathbf{BT}_{\mathsf{ID}_{|\ell}}$ for revoking capabilities and issues the key update $\mathsf{ku}_{\mathsf{ID}_{|\ell}}$ at each time period T.

Setup. Run $\mathcal{G}(1^\kappa) \to (n = p_1 p_2 p_3, \mathbf{G}, \mathbf{G}_T, \hat{e})$. Choose $\alpha, a_0, a_1, \cdots, a_L, b \xleftarrow{\$} \mathbf{Z}_n$, and a hash function $H : \mathbf{G}_T \to \{0,1\}^\kappa$. Set $\vec{w} = (a_0, a_1, \cdots, a_L, b)$, $u_i = g_1^{a_i}$ for $0 \leq i \leq L$ and $h = g_1^b$. Publish system parameters $\mathsf{pp} = \{n, \mathbf{G}, \mathbf{G}_T, \hat{e}, g_1, g_3, H\}$ and the master public key $\mathsf{mpk} = \{g_1^{\vec{w}}, \hat{e}(g_1, g_1)^\alpha\}$. Keep the master secret key $\mathsf{msk} = \{g_1^\alpha g_2^\alpha, \vec{w}\}$ secret.

KeyGen. This algorithm is differently defined according to the value ℓ.

- $\ell = 0$. Choose an unassigned leaf node η from $\mathbf{BT}_0$ at random, and store $\mathsf{ID}_{|1}$ in the node η. For all $\theta \in \mathsf{Path}(\eta_{\mathsf{ID}_{|1}}) \subset \mathbf{BT}_0$, recall S_θ in $\mathbf{BT}_0$ if it is defined. Otherwise, $S_\theta \xleftarrow{\$} \mathbf{G}_{p_1} \mathbf{G}_{p_2}$ and store it in the node $\theta \in \mathbf{BT}_0$. Choose $r_\theta \xleftarrow{\$} \mathbf{Z}_n$, set $t_1 = \log_{g_1} S_\theta$, $t_2 = \log_{g_2} S_\theta$ and

$$\vec{d}_{\mathsf{ID}_{|1}}^{(1,j)} = \mathsf{R}_3((g_1^{\mathsf{rE}(\alpha,t_1,\mathsf{ID}_{|1},\vec{w};r_\theta)}) \circ (g_2^{\mathsf{rE}(\alpha,t_2,\mathsf{ID}_{|1},\vec{w};0)})).$$

Return $\mathsf{sk}_{\mathsf{ID}_{|1}} = \{\vec{d}_{\mathsf{ID}_{|1}}^{(1,j)}\}_{j \in [1,n_u]}$, where j is the level of node θ in binary tree $\mathbf{BT}_0$.

Table 1. Notations

Symbol	Description
g, g_1, g_2, g_3	$g \in \mathbf{G}$, $g_1 \in \mathbf{G}_{p_1}$, $g_2 \in \mathbf{G}_{p_2}$ and $g_3 \in \mathbf{G}_{p_3}$
$\vec{u} \circ \vec{v}$	$(u_i v_i)_{i=1}^m$, where $\vec{u} = (u_i)_{i=1}^m \in \mathbf{G}^m$ and $\vec{v} = (v_i)_{i=1}^m \in \mathbf{G}^m$
$\vec{u}^r$	$(u_i^r)_{i=1}^m$, where $\vec{u} = (u_i)_{i=1}^m \in \mathbf{G}^m$, $r \in \mathbf{Z}_n$
$g^{\vec{x}}$	$(g^{x_i})_{i=1}^m$, where $g \in \mathbf{G}$, $\vec{x} = (x_i)_{i=1}^m \in \mathbf{Z}_n^m$
$g^{\mathsf{rE}(\alpha,t,\mathsf{ID}_{\mid \ell},\vec{w};r)}$	$(g^t g^\alpha (u_1^{l_1} \cdots u_\ell^{l_\ell} h)^r, g^r, u_0^r, u_{\ell+1}^r, \cdots, u_L^r) \in \mathbf{G}^{L-\ell+3}$
$g^{\mathsf{rE}(\alpha,t,\mathsf{T}\|\mathsf{ID}_{\mid \ell},\vec{w};r)}$	$(g^t g^\alpha (u_0^\mathsf{T} u_1^{l_1} \cdots u_\ell^{l_\ell} h)^r, g^r, u_{\ell+1}^r, \cdots, u_L^r) \in \mathbf{G}^{L-\ell+2}$
$g_1^{\mathsf{sE}(\mathsf{T}\|\mathsf{ID}_{\mid \ell},\vec{w})}$	$(g_1, (u_0^\mathsf{T} u_1^{l_1} \cdots u_\ell^{l_\ell} h)) \in \mathbf{G}^2$
$\mathsf{R}_3(\vec{u})$	$\vec{u} \circ g_3^{\vec{x}}$, where $\vec{x} \xleftarrow{\$} \mathbf{Z}_n^m$
$\log_{g_i} X$	$t_i \in \mathbf{Z}_{p_i}$, where $X = g_1^{t_1} g_2^{t_2} g_3^{t_3}$,

- $\ell > 0$. Choose an unassigned leaf η from $\mathbf{BT}_{\mathsf{ID}_{\mid \ell}}$ at random, store $\mathsf{ID}_{\mid \ell+1}$ in the node $\eta_{\mathsf{ID}_{\mid \ell+1}}$. Parse

$$\mathsf{sk}_{\mathsf{ID}_{\mid \ell}} = \{\vec{d}_{\mathsf{ID}_{\mid \ell}}^{(i,j)}\}_{(i,j)\in[1,\ell]\times[1,n_u]} = (d^{(i,j)}, d'^{(i,j)}, d_0^{(i,j)}, d_{\ell+2}^{(i,j)}, \cdots, d_L^{(i,j)}).$$

For all $(i,j) \in [1,\ell] \times [1,n_u]$, recall $S_{(i,j)}$ from $\mathsf{st}_{\mathsf{ID}_{\mid \ell}}$ if it is defined. Otherwise, $S_{(i,j)} \xleftarrow{\$} \mathbf{G}_{p_1} \mathbf{G}_{p_2}$ and store it in $\mathsf{st}_{\mathsf{ID}_{\mid \ell}}$. Choose $r_{(i,j)} \xleftarrow{\$} \mathbf{Z}_n$, set

$$\vec{d}_{\mathsf{ID}_{\mid \ell+1}}^{(i,j)} = \mathsf{R}_3((S_{(i,j)} d^{(i,j)} (d_{\ell+1}^{(i,j)})^{l_{\ell+1}} (u_1^{l_1} \cdots u_{\ell+1}^{l_{\ell+1}} h)^{r_{(i,j)}}, d'^{(i,j)}, d_0^{(i,j)},$$
$$d_{\ell+2}^{(i,j)}, \cdots, d_L^{(i,j)}) \circ (g, u_0, u_{\ell+2}, \cdots, u_L)^{r_{(i,j)}}))$$

For all $\theta \in \mathsf{Path}(\eta_{\mathsf{ID}_{\mid \ell+1}}) \subset \mathbf{BT}_\ell$, recall S_θ from the corresponding node θ in $\mathbf{BT}_{\mathsf{ID}_{\mid \ell}}$ if it is defined. Otherwise, $S_\theta \xleftarrow{\$} \mathbf{G}_{p_1} \mathbf{G}_{p_2}$ and store it in the node $\theta \in \mathbf{BT}_\ell$, choose $r_\theta \xleftarrow{\$} \mathbf{Z}_n$ and set

$$\vec{d}_{\mathsf{ID}_{\mid \ell+1}}^{(\ell+1,j)} = \mathsf{R}_3((g_1^{\mathsf{rE}(\alpha,t_1,\mathsf{ID}_{\mid \ell+1},\vec{w};r_\theta)}) \circ (g_2^{\mathsf{rE}(\alpha,t_2,\mathsf{ID}_{\mid \ell+1},\vec{w};0)}))$$

Return $\mathsf{sk}_{\mathsf{ID}_{\mid \ell+1}} = \{\vec{d}_{\mathsf{ID}_{\mid \ell+1}}^{(i,j)}\}_{(i,j)\in[1,\ell+1]\times[1,n_u]}$, where j is the level of θ in the tree $\mathbf{BT}_{\mathsf{ID}_{\mid \ell}}$.

KeyUp. This algorithm is differently defined according to the value ℓ.

- $\ell = 0$. For nodes $\theta \in \mathsf{KUNode}(\mathbf{BT}_0, \mathbf{RL}_0, \mathsf{T})$, recall S_θ from the node $\theta \in \mathbf{BT}_0$, choose $\delta_\theta \xleftarrow{\$} \mathbf{Z}_n$, and compute

$$\vec{f}_{\mathsf{ID}_{\mid 0},\theta} = \mathsf{R}_3((g_1^{\mathsf{rE}(\alpha,-t_1,\mathsf{T}\|\mathsf{ID}_{\mid 0},\vec{w};\delta_\theta)}) \circ (g_2^{\mathsf{rE}(\alpha,-t_2,\mathsf{T}\|\mathsf{ID}_{\mid 0},\vec{w};0)}))$$

Return $\mathsf{ku}_{\mathsf{ID}_{\mid 0},\mathsf{T}} = \{\emptyset, \vec{f}_{\mathsf{ID}_{\mid 0},\theta}\}$ for all $\theta \in \mathsf{KUNode}(\mathbf{BT}_0, \mathbf{RL}_0, \mathsf{T})$.

- $\ell > 0$. For all $\theta \in \mathsf{KUNode}(\mathbf{BT}_{\mathsf{ID}_{|\ell-1}}, \mathbf{RL}_{\mathsf{ID}_{|\ell-1}}, \mathsf{T})$, let Lv_i be the level of θ_i in $\mathbf{BT}_{\mathsf{ID}_{|i-1}}$ where $\theta_i = \mathsf{Path}(\mathsf{ID}_{|i}) \cap \mathsf{KUNode}(\mathbf{BT}_{\mathsf{ID}_{|i-1}}, \mathbf{RL}_{\mathsf{ID}_{|i-1}}, \mathsf{T})$. Then, parse $\mathsf{ku}_{\mathsf{ID}_{|\ell-1}, \mathsf{T}} = \{\{\mathsf{Lv}_i\}_{i\in[1,\ell-1]}, \vec{f}_{\mathsf{ID}_{|\ell-1,\theta}}\}$. For all $i \in [1,\ell]$, recall $S_{(i,\mathsf{Lv}_i)}$ from $\mathsf{st}_{\mathsf{ID}_{|\ell}}$, compute $\tilde{S} = \prod_{i=1}^{\ell} S_{(i,\mathsf{Lv}_i)}$, identify one node $\tilde{\theta}$ from the intersection between $\mathsf{Path}(\eta_{\mathsf{ID}_{|\ell}})$ and $\mathsf{KUNode}(\mathbf{BT}_{\mathsf{ID}_{|\ell-1}}, \mathbf{RL}_{\mathsf{ID}_{|\ell-1}}, \mathsf{T})$. Let Lv_ℓ be the level of $\tilde{\theta}$, the PKG can get $\vec{f}_{\mathsf{ID}_{|\ell-1,\tilde{\theta}}}$ that contains $(f_{\tilde{\theta}}, f_{\tilde{\theta}}', f_{\tilde{\theta},\ell}, \cdots, f_{\tilde{\theta},L})$. For all $\theta \in \mathsf{KUNode}(\mathbf{BT}_{\mathsf{ID}_{|\ell}}, \mathbf{RL}_{\mathsf{ID}_{|\ell}}, \mathsf{T})$, recall S_θ from $\theta \in \mathbf{BT}_{\mathsf{ID}_{|\ell}}$, choose $\delta_\theta \xleftarrow{\$} \mathbf{Z}_n$, and compute

$$\vec{f}_{\mathsf{ID}_{|\ell},\theta} = \mathsf{R}_3((S_\theta \tilde{S})^{-1} f_{\tilde{\theta}} (f_{\tilde{\theta},\ell})^{\mathsf{l}_\ell} (u_0^\mathsf{T} u_1^{\mathsf{l}_1} \cdots u_\ell^{\mathsf{l}_\ell})^{\delta_\theta}, (f_{\tilde{\theta}}', f_{\tilde{\theta},\ell+1}, \cdots, f_{\tilde{\theta},L})$$
$$\circ (g, u_{\ell+1}, \cdots, u_L)^{\delta_\theta})$$

Return $\{\{\mathsf{Lv}_i\}_{i\in[1,\ell]}, \vec{f}_{\mathsf{ID}_{|\ell},\theta}\}$ for all $\theta \in \mathsf{KUNode}(\mathbf{BT}_{\mathsf{ID}_{|\ell}}, \mathbf{RL}_{\mathsf{ID}_{|\ell}}, \mathsf{T})$.

DKG. If $\mathsf{KUNode}(\mathbf{BT}_{\mathsf{ID}_{|\ell}}, \mathbf{RL}_{\mathsf{ID}_{|\ell}}, \mathsf{T}) \cap \mathsf{Path}(\eta_{\mathsf{ID}_{|\ell}}) = \emptyset$, it returns $\bot$. Otherwise, choose a node $\theta \in \mathsf{KUNode}(\mathbf{BT}_{\mathsf{ID}_{|\ell}}, \mathbf{RL}_{\mathsf{ID}_{|\ell}}, \mathsf{T}) \cap \mathsf{Path}(\eta_{\mathsf{ID}_{|\ell}})$ and $r \xleftarrow{\$} \mathbf{Z}_n$, compute

$$\mathsf{dk}_{\mathsf{ID}_{|\ell}, \mathsf{T}} = \mathsf{R}_3(((\prod_{i\in[1,\ell]} d^{(i,\mathsf{Lv}_i)}(d_0^{(i,\mathsf{Lv}_i)})^\mathsf{T}), (\prod_{i\in[1,\ell]} d'^{(i,\mathsf{Lv}_i)}), (\prod_{i\in[1,\ell]} d_{\ell+1}^{(i,\mathsf{Lv}_i)}), \cdots,$$
$$(\prod_{i\in[1,\ell]} d_L^{(i,\mathsf{Lv}_i)})) \circ g_1^{\mathsf{rE}(0,0,\mathsf{T}\|\mathsf{ID}_{|\ell}, \vec{w}; r)} \circ (f_\theta(f_{\theta,\ell})^{\mathsf{l}_\ell}, f_\theta', f_{\theta,\ell+1}, \cdots, f_{\theta,L}))$$

Enc. The sender chooses $s \xleftarrow{\$} \mathbf{Z}_n$ and computes

$$\mathsf{ct} = (g_1^{\mathsf{sE}(\mathsf{T}\|\mathsf{ID}_{|\ell}, \vec{w})})^s = (g_1^s, (u_0^\mathsf{T} u_1^{\mathsf{l}_1} \cdots u_\ell^{\mathsf{l}_\ell})^s), \mathsf{K} = H(\hat{e}(g_1, g_1)^{\alpha s}).$$

Dec. On receiving the ciphertext ct, the receiver first parses $\mathsf{ct} = (A, B)$ and $\mathsf{dk}_{\mathsf{ID}_{|\ell}, \mathsf{T}} = (D, D', D_{\ell+1}, \cdots, D_L)$, and computes

$$\mathsf{K} = H(\frac{\hat{e}(D, A)}{\hat{e}(D', B)}).$$

Revoke. Let η be the leaf node in $\mathbf{BT}_{\mathsf{ID}_{\ell-1}}$ associated with $\mathsf{ID}_{|\ell}$. The PKG updates the revocation list by $\mathbf{RL}_{\mathsf{ID}_{|\ell-1}} \leftarrow \mathbf{RL}_{\mathsf{ID}_{|\ell-1}} \cup \{(\eta, \mathsf{T})\}$.

5 Efficiency and Security Analysis

The performance comparison of our RHIBE scheme with existing RHIBE schemes is summarized in Table 2, where $|\mathbf{G}|$ and $|\mathbf{G}_T|$ are the bit-length of an element in group $\mathbf{G}$ and $\mathbf{G}_T$, respectively. t_e is the computation cost for performing a bilinear pairing $\hat{e}(\mathbf{G}, \mathbf{G}) \rightarrow \mathbf{G}_T$, DKE means decryption key exposure and DBDH is the decisional bilinear Diffie-Hellman assumption, and SGD is the subgroup decision assumption.

Table 2. Comparison with existing RHIBE schemes

RHIBE schemes	Security model	Complexity assumption	Scalability	DKE resistance	CT size	Dec. cost
[12]	Adaptive-ID	SGD	No	No	$\ell(5|\mathbf{G}| + |\mathbf{G}_T|)$	$5\ell t_e$
[11]	Selective-ID	DBDH	Yes	Yes	$(\ell + 2)|\mathbf{G}| + |\mathbf{G}_T|$	$(\ell + 2)t_e$
Ours	Adaptive-ID	SGD	Yes	Yes	$2|\mathbf{G}| + |\mathbf{G}_T|$	$2t_e$

Theorem 3. *Our RHIBE scheme is adaptively secure against any adversary under the Assumptions 1 and 2. More precisely, we have*

$$Adv_{\mathcal{A}}^{RHIBE} \leq (L + 1)q^3 (Adv_{\mathcal{G},\mathcal{A}_1}^{SGD_1}(\kappa) + 4q Adv_{\mathcal{G},\mathcal{A}_2}^{SGD_2}(\kappa) + 2^{-\Omega(\kappa)})$$

Proof. We will provide detailed security proof in the extended version.   □

6   Conclusion

In this paper, we give formal syntax and security definitions for revocable hierarchical identity-based encryption scheme, which supports both key delegation and scalable revocation functionality. We propose the first adaptive-ID secure revocable hierarchical identity-based encryption scheme with constant ciphertext size by integrating Lewko and Waters's hierarchical identity-based encryption scheme with the complete subtree method. The proposed scheme is proved to be semantic secure in the adaptive-ID model by applying the dual system encryption methodology and predicate encoding strategy. An interesting open problem is to combine hierarchical identity-based encryption scheme with subset difference method (instead of complete subtree method).

Acknowledgments. This research is funded by National Natural Science Foundation of China (Grant No. 61173189).

References

1. Horwitz, J., Lynn, B.: Toward hierarchical identity-based encryption. In: Knudsen, L.R. (ed.) EUROCRYPT 2002. LNCS, vol. 2332, pp. 466–481. Springer, Heidelberg (2002). doi:10.1007/3-540-46035-7_31
2. Boneh, D., Franklin, M.: Identity-based encryption from the weil pairing. In: Kilian, J. (ed.) CRYPTO 2001. LNCS, vol. 2139, pp. 213–229. Springer, Heidelberg (2001). doi:10.1007/3-540-44647-8_13
3. Yan, L., Rong, C., Zhao, G.: Strengthen cloud computing security with federal identity management using hierarchical identity-based cryptography. In: Jaatun, M.G., Zhao, G., Rong, C. (eds.) CloudCom 2009. LNCS, vol. 5931, pp. 167–177. Springer, Heidelberg (2009). doi:10.1007/978-3-642-10665-1_15
4. Gentry, C., Silverberg, A.: Hierarchical ID-based cryptography. In: Zheng, Y. (ed.) ASIACRYPT 2002. LNCS, vol. 2501, pp. 548–566. Springer, Heidelberg (2002). doi:10.1007/3-540-36178-2_34

5. Lewko, A., Waters, B.: New techniques for dual system encryption and fully secure HIBE with short ciphertexts. In: Micciancio, D. (ed.) TCC 2010. LNCS, vol. 5978, pp. 455–479. Springer, Heidelberg (2010). doi:10.1007/978-3-642-11799-2_27

6. Naor, D., Naor, M., Lotspiech, J.: Revocation and tracing schemes for stateless receivers. In: Kilian, J. (ed.) CRYPTO 2001. LNCS, vol. 2139, pp. 41–62. Springer, Heidelberg (2001). doi:10.1007/3-540-44647-8_3

7. Boldyreva, A., Goyal, V., Kumar, V.: Identity-based encryption with efficient revocation. In: Proceedings of the 15th ACM Conference on Computer and Communications Security - ACM CCS, pp. 417–426 (2008)

8. Libert, B., Vergnaud, D.: Adaptive-ID secure revocable identity-based encryption. In: Fischlin, M. (ed.) CT-RSA 2009. LNCS, vol. 5473, pp. 1–15. Springer, Heidelberg (2009). doi:10.1007/978-3-642-00862-7_1

9. Seo, J.H., Emura, K.: Revocable identity-based cryptosystem revisited: security models and constructions. IEEE Trans. Inf. Forensics Secur. 9(7), 1193–1205 (2014)

10. Wang, C.J., Li, Y., Xia, X.N., Zheng, K.J.: An efficient and provable secure revocable identity-based encryption scheme. PLoS ONE 9(9), 1–11 (2014)

11. Seo, J.H., Emura, K.: Efficient delegation of key generation and revocation functionalities in identity-based encryption. In: Dawson, E. (ed.) CT-RSA 2013. LNCS, vol. 7779, pp. 343–358. Springer, Heidelberg (2013). doi:10.1007/978-3-642-36095-4_22

12. Tsai, T.T., Tsen, Y.M., Wu, T.Y.: RHIBE: constructing revocable hierarchical ID-Based encryption from HIBE. Infomatica 25(2), 299–326 (2014)

13. Seo, J.H., Emura, K.: Revocable hierarchical identity-based encryption: history-free update, security against insiders, and short ciphertexts. In: Nyberg, K. (ed.) CT-RSA 2015. LNCS, vol. 9048, pp. 106–123. Springer, Heidelberg (2015). doi:10.1007/978-3-319-16715-2_6

14. Wee, H.: Dual system encryption via predicate encodings. In: Lindell, Y. (ed.) TCC 2014. LNCS, vol. 8349, pp. 616–637. Springer, Heidelberg (2014). doi:10.1007/978-3-642-54242-8_26

15. Goyal, V., Pandey, O., Sahai, A., Waters, B.: Attribute based encryption for fine-grained access conrol of encrypted data. In: ACM conference on Computer and Communications Security, pp. 89–98 (2006)

16. Bethencourt, J., Sahai, A., Waters, B.: Ciphertext-policy attribute-based encryption. In: IEEE Symposium on Security and Privacy, pp. 321–334 (2007)

17. Waters, B.: Ciphertext-policy attribute-based encryption: an expressive, efficient, and provably secure realization. In: Catalano, D., Fazio, N., Gennaro, R., Nicolosi, A. (eds.) PKC 2011. LNCS, vol. 6571, pp. 53–70. Springer, Heidelberg (2011). doi:10.1007/978-3-642-19379-8_4

18. Waters, B.: Dual system encryption: realizing fully secure IBE and HIBE under simple assumptions. In: Halevi, S. (ed.) CRYPTO 2009. LNCS, vol. 5677, pp. 619–636. Springer, Heidelberg (2009). doi:10.1007/978-3-642-03356-8_36

19. Herranz, J., Hofheinz, D., Kiltz, E.: Some (in) sufficient conditions for secure hybrid encryption. Inf. Comput. 208(11), 1243–1257 (2010)

Understanding Networking Capacity Management in Cloud Computing

Haokun Jiang[1] and Xiaotong Sun[2(✉)]

[1] Shengli Oil Field No. 1 Middle School, Dongying 257000, Shandong, China
haokunjiang2016@yahoo.com
[2] Department of Computer Science, Pace University, New York City, NY 10038, USA
xs43599n@pace.edu

Abstract. Contemporary advances of networking technologies have enabled enormous new applications in various domains. The networking capacity has been playing a significant role in achieving a high performance of the Web-based solution. However, the challenging sector is that the diverse of the networks have brought the difficulties in establishing an efficient mechanism, since different networking types may have various performances in various operating environments. This paper focuses on understanding the fact of networking capacity management in heterogeneous cloud computing and aims to provide a comprehensive survey on the target issue for assisting enterprises in developing proper web-related strategies. In this survey, we map the crucial issues of the capacity planning, as well as the main solutions, in order to represent a holistic view of the subject.

Keywords: Cloud computing · Network management · Capacity planning · Networking optimization

1 Introduction

Recent rapid development of the networks have been improving the Web-based applications in various disciplines or fields, which is strongly impacting on current individuals' lives [1–3]. The great change in networking deployments has enabled the networking capacity to be an important role in delivering an expected service presentation [4,5]. The challenge becomes greater align with the growth of the data sizes [6]. For example, a bottleneck in a network is a critical obstacle to increase the performance of the entire network system [7], which can result in either a great latency or a networking congestion [8,9]. Therefore, a quality implementation of the networks is an urgent demand for current network-based deployments.

However, the challenging issue is that it is difficult to be aware of different networks due to numerous new technologies enabled by the growth of networking usage [10–12]. This paper addresses this topic and aims to review the main superiorities of applying networking capacity management in cloud computing,

© Springer International Publishing AG 2017
M. Qiu (Ed.): SmartCom 2016, LNCS 10135, pp. 516–526, 2017.
DOI: 10.1007/978-3-319-52015-5_53

as well as represents the virtues of the networking management for current organizations. A few aspects will be covered in this survey and the findings of this work can aid enterprises to conquer various challenges of networking deployments and operations. More specifically, we will review following aspects, which include the demands of the networking capacity management, awareness of the techniques, data collection and monitoring mechanisms, and crucial advantages and features of the networking capacity management.

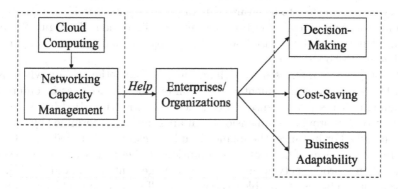

Fig. 1. Architecture of benefits received from using cloud-based networking capacity management in a perspective of strategy-making.

The significance of understanding the networking capacity management in cloud computing is great for most current enterprises or organizations that intend to utilize cloud-based solutions. Figure 1 represents three vital benefits in this field from strategic perspective, which were represented below. First, understanding the advantages of cloud-based networking capacity management can assist enterprises to make proper decisions in multiple dimensions, such as risk management [13]. The outcomes of the strategic planning need to be aligned with updated industrial trends and developments. Second, enterprises or organizations can justify the costs to reduce the financial budget based on a great understanding of new equipment, facilities, or software. Finally, an enhanced business adaptability can be retrieved when applying a cloud-based networking capacity management, since current manipulative environment is dynamic.

The main contributions of this work include:

1. This work has provided a panoramic review on the networking capacity management in cloud computing.
2. The findings of this work can be used as a reference for enterprises' future technical strategy-making.

The remainder of this paper follows the order below. Section 2 describes the structure of the cloud-based networking capacity management. Next, Sect. 3 further represent the main issues and proposes the solution framework in the discipline. Finally, we draw the conclusions in Sect. 4.

2   Structure of Networking Capacity Management in Clouds

Networking capacity management can penetrate through different layers of the business, such as employee operations, facility deployments, and production assessments. We map the crucial advantages of implementing networking capacity management techniques in this section in order to discern the topic in a perspective of business operations. One dimension of understanding cloud-based networking capacity management is to be aware of its impacts on enterprises' operations [14]. We summarize four crucial aspects concerning this dimension [15]. Figure 2 depicts a virtue structure of cloud-based networking capacity management in a perspective of enterprise operations.

The first aspect addresses the enhancement of prediction ability when using networking capacity management in cloud computing, which can assist enterprises in avoiding unnecessary expenses on technical infrastructure [16,17]. There are a variety of focuses providing dimensions that are generally determined by the business demands. For instance, the financial budgets can be controlled, maintained, and adjusted in terms of the economic conditions, either in a growth or a recession period, when a proper networking capacity management is achieved [18,19]. The achievement can be also favorable to creating an effective technical facility budget based on the networking resource usage.

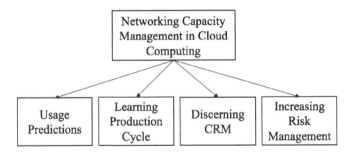

Fig. 2. Virtue structure of using cloud-based networking capacity management in a perspective of enterprise operations.

The next aspect is that having an effective cloud-based networking capacity management strategy can help enterprises to obtain a proper production cycle, even though a great deal of maintenance work needs to be carried [20]. Fully understanding the production cycle addressing the networking capacity dimension is an important fundamental for enterprises to make a strategy on resource distribution and planning [21]. Enterprises can adjust their technical or networking tactics based on their positions within the whole production cycle as considering using network-based solutions. The utilization of the cloud resources can be scaled up and down based on the continuous changes in the production cycles [22]. The changes can be designed and implemented by aligning with both internal operations and external conditions.

Furthermore, enterprises need to discern the method of increasing *Customer Relationship Management* (CRM) by optimizing cloud-based networking capacity management [18,23] in order to adapt themselves for continuous changing requirements of customers. The scalability and adaptability can be beneficial for increasing the quality of the service deliveries, which can increase the satisfaction of customers [24,25]. For instance, cloud resource manager can assist customers to obtain the computing resources from the nearest cloud servers in order to speed up the efficiency of the resource acquisitions. This benefit becomes expressive and significant when the enterprise encounters unexpected external changes, such as a dramatical marketing transformations or a quick networking service level update [26–28]. In the financial industry, this feature has become an important role in connecting customers with financial services, such as mobile payments. New financial services need to be not only scalable but also efficiently reachable.

Finally, it is also a significant component of the networking management for an enterprise's risk management [29–32]. The networking capacity within a cloud computing environment is a new issue in the field of networking management. A great loss can occur if the enterprise does not have a prepared solution. However, this issue is still a challenging task for most current enterprises, since understanding most cyber risks and preparing solutions to all potential risks are not easy [33,34]. The cyber risks derive from not only capacity limitations but also adversarial attacks [35], such as network abuse or malicious codes [36].

In summary, it is important to understand that creating a long-term plan of networking capacity management plan in cloud computing for most current companies that highly rely on the implementations of networks. Having an effective cloud-based networking capacity management can bring various advantages based on the reviews above. The next section will present the main issues of applying networking capacity management in cloud computing, as well as the corresponding potential solutions.

3   Main Issues and Solutions

3.1   Main Concepts

The concept of networking capacity management in cloud computing is a series of procedures forming a whole process in which the network system's usage demand can be predicted for a certain purpose configured by the organization's demand, such as avoiding networking resource wastes or reducing latency. By using networking capacity management in cloud computing, the organization or enterprise can plan the networking usage ahead in order to achieve the optimizations of the network system.

In addition, the crucial issue in networking capacity management is to obtain a quality ability of predictions [37], so that the risks or obstacles can be avoided or solved by the planned solutions. This process is usually implemented by an application that examines and controls the networking configurations in terms of the real-time networking requirements [38,39]. To reach this goal, an examination of the workloads is required for obtaining the estimated values of various parameters, such as the networking service schedules, service response time, and estimated service frequency [40]. Therefore, the forecast techniques for predicting networking conditions are considered one of the crucial required conditions in the establishment of the efficient networking capacity management [41].

3.2 Bottleneck Issues

The bottleneck issue in networking capacity management mainly refers to the network node that takes majority time consumptions. Thus, it is one of the most important aspects that can improve the performance of the entire network system from the perspective of reducing total response time [42,43]. There are a variety of elements that can influence the outcomes of web performances. One element is that there exist some trade-off in the network system, such as the contradiction between response time and workloads. Another common element is the saturation of the network system [9,44]. The saturation refers to the usage level of the network system with a limitation. Figure 3 illustrates a high level framework of our proposed solution, which shows the client/server-based networking environment for capacity management in cloud computing. The centralized node monitors the operations of the entire network in order to ensure the saturation level of the network is attached to a proper position.

In Fig. 3, client-end refers to the representation of the service deliveries. For each client-end, a server is required for the purpose of data storage or processing [45]. Meanwhile, the data can be also transferrable between local and remote servers, to which the authenticated users have the access. It means that each client-end

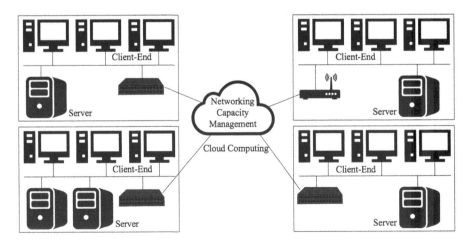

Fig. 3. Example of client/server-based networking environment for capacity management in cloud computing.

location has a local network that establishes the sub-level network system. In general, a router is required to connect the local network to the Internet or other networks. Our solution is to use a added layer that is called *Networking Capacity Management Controller* (NCMC) to monitor, control, and schedule the networking resources based on the real-time demands. The purpose of adding this layer in the cloud-based network system is to make sure the networking resources are manageable for avoiding bottleneck issues.

Moreover, prior researches have also pointed out that the crucial factors in improving networking performances include clients, network, and servers. In these three factors, a server is the critical component in the Web by which the service requests are delivered [46]. Web services are "carriers" for delivering services between computing and networking nodes [47]. In some situations, the main obstacles of the networks are mainly caused by the limitations of information reaching, such as unpredictable service requests or retrievals. We summarize that the networking problems derive from a few aspects, which include data, applications, infrastructure, operating systems, and networking bandwidth [48,49]. Any improper operations or unexpected mechanical problems can create networking bottlenecks.

In summary, the bottleneck issues in cloud computing mainly exist when multiple service requests are reaching the same computing source in clouds. The jams can be caused by either networking collisions or cloud servers' overloads.

3.3   State Estimations

In the cloud context, the capacity statuses of the networks can be considered the "states". Most current active task schedulers have fixed task scheduling algorithms so that the task forwards are usually not changed. This method can avoid networking bottlenecks only when the networking conditions are stable and continuous, which means the bottleneck issues will occasionally take place within the dynamic networking usage environment. Therefore, estimating the networking states is a fundamental for task schedulers to create a dynamic task forwarding method. The support of the state estimations brings the dimension of using a variety of mature techniques, such as parallel computing, state optimizations, or networking capacity planning.

Some recent studies also addressed the explorations of forecasting or analyzing networking states. The purposes might be varied due to different demands, such as security solutions and efficient performances [50]. A common description of state forecasts can be defined as the activities of predicting the upcoming statuses of the states based on the applied prediction techniques. The methods of state estimations are generally based on the historical data sets or real-time state condition updates. For example, one of the approaches for predicting cloud servers' conditions is using time-series prediction methods [51], which normally utilize the state condition at the precedent time slot to forecast the state condition at the succeeding time slot. The other common approach is using machine-learning techniques for decision-making assistance [52,53].

Based on the reviews and syntheses above, we propose a solution framework that is designed to deal with the common constraints caused by networking bottlenecks. The proposed solution is represented in Sect. 3.4.

3.4   Solution Framework

Figure 4 illustrates our proposed framework that consists of a few components. There are mainly three phases in the framework, which are data collections, data analyses, and plan generations. At the phase of data collections, the system needs to gather the data that are required for scheduling tasks, such as the earliest arrival time, the earliest finish time, bandwidth, and required processing time. Next, the phase of data analyses consists of two steps, namely data re-processing and scheduling tasks. The step of data pre-processing is for estimating networking states by forecasts. In the figure, *State 1* means the precedent state and *State 2* means the succeeding state that is forecasted from *State 1*. There are a few dimensions of forecasts, which are determined by the demands. A number of examples are shown in the figure, including security, energy consumptions, networking node status, and *Virtual Machine* (VM) availabilities. The other step at the phase of data analyses is scheduling tasks, which arrange the tasks to the proper cloud sources for processing workloads. The parameters used at this step depend on the outcomes produced from data pre-processing. Finally, a plan will be created from using the task scheduler.

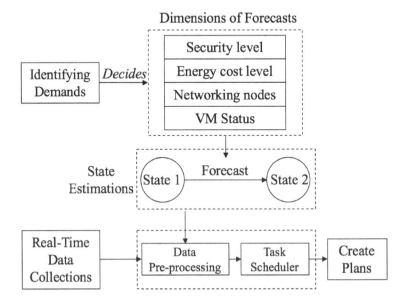

Fig. 4. Networking capacity management framework.

4 Conclusions

This paper focused on the issue of networking capacity management in cloud computing and completed a survey on the corresponding topics. The study concentrated on the main issues of networking limitations and proposed potential solutions based on the prior researches. Future work will address the implementations of the networking capacity management in cloud computing and examine the proposed approach in various applications.

References

1. Jeong, H., Tombor, B., Albert, R., Oltvai, Z., Barab, A.: The large-scale organization of metabolic networks. Nature **407**(6804), 651–654 (2000)
2. Wu, L., Garg, S., Buyya, R.: SLA-based admission control for a software-as-a-service provider in cloud computing environments. J. Comput. Syst. Sci. **78**(5), 1280–1299 (2012)
3. Gai, K., Qiu, M., Zhao, H., Tao, L., Zong, Z.: Dynamic energy-aware cloudlet-based mobile cloud computing model for green computing. J. Netw. Comput. Appl. **59**, 46–54 (2015)
4. Gai, K., Li, S.: Towards cloud computing: a literature review on cloud computing and its development trends. In: 2012 Fourth International Conference on Multimedia Information Networking and Security, Nanjing, China, pp. 142–146 (2012)
5. Chen, L., Duan, Y., Qiu, M., Xiong, J., Gai, K.: Adaptive resource allocation optimization in heterogeneous mobile cloud systems. In: The 2nd IEEE International Conference on Cyber Security and Cloud Computing, pp. 19–24. IEEE, New York (2015)
6. Gai, K., Qiu, M., Chen, L., Liu, M.: Electronic health record error prevention approach using ontology in big data. In: 17th IEEE International Conference on High Performance Computing and Communications, New York, USA, pp. 752–757 (2015)
7. Gai, K., Qiu, M., Zhao, H.: Security-aware efficient mass distributed storage approach for cloud systems in big data. In: The 2nd IEEE International Conference on Big Data Security on Cloud, New York, USA, pp. 140–145 (2016)
8. Sun, H., Chen, Y., Lin, Y.: oPass: a user authentication protocol resistant to password stealing and password reuse attacks. IEEE Trans. Inf. Forensics Secur. **7**(2), 651–663 (2012)
9. Gai, K., Qiu, M., Zhao, H., Liu, M.: Energy-aware optimal task assignment for mobile heterogeneous embedded systems in cloud computing. In: 2016 IEEE 3rd International Conference on Cyber Security and Cloud Computing (CSCloud), pp. 198–203. IEEE, Beijing, China (2016)
10. Gai, K., Qiu, M., Elnagdy, S.: Security-aware information classifications using supervised learning for cloud-based cyber risk management in financial big data. In: The 2nd IEEE International Conference on Big Data Security on Cloud, New York, USA, pp. 197–202 (2016)
11. Jean-Baptiste, H., Qiu, M., Gai, K., Tao, L.: Meta meta-analytics for risk forecast using big data meta-regression in financial industry. In: The 2nd IEEE International Conference on Cyber Security and Cloud Computing, pp. 272–277. IEEE, New York (2015)

12. Jean-Baptiste, H., Tao, L., Qiu, M., Gai, K.: Model risk management systems-back-end, middleware, front-end and analytics. In: The 2nd IEEE International Conference on Cyber Security and Cloud Computing, pp. 312–316. IEEE, New York (2015)

13. Gai, K., Qiu, M., Elnagdy, S.: A novel secure big data cyber incident analytics framework for cloud-based cybersecurity insurance. In: The 2nd IEEE International Conference on Big Data Security on Cloud, New York, USA, pp. 171–176 (2016)

14. Hu, J., Holt, S., Marques, J., Camillo, A.: Marketing channels and supply chain management in contemporary globalism: E-commerce development in China. In: Handbook of Research on Effective Marketing in Contemporary Globalism, p. 325 (2014)

15. Gai, K., Steenkamp, A.: Feasibility of a Platform-as-a-Service implementation using cloud computing for a global service organization. In: Proceedings of the Conference for Information Systems Applied Research ISSN, vol. 2167, p. 1508 (2013)

16. Elnagdy, S., Qiu, M., Gai, K.: Understanding taxonomy of cyber risks for cybersecurity insurance of financial industry in cloud computing. In: 2016 IEEE 3rd International Conference on Cyber Security and Cloud Computing, The 2nd IEEE International Conference of Scalable and Smart Cloud, pp. 295–300. IEEE (2016)

17. DeStefano, R., Tao, L., Gai, K.: Improving data governance in large organizations through ontology and linked data. In: 2016 IEEE 3rd International Conference on Cyber Security and Cloud Computing, The 2nd IEEE International Conference of Scalable and Smart Cloud, pp. 279–284. IEEE (2016)

18. Steenkamp, A., Alawdah, A., Almasri, O., Gai, K., Khattab, N., Swaby, C., Abaas, R.: Teaching case enterprise architecture specification case study. J. Inf. Syst. Educ. **24**(2), 105 (2013)

19. Hu, Z., Lin, H., Zhou, Y.: A fuzzy reputation management system with punishment mechanism for P2P network. J. Netw. **6**(2), 190–197 (2011)

20. Gai, K., Steenkamp, A.: A feasibility study of Platform-as-a-Service using cloud computing for a global service organization. J. Inf. Syst. Appl. Res. **7**, 28–42 (2014)

21. Elnagdy, S., Qiu, M., Gai, K.: Cyber incident classifications using ontology-based knowledge representation for cybersecurity insurance in financial industry. In: 2016 IEEE 3rd International Conference on Cyber Security and Cloud Computing, The 2nd IEEE International Conference of Scalable and Smart Cloud, pp. 301–306. IEEE (2016)

22. Gai, K.: A review of leveraging private cloud computing in financial service institutions: Value propositions and current performances. Int. J. Comput. Appl. **95**(3), 40–44 (2014)

23. Gai, K.: A report about suggestions on developing e-learning in China. In: IEEE 2010 International Conference on E-Business and E-Government, pp. 609–613. IEEE (2010)

24. Zhang, Y., Bian, J., Zhu, W.: Trust fraud: A crucial challenge for China's e-commerce market. Electron. Commer. Res. Appl. **12**(5), 299–308 (2013)

25. Tian, G., Peng, H., Sun, C., Li, Y.: Analysis of reputation speculation behavior in china's C2C e-commerce market. J. Comput. **7**(12), 2971–2978 (2012)

26. Gai, K., Qiu, M., Zhao, H., Dai, W.: Anti-counterfeit schema using monte carlo simulation for e-commerce in cloud systems. In: The 2nd IEEE International Conference on Cyber Security and Cloud Computing, pp. 74–79. IEEE, New York, USA (2015)

27. Liang, H., Gai, K.: Internet-based anti-counterfeiting pattern with using big data in China. In: The IEEE International Symposium on Big Data Security on Cloud, pp. 1387–1392. IEEE, New York, USA (2015)
28. He, X., Wang, C., Liu, T., Gai, K., Chen, D., Bai, L.: Research on campus mobile model based on periodic purpose for opportunistic network. In: 2015 IEEE 17th International Conference on High Performance Computing and Communications, pp. 782–785. IEEE, New York, USA (2015)
29. Gai, K., Qiu, M., Tao, L., Zhu, Y.: Intrusion detection techniques for mobile cloud computing in heterogeneous 5G. Secur. Commun. Netw. 9(16), 1–10 (2015)
30. Li, Y., Gai, K., Qiu, L., Qiu, M., Zhao, H.: Intelligent cryptography approach for secure distributed big data storage in cloud computing. Inf. Sci. PP(99), 1 (2016)
31. Ma, L., Tao, L., Gai, K., Zhong, Y.: A novel social network access control model using logical authorization language in cloud computing. Concurrency Comput. Pract. Experience PP(99), 1 (2016)
32. Jean-Baptiste, H., Tao, L., Gai, K., Qiu, M.: Understanding model risk management - model rationalization in financial industry. In: The 2nd IEEE International Conference on Cyber Security and Cloud Computing, pp. 301–306. IEEE, New York, USA (2015)
33. Li, Y., Gai, K., Ming, Z., Zhao, H., Qiu, M.: Intercrossed access control for secure financial services on multimedia big data in cloud systems. ACM Trans. Multimedia Comput. Commun. Appl. 12(4s), 67 (2016)
34. Yu, X., Pei, T., Gai, K., Guo, L.: Analysis on urban collective call behavior to earthquake. In: The IEEE International Symposium on Big Data Security on Cloud, pp. 1302–1307. IEEE, New York (2015)
35. Gai, K., Qiu, M., Zhao, H., Xiong, J.: Privacy-aware adaptive data encryption strategy of big data in cloud computing. In: 2016 IEEE 3rd International Conference on Cyber Security and Cloud Computing (CSCloud), The 2nd IEEE International Conference of Scalable and Smart Cloud (SSC 2016), pp. 273–278. IEEE, Beijing, China (2016)
36. Gai, K., Qiu, M., Thuraisingham, B., Tao, L.: Proactive attribute-based secure data schema for mobile cloud in financial industry. In: The IEEE International Symposium on Big Data Security on Cloud; 17th IEEE International Conference on High Performance Computing and Communications, pp. 1332–1337, New York, USA (2015)
37. Yin, H., Gai, K., Wang, Z.: A classification algorithm based on ensemble feature selections for imbalanced-class dataset. In: The 2nd IEEE International Conference on High Performance and Smart Computing, New York, USA, pp. 245–249 (2016)
38. Liu, S., Papageorgiou, L.: Multiobjective optimisation of production, distribution and capacity planning of global supply chains in the process industry. Omega 41(2), 369–382 (2013)
39. Oh, S., Özer, Ö.: Mechanism design for capacity planning under dynamic evolutions of asymmetric demand forecasts. Manage. Sci. 59(4), 987–1007 (2013)
40. Ren, C., Wang, D., Urgaonkar, B., Sivasubramaniam, A.: Carbon-aware energy capacity planning for datacenters. In: 20th International Symposium on Modeling, Analysis and Simulation of Computer and Telecommunication Systems, Washington, DC, USA, pp. 391–400 (2012)
41. Gai, K., Qiu, M., Chen, M., Zhao, H.: SA-EAST: security-aware efficient data transmission for ITS in mobile heterogeneous cloud computing. ACM Trans. Embed. Comput. Syst. PP(1) (2016)

42. Gai, K., Qiu, M., Zhao, H.: Cost-aware multimedia data allocation for heterogeneous memory using genetic algorithm in cloud computing. IEEE Trans. Cloud Comput. **PP**(99), 1–11 (2016)
43. Qiu, M., Cao, D., Su, H., Gai, K.: Data transfer minimization for financial derivative pricing using Monte Carlo simulation with GPU in 5G. Int. J. Commun. Syst. **29**(16), 2364–2374 (2015)
44. Dehghanian, P., Fotuhi-Firuzabad, M., Aminifar, F., Billinton, R.: A comprehensive scheme for reliability-centered maintenance in power distribution systems, Part II: Numerical analysis. IEEE Trans. Power Delivery **28**(2), 771–778 (2013)
45. Ou, C., Pavlou, P., Davison, R.: Swift guanxi in online marketplaces: the role of computer-mediated communication technologies. MIS Q. **38**(1), 209–230 (2014)
46. Li, Y., Gai, K., Qiu, M., Dai, W., Liu, M.: Adaptive human detection approach using FPGA-based parallel architecture n reconfigurable hardware. Concurrency Comput. Pract. Experience **PP**(99), 1 (2016)
47. Tao, L., Golikov, S., Gai, K., Qiu, M.: A reusable software component for integrated syntax and semantic validation for services computing. In: 9th International IEEE Symposium on Service-Oriented System Engineering, San Francisco Bay, USA, pp. 127–132 (2015)
48. Qiu, M., Gai, K., Thuraisingham, B., Tao, L., Zhao, H.: Proactive user-centric secure data scheme using attribute-based semantic access controls for mobile clouds in financial industry. Future Gener. Comput. Syst. **PP**(1) (2016)
49. Ma, L., Tao, L., Zhong, Y., Gai, K.: RuleSN: Research and application of social network access control model. In: IEEE International Conference on Intelligent Data and Security, New York, USA, pp. 418–423 (2016)
50. Zhao, H., Qiu, M., Gai, K., Li, J., He, X.: Maintainable mobile model using precache technology for high performance android system. In: The 2nd IEEE International Conference on Cyber Security and Cloud Computing, pp. 175–180. IEEE, New York (2015)
51. Gai, K., Du, Z., Qiu, M., Zhao, H.: Efficiency-aware workload optimizations of heterogenous cloud computing for capacity planning in financial industry. In: The 2nd IEEE International Conference on Cyber Security and Cloud Computing, pp. 1–6. IEEE, New York (2015)
52. Suthaharan, S.: Big data classification: Problems and challenges in network intrusion prediction with machine learning. ACM SIGMETRICS Perform. Eval. Rev. **41**(4), 70–73 (2014)
53. Yin, H., Gai, K.: An empirical study on preprocessing high-dimensional classimbalanced data for classification. In: The IEEE International Symposium on Big Data Security on Cloud, New York, USA, pp. 1314–1319 (2015)

A PSO-Based Virtual Network Mapping Algorithm with Crossover Operator

Ying Yuan[1], Sancheng Peng[2(✉)], Lixin Zhou[3], Cong Wang[1],
Cong Wan[1], and Hongtao Huang[2]

[1] School of Computer and Communication Engineering, Northeastern
University at Qinhuangdao, Qinhuangdao 066004, China
[2] School of Informatics, Guangdong University of Foreign Studies,
Guangzhou 510420, China
psc346@aliyun.com
[3] Department of Science and Technology Information, Public Security Bureau of
Qinhuangdao, Qinhuangdao 066004, China

Abstract. Virtual network mapping (VNM) is a crucial technology for network virtualization to allocate network resource. One of the major challenges for virtual network mapping is the efficient allocation of substrate resources to the virtual networks. In order to further improve the efficiency of the previous algorithms in large scale network, the crossover operator is introduced into discrete particle swarm optimization algorithm, and a hybrid intelligent algorithm is designed. The algorithm can solve the problem that the traditional particle swarm algorithm is easy to fall into local optimal point and is difficult to reach the global optimal. Experimental results show that the algorithm consumes the lowest cost in the case of mapping the same virtual network requests than existing ones, and has higher revenue/costs ratio.

Keywords: Network virtualization · Virtual network · Virtual network mapping · Revised discrete particle swami optimization

1 Introduction

With the rapid development of Internet, there is difficulty in meet the development of new applications for existing network architecture. Network virtualization is an effective method to solve the problem caused by ossification of Internet architecture [1]. In network virtualization, infrastructure providers (InPs) and service providers (SPs) play different roles: InPs manage the physical infrastructure while SPs create VNs and offer end-to-end services [2]. The virtual network (VN) embedding problem deals with finding a mapping of a virtual network request onto the substrate network [3]. Virtual network Embedding problem is known to be NP-hard even in the offline case. Even if all the virtual nodes are mapped, it is still NP-hard to mapped virtual links [4].

In this paper, we formulate the virtual network mapping problem using integer linear programming, present a discrete particle swarm optimization algorithm with crossover operator to avoid premature phenomenon in traditional PSO algorithm. The algorithm has stronger global searching ability so as to be more close to the optimal solution.

© Springer International Publishing AG 2017
M. Qiu (Ed.): SmartCom 2016, LNCS 10135, pp. 527–536, 2017.
DOI: 10.1007/978-3-319-52015-5_54

The remainder of this paper is organized as follows. In Sect. 2, we provide an overview on the related work. In the Sect. 3, we describe the virtual network mapping model, and present an algorithm GADPSO-VNM in Sect. 4. In Sect. 5, we conduct the simulation and performance analysis, and conclude this paper in Sect. 6.

2   Related Work

In the existing work, the VNM problem was solved by formulating the VNM as an optimization problem with the mapping cost as the objective. Recently, some researchers have presented some algorithms or customized algorithms in [4–10].

Minlan et al. in [5] provided a two stage algorithm for mapping the VNs. Firstly, they embedding the virtual nodes. Secondly they proceed to map the virtual links using shortest paths and multi-commodity flow (MCF) algorithms. In order to increase the acceptance ratio and the revenue, D-ViNE and R-ViNE were designed in [6]. The authors formulated the virtual network mapping problem as a mixed integer program through substrate network augmentation, and then relaxed the integer constraints to obtain a linear program. VNE-AC algorithm in [7] is a new VNE algorithm based on the ant colony meta-heuristic. They do not restrain the virtual network mapping problem by assuming unlimited substrate resources, or specific virtual network topologies or restricting geographic locations of the substrate core node. Hong et al. in [8] used a joint node and link mapping approach for the VI mapping problem and develop a virtual infrastructure mapping algorithm. Jiang et al. proposed a time-based VN embedding algorithm in [9]. They built a probability model to obtain the maximum probability that the available resources of substrate network can be used by succeeding VN requests. Then separate the VN mapping into two independent phases: node mapping and link mapping, and use greedy algorithm to embed the virtual nodes and a shortest path algorithm to embed the virtual links. Zhang and Gao presented a locality awareness approach in [10] with topological potential ranking and influence choosing node algorithm. Such mechanism can reflect the relative importance of nodes, and considers the mutual influence between a mapped node and its candidate mapping nodes. So it can improve the integration of node and link mapping. Considering time-varying resource requirements of virtual networks, Sheng et al. presented an opportunistic resource sharing-based mapping framework [11], in which the substrate resources are opportunistically shared among multiple virtual networks. They modeled the VNE problem as a bin packing problem. By adopting the core idea of first-fit, two practical solutions are proposed: first-fit by collision probability (CFF) and first-fit by expectation of indicators' sum (EFF). The embedding framework consists of the macro-level node-to-node/link-to-path embedding and the micro-level time slot assignment.

3   Network Model and Problem Description

In this section, we first provide the network model, including substrate network, virtual network request, and then introduce the virtual network mapping problem, including basic definitions and object.

3.1 Substrate Network and Virtual Network Request

We model the substrate network as a weighted undirected graph and denote it by $P = (N^S, L^S)$, where N^S and L^S is the set of substrate nodes and substrate link, respectively; We denote the set of loop-free substrate paths by P^S.

Similar to the substrate network, We denote a virtual network request can be defined by $VNR = (V, t)$, where V is a weighted undirected graph; t the duration of VN staying in the substrate network; the virtual network will be modeled $V = (N^V, L^V)$, where N^V and L^V denote the set of virtual nodes and virtual link, respectively.

In this paper, each substrate node $n^s \in N^S$ is associated with the CPU. Each substrate link $l^s(m, n) \in L^S$ between two substrate nodes m and n is associated with the bandwidth.

3.2 Problem Statement

When a VN request arrives, the substrate network has to determine whether to accept the request or not. If the request is accepted, the substrate network has to perform a suitable Virtual network embedding and allocate substrate resource.

The virtual network embedding problem is defined by an embedding Γ from a V to a subset of the substrate network P. The embedding action can be expressed as follows:

$$\Gamma : V \mapsto (N^{S*}, P^{S*}, \delta_N, \delta_P) \tag{1}$$

where $N^{S*} \subseteq N^S$, $P^{S*} \subseteq P^S$ δ_N and δ_P denote the attributes of the node and link, respectively.

Each virtual node and link from a VN is mapped to substrate node and link by a one-to-one mapping:

$$\Gamma_N : N^V \mapsto N^{S*} \tag{2}$$

$$\Gamma_L : L^V \mapsto P^{S*} \tag{3}$$

Such that $\forall n^v \in N^V$, $\Gamma_N(n^v) \in N^{S*}$, subject to the CPU constraint:

$$Rcpu(n^v) \leq Ccpu(\Gamma_N(n^v)) \tag{4}$$

$$Ccpu(\Gamma_N(n^v)) = cpu(\Gamma_N(n^v)) - \sum_{h \in N^V \mapsto \Gamma_N(n^v)} Rcpu(h) \tag{5}$$

Such that $\forall l(i,j) \in L^V$, and P^S is the set of simple paths of S. We have $\Gamma_L(l^v) \supseteq P^S(\Gamma_{N(i)}, \Gamma_{N(j)})$, subject to the bandwidth capacity constraints:

$$Rbw(l^v) \leq Cbw(p), \forall p \in \Gamma_L(l^v) \tag{6}$$

$$Cbw(p) = \min_{l^s \in p} Cbw(l^s) \tag{7}$$

$$Cbw(l^s) = bw(l^s) - \sum_{k \in L^V \mapsto l^s} Rbw(k) \tag{8}$$

$$DIS(n^v, \Gamma_N(n^v)) \leq \gamma \tag{9}$$

where $Rcpu(n^v)$ is requested CPU constraints for the virtual node n^v; $Ccpu(\Gamma_N(n^v))$ denote the available CPU capacity of substrate node; $cpu(\Gamma_N(n^v))$ denote the available total CPU capacity of substrate node; $Rbw(l^v)$ is requested bandwidth constraints for the virtual link l^v; $Cbw(p)$ denote the available bandwidth capacity of substrate link; $bw(l^s)$ denote the available total bandwidth capacity of substrate link.

3.3   Object

The main objective of virtual network embedding is to make efficient use of the substrate network resources when mapping the virtual network into the substrate network. In this paper, we aim to decrease cost of the InP so as to support more virtual networks.

Similar to the early work in [5–7], Firstly, we define the revenue and cost.

Definition 1: Revenue the sum of total an virtual network request gain from InP at time t as

$$Rev(G^V(t)) = \sum_{l^v \in E^V} bw(l^v) + \sum_{n^v \in N^V} cpu(n^v) \tag{10}$$

where $bw(l^v)$ is the bandwidth of the request link, $cpu(n^v)$ is the CPU capacity of the request node.

Definition 2: Cost the sum of total substrate resources allocated to that virtual network at time t as

$$Cos(G^V(t)) = \sum_{l^v \in E^V} bw(l^v)Length(P) + \sum_{n^v \in N^V} cpu(n^v) \tag{11}$$

Where $Length(P)$ is the hop count of the virtual link l^v when it is assigned to a set of substrate links.

Definition 3: Revenue to Cost ratio the long-term revenue-to-cost ratio, which is formulated by

$$Rev/Cos = \lim_{T \to \infty} \frac{\sum_{t=0}^{T} Rev(G^V(t))}{\sum_{t=0}^{T} Cos(G^V(t))} \tag{12}$$

The long-term revenue to cost ratio indicates the efficiency of substrate network resource usage, which is an important factor to judge the performance of a virtual network embedding algorithm.

4 DPSO-Based VNE Algorithm with Crossover Operator

4.1 System Model

We model the problem of embedding of a virtual network as a mathematical optimization problem using integer linear programming (ILP). In order to minimize the usage of the substrate resources, the objective of our optimization problem is defined as follows:

ℓ_{ij}^w is a binary variable. $\forall w \in l^v$ $\forall i,j \in N^s$ $\ell_{ij}^w \begin{cases} 0 & i=j \\ 1 & i \neq j \end{cases}$

Object:

$$Minimize \sum_{(i,j) \in P^s} \ell_{ij}^w \times bw(l^w) \tag{13}$$

Constraints

$$\forall u \in N^V, \forall i \in N^S$$
$$Ccpu(i) = cpu(i) - \sum_{n^v \to i} cpu(n^v) \geq Rcpu(u) \tag{14}$$

$$\forall i,j \in N^s, P^{ij} \in P^s, \forall w \in l^v$$
$$Cbw(P^{ij}) = \min_{l^s \in P^{ij}} Cbw(l^s) \geq Rbw(w) \tag{15}$$

$$Cbw(l^s) = bw(l^s) - \sum_{l^v \to l^s} bw(l^s) \tag{16}$$

where $\sum_{n^v \to i} cpu(n^v)$ is the total amount of CPU capacity allocated to different virtual nodes hosted on the substrate node; $\sum_{l^v \to l^s} bw(l^s)$ is the total amount of bandwidth capacity allocated to different virtual links hosted on the substrate link.

4.2 Revised Discrete PSO for Virtual Network Mapping

Particle swarm optimization (PSO) is a population based stochastic optimization technique developed by Dr. Eberhart and Dr. Kennedy in 1995, inspired by social behavior of bird flocking or fish schooling [12].

During the evolutionary process, the velocity and position of particle updated as follows:

$$V^{k+1} = \omega V^k + c_1 r_1 \left(X_p^k - X^k \right) + c_2 r_2 \left(X_g^k - X^k \right) \tag{17}$$

$$X^{k+1} = X^k + V^{k+1} \tag{18}$$

Where V^k is the velocity vector; X^k is the position vector; ω denote the inertia weight; r_1 and r_2 denote two random variables uniformly distributed in the range of $(0, 1)$; c_1 and c_2 denote the accelerator of particle; X_p^k denote the position with the best fitness found so far for the kth particle; X_g^k denote the best global position in the swarm.

Standard PSO is not directly applicable to the optimal virtual network mapping problem, so we used variants of PSO for discrete optimization problems to solve the optimal virtual network mapping problem.

Redefine the position and velocity parameters for discrete PSO as follows:

Definition 4: Position $X^k = [x_1^k, x_2^k, \cdots, x_m^k]$ a possible virtual network mapping solution, where x_i^k is the number of the substrate node the ith virtual node embedding to. m denotes the total number of nodes in virtual network k.

Definition 5: Velocity $V^k = [v_1^k, v_2^k, \cdots, v_m^k]$ makes the current virtual network mapping solution to achieve a better solution, where v_i^k is a binary variable. For each v_i^k, if $v_i^k = 1$, the corresponding virtual node's position in the current virtual network mapping solution should be remains; otherwise, should be adjusted by selecting another substrate node.

The operations of the particles are redefined as follows:

Definition 6: Addition of Position and Velocity $X^k + V^k$ a new position that corresponds to a new virtual network embedding solution. If the value of v_i^k equals to 1, the value of x_i^k will be kept; otherwise, the value of x_i^k should be adjust by selecting another substrate node.

Definition 7: Subtraction of Position $X^m - X^n$ a velocity vector. It indicates the differences of the two virtual network embedding solutions X^m and X^n. If X^m and X^n have the same values at the same dimension, the resulted value of the corresponding dimension is 1, otherwise, the resulted value of the corresponding dimension is 0.

Definition 8: Multiple of Velocity $\psi * V^m$ keep V^m with probability ψ in the corresponding dimension.

Definition 9: Addition of Multiple $\psi_1 * V^m + \psi_2 * V^n$ a new velocity that corresponds to a new virtual network embedding solution, where $\psi_1 + \psi_2 = 1$. If V^m and V^n have the same values at the same dimension, the value of the corresponding dimension will be kept; otherwise, keep V^m with probability ψ_1 and keep V^n with probability ψ_2.

Because of the specificity of discrete quantity operation, we modify the particle motion equation and cancel the original inertia item. The position and velocity of particle k are determined according to the following velocity and position update recurrence relations:

$$V^{k+1} = \psi_1 * \left(X_p^k - X^k \right) + \psi_2 * \left(X_g^k - X^k \right) \tag{19}$$

$$X^{k+1} = X^k + V^{k+1} \tag{20}$$

where ψ_1 and ψ_2 are set to constant values that satisfy the inequality $\psi_1 + \psi_2 = 1$.

For multidimensional optimization problem, Discrete PSO (DPSO) algorithm is easy to fall into local optimal solution in later stage, unable to achieve the global optimal solution. And the speed is not fast when solve some particular problems. In order to overcome above defect, we proposed a revised DPSO with crossover operator (GADPSO). The algorithm introduces the idea of crossover operation in genetic algorithm during search process. Crossover method is as follows: In each iteration, select the first half particles with good fitness and make them directly into the next generation; put the after half particles into a pool and pairwise coupling, make a crossover operation like in genetic algorithm. Produce child particles with the same number of theirs parent, and compare them to theirs parent to select the first half with good fitness into the next generation to maintain the same population number of particles. Such mechanism can increase the diversity of particles, jump out of local optimum, and can also speed up the convergence speed.

4.3   GADPSO-VNE Algorithm

Specific steps of GADPSO-VNM Algorithm are as follows:

1. For substrate network, sort the substrate nodes *NList* according to their CPU capacity $Ccpu(n^s)$ in descending order sort the substrate edges *LList* according to their bandwidth capacity $Cbw(l^s)$ in descending order;
2. For a virtual network request, sort the virtual nodes *VNList* according to their required CPU resources $Rcpu(n^v)$ in ascending order, then $\min Rcpu(n^v) \leftarrow$ required CPU resources of the first node of *VNList*; sort the virtual edges *VLList* according to their required bandwidth resources $Rbw(l^v)$ in ascending order, then $\min Rbw(l^v) \leftarrow$ required bandwidth resources of the first edge of *VLList*;
3. Obtain $NList' \leftarrow NList|\{Ccpu(n^s) < \min Rcpu(n^v)\}$;

$$LList' \leftarrow LList|\{Cbw(l^s) < \min Rbw(l^v)\}.$$

4. Initialize n = Particle Count, m = Max Iteration Count;
5. Randomly generated X^k and V^k;
6. Each X^k corresponds to link embedding with shortest path. Set X_p^k and X_g^k to these particles according to their fitness values of $\phi(x)$. If X^k is an unfeasible position, the fitness value $\phi(x)$ of this particle will be set to $+\infty$;
7. If $\phi(x)$ of the particle equal to $+\infty$, re-initialize its X^k and V^k. Otherwise, if $\phi\left(x_p^k\right) \geq \phi\left(x^k\right)$, then set X^k to be the x_p^k of the particle. If $\left(\phi\left(x_g^k\right) \geq \phi\left(x^k\right)\right)$, then set X^k to be the x_g^k of the particle;
8. Use formula (19) and (20) to update X^k and V^k;
9. For each particle, calculate its fitness according to its current position and sort them from big to small order. Select the first half particles with good fitness and make them directly into the next generation; put the after half particles into a pool and pairwise coupling. For the after half, randomly generate a cross position and do selection and crossover operation like in the genetic algorithm to generate the same number of offspring.

10. When cross over, update and calculate fitness of the offspring and compare to their parent. Select the first half particles with good fitness in both generations and put them into the next generation. Note that the number of particles doesn't change.

11. Evaluate each particle's fitness, if the current fitness value of the new position is superior to X_p^k which the particle has experienced, then update X_p^k, i.e., use its current position as its new best position; if the current fitness value of the new position is superior to X_g^k, then update X_g^k.

12. Check termination conditions, whether algorithm reached the maximum number of iterations or reached the best fitness (X_g^k does not change during a given number of rounds). If reached such conditions, the algorithm will be ended. Otherwise, will jump back to step 8.

5 Experimental Results and Analysis

We implemented the RBVNM algorithm using the CloudSim 3.0.1 on a PC which has one Intel Core i7-3770 CPU and 20G DDR3 1600 RAM. We design a random topological generator in java to generate topologies for the underlying substrate networks and virtual networks in CloudSim. Substrate networks in our experiments have 60 nodes, each node connect to other nodes with probability 0.2, so there are about 300 links in the networks. The substrate nodes and links were assigned resources by generating a uniform random number between 50 and 100. For each VN request, the number of virtual nodes was determined by a uniform distribution between 2 and 12, with the probability of a virtual link between any two virtual nodes set to 0.5. The CPU and bandwidth requirements of virtual nodes and links are real numbers uniformly distributed between 3 and 30 units. Each virtual network's living time uniformly distributed between 100 and 1000 time unit. We analyze the performance of the new algorithm by comparing it with the D-ViNE-SP, M-VNE-DPSO [13] and MLB-VNE-SDPSO [14].

Figure 1 shows the Cost of substrate network changing with different number of virtual network requests. From the results we can see, when running 100 virtual network request, D-ViNE-SP algorithm is the highest Cost, the GADPSO-VNM algorithm less than M-VNE-DPSO and MLB-VNE-SDPSO algorithm, along with the increase of the number of virtual network, their difference is more and more obvious. The reason is that in D-ViNE-SP, its relaxation-based approach weakens the coordination between node mapping and link mapping, which results in poor performance. The other three algorithms use repeatable embedding over substrate nodes that saves the substrate link cost and makes more virtual network embedded. The GADPSO-VNM algorithm adopts the revised discrete particle swarm optimization algorithm to avoid premature phenomenon in traditional PSO algorithm and has stronger global searching ability, the answer is closer to the optimal.

From Fig. 2 we can see that the Revenue-Cost (Rev/Cos) ratio of the GADPSO-VNM algorithm was larger. Along with the increase of the number of virtual network, their Rev/Cos ratio becomes more balanced. The reason is that the GADPSO-VNM algorithm adopted the discrete particle swarm optimization algorithm with crossover operator to avoid premature phenomenon in traditional PSO algorithm. The higher Rev/Cos ratio

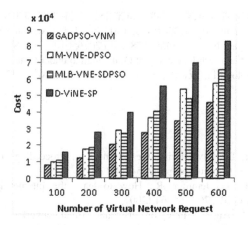

Fig. 1. Comparison of embedding cost for different number of VN request

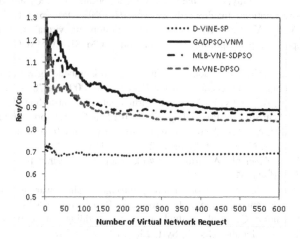

Fig. 2. Revenue-Cost ratio

indicates higher resource utilization, which further results in accepting more VNs at certain infrastructure resources.

6 Conclusion

The problem of efficiently allocating network resources of multiple VNs coexist on a shared substrate network is challenging. This paper introduced a novel VNE algorithm which adopted the discrete particle swarm optimization algorithm with crossover operator to avoid premature phenomenon in traditional PSO algorithm. Simulation results showed that the algorithm consumes the lowest cost in the case of mapping the same virtual network requests than existing ones and has higher revenue/costs ratio.

Acknowledgment. This work was partially supported by the National Natural Science Foundation of China under Grant Nos. 61300195 and 61379041, the Natural Science Foundation of Hebei Province under Grant Nos. F2014501078 and F2016501079, the Science and Technology Support Program of Northeastern University at Qinhuangdao under Grant No. XNK201401, and the Science and Technology Project of Guangzhou under Grant No. 2013Y2-00069.

References

1. Turner, J., Taylor, D.: Diversifying the internet. In: Proceedings of the IEEE Global Telecommunications Conference, pp. 755–760 (2005)
2. Feamster, N., Gao, L., Jennifer, R.: How to lease the Internet in your spare time. ACM SIGCOMM CCR **37**(1), 61–64 (2007)
3. Herker, S., Khan, A., An, X.: Survey on survivable virtual network embedding problem and solutions. In: Proceedings of the Ninth International Conference on Networking and Services, pp. 99–104 (2013)
4. Ilhem, F., Nadjib, A., Guy, P., Hubert, Z.: VNR algorithm: a greedy approach for virtual networks reconfigurations. In: Proceedings of the GLOBECOM, pp. 1–6 (2011)
5. Minlan, Y., Yung, Y., Jennifer, R., Mung, C.: Rethinking virtual network embedding: substrate support for path splitting and migration. ACM SIGCOMM CCR **38**(2), 17–29 (2008)
6. Mosharaf, C., Muntasir, R.R., Raouf, B.: Virtual network embedding with coordinated node and link embedding. In: Proceedings of the INFOCOM 2009, pp. 783–791 (2009)
7. Ilhem, F., Nadjib, A.S., Guy, P.: VNE-AC: virtual network embedding algorithm based on ant colony metaheuristic. In: Proceedings of the IEEE ICC 2011, pp. 1–6 (2011)
8. Hong, F.Y., Vishal, A., et al.: A cost efficient design of virtual infrastructures with joint node and link mapping. J. Netw. Syst. Manage. **20**(1), 97–115 (2012)
9. Jiang, M., Zhao, Z., et al.: A virtual network mapping algorithm based on time. Chin. J. Electron. **23**(CJE-1), 31–36 (2014)
10. Zhang, D., Gao, L.: Virtual network mapping through locality-aware topological potential and influence node ranking. Chin. J. Electron. **23**(CJE-1), 61–64 (2014)
11. Sheng, Z., Zhuzhong, Q., et al.: Virtual network embedding with opportunistic resource sharing. IEEE Trans. Parallel Distrib. Syst. **25**(3), 816–827 (2014)
12. James, K., Russell, E.: Particle swarm optimization. In: Proceedings of the International Conference on Neural Networks, pp. 1942–1948 (1995)
13. Yuan, Y., Wang, C.-R., Wan, C., Wang, C., Song, X.: Repeatable optimization algorithm based discrete PSO for virtual network embedding. In: Guo, C., Hou, Z.-G., Zeng, Z. (eds.) ISNN 2013. LNCS, vol. 7951, pp. 334–342. Springer, Heidelberg (2013). doi:10.1007/978-3-642-39065-4_41
14. Yuan, Y., Wang, C., Wang, C.: Load controllable virtual network embedding algorithm based on discrete particle swarm optimization. J. Northerstern Univ. Nat. Sci. **35**(1), 10–13 (2014)

WiHumidity: A Novel CSI-Based Humidity Measurement System

Xiang Zhang, Rukhsana Ruby, Jinfeng Long, Lu Wang, Zhong Ming, and Kaishun Wu$^{(\boxtimes)}$

College of Computer Science and Software Engineering,
Shenzhen University, Shenzhen, China
zhangxiangdavid@126.com, rukhsana@ece.ubc.ca, jinfeng.long@outlook.com,
{wanglu,mingz,wu}@szu.edu.cn

Abstract. Atmospheric humidity is one of the most important environmental attributes for weather condition. It affects the economy of nature as well as human life. Many environmental processes are affected by this attribute. For example, rice has the most powerful photosynthesis when the atmospheric humidity is in between 50% and 60%. For most of the human being, the humidity in between 20% and 80% is good to have a healthy life. Consequently, humidity measurement methods are urgently required. The existing methods are neither convenient for large scale deployment due to the high cost nor accurate enough to use. Recently, researchers found that humidity has a direct effect on radio propagation. This observation is undoubtedly useful to measure humidity in the environment. However, the humidity estimation based on received signal strength indicator (RSSI) is easily affected by the temporal and spatial variance due to multipath effect. Meanwhile, the change of radio signals incurred by RSSI-based systems is not that much obvious when the transmitter and receiver are in close distance. As a result, it is challenging to measure humidity in indoor environments. In this work, we provide a novel system, namely WiHumidity, to tackle this problem. The system utilizes the special diversity of channel state information (CSI) to alleviate multipath effect at the receiver. Extensive experiments have been conducted to verify the effectiveness of WiHumidity. The experimental results verify that on average, WiHumidity can achieve 79% measurement accuracy.

Keywords: Humidity measurement · Channel state information · Received signal strength indicator

1 Introduction

Wireless network technologies [1–3] have been developing rapidly in recent years. At the same time, people are increasingly concerned about the relationship between their health and nature. Atmospheric humidity strongly influences economic sectors as well as plays an important role in a variety of environmental processes.

© Springer International Publishing AG 2017
M. Qiu (Ed.): SmartCom 2016, LNCS 10135, pp. 537–547, 2017.
DOI: 10.1007/978-3-319-52015-5_55

Besides hygrometer [4] and weather satellites [5], the existing humidity measurement methods are typically developed based on the idea of wireless signal attenuation. The accuracy of hygrometers is very high. However, due to the high cost, hygrometers are not suitable for large-scale deployment. On the other hand, meteorological measurement near the Earth surface is not accurate. The humidity measurement methods based on wireless signal attenuation require the use of several tens of GHz band and special equipments. These special equipments are quite expensive, and hence these methods are not suitable for large-scale deployment. Consequently, the necessity of an accurate and low-cost method to measure atmospheric humidity is high.

Wireless technologies are capable to identify the changes of environments. One application that utilizes this idea is [6]. This system exploits the change of wireless signal to measure rainfall. The heavier the rainfall the more attenuation is caused to the wireless channel. Based on the same principle, wireless network technologies can be used to measure atmospheric humidity [7]. However, there are still two problems with this idea. First, due to the small attenuation caused by water vapor, measurement over a long distance is required to have a significant amount of attenuation. A typical experimental distance can be upto several kilometers. It is inconvenient to conduct measurements over such long distance in one shot. Second, a very high frequency and special device have been used, which is not common in life.

There are some humidity measurement systems [8,9], which are developed based on the received signal strength indicator (RSSI) concept of radio signals. There are many problems associated with RSSI-based measurement systems. First, RSSI is measured from the RF signal at a per-packet level, the accurate value of which is difficult to obtain. Some surveys [10–12] show that the variance of RSSI can be upto 5 dB in 1 min. Second, multipath effect of the radio signal always exists, especially in indoor environments. RSSI is easily affected by this effect. In theory, a model to measure the humidity using the received power can be established. However, in practice, RSSI value is not monotonic due to multipath effect. Consequently, the RSSI-based propagation model is invalid in short distance.

In order to tackle these challenges, we utilize channel state information (CSI) instead of RSSI of wireless signals in designing our system. And, unlike in [8,9], we redefine a new propagation model to accommodate these issues. Our proposed WiFi signal-based humidity measurement system is based on the new propagation model, and we name our system as WiHumidity. In the system, two Commercial Off-The-Shelf (COTS) WiFi devices are used, one of which is a sender (e.g., a router) and another one is a receiver (e.g., a laptop). The sender continuously emits signals and the receiver continuously receives signals. We conduct the humidity measurement experiments over a short distance on 5 GHz radio band.

Our refined propagation model has an unknown parameter, the value of which is greatly dependent on the environment. Since the model parameter is unknown, we cannot directly apply the new model to predict the humidity of a certain place. Consequently, we adopt a supervised machine learning algorithm, i.e., Support Vector Machine (SVM) to build a learning model that infers the relationship between CSI samples and humidity. Once we establish the model using training data, we verify the effectiveness of our model using test data. The main contributions of this paper are summarized as follows.

- We develop a supervised learning model that concludes the relationship between CSI samples and humidity of a certain place, and can measure humidity accurately. To the best of our knowledge, this is the first work that has used fine-gained PHY layer information in orthogonal frequency division multiplexing (OFDM)-based systems to build a propagation model so as to improve the performance of atmospheric humidity measurement.
- Since the refined propagation model has an unknown parameter, it cannot be directly applied to measure humidity of a certain environment. Consequently, we develop a supervised learning model that concludes the relationship between CSI samples and humidity of a certain place.
- We implement our method in commercial 802.11 Network Interface Cards (NICs), and conduct extensive experiments in typical indoor environments to show the feasibility of our design.
- Experimental results demonstrate that WiHumidity can measure atmospheric humidity effectively and accurately. The average measurement accuracy can be upto 79%.

The rest of this paper is organized as follows. We first provide some background information related to the content of this paper in Sect. 2, which includes the detailed explanation of CSI and the traditional wireless signal propagation model. In Sect. 3, the refined propagation model based on CSI is introduced. In Sect. 4, we present the detailed design description of our CSI-based humidity measurement system. Followed by the experimental setup, we evaluate the performance of our system in Sect. 5. Finally, Sect. 6 concludes the paper with some directions on future research.

2 Preliminaries

In this section, we provide a detailed description of CSI and the general wireless signal propagation model.

2.1 Elaboration of CSI

CSI represents the channel properties of a communication link. In wireless communications, the radio signal is affected by surrounding physical environments. The combined effect of reflections, diffractions and scattering is revealed by CSI. In frequency domain, the narrow band flat-fading channel model is

$$y = Hx + n, \tag{1}$$

where y and x are the received and transmitted signal vectors, respectively. n is the additive white Gaussian noise vector, and H is the CSI matrix. The noise is often modeled as to have circular symmetric complex normal distribution with $n \approx cN(0, S)$. Thus, H in the above formula can be estimated as $H = \frac{y}{x}$.

2.2   General Radio Propagation Model

Previous studies [8,9,13] have indicated that many weather attributes can affect the transmission of electromagnetic waves. Precipitation, water vapor, oxygen, snow, mist and fog are the typical notions of weather attributes. Diffraction, refraction, absorption and scattering caused by these weather factors can affect the electromagnetic waves and cause attenuation to radio signals. Hence, the current widely distributed wireless communication network technologies have opened a new door to measure environmental attributes. Quite a lot of research have been conducted in this context for many different purposes, such as rainfall measurement [6,14–16], etc.

The research results in the field of Physics show that oxygen and water vapor are the main absorbing gases, especially in lower atmosphere [17]. Although other atmospheric molecules have definite effect on radio signals, their impacts are too small to be ignored. Hence, the attenuation γ due to dry air and water vapor can be described as follows.

$$\gamma = A_w + A_o = 0.1820 f_{GHz} N^{''}(f), \tag{2}$$

where A_w is the attenuation amount caused by the water vapor, and A_o is the attenuation amount caused by the dry air. $N^{''}(f)$ is the imaginary part of the frequency-dependent complex refractivity, which is a function of pressure, temperature and water vapor density. And, $N^{''}(f)$ is given by

$$N^{''}(f) = \sum_i S_i(.)F_i(.) + N_D^{''}(f), \tag{3}$$

where $S_i(.) = S_i(p, T)$ (the function of pressure and temperature) is the strength of the ith frequency line; $F_i(.) = F_i(p, T, f)$ (the function of pressure, temperature and frequency) is the line shape factor; and $N_D^{''}(f)$ represents the dry continuum due to pressure induced nitrogen absorption and the Debye spectrum [18]. Hence, according to (2) and (3), we can obtain

$$\gamma = 0.1820 f(\sum_i S_i(.)F_i(.) + N_D^{''}(f)). \tag{4}$$

Reorganizing (4), we obtain

$$\sum_i S_i(.)F_i(.) = \frac{\gamma}{0.1820 f} - N_D^{''}(f). \tag{5}$$

The right hand part of (5) does not have any variable related to humidity. However, both $S_i(.)$ and $F_i(.)$ have parameters, which are related to humidity. As for $S_i(.)$ and $F_i(.)$, the precise expression can be found in [18].

3   Proposed Refined Radio Propagation Model

In this section, we utilize the fine grained CSI instead of RSSI of radio signals to build a propagation model, and accommodate the issues relevant to humidity measurement in the model. Before proposing the new refined model, we discuss about the deficiencies of the existing RSSI-based propagation model.

3.1   Deficiencies of the RSSI-based Model

According to (4), we have plotted some figures, that characterize this model. As illustrated in Fig. 1(a), maximum attenuation occurs at around 22 GHz band within the range of $[0 \sim 40]$ GHz frequency band. In theory, water vapor has a resonance line at 22.235 GHz channel. Hence, at $\sim$22 GHz band, the signal loss is caused predominantly by the water vapor. Consequently, $[22 \sim 23]$ GHz electromagnetic wave was used for the experiments in [8]. Figure 1(b) shows us the relationship between attenuation and temperature. Below 100°C, the higher the temperature, the smaller the attenuation.

Figure 1(c) shows us the important information in this context. The functional relationship between the attenuation and humidity is monotonic. The higher the humidity, the larger the attenuation. At the same time, one important key point is, the attenuation ratio is relatively low (in the order of 10^{-3}) at 2.4 GHz frequency or at 5 GHz frequency band compared to that in Fig. 1(a). Consequently, in order to obtain obvious and measurable attenuation value, the measurement distance is often several kilometers long even at $\sim$5 GHz frequency band[1].

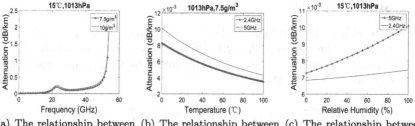

(a) The relationship between frequency and attenuation ratio (dB/km).

(b) The relationship between temperature and attenuation ratio (dB/km).

(c) The relationship between humidity and attenuation ratio (dB/km).

Fig. 1. The RSSI-based model.

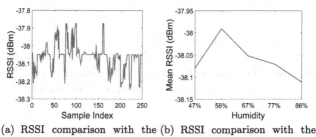

(a) RSSI comparison with the varying humidity samples.

(b) RSSI comparison with the varying humidity.

Fig. 2. The relationship between humidity and RSSI of the signal.

---

[1] The higher the frequency band, the larger the attenuation value.

We chose a confined space, which is 4 m long and 2 m high. We used humidifier to adjust the humidity in this small space. Then, we collected data for five different humidity values. 50 samples were collected at each humidity value. In total, there were 5 different humidity points and their corresponding 250 samples. The distance between TX and RX was set to 3 m. Figure 2(b) describes the changing trend of RSSI for different humidity values. Whereas, in Fig. 2(a), we show the RSSI variance w.r.t. different humidity samples.

We conducted some experiments to emphasize the deficiencies of this model further. The distance between TX and RX was set to 3 m. In total, we collected 5 different humidity values and 50 samples for each humidity value. Figure 2(b) describes the changing trend of RSSI for different humidity values. Whereas, in Fig. 2(a), we show the RSSI variance w.r.t. different humidity samples.

To summarize, the greater the humidity, the greater the attenuation ratio. In theory, the RSSI value is monotonically decreasing with humidity. However, in Fig. 2, we see the fluctuations in terms of this trend. There are two reasons which contribute to this observation. First, the attenuation ratio is relatively low even at 5 GHz band, which is obvious in Fig. 1(c). Second, multipath effect always exists, especially in the indoor environment.

3.2   CSI-Based Model

In order to address the problems with the RSSI-based model, we refine the propagation model here. Comparing with RSSI, CSI can better reflect the quality of the channel. Denoting the CSI-based attenuation by CSI_{eff} and replacing γ with CSI_{eff} in (5), we can obtain

$$\sum_i S_i(.)F_i(.) = \frac{\alpha \cdot |CSI_{eff}|}{0.1820f} - N_D^{''}(f), \tag{6}$$

where CSI_{eff} is given by

$$\text{CSI}_{eff} = \frac{1}{K} \sum_{k=1}^{K} (\frac{f_k}{f_0} \times |H_k|), \quad k \in (-15, 15). \tag{7}$$

α is an unknown factor which is greatly dependent on the environment. f_0 is the central frequency. Respectively, f_k and $|H_k|$ are the frequency and amplitude of the kth subcarrier. The reason to replace γ by CSI_{eff} is that it can exploit the frequency diversity to compensate the small-scale fading effect.

4   System Design

Using the refined CSI-based propagation model described in the previous section, we propose a system that measures humidity in a certain place effectively. We name our system as WiHumidity. Although the model provides the relationship between the humidity of a certain place and CSI of wireless signal, our system is

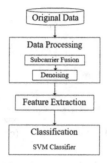

Fig. 3. The system flow diagram.

based on the supervised machine learning algorithm. Supervised learning algorithm requires some training data to build a prediction model, and test data to validate the model. In the following discussions, we provide the detailed description of our system including the methodology. Figure 3 shows the functional flow diagram of our system. There are mainly two components in the system, the functionalities of which are described as follows.

4.1   Data Collection and Processing

In a typical environment, CSI is easily influenced by human activities, and hence CSI needs filtering to reduce noise. We choose a confined space in the indoor environment, in which there are N_T number of transmit antennas and N_R number of receive antennas. WiFi NIC(s) of the receiver(s) report(s) CSI values over 30 OFDM subcarriers of the 20 MHz wide WiFi channel. This leads to 30 CSI samples with dimension $N_R \times N_T$ per humidity value. Consequently, the number of collected CSI samples for one humidity value is $30 \times N_R \times N_T$.

4.2   Feature Extractioin and Classification

In order to cope with the randomness of data, we collect a set of samples for each humidity value. Seven features of the collected $30 \times N_R \times N_T$ CSI samples (for per humidity sample) are used to build a classification model. These seven features are: (1) mean value, (2) normalized standard deviation (NSTD), (3) median absolute deviation (MAD), (4) interquartile range (IR), (5) maximum value, (6) skewness, and (7) signal entropy. As shown in the next section, none of the features alone can determine the corresponding humidity value effectively. Taking all features jointly into account, we apply a supervised machine learning algorithm, i.e., SVM to make a definite decision about the corresponding humidity.

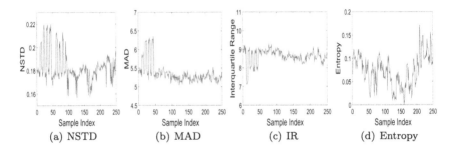

Fig. 4. The change of individual feature value w.r.t. different humidity samples.

5   Experiments and Performance Evaluation

In this section, we first describe the details of the experimental settings and the implementation of our system, WiHumidity. Then, to justify the effectiveness of WiHumidity, we show the results of our experimentation.

5.1   Experimental Settings

For the experiment, we design a confined space, the size of which is 4 m long and 2 m high. We use humidifier to adjust the humidity of environment in this small space. One TP-LINK router acts as the TX. A desktop of DELL with 2.4 GHz dula-core CPU as the RX, and implements WiHumidity. The desktop is equipped with Intel WiFi Link 5300(iwl5300) 802.11n NICs. In order to avoid the interference from the neighboring bands, we conduct our experiment on 5 GHz band. Moreover, we implement our system on the MATLAB platform, and evaluate the system in a typical laboratory room. Then, we collect data to evaluate the performance of our system. In order to ensure the diversity of data, 50 samples are collected for each humidity value.

5.2   Experimental Results

5.2.1   Signal-Level Analysis of the Collected Data

Previous works [9,13] have proved that water vapor can cause attenuation to radio signals, which leads to the change in RSSI or CSI values. However, to the best of our knowledge, no model based on CSI has been established before. This is the first work that has established the relationship between humidity and CSI. The key of the humidity measurement problem is to extract features from the CSI samples to discriminate different humidity values. We select four different features of the CSI samples to show the variation with different humidity values. From Fig. 4, for different humidity values, we can find obvious difference in terms of a feature value. It implies that different humidity values have different variation of CSI.

5.2.2   The Final Measurement Performance

As mentioned previously, a total of 250 CSI-humidity samples are collected. To establish a classification model, 120 of them are used for training, the rest (130) are to test. Finally, we run the classification process for 10000 times. As demonstrated in Fig. 5, WiHumidity can achieve detection accuracy 79% on average. The highest accuracy can be even upto 91%. Figure 5(a) shows comparison of three features combination. Figure 5(b) is the pdf plot for the measurement accuracy when all seven features are taken into account in building the SVM classification model. One of the outstanding works, in this context, is to predict the humidity following the relation in (6). However, there is an unknown factor α in this equation, which can be determined using the relations in (5) and (6). Then, the results of the proposed classification model can be verified with that of the proposed CSI-based propagation model. We would like to leave this work as our future study. Furthermore, we have not compared the results of our classification model with that of a RSSI-based measurement model. This is because the short distance over which we have conducted our experiments, and the RSSI-based models are not applicable for the short distance. Moreover, it requires specialized expensive equipments which are not easy to obtain.

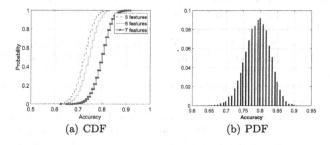

Fig. 5. Accuracy of our SVM classification model.

6   Conclusions and Future Work

In this paper, we proposed an effective and low-cost humidity measurement system. Having noticed several drawbacks of the existing RSSI-based humidity measurement systems, we proposed a refined CSI-based propagation model. To the best of our knowledge, this is the first work that used fine-gained PHY layer information (CSI) in OFDM-based systems to build a propagation model for improving the performance of atmospheric humidity measurement. Since the refined model has an unknown parameter and it is greatly dependent on the environment, we used SVM learning model to predict the humidity. Comprehensive experiments were conducted to verify the effectiveness of the learning model in predicting humidity. The average measurement accuracy that we obtained was around 79%. The accuracy of the prediction can be further improved by using better machine learning algorithm.

As a continuation of this work, our first objective is to improve the measurement accuracy of the system. Perhaps, we can use higher frequency radio signals to improve the accuracy. Furthermore, we can conduct more comprehensive experiments at different environments while considering different distance between the TX and the RX to verify the new CSI-based propagation model further.

Acknowledgements. This research is supported in part by the Shenzhen Science and Technology Foundation (No. KQCX20150324160536457, JCYJ20150324141711621.), China NSFC Grant 61472259, 61502313, Program for New Century Excellent Talents in University (NCET-13-0908), Guangdong Talent Project 2014TQ01X238, 2015TX01X111 and GDUPS (2015). Tencent "Rhinoceros Birds" - Scientific Research Foundation for Young Teachers of Shenzhen University.

References

1. Wu, K., Tan, H., Liu, Y., Zhang, J., Zhang, Q., Ni, L.M.: Side channel: bits over interference. IEEE Trans. Mob. Comput. **11**(8), 1317–1330 (2012)
2. Wu, K., Tan, H., Ngan, H., Liu, Y., Ni, L.M.: Chip error pattern analysis in IEEE 802.15.4. IEEE Trans. Mob. Comput. **11**(4), 543–552 (2012)
3. Kaishun, W., Haochao Li, L., Wang, Y.Y., Liu, Y., Chen, D., Luo, X., Zhang, Q., Ni, L.M.: hJam: attachment transmission in WLANs. IEEE Trans. Mob. Comput. **12**(12), 2334–2345 (2013)
4. Knowles Middleton, W.E.: A History of the Thermometer. Johns Hopkins Press, Baltimore (1966). ISBN 0-8018-7153-0
5. Sauvageot, H.: Radar Meteorology. Artech House Publishers, Norwood (1992)
6. Zamora, J.L.F., Kashihara, S., Yamaguchi, S.: Radio signal-based measurements for localized heavy rain detection using smartphones. In: Proceedings of the IEEE GHTC (2013)
7. David, N., Alperta, P., Messer, H.: Technical note: novel method for water vapor monitoring using wireless communication networks measurements. Atmos. Chem. Phys. **9**(7), 2413–2418 (2009)
8. Liebe, H.J.: An updated model for millimeter wave propagation in moist air. Radio Sci. **20**(5), 1069–1089 (1985)
9. Kobayashi, H.K.: Atmospheric effects on millimeter radio waves. DTIC Document (1980)
10. Kaishun, W., Xiao, J., Yi, Y., Chen, D., Luo, X., Ni, L.M.: CSI-based indoor localization. IEEE Trans. Parallel Distrib. Sys. **24**(7), 1300–1309 (2013)
11. Wang, G., Zou, Y., Zhou, Z., Kaishun, W., Ni, L.M.: We can hear you with WiFi!. IEEE Trans. Mob. Comput. **15**(11), 2907–2920 (2016)
12. Wang, L., Qi, X., Xiao, J., Kaishun, W., Hamdi, M., Zhang, Q.: Exploring smart pilot for wireless rate adaptation. IEEE Trans. Wireless Commun. **15**(7), 4571–4582 (2016)
13. Zhevakin, S.A., Naumov, A.P.: The propagation of centimeter, millimeter, and submillimeter radio waves in the earth's atmosphere. Radiophys. Quantum Electron. **10**(9–10), 678–694 (1967)

14. Messer, H.: Rainfall monitoring using cellular networks. IEEE Sig. Process. Mag. **24**(3), 142–144 (2007)
15. Zinevich, A., Alpert, P., Messer, H.: Estimation of rainfall fields using commercial microwave communication networks of variable density. Adv. Water Resour. **31**(11), 1470–1480 (2008)
16. Zinevich, A., Messer, H., Alpert, P.: Frontal rainfall observation by a commercial microwave communication network. J. Appl. Meteorol. Climatol. **48**(7), 1317–1334 (2009)
17. Van Vleck, J.H.: The absorption of microwaves by uncondensed water vapor. Phys. Rev. **71**(7), 425 (1947)
18. Ippolito, L.J., Jr.: Attenuation by Atmospheric Gases. In: Radiowave Propagation in Satellite Communications, pp. 25–37. Springer, Heidelberg (1986)

Bike-Sharing System: A Big-Data Perspective

Zhili Jia[1], Gang Xie[1(✉)], Jerry Gao[1,2], and Shui Yu[3]

[1] College of Information Engineering, Taiyuan University of Technology,
79 Yingze West Street, Taiyuan 030024, China
jiazhili0276@link.tyut.edu.cn, xiegang@tyut.edu.cn
[2] Computer Engineering Department, San Jose State University,
1 Washington Square, San Jose, CA 95192, USA
jerry.gao@sjsu.edu
[3] School of Information Technology, Deakin University,
221 Burwood HWY, Burwood, VIC 3125, Australia
syu@deakin.edu.au

Abstract. Bike-sharing systems, as a green travel way, recently have been widely spreading over 1000 cities around the world. How to plan and optimize such systems receive attention in academia as well as in practice. However, scientific literature about planning, usage prediction, pattern analysis, and system operation in this field is still rather scarce and full of challenges. And the solutions of these articles can hardly meet the increasing the demands of users and management of bike-sharing systems in the big-data era. To this end, a comprehensive literature comparison and analysis has been given focused on four topics. Then, a bike-sharing systems process framework from a big-data analysis perspective is proposed in the paper.

Keywords: Bike-sharing system · Big-data · 5Vs model · Framework

1 Introduction

Bike-Sharing System (BSS) refers to a public transportation for short distance trips either free or at an affordable price in the first half or one hour. According to the bike-sharing blog [1], over 1,000 cities or districts now provide the bike-sharing programming, with an estimated 1,270,000 bikes by the end of 2015. From the first generation the White Bicycle Plan in 1960s in Amsterdam, to the third generation integrated with information technology [2], the challenges of programs come along with the massive increasing number of subscribers or short-term consumers. How to analyze from such a huge volume of data generated by BSS and meet the realtime demands from users became the main problems constraint the performance of the BSS.

Owing to the big-data technologies springing out recently, this gives us a clue to solve the above-mentioned problems. This paper gives a big-data perspective on the BSS problems. Our contributions are: (1) make a comprehensive literature comparison on the topics that most talked in recently, including the BSS

© Springer International Publishing AG 2017
M. Qiu (Ed.): SmartCom 2016, LNCS 10135, pp. 548–557, 2017.
DOI: 10.1007/978-3-319-52015-5_56

planning, BSS pattern analysis, BSS demands or trip prediction, and repositioning problems on BSS. (2) talk BSS' 5Vs features and propose a BSS framework by big-data technologies to solve the problems faced by the system. (3) analyze the challenges faced by the BSS.

2 Studies on BSS

BSS has received an increasing attention with the rapid expansion over the world. DeMaio [2] and Shaheen et al. [3] studied the history, system structures, influence and future. Other studies on BSS are mainly summarized to four parts: system planning [4–7], usage prediction [8–13], pattern analysis [14–18], and system operation (repositioning problem) [19–22]. A comparison list of studies in patterns and predictions on BSS is showed in Table 1.

2.1 BSS Planning

To launch a BSS, the planner should identify goals and expectations of what the system will look like. Kim et al. [4] analyzed factors influencing travel behaviors, argued that appropriate scale of each station must be calculated in consideration of nearby land use and facilities. Buck et al. [5] found a significant correlation between the presence of bicycle lanes and Capital Bikeshare usage, and highlighted the importance of population density and mixed-uses in encouraging ridership. Martinez et al. [6] proposed an algorithm to optimise the location of stations by using the Mixed Integer Linear Program algorithm. Lin et al. [7] developed a mathematical model to determine the number and locations of bike stations.

2.2 BSS Prediction

Prediction model infers station demands and trips (duration and destination) ahead of time. Froehlich [8], the first to study the prediction of BSS for station demands, introduced four probability model Last Value (LV), Historic Mean (HM), Historic Trend (HT) and Bayesian Network (BN) to predict the available bikes for each station. Borgnat et al. [9] and Singhvi et al. [10] predicted bike demands using regression with many factors. Li et al. [11] proposed a hierarchical prediction model to predict the check-out/in of each station cluster in a bike-sharing system, based on the historical bike data and meteorology data.

Trip prediction is an interesting but important problem in BSS operation. Chen et al. [12] formulated the trip inference problem as an ill-posed inverse problem, and proposed a regularization technique to incorporate the a priori information about bike trips to solve the problem. Zhang et al. [13] adopted two models: trip destination prediction and trip duration inference on individual trip prediction problem, using MART and Lasso regression model, respectively.

Table 1. Comparison of studies on patterns and predictions of BSS

City/Program	Focuses	Method	Value	Index
Barcelona/Bicing	Station-level demands prediction	Probability model (LV, HM, HT, BN)	First to study the prediction, classify station state with 80% accuracy up to two hours into the future	[8]
Lyon/Vélo'v	Global activity prediction, spatial patterns	Signal processing tools for prediction, data mining for patterns	Used methods from signal processing and data analysis to study the Vélo'v system	[9]
NYC/Citi Bike	Bike demands prediction	Regression models with covariates including population, weather and taxi usage	First to use taxi data to predict bike trip volume	[10]
NYC/Citi Bike; DC/CaBi	Prediction on station cluster level	Offline: clustering, GBRT, trip duration learning and inter-cluster transition learning, online: inter-cluster transition prediction, check-out prediction	Proposed a hierarchical model to predict check-out/in of each cluster of stations	[11]
DC/CaBi	Trip prediction	Treated as an ill-posed inverse problem	Proposed a method infer bike trip patterns directly from the public station status data	[12]
Chicago/Divvy	Trip prediction	MART, Lasso regression model	Inferred the potential trip destination station and trip duration	[13]
Brisbane/CitiCycle	Spatio-temporal patterns	Multivariate regression modelto capture the effect of factors, Flow-comap to depict the usage patterns	Developed a tool to quantify salient factors influencing the system	[14]
Lyon/Vélo'v	Station patterns	Spatial clustering of the traffic flows, PCA, K-means, and visualization	Introduced the notion of spatial analysis for bike-sharing	[15]
Boston/Hubway; DC/CaBi; Chicago/Divvy	Station-level usage patterns between cities	Maximal Clique, Louvain Modularity, and ST-DBSCAN to clustering	Provided an accessible interface for examining bikeshare programs across cities	[16]

2.3   BSS Pattern

Spatio-temporal patterns and station-level patterns are two main part of the BSS patterns. While spatio-temporal reveals the flow of the city over time, station-level patterns focus on the similarity of the bike rental behavior among the stations. Corcoran et al. [14] developed a visualization analysis technique flow-comap to depict the subtle changes in spatio-temporal usage patterns under different weather conditions and various calendar events in Brisbane's CityCycle Scheme.

Other researchers did more on the station-level analysis. Borgnat et al. [15] proposed an approach based on spatial clustering of the flows between stations, highlighting the main features of bicycle movements all along the week. Bargar et al. [16] adopted ST-DBSCAN to reveal different patterns between cities. They performed graph theoretic algorithms, including maximal clique and Louvain modularity, to select a subset of the trip data that the stations are more connected before clustering. Sarkar et al. [17] compared 10 cities Bike-Sharing system patterns by using a hierarchical clustering algorithm to run a cross-city occupancy clustering and activity clustering. Keenan [18] leveraged big-data technology tools like Hadoop and Cloudera to handle large datasets, and got some statistics relationship between trips and weather-related data.

2.4   BSS Repositioning

Repositioning is an operation which redistribute the bicycles to meet the demands of BSS, focusing on the decision making of the vehicles route and regulating of the inventory for each station. Most of those studies are based on the assumption of knowing the inventory of stations, treated as a vehicle scheduling problem. Vogel et al. [19] proposed a model to assesses the prospects of operational repositioning services by an aggregate feedback loop model with Karmarkar algorithm. Chemla et al. [20] described the problem as a Single Vehicle One-commodity Capacitated Pickup and Delivery Problem, adopting a ranch-and-cut algorithm to solve the relaxation problem, and a tabu search to get the upper bound of the optimal solution. Dell' Amico [21] presented four mixed integer linear programming formulations for the same problem. Schuijbroek et al. [22] presented a cluster-first route-second approach to decomposes the multi-vehicle rebalancing problem into separate single-vehicle problems by using mixed integer programming based on the clustering problem.

3   BSS with Big-Data

From last section, we infer the fact that studies on BSS top four topics are most achieved in the conventional method with a great numbers of assumptions. We also found that BSS with big-data is scarcely ever mentioned in those papers. The problems that system data accumulated in a rapid rate of speed and volume, that realtime demands from users are still there, and big-data can save us a life. In the following part, we will get an in-depth analysis and study to the big-data technology and the solution by a big-data perspective.

3.1 Brief Talk About Big-Data

Big-data has been a hot buzzword since 2014, yet with no rigorous definition. The definition is intentionally subjective and incorporates a moving definition of how big a dataset needs to be in order to be considered big-data said by McKinsey researchers [23]. Here is a generally definition from Wikipedia: a term for data sets that are so large or complex that traditional data processing applications are inadequate [24]. As a fact, 3Vs model (early definition of big-data: Volume, Velocity, and Variety) was firstly introduced by Doug Laney in 2001 [25], although such a model was not originally used to define big data. IDC (International Data Corporation) proposed a 4Vs model (3Vs plus Value): extracting value from very large volumes of a wide variety of data, by enabling the high-velocity capture, discovery, and/or analysis [26]. At the same time, IBM also proposed their 4Vs model: with Veracity instead, and later the 5Vs model: with both Veracity and Value included.

According to MGI's (McKinsey Global Institute) report on big-data [23] there will be billions of big-data related market. And lots of country administrations and economic commissions like U.S., European Commission, Australia, and China have released their official documents on the promotion of big-data in order to embrace the big-data era.

3.2 Deep into Big-Data

The point of Big-data is to develop a system moves data along the path – raw data to actionable insights [27]. A big data solution typically comprises four logical layers: big data sources, data massaging and store layer, analysis layer, and consumption layer [28]. Figure 1 shows the architecture of the Big-data solution.

Big data sources: data that available for analysis.

Data massaging and store: it is where the source data lives, using a distributed file system like HDFS (Hadoop Distributed File System) or GFS (Google File System). Data is broken down into smaller pieces (called blocks) and distributed throughout the cluster.

Analysis layer: reads the data from the store layer, then process and analyze the data using a MapReduce tool such as Pig or Hive.

Output layer: gets the output of the analysis layer. The output can take the form of charts, figures, and recommendations, etc.

When comparing with the big-data analysis at extreme scale, conventional methods could be out at elbows not only on the storage but the CPU processing performance. Big-data can easily scale-out rather than scale-up owing to the distributed idea. Big-data analysis can also take as much covariates as you want into the system to give a better understanding of influences for each factor. Duing to such advantages big-data have, we will propose a framework later to cope with 5Vs in BSS.

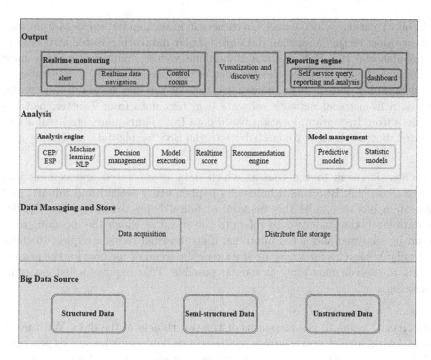

Fig. 1. Architecture of big-data

3.3   5Vs in BSS

Big-data is featured with 5Vs: Volume, Variety, Velocity, Veracity, and Value. As a data related service, BSS faces the same problems most service-oriented companies do. In the following, we will talk the BSS' 5Vs in detail.

Volume. It refers to the amount of data. A typical BSS generates amounts of trips. For example, the world biggest BSS program: Hangzhou Public Bicycle Service. There are 84,100 bicycles scattered in 3574 stations, more than 150 million trips generated in 2015, 448,600 highest daily usages, more than 310,000 on average. Another successful scheme in Taiyuan (China), with 41,000 bicycles in 1258 stations, less than half of the Hangzhou's, reached a record-breaking 568,500 highest daily usages, and more than 400,000 on average. Vélib' in Paris, the largest BSS outsides of China, with around 14,500 bikes and more than 1,230 stations in their fleet, with an average daily ridership of 110,000 (75 hires per minute), and more than 283 million rentals since the launch (2007).

Variety. It refers to the number of types of data. The data is structurally heterogeneity nowadays. In a BSS, data is generated with each check-in or check-out behavior: timestamps, serial number of the slot and station, duration, GPS (trip path, bike location), video data from camera in kiosk; gender, age data from subscribers. Most of this operation data are tabular data being stored structurally in spreadsheets. But BSS is not a standalone system, many factors may have influence on the system also collected such as weather-related data, resident density, geographic data (streets, elevation), time of the day, holidays and events, even data from social network software (e.g. text data from Twitter, pictures or videos from Instagram, location-based data from Foursquare, etc.). All above included, but not limited to, contribute to the BSS for big-data analysis.

Velocity. It refers to the rate at which data is generated and coming into the database [29]. BSS rental behaviors happen every second and minute, and the system has to handle this tsunami of rental behaviors every day. Besides the data-level, the velocity also refers to the speed at which the incoming data should be analyzed and acted upon, i.e. if a subscriber sends a request to check the available bicycles for a specific station or the trend of the flow for the nearest station, the server must reply as soon as possible. That requires a fast real-time processing.

Veracity. It refers to the messiness or trustworthiness of the data. With many forms of data, quality and accuracy are less controllable, which requires processes to keep the bad data from accumulating in your systems. i.e. if a trip duration is less than 90 s, it is an invalid record in the database, should be set aside in some processing like ridership prediction.

Value. It is the end game. The purpose of the big-data analysis is to extract the potential value behind the data. The value of the BSS is to find the patterns of the system, improve the performance of the system, and reduce the cost for both venders and subscribers.

3.4   BSS Framework

This framework shared the insights from an air quality inference model proposed by Zheng et al. [30]. The framework give a clear flow path on how the system works. As shown in Fig. 2, it consists two main part: the offline learning and online inference. Three main process flows are processed in the system: preprocessing data flow, learning data flow, and inference data flow.

Preprocessing Data Flow. We receive the kinds of data like the bicycle records: timestamps, station and slot serial numbers bike borrowed from or returned to, trip duration, etc. the preprocessing is an important part before the offline learning, dirty data like trip duration less than 90 s is set aside or deleted in this process. Others like weather-related data were emerged with the operation data. So the data stored in the database are usually not the raw data, means the data is preprocessed.

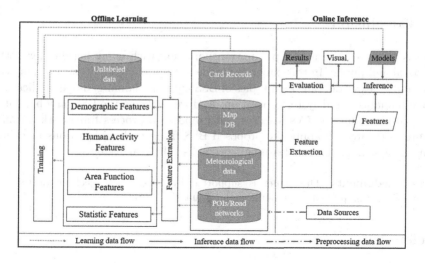

Fig. 2. BSS Framework

Learning Data Flow. Multi-factors like rental records, geography information, metrological, and POIs or road networks data are included in the system. In this part of the system, the factors that have an influence on the BSS are extracted as features to get the spatio-temporal patterns of bicycle flows. Demographic, human activity, area function, and the BSS itself are taken into consideration. By a series of training (e.g. machine learning, semi-supervised learning, deep learning, and community detection, etc.), we can get knowledge from the complex system. After the offline training, each station will get a label to be clustered by similarity.

Inference Data Flow. While offline learning cares about the historical data and correlation between them, inference flow focus on the streaming data, and infer the demands of the station in the future. The inference results need a

3.5   Challenges

There are two main challenges about the data. The first one is how to break the data source barriers [31], e.g. not all programs put their operation data on a public website, especially in China where the largest and non-profit programs are in there. The ridership can be times more than other countries, and worth a study in-depth. Besides the trip data, the types of data we can access are also limited, like residential density data. Second is the trustiness of the data: fake data and outdated data can have nothing to do with the system but have negative effects. So, a data quality examine tool in big-data should be introduced and add into the framework. Another serious problem is the security of the data [32,33], every single trip contributes to the way users' living habits in some ways. How to protect the users' information from being hacked should be treated at a priority level.

4 Conclusion

BSS is a perfect alternative public transportation with many benefits for both personal and public. In this paper, we made a comprehensive literature comparison and analysis on four main topics of BSS: system planning, usage prediction, pattern analysis, and system operation. And we analyzed the BSS on a big-data perspective about the 5Vs features and proposed a process framework. In our future work, we will apply the proposed BSS framework with various machine leaning or semi-supervisor learning to establish the real system.

Acknowledgments. This paper is supported by Research Project Supported by Shanxi Scholarship Council of China under grant No. 2016-044.

References

1. The bike-sharing blog. http://bike-sharing.blogspot.com/
2. DeMaio, P.: Bike-sharing: history, impacts, models of provision, and future. J. Public Transp. **12**(4), 41–56 (2009)
3. Shaheen, S., Guzman, S., Zhang, H.: Bikesharing in Europe, the Americas, and Asia. Transp. Res. Rec. J. Transp. Res. Board **2143**, 159–167 (2010)
4. Kim, D.J., Shin, H.C., Im, H., et al.: Factors influencing travel behaviors in bike-sharing. In: Transportation Research Board 91st Annual Meeting (2012)
5. Buck, D., Buehler, R.: Bike lanes and other determinants of capital bikeshare trips. In: Transportation Research Board 91st Annual Meeting (2012)
6. Martinez, L.M., Caetano, L., Eiró, T., Cruz, F.: An optimisation algorithm to establish the location of stations of a mixed fleet biking system: an application to the city of Lisbon. Procedia-Soc. Behav. Sci. **54**, 513–524 (2012)
7. Lin, J.R., Yang, T.H.: Strategic design of public bicycle sharing systems with service level constraints. Transp. Res. Part E Logistics Transp. Rev. **47**(2), 284–294 (2011)
8. Froehlich, J., Neumann, J., Oliver, N.: Sensing and predicting the pulse of the city through shared bicycling. In: Proceedings of the 21st International Joint Conference on Artificial Intelligence (IJCAI), pp. 1420–1426. Morgan Kaufmann, San Francisco (2009)
9. Borgnat, P., Abry, P., Flandrin, P., et al.: Shared bicycles in a city: a signal processing and data analysis perspective. Adv. Complex Syst. **14**(3), 415–438 (2011)
10. Singhvi, D., Singhvi, S., Frazier, P.I., et al.: Predicting bike usage for New York Citys bike sharing system. In: AAAI 2015 Workshop on Computational Sustainability (2015)
11. Li, Y., Zheng, Y., Zhang, H., Chen, L.: Traffic prediction in a bike-sharing system. In: Proceedings of the 23rd SIGSPATIAL International Conference on Advances in Geographic Information Systems (GIS 2015) (2015)
12. Chen, L., Jakubowicz, J.: Inferring bike trip patterns from bike sharing system open data. In: 2015 IEEE International Conference on Big Data (Big Data), pp. 2898–2900. IEEE, New York (2015)
13. Zhang, J., Pan, X., Li, M., Yu, P.S.: Bicycle-sharing system analysis and trip prediction. In: 17th IEEE International Conference on Mobile Data Management (MDM) (2016)

14. Corcoran, J., Li, T., Rohde, D., Charles-Edwards, E., et al.: Spatio-temporal patterns of a public bicycle sharing program: the effect of weather and calendar events. J. Transport Geogr. **41**, 292–305 (2014)
15. Borgnat, P., Robardet, C., Rouquier, J.B., et al.: Shared bicycles in a city: a signal processing and data analysis perspective. Adv. Complex Syst. **14**(3), 415–438 (2011)
16. Bargar, A., Gupta, A., Gupta, S., et al.: Interactive visual analytics for multicity bikeshare data analysis. In: 3rd International Workshop on Urban Computing (2014)
17. Sarkar, A., Lathia, N., Mascolo, C.: Comparing cities' cycling patterns using online shared bicycle maps. Transportation **42**(4), 541–559 (2015)
18. Keenan, C.D.: Chicago's shared bikes: how big data technology can assess ridership. In: 2016 Escape (2016)
19. Vogel, P., Mattfeld, D.C.: Modeling of repositioning activities in bike-sharing systems. In: 12th World Conference on Transport Research (WCTR) (2010)
20. Chemla, D., Meunier, F., Calvo, R.W.: Bike sharing systems: solving the static rebalancing problem. Discrete Optim. **10**(2), 120–146 (2013)
21. Dell'Amico, M., Hadjicostantinou, E., Iori, M., Novellani, S.: The bike sharing rebalancing problem: mathematical formulations and benchmark instances. Omega **45**, 7–19 (2014)
22. Schuijbroek, J., Hampshire, R., Hoeve, W.J., et al.: Inventory rebalancing and vehicle routing in bike sharing systems (2013)
23. Manyika, J., Chui, M., Brown, B., et al.: Big data: the next frontier for innovation, competition, and productivity. The McKinsey Global Institute (2011)
24. Big data. https://en.wikipedia.org/wiki/Big_data
25. Laney, D.: 3D data management: controlling data volume, velocity and variety. META Group Res. Note **6**, 70 (2001)
26. Gantz, J., Reinsel, D.: Extracting value from chaos. IDC iView **1142**, 1–12 (2011)
27. Big data: the 4 layers everyone must know. http://www.linkedin.com/pulse/20140918062736-64875646-big-data-the-4-layers-everyone-must-know
28. Understanding the architectural layers of a big data solution. http://www.ibm.com/developerworks/library/bd-archpatterns3/bd-archpatterns3-pdf.pdf
29. Gandomi, A., Murtaza, H.: Beyond the hype: big data concepts, methods, and analytics. Int. J. Inf. Manage. **35**(2), 137–144 (2015)
30. Zheng, Y., Liu, F., Hsieh, H.P.: U-Air: when urban air quality inference meets big data. In: Proceedings of the 19th ACM SIGKDD International Conference on KDD, pp. 1436–1444. ACM, New York (2013)
31. Allen, A.L.: Privacy law: positive theory and normative practice. Howard Law J. **56**(3), 241–251 (2013)
32. Kshetri, N.: Big data's impact on privacy, security and consumer welfare. Telecommun. Policy **38**(11), 1134–1145 (2014)
33. Qiu, M., Gai, K., Thuraisingham, B., et al.: Proactive user-centric secure data scheme using attribute-based semantic access controls for mobile clouds in financial industry. Future Gener. Comput. Syst. (2016)

Process Mining of Event Log from Web Information and Administration System for Management of Student's Computer Networks

Radim Dolak[✉], Dominik Musil, and Jan Kolesar

School of Business Administration in Karvina, Department of Informatics and Mathematics,
Silesian University in Opava, Karvina, Czech Republic
{dolak,0150860,160358}@opf.slu.cz

Abstract. Process mining is relatively new approach which is often using for performance managing and optimizing of the most important business processes. Process mining analysis allows extracting information from event logs. The main purpose of this paper is to describe advantages of process mining, current trends and provide process mining analysis of event log from web information and administration system for management of student's computer networks in case study. The case study deals with using Disco software tool for process mining of mentioned event log. Case study will provide information about processes such as automatic discovery of process model based on imported data, process map with detail information about activities and paths (frequency, repetitions and duration), number of events, overview about events and active cases over time and finally also using resources.

Keywords: Process mining · Events log · Management of computer networks · Disco software tool

1 Introduction

There are many web information and administration systems that store the information about events in the log files. We will discuss the issue of analyzing events log from web information and administration system for management of student's computer networks. If we want to provide analysis of events log we will need to use process mining techniques to extract event logs from data sources such as information systems databases, transaction logs, audit trails etc. We will use events log that are saved in MySQL database of Pawouk information system for administration and management of student's computer networks.

The structure of this paper is divided into the following chapters: Process mining, Management of student´s computer networks at School of Business Administration in Karvina, Process mining software, Case study: Process mining of events log from information system for management of network users and Conclusion.

© Springer International Publishing AG 2017
M. Qiu (Ed.): SmartCom 2016, LNCS 10135, pp. 558–567, 2017.
DOI: 10.1007/978-3-319-52015-5_57

2 Process Mining

Process mining is defined by van der Aalst [1] as a relatively young research discipline that sits between computational intelligence and data mining on the one hand, and process modeling and analysis on the other hand. Process mining is according to [12] closely related to BAM (Business Activity Monitoring), BOM (Business Operations Management), BPI (Business Process Intelligence), and data/workflow mining.

The main idea is to discover, monitor and improve real processes by extracting knowledge from event logs. We can find many sources of event logs such as audit trails of a workflow management system or the transaction logs of an enterprise resource planning system. Why is process mining so important? Process mining techniques allow extracting information from event logs [4]. We can find at least two reasons that are described in [5]: first of all, it could be used as a tool to find out how people and/or procedures really work. Second, process mining could be used for Delta analysis, i.e., comparing the actual process with some predefined process.

Starting point for process mining is according to van der Aalst et al. [2] an event log and all process mining techniques assume that it is possible to sequentially record events such that each event refers to an activity and is related to a particular case. Process Mining Manifesto [11] discusses important aspects that process mining includes such as: process discovery (i.e., extracting process models from an event log), conformance checking (i.e., monitoring deviations by comparing model and log), social network/ organizational mining, automated construction of simulation models, model extension, model repair, case prediction and also history-based recommendations. Process mining is used for example in software development and testing process [16], application of process mining in healthcare [9], for analyzing inventory processes [14]. We can find more case studies about using process mining for example in academic sphere. Department of Mathematics and Computer Science at University of Technology Eindhoven is engage in an extensive research on the application of process mining in many research areas. We can find a list of process mining case studies such as: improved invoicing, support of operational excellence and sales, refund service process of an electronics manufacturer, inefficiencies in city government, process mining for auditing, process mining in invoice handling process are available here [13].

3 Management of Student's Computer Networks at School of Business Administration in Karvina

During the last ten years of administration of student's computer networks in dormitories of The School of Business Administration in Karvina was introduced and updated web information system called Pawouk. Pawouk is a web information and administration system for managing student's computer networks. Information and administration system Pawouk currently provides complete operation of student's computer networks in dormitories of The School of Business Administration in Karvina.

Pawouk is based on using open source technologies and it was designed as a three-layer web database application. Architecture and structure is more detailed described by Petránek [10] who is author and developer of this system. There is used MySQL database for storing the data and to ensure running applications is used layer of Apache Web server module for PHP. As the main development environment has been chosen PHP. For communication between the application and data layer is used the PHP Data Objects (PDO).

The web application of Pawouk system is divided into two separate interfaces: user and administrator. The user interface is designed for both network users and visitors to the website portal. Website portal provides general information and technical support for users of the network. Admin interface is designed only for network administrators. It is a tool for network administrators to perform most tasks related to network management: to register new users, access to information about current network users and their devices (PC, notebooks, tablets, smartphones), administration of evidences such as charges for internet connection and payments for print accounts, setting the rules, configuring the server, providing technical support to users etc.

There are many common activities and processes associated with management of student's computer networks which are support by Pawouk information system. We can recognize main following types of functionalities defined by Petránek [10]: website (information about the network, documents, manuals and contacts), technical support system for network users, registration of new users of the network via a web form, user authentication system, list of users and connected devices, processing and recording of payments.

4   Process Mining Software

There are many tools that provide process mining. The most famous is open source project framework called ProM which has been developed at the Eindhoven University of Technology by team of professor van der Aalst. There are also commercial tools such as Disco by Fluxicon, Interstage Automated Process Discovery by Fujitsu, ARIS Process Performance Manager by Software AG etc.

ProM. ProM is an extensible framework that supports a wide variety of process mining techniques in the form of plug-ins. It is platform independent as it is implemented in Java, and can be downloaded free of charge [15]. The ProM framework has been developed as a completely plug-able environment. The most interesting plugins are the mining plugins and the analysis plugins [3]. The architecture of ProM allows different types of plugins: mining plug-ins, export plug-ins, import plug-ins, analysis plug-ins and conversion plug-ins. The functionality of ProM is according to Aalst [1] unprecedented and there is no product offering a comparable set of process mining algorithms. Some of the process mining plug-ins present used in ProM 6 are following: alpha miner, heuristic miner, genetic miner, fuzzy miner, simple log filter, guide tree miner and social network miner.

Disco by Fluxicon. Disco is complete process mining software developed by Fluxicon. Disco provides very fast process mining algorithms and also includes tools such as efficient log management and filtering framework. We can easily add your Excel or CSF file with log events from information system or other data storages.

Professor van der Aalst from Eindhoven University of Technology described Disco as a great process mining tool that simply works and it is able to deal with large event logs and complex models and conversion and filtering are made easy. Performance metrics are shown in a direct and intuitive manner and the history can be animated on the model [6].

Disco provides according to official website [6] useful functionality for optimize performance, control deviations, or explore variations such as: automated process discovery, process map animation, detailed statistics, cases, filters, import and export and project management.

Günther and Rozinat [8] deals with idea that the core functionality of process mining is the automated discovery of process maps by interpreting the sequences of activities in the imported log file. Disco provides this functionality in an intuitively understandable way of process map visualization. Visualization is user friendly with good possibility of visual discovering the main paths of the process flows (coloring and thickness of paths) and also wasteful rework loops.

5   Case Study: Process Mining of Events Log from Information System for Management of Network Users

In this case study is utilized a commercial process mining tool called Disco which is used at our faculty under academic license for research in process mining area.

5.1   Preparing Data from Events Log

We have obtained events log from Pawouk information and administration system which contains 4314 events and we can analyze behavior of 361 internet users during academic year 2014–2015. There are also resources such as 3 network administrators which have consultation hours at student's dormitory, 2 IT specialists from faculty Institute of Information Technology, WWW interface of web portal which is connected with Pawouk system, resource SYSTEM means automatic termination of network traffic when academic year ends. There are also student's users as resources in cases of their own user and device registration or updating device information through web portal of Pawouk system. These registration and updates are in all cases subsequent approved or rejected by network administrators. There are the following basic activities (processes) during academic year: new user registration (WWW), registration of new user device (WWW), user authentication, updating the user detail, updating the device detail, restart password for recovery of user registration, restart the user's password, changing user password, added a new fee, added a new fee-cancellation, inserting a new printing bill, update of printing bill, termination of network traffic.

We can see an example of default events log from Pawouk system which is available in module called "logging system". (Table 1)

Table 1. Default events log

Date and time	Source	Network	Text	IP address
21.9.2014 11:12	WWW [ID: 0]	Z@vináč	new user registration WWW (o0923xx – Name_Surname)	85.70.56.xxx
21.9.2014 11:16	ANONYM [ID: 2000000000]	N/A	user authentication (o0923xx – Name_Surname) [id: 90]	85.70.56.xx
21.9.2014 20:37	Tomas01 [ID: 11]	Z@vináč	updating the user detail (o0923xx – Name_Surname) [id: 90]	193.84.220.x
7.10.2014 18:31	radim_dolak [ID: 4]	Z@vináč	updating the user detail (o0923xx - Name_Surname) [id: 90]	193.84.220.x
7.10.2014 18:31	radim_dolak [ID: 4]	Z@vináč	updating the device detail (o0923xx) [id: 98]	193.84.220.x
7.10.2014 20:51	Miro [ID: 10	Z@vináč	added a new fee (ID number: 259) [id: 259]	193.84.220.xx
...		...	...	...
31.8.2015 23:59	SYSTEM	Z@vináč	termination of network traffic	N/A

User must be extracted from column text, where is information about user identification. It is primarily key student´s number in the following form oYYXXXXXX, where YY represents last two numbers from year when student starts to study, XXXXXX represents six specific numbers. There are many data storage in Pawouk events log for managing student computer networks. There is following structure of columns for transformed events log: date and time, user, source, action. We can see an example of event log after data transformation for one internet user with o0923xx from Pawouk system in the following Table 2 called Transformed events log.

There are 7 basic different actions during academic year which is typical period for internet services for students who are accommodate in students dormitory of our faculty. The first action is typically new user registration (WWW) following by user authentication, updating the user and device detail. Sometimes are students changing their user password. Internet services are not free of charge so there is also possibility for adding a new fee. The last action is termination of network traffic associated with ending of summer academic semester or student's accommodation contracts.

Table 2. Transformed events log

Date and time	User	Source	Action
21.9.2014 11:12	o0923xx	WWW [ID: 0]	new user registration (WWW)
21.9.2014 11:16	o0923xx	ANONYM [ID: 2000000000]	user authentication
21.9.2014 20:37	o0923xx	Tomas01 [ID: 11]	updating the user detail
21.9.2014 20:37	o0923xx	Tomas01 [ID: 11]	updating the user detail
7.10.2014 18:31	o0923xx	radim_dolak [ID: 4]	updating the user detail
7.10.2014 18:31	o0923xx	radim_dolak [ID: 4]	updating the device detail
7.10.2014 18:44	o0923xx	radim_dolak [ID: 4]	updating the user detail
7.10.2014 18:44	o0923xx	radim_dolak [ID: 4]	changing user password
7.10.2014 20:49	o0923xx	Miro [ID: 10]	updating the user detail
7.10.2014 20:49	o0923xx	Miro [ID: 10]	changing user password
7.10.2014 20:51	o0923xx	Miro [ID: 10]	updating the user detail
7.10.2014 20:51	o0923xx	Miro [ID: 10]	added a new fee
4.1.2015 20:34	o0923xx	Tomas01 [ID: 11]	added a new fee
26.2.2015 15:31	o0923xx	Miro [ID: 10]	added a new fee
31.8.2015 23:59	o0923xx	SYSTEM	termination of network traffic

Text column contains information about actions and also about users which are identified according to their student numbers. It is necessary to extract information about actions and users to the separated columns. The easiest way is using for example Microsoft Excel text functions.

Fluxicon, Disco User´s Guide [7] is defining the minimum requirements for an event log. The data columns determine the analysis possibilities. We need to identify at least the following three elements in our event log for providing process mining analysis in Disco: Timestamp, Case ID and Activity. Other elements in Disco terminology are for example resources, costs etc. We can see an example of transformed events log from Pawouk system in previous Table 2. It is prepared to be imported into Disco process mining tool. We can define analyzed columns in the following bullet item list.

- Date and time = Timestamp in Disco
- User = Case ID in Disco
- Source = Resource in Disco
- Action = Activity in Disco

There are also other columns which are possible to analyze but we made a decision to analyze them in a future work. Other columns in Pawouk system are the following: network, IP, Amount (internet connection payment or payment for printing tasks)

5.2 Process Mining Analysis in Disco

In this section are presented the results of process mining analysis using Disco software. We have used for the verification, 361 cases ad 4314 events which were generated by Pawouk information and administration system.

Process model map. We can see only the most frequent paths in the flow if we want. It is possible to change the level of detail of the shown process map by using the interactive setting in the software (setting detail of paths in percent). The some situation is in number of activities (setting detail of activities in percent). We can easily create with Disco animations for visualizing our processes. Animations can be beneficial and helpful because it can help us find out spot bottlenecks where work is piling up. The following Fig. 1 shows the discovered process model using DISCO software tool.

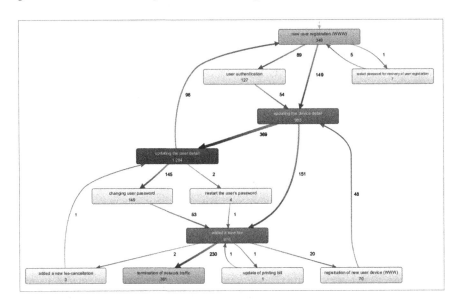

Fig. 1. Disco process model map. (Source: Own analysis in Disco)

There are more important results than only display of paths and actions in process map. We can use possibilities of advanced filtering allows us to set the focus to certain process variants or processes for further analysis.

There are three ways how can be user added to the Internet network: "user authentication" using confirmation email, verification by network administrators ("updating the device detail" together with "updating the user detail") when new user must visit them during consultation hours in internet office on student´s dormitory and the last direct import of users from the users database from the last academic year. Only the last variant is possible without "new user registration".

We have used mean, median, maximum and minimum duration for analysis of the most important activities related to user management. Median duration between new user registration using web portal and user authentication is 3 min. Mean duration is 9.9 h. It is caused by some users which did not use confirmation email immediately but after several days. We can compare also minimum (60 s) and maximum (11.3 d) duration. We can see median duration for some basic activities in the following Fig. 2.

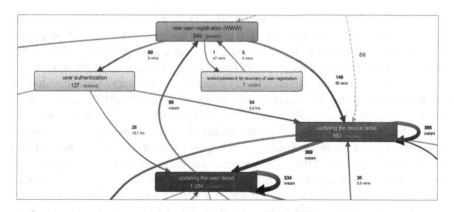

Fig. 2. Median duration of some processes (Source: Own analysis in Disco)

We can see in the previous Fig. 2 that median duration between "new user registration" and "updating the device detail" together with "updating the user detail" by network administrators is only 60 s. This good result is due to the fact that at the beginning of the academic year are all three network administrators in the internet office ready to help with the registration of first-year students, who are almost immediately after registration authenticated and connected to the network ("updating the device detail" together with "updating the user detail").

We can see the loop at some activities such as "updating the device detail" and "updating the user detail". It is due to the fact that these records contain many items sometimes in different tabs and are not always filled at once. Another reason is that there are errors in entering data such as MAC address and so on. And these errors are subsequently corrected.

Activity Event Classes. We have found activity event classes based on DISCO analysis. The most common activities are: updating the user detail, updating the device detail and added a new fee.

Events Over Time. We can see an overview of events during academic year from 1.9.2014 to 31.8.2015 in the following Fig. 3. There are some periods with an increased occurrence of events. There are total 4314 events, 361 cases and 12 activities.

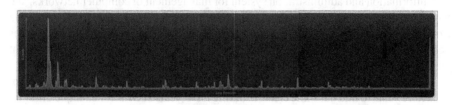

Fig. 3. Events over time (Source: Own analysis in Disco)

The third week of September is characterized by the arrival of students at university dormitory and there are the following activities: new user registration (WWW), registration of new user device (WWW), user authentication, updating the user and the device detail. Some students forgot the password from last year for renewal of registration, and in this case there are these activities: restart password for recovery of user registration, restart the user´s password, changing user password. Fees for Internet access are set from the month of October. There is the choice of payment for 1 month, 3 months and 10 months. Another increase in events during the year, according to the analysis presented payment for Internet access in the event of short-term payments for 1 or 3 months at the beginning of the month for which the fee is necessary to be paid together with payment for printing tasks on network printer. We can distinguish these activities: added a new fee, added a new fee-cancellation, inserting a new printing bill and update of printing bill. The last significant activity is the closure of a student´s computer networks at the end of the academic year - 31 August 2015. It is necessary to make backup of the user database and create a new empty database for the next academic year, which is connected to the web portal to register new users or renewal of registrations from the previous academic year. It is also possible to directly export some users from the past to the current database.

Overview of Using Resource. There are 3 network administrators which have consultation hours at student's dormitory (Miro, radim_dolak, and Tomas1). We can see that the workload of network administrators is nearly almost evenly in all. It is therefore not necessary to make changes in the distribution of their consulting hours.

SYSTEM provides automatic command to terminate traffic of student´s computer network by disconnecting all current users.

WWW represents all activities that are carried through the information web portal, which allows particularly user registration and adding equipment.

ANONYM represents activities that can be conducted before a portal or via email registered users. These are the following activities: restart password for recovery of user registration, user authentication and restart the user's password.

6   Conclusion

The main purpose of this paper was to introduce process mining in a specific area of web information and administration system for management of computer networks. We have used Disco process mining tool to provide process mining analysis of events log from web information and administration system for management of student's computer networks. We have obtained the following results: process model map, activity event classes, events over time and also overview of using resources.

As a future work, we are planning to compare process mining analysis from different academic years to find out some changes during last 5 years such as for example different behavior of users (more automatic registration through web portal with user authentication by confirmation email, payments for longer periods etc.), using of resources and times (mean, median, min and max) for the most important activities.

Acknowledgement. The work was supported by SGS/19/2016 project called "Advanced mining methods and simulation techniques in business process domain".

References

1. van der Aalst, W.M.P.: Process Mining: Discovery: Conformance and Enhancement of Business Processes. Springer, Berlin (2011)
2. Aalst, W., et al.: Process mining manifesto. In: Daniel, F., Barkaoui, K., Dustdar, S. (eds.) BPM 2011. LNBIP, vol. 99, pp. 169–194. Springer, Heidelberg (2012). doi: 10.1007/978-3-642-28108-2_19
3. van der Aalst, W.M.P., Reijers, H.A., Weijters, A.J.M.M., van Dongen, B.F., Aalves de Mediros, A.K., Song, M.S., Verbeek, H.M.W.: Business process mining: an industrial application. Inf. Syst. **32**, 713–732 (2007)
4. van der Aalst, W.M.P.: Process mining in CSCW systems. In: Proceedings of the Ninth International Conference on Computer Supported Cooperative Work in Design, pp. 1–8. IEEE Press, New York (2005)
5. van der Aalst, W.M.P., Weijters, A.J.M.M., Maruster, L.: Workflow mining: discovering process models from event logs. IEEE Trans. Knowl. Data Eng. **16**(9), 1128–1142 (2004)
6. Fluxicon. http://www.fluxicon.com/disco/
7. Fluxicon Disco User's Guide. https://fluxicon.com/disco/files/Disco-User-Guide.pdf
8. Günther, C.W., Rozinat. A.: Disco: discover your processes. In: Proceedings of the Demonstration Track of the 10th International Conference on Business Process Management, pp. 40–44. Springer, Heidelberg (2012)
9. Mans, R.S., Schonenberg, M.H., Song, M., Aalst, W.M,P., Bakker, P,J,M.: Application of process mining in healthcare – a case study in a dutch hospital. In: Fred, A., Filipe, J., Gamboa, H. (eds.) BIOSTEC 2008. CCIS, vol. 25, pp. 425–438. Springer, Heidelberg (2008). doi: 10.1007/978-3-540-92219-3_32
10. Petránek, T.: P@wouk – nástroj pro jednoduchou správu a vedení agendy studentských počítačových sítí na kolejích SU OPF Karviná. Open Source řešení v sítích (2011)
11. Process Mining Manifesto. http://www.win.tue.nl/ieeetfpm/lib/exe/fetch.php? media=shared:process_mining_manifesto-small.pdf
12. Process mining. http://www.processmining.org/research/start
13. Process Mining Case Studies. http://www.win.tue.nl/ieeetfpm/doku.php?id=shared: process_mining_case_studies
14. Process Mining for Analyzing Inventory Processes. https://fluxicon.com/blog/2014/01/case-study-process-mining-for-analyzing-inventory-processes/)
15. ProM. http://www.promtools.org/doku.php
16. Saylam, R., Sahingoz, O.K.: A process mining approach in software development and testing process: a case study. In: Proceedings of the World Congress on Engineering, pp. 407–411, Newswood Limited, Hong Kong (2014)

A Virtual Network Embedding Algorithm Based on Hybrid Particle Swarm Optimization

Cong Wang[1], Yian Su[1], Lixin Zhou[2], Sancheng Peng[3(✉)], Ying Yuan[1], and Hongtao Huang[3]

[1] School of Computer and Communication Engineering,
Northeastern University at Qinhuangdao, Qinhuangdao 066004, China
[2] Department of Science and Technology Information, Public Security Bureau of Qinhuangdao, Qinhuangdao 066004, China
[3] School of Informatics, Guangdong University of Foreign Studies,
Guangzhou 510420, China
psc346@aliyun.com

Abstract. Allocating the underlying physical substrate network resources for users reasonably is the target of virtual network embedding (VNE), which is a hot issue in virtual resource allocation field. In order to prevent the premature convergence and poor performance of local optimization during mapping procedure, in this paper, we combine DPSO, taboo-search technology and simulated annealing algorithms to solve premature convergence problem by using taboo list and annealing process, then propose a virtual network embedding algorithm based on hybrid particle swarm optimization. Simulation results show that our algorithm can improve the revenue to cost ratio and the acceptance ratio.

Keywords: Virtual network · Virtual network embedding · DPSO · Taboo-search · Simulated annealing

1 Introduction

Network virtualization technology [1] allows the existence of multiple heterogeneous virtual network above the sharing underlying physical substrate network. The problem of allocating the request for asking underlying network resources from users in nodes (CPU, Memory, Storage) and link resources with restrictions of virtual network is called Virtual Network Embedding (VNE) problem [2]. VNE has been proved to be a NP-hard problem. Even if all the virtual nodes have been mapped, virtual links with restrictions on bandwidth resources are also NP-hard [3]. Therefore, most of the existing work used intelligent algorithm to solve VNE problem [4].

Discrete particle swarm optimization (DPSO) is a global optimization technique based on swarm intelligence. Each particle represents a candidate solution. The algorithm finds the optimization area in searching space by interactions between two particles. It has some advantages, such as fast convergence, simple algorithm and high searching efficiency [5]. However, in practice, the use of intelligent algorithm to solve VNE problem is not efficient, because there is be a convergence to local optimization problem after running for a time in mapping scheme based on DPSO. Thus, in this

© Springer International Publishing AG 2017
M. Qiu (Ed.): SmartCom 2016, LNCS 10135, pp. 568–576, 2017.
DOI: 10.1007/978-3-319-52015-5_58

paper, we propose an algorithm, called HPTS-VNE-PSO (Hybrid of Particle swarm optimization, Taboo search and Simulated annealing about Virtual Network Embedding Based on Particle Swarm), to solve premature convergence problem in finding optimal solutions, by utilizing the taboo list in taboo search and the annealing process in simulated annealing algorithm.

2 Related Work

Without any restriction above the problem space, by allowing the substrate network to support path splitting and migration, Yu et al. [5] advocated a different approach on the design of the substrate network to enable simpler embedding algorithms and more efficient use of resources. They simplified virtual link embedding by allowing the substrate network to support path splitting and migration, and proposed VN embedding algorithms without restricting the VN embedding problem space.

Lischka and Karl [6] modeled the topology of the substrate and the VN as a directed graph and propose a VN embedding algorithm based on subgraph isomorphism that maps nodes and links during the same stage. Their algorithm can be seen as an extended version of the classic VF graph matching algorithms [7], where link-on-link mapping has been relaxed. However, this algorithm cannot work when the location constraints on the virtual nodes are taken into consideration.

Xiang et al. [8] applied the particle swarm optimization to solve the VNE problem. For the particle swarm algorithm is simple, the virtual network mapping problem becomes easier to solve. The algorithm defined the virtual nodes mapping targets as positions of particles and uses resource consumption of substrate network as fitness function to evaluate the current scenario. In the process of iteration, each particle adjusts their position based on the individual and global optimal information to obtain optimal virtual network mapping. The proposed algorithm can improve VNR acceptance rate and reduce the time to gain VNE solutions.

Ying et al. [9] proposed a DPSO-based load balancing algorithm to solve the VNE problem also used particle swarm optimization algorithm. And put forward a load balancing MLB-VNE-SDPSO algorithm which can reuse physical nodes. The algorithm considered the CPU resource utilization, saved bandwidth resources of physical links and reduced the virtual links mapping times. However, deficiencies are not taken the premature problem and resource consumptions in link nodes into consideration.

PSO algorithm is independent of the problem, which is to say that without knowing too much specific information related to the problem, just knowing the fitness value of each solution, we can easily proceed the optimizing operation, which makes PSO algorithm become much stronger than other search algorithms. However, since the particle swarm algorithm is a random search algorithm and has been proved to be an uncertain global search. If the solving process to the problem is very difficult and complicated, PSO algorithm may not find the best solution needed. Tabu Search (TS) and Simulated Annealing (SA) algorithms can avoid of falling into local optimization to some extent, and the search process can be controlled by the cooling strategy. By designing the neighborhood structure and cooling strategy of TS and SA algorithms to control the search process, individuals can effectively avoid of falling into local optimization. Therefore, we propose HPTS-VNE-PSO algorithm and expand search rang from the horizontal and vertical levels.

3   Hybrid VNE Approach

TS algorithm [10], as a meta-heuristic algorithm, to some degree is an intelligent search of simulating human's behaviors. Different from swarm intelligence algorithm like genetic algorithm and simulated annealing algorithm, the TS algorithm overcomes the premature convergence in searching process by using the tabu strategy to achieve global optimization. This paper makes use of this characteristic of TS algorithm and references the tabu list to avoid repetitive search and improve the convergence efficiency. SA algorithm [11] introduced the natural mechanism of the annealing process from the physical system. In the iterative process, it not only accepts the tentative point which makes the fitness value become "good", but also accepts the point that makes the fitness value become "bad" with a certain probability. Decrease in temperature will also reduce the probability of acceptance. Such a search strategy can effectively avoid falling into the scenario that the searching process cannot escape because of the local optimal solution and improve the reliability of receiving the global optimal solution.

3.1   Problem Formalization

There are several key technologies in the design of HPTS-VNE-PSO algorithm, such as (1) search space, (2) PSO algorithm framework, (3) neighbor search, (4) taboo list, (5) fitness function, and (6) cooling control, which are described as follows:

(1)   Search space

Assume the number of virtual nodes is N, then the searching range of particle swarm is an N-dimensional space. Set the initial number of particle swarms to m. Since the VNE problem is a discrete problem, we use the DPSO algorithm and redefine the position, velocity, individual optimal position and global optimal position of particle i as follows.

$$\text{Position} : X_i = [x_i^1, x_i^2, \ldots, x_i^N], x_i^j \geq 0, j \in (1, N), i \in \{1, 2, \ldots, m\}$$

N denotes the number of virtual nodes in the virtual network request, x_i^j sets to a positive integer, the value of which denotes the number of underlying physical node to which the j-th virtual node is mapped.

$$\text{Velocity} : V_i = [v_i^1, v_i^2, \ldots, v_i^N], i \in \{1, 2, \ldots, m\}$$

v_i^j is a binary number. If its value is 0, means it need to select another node for mapping from the candidate node. Otherwise, there is no adjustment to make.

(2) PSO algorithm framework

The inertia weight of the PSO algorithm is:

$$w = w_{max} - \frac{w_{max} - w_{min}}{iter \max} \times iter \tag{1}$$

where w_{max} is the initial weight, w_{min} is the final weight, *itermax* is the maximum number of iterations, *iter* represents the number of current iteration. The inertia weight w is very critical to the searching ability of the PSO algorithm. The bigger the value is, the further space is more inclined to be searched. On the opposite, the smaller the value is, the closer space is more inclined to be searched. Therefore, the initial weight value used in this paper linearly decreases from a higher value to a lower value, who tends to have strong global search ability from the outset, and then tends to the local search."

$$V_{i+1} = \varphi\{wV_i \oplus c_1 \times Rand() \times (X_p^i \ominus X_i) \oplus c_2 \times Rand() \times (X_g \ominus X_i)\} \tag{2}$$

$$\varphi = \frac{2}{|2 - c - \sqrt{c^2 - 4c}}, c = c_1 + c_2, c > 4 \tag{3}$$

In a particle swarm, "cognitive" weight c_1 and "social" weight c_2 control the range of particles' movements, usually we set it to 2.0. In this paper, the formula for updating the velocity of particle is described in (2) and (3).

(3) Neighbor search

In order to continuously expand the search space, taboo search needs to keep neighbor moving constantly. Neighbor movement is based on the current solution, according to a certain mobile strategy to produce a certain number of new solutions, known as the neighbor solution. Neighbor search is very effective for local search, and generally, unnecessary and infeasible movements must terminate. Currently the most famous neighbor structure is based on "block". The neighbor structure of our algorithm is set to be an array.

(4) Taboo list

Taboo objects, which are stored in the taboo list, cannot be re-searched before being lifted. The taboo list used in this article is an array, and taboo objects in this list have been searched before.

Taboo length refers to the survival time of a taboo object in the taboo list. When a taboo object is added to the taboo list, sets its term to a taboo length value, and the HTPS-VNE-PSO algorithm sets the population size to the taboo length. The taboo length is dynamically changed during the searching process. In this paper, the term of the taboo object is automatically decremented by 1 for each iteration. When the term of a taboo object is 0, then remove it from the taboo list. A taboo object with a term higher than 0 is set to taboo state, during the searching process it cannot be selected as a new solution.

(5) Fitness function

The fitness function can calculate the corresponding evaluation for solutions, and evaluate the solution from the taboo search space by the evaluation function. The value of evaluation function represents the degree of merit of the solution, which is a performance index of the particle swarm. By comparing the value of evaluation function (i.e., fitness function), we can determine whether the current solution is better or not. We use the minimization of bandwidth as the fitness function: $Min \sum_{l_{uv} \rightarrow l_{ij}} f_{ij}^{uv} b(l_{uv})$. The

smaller the value is, the better it will be. This is also a typical goal to the VNE problem.

(6) Cooling control

The SA algorithm module can be controlled by a cooling strategy. The initial temperature of the cooling strategy $T_0 = \Delta f_{max}$, in which Δf_{max} represents the maximum different value between the fitness values of two neighbors. The formula of temperature changing is: $T_k = B \times T_{k-1} (k = 1, 2, 3 \ldots)$, descent factor B is less than 1.

3.2 HPTS-VNE-PSO Algorithm

The main idea of HPTS-VNE-PSO is to use PSO algorithm to generate the position particle in each turn, which is a possible mapping plan; and then use the shortest path algorithm to find the shortest path for the virtual link and allocate links for it; add the initial particle information to the taboo list; for each particle, calculate the fitness value, *gbest*, *pbest* and find the optimal solution; then create a neighbor solution for the optimal solution according to the specific rule and store the best neighbor solution from the neighbor solution set, which is not in the taboo list as the current solution; update the taboo list. In the simulated annealing section, we first select the initial temperature, when the temperature is not high enough, continue to generate the neighbor solution with the current solution, and calculate the fitness value. If it reaches the terminal condition, then update *gbest* and add it to the taboo list, then reduce the temperature. When the temperature decreases to a certain degree, return the optimal solution, or continue with the cooling process.

The PSO algorithm provides the initial particles for the HPTS-VNE-PSO algorithm and forms the outer frame of the algorithm. The pseudo-code is shown as follows:

```
Algorithm 1. PSO algorithm
Generate initial particle swarm by using particle ini-
tialization allocation strategy, use Freud shortest path
algorithm to allocate links for them;
While not reaching the maximum iteration do
     Iteration++;
     Update the particle position and velocity to generate
the next swarm;
     Use TS-SA algorithm to find pbest and gbest;
     Update pbest and gbest;
End while
```

Algorithm 2 is the operation process of the TA(SA) algorithm, including the neighbor operation and the use of taboo list of TS algorithm. Meanwhile, the cooling process of SA algorithm is adopted. When the temperature reaches a certain value, break the loop and output the optimal solution, which constitutes the core of HPTS-VNE-PSO algorithm.

```
Algorithm 2. TS-SA algorithm
Iter++, generate a neighbor solution according to the
neighbor structure of the cur-rent optimal solution x,
select several candidate solutions from the neighbor so-
lution;
Select and store the best neighbor solution which is not
in the taboo list or the one that meets the receiving
criteria as a newly current solution. Update the taboo
list.
```

For $gbest$ particles S, set temperature $T_k = T_0$;

While $(T_k \, {}^3 \, Tend)$ do

Generate a neighbor solution $S*$ of S through pair-exchange method;
Calculate the fitness value of $S*$;
Evaluate $S*, D = f(S*) - f(S)$;

If $\min[1, \exp(-D/T_k)] < random[0,1]$ then

Accept $S*$;
Update $gbest$;
End if;
$T_k = B * T_{k-1}$;
End while
End for
If $iter \, £ \, \max Iter$ then goto 4;
Return the optimal solution;

4 Experimental Results and Analysis

We compare our proposed algorithm to VNE-R-PSO [9] and D-ViNE-SP [2] which is a typical PSO-based VNE approach in dealing with online VNRs. Both algorithms are implemented in CloudSim [12]. The control granularity of CloudSim has been modified from virtual machines to virtual networks. A topology generator is developed to generate virtual and physical topologies. The parameters of the generator contain number of nodes, connected probability and task duration of VNRs. The substrate network is set to have 100 nodes and 500 links. The physical node CPU capacity and physical link bandwidth are uniform distribution from 50 to 100. For online virtual network requests, assume that the arrival process of them follows Poisson process, with time unit 100 and strength 5, and the lifetime of each virtual network follows exponential distribution, with parameter 400. For every virtual network request, the number of virtual nodes uniform distribute from 2 to 20. The probability of link generation between nodes in virtual network is 50%. Both virtual network node CPU request and virtual bandwidth request are uniform distribute from 3 to 50. There are 50,000 time units running in each simulation experiment, and average 2500 virtual network requests are operated. The maximum number of iterations is 20.

Figure 1 shows the changing curve of the acceptance rate of the virtual network request with the clock tick of the algorithm increases, the value of load factors is 1. It can be seen from the figure that the acceptance rate of virtual network request decreases gradually and tends to become flat over time, and it's better than other algorithms.

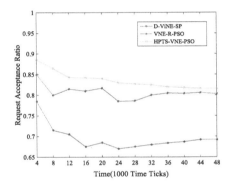

Fig. 1. The comparison of virtual network request acceptance ratio by HPTS-VNE-PSO algorithm

Figure 2 shows the comparison of average cost-to-benefit ratio in the long-term among the HPTS-VNE-PSO algorithm with other two algorithms, the value of load factor is 1. It can be seen from the figure that fluctuations of average long-term

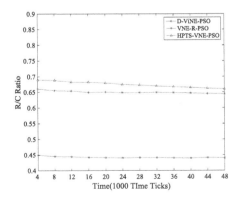

Fig. 2. The comparison of R/C ratio by HPTS-VNE-PSO algorithm

cost-to-benefit ratio from several algorithms are less than our proposed algorithm. However, the HPTS-VNE-PSO algorithm performs better than the other two algorithms over time.

5 Conclusion

In this paper, we proposed the HPTS-VNE-PSO algorithm by using taboo search algorithm and annealing technology of simulated annealing algorithm. The taboo list can avoid repetitive search and extend the horizontal search range, while the stimulated annealing algorithm can avoid the dilemma of local optimization from the vertical level. The combination of them can solve the premature convergence problem of VNE problem, and can also obtain better virtual net-work cost-to-benefit ratio.

Acknowledgement. This work was partially supported by the National Natural Science Foundation of China under Grant Nos. 61300195 and 61379041, the Natural Science Foundation of Hebei Province under Grant Nos. F2014501078 and F2016501079, the Science and Technology Support Program of Northeastern University at Qinhuangdao under Grant No. XNK201401, and the Science and Technology Project of Guangzhou under Grant No. 2013Y2-00069.

References

1. Chowdhury, N.M.M.K., Boutaba, R.: Network virtualization: state of the art and research challenges. IEEE Commun. Mag. **47**(7), 20–26 (2004)
2. Cheng, X., Zhang, Z., Su, S., Yang, F.: Survey of virtual network embedding problem. J. Commun. **32**(10), 141–143 (2011)
3. Zhi-ping, C., Qiang, L., Pin, L., et al.: Virtual network mapping model and optimization algorithms. J. Softw. **23**(4), 864–877 (2012)
4. Beck, M.T., Fischer, A., Botero, J.F., et al.: Distributed and scalable embedding of virtual networks. J. Netw. Comput. Appl. **56**, 124–136 (2015)
5. Ma, X., Liu, Q.: Particle swarm optimization for multiple multicast routing problem. J. Comput. Res. Dev. **50**(2), 260–268 (2013)
6. Lischka, J., Karl, H.: A virtual network mapping algorithm based on subgraph isomorphism detection. In: Proceedings of the 1st ACM Workshop on Virtualized Infrastructure Systems and Architectures, pp. 81–88 (2009)
7. Cordella, L.P, Foggia, P., et al.: An improved algorithm for matching large graphs. In: 3rd IAPR-TC15 Workshop on Graph-Based Representations in Pattern Recognition, pp. 149–159 (2001)
8. Xiang, C., Baozhong, Z., et al.: Virtual network embedding based on particle swarm optimization. Acta Electronica Sin. **39**(10), 2240–2244 (2011)
9. Ying, Y., Cuirong, W., et al.: Load controllable virtual network embedding algorithm based on discrete particle swarm optimization. J. Northeast. Univ. (Nat. Sci.) **35**(1), 10–14 (2014)
10. Glover, F., et al.: Future paths for integer programming and links to artificial intelligence. Comput. Oper. Res. **13**, 533–549 (1986)

11. Kirkpatrick S., Jr., G.C., Vecchi, M.P.: Optimization by simulated annealing. Sciennce **11**, 650–671 (1983)
12. Calheiros, R.N., Ranjan, R., et al.: Cloudsim: a toolkit for modeling and simulation of cloud computing environments and evaluation of resource provisioning algorithms. Softw. Pract. Exp. **41**(1), 23–50 (2011)

Research on Content Distribution of P2P VoD with Cloud Assisting

Hongfang Guo[(✉)] and Tingting Ma

Beijing Wuzi University, No. 1 Fuhe Street, Tongzhou District,
Beijing 101149, China
guohongfang@bwu.edu.cn

Abstract. The large-scale application of VoD still remains to be restricted due to users' dynamics and time sensitivity with popular P2P technology. Driven by this demand, a hybrid streaming distribution overlay is proposed, called CAPMedia, which supports the P2P VoD service with on-demand cloud assisting with its high reliability, storage, and processing ability. In this paper, the mechanism of segments selected is proposed which combines the segments popularity and nodes failure. Then, the formalized cost model is detailed to guide the VoD service to reduce the deployment cost. The extensive simulation experiments validate CAPMedia's efficiency. Compared with traditional solution, CAPMedia improves the user's satisfaction, bandwidth saving significantly, and reduce the bandwidth consumption.

Keywords: VoD · P2P · Cost model · Cloud service

1 Introduction

With the rapid advances of network and multimedia technologies, Video-on-Demand has become an important part of Internet life. Online video streaming has proliferated rapidly, it emerges a dominant form of big data on Internet world. According to Cisco Systems, Internet videos accounted for 78% of all U.S. Internet traffic in 2014, and is expected to rise to 84% in 2018 [1]. However, the growing user scale, huge volume multimedia data and high dynamic of user challenge VOD service, how to support interactive operations with low cost on the current physical network is still a key issue [2].

Among of all content delivery network, peer-to-peer (P2P) overlay has become an promising choice for efficient and low-cost delivery for VoD, however its successful application remains to be restricted by the dynamic, view-independent behaviors of P2P users. In addition, user's demand varies with the time in one-day or in a special period, such as a big events or game and et al. It is reported that it consumes 60% bandwidth with the top 25 popular videos. VoD application is of time-constraint, it has to support 95% value for golden time however it must sustain a few hours leads to bandwidth idle at other time in one day [3] which leads to a big waste.

© Springer International Publishing AG 2017
M. Qiu (Ed.): SmartCom 2016, LNCS 10135, pp. 577–586, 2017.
DOI: 10.1007/978-3-319-52015-5_59

On-Demand service of cloud supporting enough caching and high-speed computing exploits a new opportunity for VoD application. However, it will become the traditional C/S model with all the VoD application migrating to the cloud platform, VoD provider has to spend enormous amounts for user's frequent accesses. It is reported that it will spend 130 thousands dollars a month for bandwidth of MMVE service [4]. For numerous video data, the cost will be more unimaginable.

The issues which are described above are incentives to investigate two mechanisms: cloud computing and P2P networks. In this paper, we targeted to design a hybrid topology combining P2P and cloud service, called CAPMedia. It makes full use of advantages of P2P and cloud, meanwhile avoiding the long jumping latency of playback and high cost of bandwidth consumption. Then an effective content-selection mechanism was proposed considering the popularity and user's leaving behaviors. In order to evaluate them, we compared it with state-of-the-art solutions, and the results show that it outperforms the traditional mechanisms by high user satisfactory and low cost significantly.

The remainder of this paper is organized as follows. Section 2 gives the related work. The cloud-assisted P2P video content distribution model is given in Sect. 3. The video segment uploading Mechanism and cost estimating of CAPMedia system is detailed in Sects. 4 and 5 respectively. Then, in Sect. 6, we present the evaluation results. Finally, Sect. 7 concludes the paper.

2   Related Work

As two popular mechanism of streaming content distribution, P2P and cloud draw more and more attentions, works focusing on the integrating of P2P and cloud can be classified into two categories, we called them P2P-in-Cloud and Cloud-in-P2P.

P2P-in-Cloud model: for the VoD supporter, the cloud platform can be considered as a central server which contains many data centers deploying in various positions, however, the single point failure is fatal shortcoming of the cloud-like topology service. A lot of works on cloud server optimization with P2P technology are introduced [5–8]. For GFS file storing systems, in order to avoid bottleneck of master directory servers. Structured DHT is introduced to organize Trunk Servers [6]. Duplicated deployment of more master directory servers is cited to avoid single point of failure [7]. Rayjan and et al. proposed the approach with a structured peer-to-peer network model to connect cloud system components [8]. P2P-in-Cloud focus on cloud servers, which is different with application of migrating from users' aspect, called Cloud-in-P2P model.

Cloud-in-P2P model: With P2P and cloud technology, a content distribution overlay was proposed for mobile devices. Storing and computing are processed with clouds, otherwise load-balance of content distribution is finished with P2P among mobile devices [9]. Paper [4] introduce the application of p2p and cloud in Massively Multiuser Virtual Environments(MMVE). Feng Wang and et al. proposed the cloud-assisted live media streaming [10]. P2Pcloud platform is proposed to process reliable resource reservations, while our work focuses on Video-on-Demand application, especially on bandwidth cost optimization of application.

The tasks of this paper conclude: (1) Exploiting cloud-assisted VoD content distribution model to solve the problems of high-cost and low utilization; (2) Then proposing the key issues of content segment selection mechanism and cost estimation model; (3) Evaluation and analysis of simulation results.

3 Model Description

3.1 Prerequisites of Model

Firstly, we introduced the basis of thoughts from the following three aspects: (1) P2P model is not perfect solution for VoD application due to its unreliable nodes, low scalability of servers. With the storing, computing, reliable service, content will be distributed to cloud platform to migrate the load which leads to reduce the load and the cost; (2) It was reported that 20% super nodes serve 80% user demand. Super nodes in P2P overlay is crucial, therefore, we can consider cloud server as super node in cloud-assisted P2P overlay; (3) For cloud service model with renting resource, P2P is a promising candidate to reduce the cost.

Migrating video content to cloud platform takes the storing and reliable service advantages of cloud, by shifting load to reduce server's cost of VoD suppliers. The key issues of solution are segments selection and the amount of traffic which are related to the cost of VoD supplier spending to cloud supplier.

In this paper, we introduced cloud server as super node cooperating with P2P node to solute the problem of high volume streaming content and high bandwidth cost. The key issues of solution contain the deployment of cloud nodes and P2P nodes, placement of content segment, optimizing of bandwidth cost.

3.2 CAPMedia Model

With the analysis above, we proposed the CAPMedia (Cloud Assisted P2P Media) overlay Model, see Fig. 1. In P2P overlay, we took advantage of super node, Cloud CDN performs super nodes' task by Internet. Uploading and downloading streaming are finished by super nodes.

In CAPMedia model, there are three participant Objects including video provider (VP), cloud provider (CP) and users (US). VP stores and supports video content source, it also manages users, content distribution and interactive operations with CP. CP is introduced to support user demand by cooperating with CP. US is the terminal serving user, their demands are the goal of CAPMedia achieving. CP, VP and other US all can provider content service for a US.

We divided these participant nodes into six categories which are showed in Fig. 2. Cloud platform providers cloud storage servers (CSS) which store big data of streaming, it also provides cloud content distribution servers (CCDN) which locate the optimal CCDN to server users by segment selection mechanism.

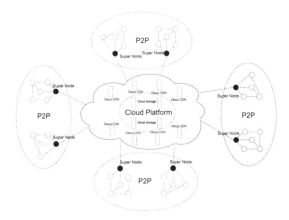

Fig. 1. The overlay of CAPMedia

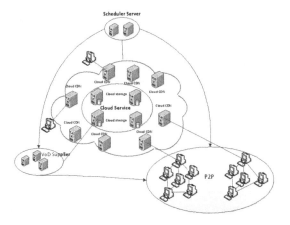

Fig. 2. Node deployment of CAPMedia model

Source servers (SS) and scheduler server (SCHS) are provided by VoD provider. SCHS is the interface between P2P and cloud platform, it also manages segment uploading mechanism. In P2P model, it concludes users (US) and super nodes (SN). In CAPMedia, with a user's request, the target content indexing will be triggered. Firstly it searches target content in P2P overlay, then in stead of resorting to VP, it will satisfy user's demand by requesting CP by cost estimating mechanism. In order to meet user's experience and reduce the service cost of VoD, user's request can be satisfied by the following three approaches: cooperation among users, VoD provider, cloud provider, which are showed in Fig. 3.

3.3   Interface of P2P and Cloud Service

As Fig. 4 shows, broker is the communication interface of cloud platform and VoD application. Cloud provider and VoD provider will decide the service level

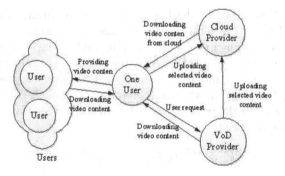

Fig. 3. The model of user request of CAPMedia

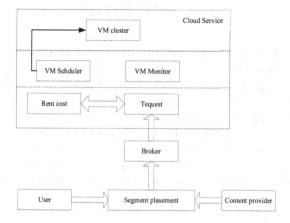

Fig. 4. Interface of P2P and Cloud service

and type basing on the cost, then arrange VM scheduler to schedule VM cluster's resource. It needs to be put forward that VoD provider and users all can upload video content to cloud platform which can reduce the private servers' load of VoD provider. It is obvious that two key problems are needed to solve, one is segment selection of uploading, the other is how to estimate the right cost to rent proper resource.

4 Video Segment Uploading Mechanism

It is reported that it consumed 60% bandwidth with the top 25 popular video. Obviously, the popular video can not cooperate to save bandwidth effectively as, though they have more users. Users who like the video do not mean they are willing to share their resource, they may leave after playing video.

For video segment selection mechanism, the video segment popularity is cited to be a key factor. The streaming service will be migrating to cloud when user's demand can not be satisfied by sharing among users. We firstly introduced a concept, called availability which takes segment popularity and node failure into consideration. Then uploading mechanism was proposed basing on availability value.

We defined m as the total of video segments in a video, users can request for any segments. We assumed the caching state of every segment is $S(T) = (n_1(t), n_2(t), n_3(t), ...n_m(t))$, which $n_i(t)$ represents the nodes of caching i_{th} segment at time slot t. On the time period $[t, t+\Theta]$, the leaving (failure) probability of nodes can be formalized:

$$F^{\Theta}(T) = (f_1^{\Theta}(t), f_2^{\Theta}(t),)f_3^{\Theta}(t),)...f_m^{\Theta}(t))) \tag{1}$$

Then the availability of node can be defined as follows:

$$\frac{\sum_{i=1}^{m}(1 - \prod_{j=1}^{n_i(t)} f_{ij}^{\Theta}(t))}{m} \tag{2}$$

The above definition and analysis describe the availability from global view which need the whole history log information. It introduced more new traffic of control information leading to new bottleneck problem. In order to solve the problem, we selected uploading segment by exchanging information among users which spread the task to every user.

We assumed the neighbor of i node is $partner_i(t)$ at time slot t, neighbor caching k segment is $partner_i^k(t)$. Caching k segment in local buffer is $localcache_i(t)$. $q_k^{\Theta}(t)$ is the probability of requesting segment k. Our mechanism choosing minimum availability value see the following equation. Minimize

$$(1 - \prod_{p \in patner_i^k(t)} f_p^{\Theta})q_k^{\Theta}(t) \qquad where \qquad k \in localcache_i(t) \tag{3}$$

If the $k \notin localcache_i(t)$, the availability value is 0.

5   Migrating Cost Estimation Mechanism

In this paper, we utilized the streaming traffic as the parameter to analyse the cost of CAPMedia model. Based on the topology of CAPMedia, we induced the

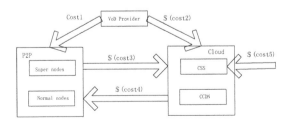

Fig. 5. Cost model of CAPMedia

Table 1. List of Notations

Notation	Description
$cost1$	Cost of video provider spending for serving user's demand
$cost2$	Cost of video provider spending to cloud platform
$cost3$	Cost of uploading video of normal nodes
$cost4$	Cost of downloading from cloud platform
$cost5$	Cost of content storing of cloud platform
$s(i)$	The size of segment
$tocs^{\varepsilon}(t)$	Segments of VP servers uploading to cloud at ε time slot
$tocp^{\varepsilon}(t)$	Segments of nodes uploading to cloud at ε time slot
$VPCapacity$	The total bandwidth of VP servers
$R^{\varepsilon}(t)$	The average request at ε time slot
$VPtoNode^{\varepsilon}(t)$	The average bandwidth of VP to nodes
$P2P^{\varepsilon}(t)$	The average bandwidth of nodes sharing
$CSS(t)$	The renting storing space of cloud
$cloud^{\varepsilon}(t)$	The average bandwidth of cloud providing to users at ε time slot

cost distribution as Fig. 5. It is obvious that the total cost of VoD application conclude five aspects. The formal representation is detailed as follows. First, we summarized some notations in Table 1. At the time period $[t, t + \varepsilon]$, the total cost is defined:

$$cost = cost1 \times VPCapacity + \sum_{i \in tocs^{\varepsilon}(t)} cost2 \times s(i) + \sum_{j \in tocp^{\varepsilon}(t)} cost3 \times s(j)$$
$$+ cost4 \times [R^{\varepsilon}(t) - VPtoNode^{\varepsilon}(t) - P2P^{\varepsilon}(t)] + cost5 \times css(t) \qquad (4)$$

Based on cost estimation above, we give the optimal approach by handling the request among VP, users and cloud servers adaptively. Assuming λ_1, λ_2 representing the resource proportion of VP and cloud supporting respectively. IT can be formalized in to minimum problems. Minimize:

$$C = \lambda_1 \times VPtoNode^{\varepsilon}(t) \times cost1 + \lambda_2 \times cloud^{\varepsilon}(t) \times cost4 \qquad (5)$$

Constraints:
$$P2P^{\varepsilon}(t) + VPtoNode^{\varepsilon}(t) + cloud^{\varepsilon}(t) \geq R^{\varepsilon}(t)$$

6   Performance Evaluation

In order to evaluate the performance of CAPMedia model, we implemented an event-driven simulator to conduct a series of simulations. We also performed extentive simulations on P2P, C/S model as comparison.

Simulation settings: the whole video contains 720 chunks and 16000 users in all. Initially, the start playback point of each user is evenly distributed between 1 to 720. Segment popularity we adopt log-normal distribution with the value $\mu = 0.0159$, $\sigma = 1.35$, The skew factor of the Zipf distribution is set to be 0.57 [11,12]. The main metrics investigated are showed as follows.

The Proportion of Failure Request (**PFR**): the ratio of failure requests and total requests. The Proportion of Saving Bandwidth (**PSB**): the ratio of request satisfied by cloud and total request. The Proportion of Cost (**PC**): the cost ratio of CAPMedia and C/S model.

6.1   The Proportion of Failure Request (PFR)

As Fig. 6 shows, the comparison of C/S, P2P and CAPMedia is represented. Assume the request of users are more than VoD servers' supporting. CAPMedia perform excellently without influence of high peer failure, even 90% nodes leave, there is no segment in whole overlay, CAPMedia also meet user's demand. This figure verifies the robustness of CAPMedia. For C/S Model, initially it can server 40% user's request. When nodes failure ratio achieves 60%, C/S can server users well. However, with P2P model, it is not robust to nodes failure, when nodes failure achieves 90%, the PFR increases quickly.

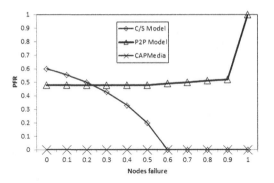

Fig. 6. Performance of nodes failure

6.2   Proportion of Saving Bandwidth

Figure 7 shows performance of video uploading mechanism, the left shows the PSB and the right shows the PFR. According to the 25 popular video consuming 60% bandwidth, we introduce the 20 video segments as research objects. Obviously, PSB is dependent to number of uploading to cloud platform. The first 8 video segments make great impact to PSB. As the availability value grows, PSB tread become gentle which verify the efficient of proposed video uploading mechanism. 32% users can't be satisfied without cloud assisting, when uploading first 3 segments (15% of total video segment), PFR reduces greatly. It improve the experience of users and scalability.

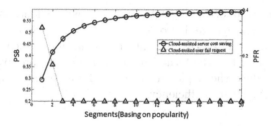

Fig. 7. Performance of video uploading mechanism

6.3   Cost Evaluation

PC is a key parameter of CAPMedia, it is crucial to decide whether adopt cloud service or not. All the value of parameters in simulation is normalized, also the traffic are counted by number of segments. Uploading or Downloading one segment charges 1 cost. Figure 8 shows the cost under different charging standards. Model1 represents that VoD provider and cloud provider have the same charging standard. The average performance of cost is dependent on VP'S bandwidth. Model2 represents that cloud provider charge standard is 1.5 for one segment, which takes the storing service into consideration. It is higher than VoD provider. It is obviously that PC reduces with the bandwidth of VP server growing. When VP can supported 60%–70% or higher, the PC will increased. Intuitively, VP can finish most of request, Cloud service take little impact on cost. Obviously, we can induce that server's load is one of crucial issue in cloud-assisted streaming system.

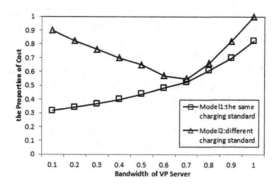

Fig. 8. Cost estimation

7   Conclusion

Due to problem of storing big data of streaming and unreliable P2P application, we introduced the cloud to P2P model and proposed a hybrid topology called CAPMedia, supporting P2P VoD service with Cloud assisted which is of high reliability, storage, and processing ability. The video segment uploading mechanism takes nodes's failure into consideration, which select the right segments effectively. Then the formalized cost model was proposed to guide the VoD service to reduce the deployment cost. The extensive simulation experiments validate CAPMedia's efficiency.

References

1. Videos may make up 84 percent of internet traffic by 2018: Cisco. http://www.reuters.com/article/2014/06/10/us-internet-consumers-cisco-systems-idUSKBN0EL15E20140610
2. Chen, Y., He, W., Hua, Y., Wang, W.: CompoundEyes: near-duplicate detection in large scale online video systems in the cloud. In: Proceedings of the IEEE INFOCOM (2016)
3. Li, H., Zhong, L., Liu, J., et al.: Cost-effective partial migration of VoD services to content clouds. In: Proceedings of the IEEE Cloud (2011)
4. Chen, K.-T., Huang, P., Lei, C.-L.: Game traffic analysis: an MMORPG perspective. Comput. Netw. **50**(16), 3002–3023 (2006)
5. Chen, Z., Zhao, Y., Miao, X., et al.: Rapid provisioning of cloud infrastructure leveraging peer-to-peer networks. In: Proceedings of ICDCS (2009)
6. Zhang, X., Song, J., Xu, K., Song, M.: A cloud computing platform based on P2P. In: Proceedings of the ITIME (2009)
7. Yang, J., Zhao, G., Wang, K., et al.: A modern service-oriented distributed storage solution. J. China Univ. Posts Telecommun. **16**(1), 120–126 (2009)
8. Ranjan, R., Zhao, L., Wu, X., Liu, A., Quiroz, A., Parashar, M.: Peer-to-peer cloud provisioning: service discovery and load-balancing. In: Antonopoulos, N., Gillam, L. (eds.) Cloud Computing. Computer Communications and Networks, pp. 195–217. Springer, London (2010). doi:10.1007/978-1-84996-241-4_12
9. Jin, X., Kwok, Y.-K.: Cloud assisted P2P media streaming for bandwidth constrained mobile subscribers. In: Proceedings of the 16th International Conference on Parallel and Distributed Systems (2010)
10. Wang, F., Liu, J., Chen, M.: CALMS: migration towards cloud-assisted live media streaming. In: Proceedings of the IEEE INFOCOM (2012)
11. Cheng, X., Liu, J.: NetTube: exploring social networks for peer-to-peer short video sharing. In: Proceedings of the IEEE INFOCOM (2009)
12. Guo, H., Liu, J., Wang, Z.: Frequency-aware indexing for peer-to-peer on-demand video streaming. In: Proceedings of the IEEE ICC (2010)

Author Index

Printed in the United States
By Bookmasters